D1442551

NORTH
AMERICAN
HORTICULTURE

NORTH AMERICAN HORTICULTURE

A REFERENCE GUIDE

SECOND EDITION

COMPILED BY THE AMERICAN HORTICULTURAL SOCIETY

THOMAS M. BARRETT
EDITOR

Macmillan Publishing Company
NEW YORK
Maxwell Macmillan Canada
TORONTO
Maxwell Macmillan International
NEW YORK   OXFORD   SINGAPORE   SYDNEY

Macmillan Publishing Company
866 Third Avenue, New York, N.Y. 10022

Maxwell Macmillan Canada, Inc.
1200 Eglinton Avenue East, Suite 200
Don Mills, Ontario M3C 3N1

Macmillan Publishing Company is part of the Maxwell
Communication Group of Companies

Library of Congress Catalog Card Number: 90-20435

Printed in the United States of America

printing number
1   2   3   4   5   6   7   8   9   10

Library of Congress Cataloging-in-Publication Data

North American horticulture : a reference guide / compiled by the
American Horticultural Society ; Thomas M. Barrett, editor. — 2nd
ed.
        p.   cm.
    Includes index.
    ISBN 0-02-897001-2
    1. Horticulture—United States—Societies, etc.—Directories.
2. Gardening—United States—Societies, etc.—Directories.
3. Conservation of natural resources—United States—Societies,
etc.—Directories.   4. Horticulture—Canada—Societies, etc.—
Directories.   5. Gardening—Canada—Societies, etc.—Directories.
6. Conservation of natural resources—Canada—Societies, etc.—
Directories.   I. Barrett, Thomas M.   II. American Horticultural
Society.
SB317.56.U6N67   1991
635'.02573—dc20   90-20435
                                                              CIP

The paper used in this publication meets the
minimum requirements of American National
Standard for Information Sciences—Permanence of
Paper for Printed Library Materials. ANSI Z39.48-1984.
∞

Contents

Introduction

This is the fourth edition of a book initiated by the American Horticultural Society in 1971 as the *Directory of American Horticulture*. The most recent edition was *North American Horticulture: A Reference Guide* (New York: Charles Scribner's Sons, 1982). Many changes have been made in the current edition, some of which reflect changes in the field of horticulture since 1982, and some of which are attempts to make the guide a more useful reference.

The guiding philosophy of this book has been to be as inclusive as possible. With the cooperation of other associations and individuals, we have tried to present a thorough listing of horticultural institutions. Nevertheless, no reference work of this sort can pretend to be all-inclusive. There is simply no way possible to root out all of the organizations involved in horticulture in North America.

Some readers might quibble that we have been *too* inclusive. It may seem odd that Marvin Johnson's Gourd Museum and the Ottawa Research Station Herbarium appear in the same book, and that both are in a reference on horticulture. If this were a theoretical treatise on horticulture, we undoubtedly would have left them out. But this is a book to be used, and we have found that the most valuable reference works cut their swath broadly. Besides, horticulture is an expansive discipline, not a narrow one—the boundaries between horticulture, botany, agriculture, environmental science, and forestry are fuzzy at best. And who knows where amateur gardening stops and horticultural science begins? Thus, we have not attempted to define horticulture as much as we have tried to compile a book that will be useful to anyone interested in plants and gardening.

We also thought it important to include more local information than the previous editions contained. Horticulture is necessarily a local pursuit—we all work within the limits of our hardiness zone, our soil type, and other local climatic conditions. This means that a state tomato growers group, for example, might have more pertinent information for the average gardener than the national Solanaceae Enthusiasts would. We have included an entire chapter on state and local horticultural groups.

The two most striking changes in content since the 1982 edition are the explosion of native plant societies and the increased concern with seed preservation and historical horticulture. In 1982, few native plant organizations existed. Now there are several national organizations devoted to wildflowers and native plants, and almost every state has its own native plant society. Similarly, there has been a dramatic increase in the number of groups involved in the preservation of heirloom plant varieties, and historic gardens and landscapes. A chapter has been devoted to both native plant societies and historical horticulture.

Horticultural therapy—another burgeoning area of horticulture—is also better represented in this edition of *North American Horticulture*. We have expanded our coverage to include regional chapters of the American Horticultural Therapy Association, educational programs and internships, and horticultural therapy programs.

This book is a part of the American Horticultural Society's mission to forge connections between the different sections of a very diverse horticultural community and to act as a clearinghouse for information on North

American horticulture. Certainly there are additional organizations which merit inclusion in this listing. We urge any group that was not included to photocopy and fill out the form on page xv or call the American Horticultural Society (800-777-7931). AHS is maintaining a computer database of this information and would like to keep it as complete and up-to-date as possible.

How to Use This Book

There are three types of entries in the following chapters—full entries, ancillary entries, and abbreviated entries. The full entry is the main entry for the organization (for example, the entry for the Denver Botanic Garden in chapter 16, "Botanical Gardens, Arboreta, Conservatories, and Other Public Gardens"). An ancillary entry is one that is cross-referenced to and from the full entry. For example, the ancillary entry for Denver Botanic Garden's horticultural therapy program is cross-referenced to the full entry in chapter 16, and vice versa.

Every full entry and ancillary entry has been given a two-part number, the first part corresponding to the chapter number and the second to the entry number in that chapter. For example, 1.60 (Herb Society of America) is the sixtieth full or ancillary entry in chapter 1.

Abbreviated entries (name, city, state, cross-reference number) without an entry number have been inserted to guide readers to related subject matter in other chapters. For example, the full entry for All-America Rose Selections is in chapter 1 ("National Organizations"). This abbreviated entry has been inserted in chapter 2 ("Plant Societies") to refer readers back to the full entry (1.4):

All-America Rose Selections

Chicago, IL

Cross-reference: 1.4

We have tried to limit full cross-reference information to the full entry as much as possible. For example, the Denver Botanic Garden (16.65) has six cross-references, one to the "Horticultural and Botanical Libraries" chapter (14.14), one to "Conservation" (8.102), two to "Horticultural Therapy" (13.16, 13.66), and two to "Test and Demonstration Gardens" (20.85, 20.272). The ancillary entries for the Denver Botanic Garden in these other chapters only cross-reference back to the full entry (16.65) or to other entries in the same chapter. Thus, entry 13.66 for horticultural therapy programs is as follows:

13.66

Denver Botanic Gardens

Denver, CO

Cross-references: 13.16, 16.65

In some cases there are only ancillary entries cross-referenced to each other and *no* full entry. In this case, each entry contains a good deal of cross-reference information. This type of cross-referencing was necessary to connect distinct parts of the same institution. For example, several different entries relating to Penn State University are all cross-referenced to each other, rather than to one full entry: two entries in chapter 12 ("Education"), the Department of Horticulture (12.278) and the Department of Landscape Architecture (12.447); the Penn State Herbarium (15.169); and four entries in chapter 20 ("Test and Demonstration Gardens"), All-America Selections flower trial ground (20.25), All-America Selections vegetable trial ground

(20.56), All-America Selections display garden (20.203), and All-America Rose Selections test garden (20.288).

Separate chapters have been given to international registration authorities, educational programs, horticultural therapy, libraries, herbaria, community gardens, and test and demonstration gardens, but entries in these chapters are often ancillary entries, cross-referenced to and from a full entry in another chapter.

Please note that in several chapters (such as chapter 3, "Native Plant Societies and Botanical Clubs") the entries are alphabetized by the postal abbreviation of the state. Thus, Alaska (AK) comes before Alabama (AL), Massachusetts (MA) comes before Maine (ME), and so on.

The American Horticultural Society

The American Horticultural Society, founded in 1922, is dedicated to excellence in horticulture—the art and science of growing plants. Through varied programs, AHS seeks to inspire and educate both amateur and professional horticulturists; to provide accurate and current horticultural information to individuals, scientific organizations, educational institutions, and commercial enterprises; and to preserve and enhance our environment.

Members of the American Horticultural Society receive the following benefits:

American Horticulturist magazine is printed in full color and published six times a year. In-depth articles on unique gardens, plants of special interest, and the personalities behind horticulture are all illustrated with magnificent photography.

American Horticulturist News Edition, also published six times yearly, is packed with the latest gardening tips, environmental information, research reports, and public garden news, and contains a calendar of major national and international horticultural events.

The *Gardeners' Information Service* provides answers to gardening problems, information on public gardens, horticultural organizations and programs, plant and seed sources, and other horticultural resources, and publishes a variety of informational brochures. Members may use this service by writing or calling the GIS toll-free phone number.

The *AHS Book Program* brings members the latest and best in gardening books through the mail at substantial discounts.

The *Annual Seed Program* offers members more than 140 varieties of seeds, many of them uncommon and hard to find.

Members are also entitled to attend the annual membership meeting, travel programs, garden symposia, and various special events. For more information on membership in the American Horticultural Society, write to AHS, 7931 East Boulevard Drive, Alexandria, VA 22308.

Entries for the Next Edition of
North American Horticulture

Name _____

Address _____

_____

City _____ State _____ Zip _____

Phone _____ Fax _____

Contact Person _____

Type of organization:

_____ National Horticultural Organization

_____ National Grower Trade Association

_____ Plant Society

_____ Native Plant Society or Botanical Club

_____ Scholarly or Scientific Organization

_____ National Garden Club Organization

_____ State or Provincial Horticultural Society

_____ Nonprofit Garden Center

_____ Conservation Organization

_____ State, Provincial, or Local Horticultural Organization

_____ International Registration Authority

_____ U.S. or Canadian Governmental Program

_____ Educational Program

_____ Horticultural Therapy Program

_____ Horticultural or Botanical Library

_____ Herbarium

_____ Botanical Garden, Arboreta, Conservatory, or Other Public Garden

_____ Historical Horticultural Organization

_____ Museum, Estate, or Historic Garden, Living Historical Farm, or Agricultural Museum

_____ Zoological Park

_____ Cemetery with a Notable Garden

_____ Community Garden

_____ Test or Demonstration Garden

_____ Horticultural Periodical

Briefly describe the purpose of the organization:

Please send to: North American Horticulture, American Horticultural Society, Dept. MN 078, 7931 East Boulevard Drive, Alexandria, VA 22308.

Acknowledgments

This book was a truly collective effort. It would not have been possible to publish *North American Horticulture* without the help of countless librarians, extension agents, professors, and others who directed me towards organizations and sources of information. Many spent quite a bit of effort compiling lists of groups to be included in our book. Perhaps most heartening were the numerous unsolicited cards and letters from people across North America informing us about gardens and horticultural groups that we did not know about. All of you have made this a much more useful reference than it otherwise might have been.

Thanks also go to the hundreds of people who filled out questionnaires for their organizations and sent them in on time. Queries such as ours are inevitably tiresome, and they always seem to arrive en masse, demanding to be returned immediately. We appreciate the trouble that everyone took.

Several individuals went way beyond the call of duty to help with this book. Ina Vrugtman of the Royal Botanical Gardens and Bob McClelland of Agriculture Canada provided much information on Canada—if Canada is underrepresented in these pages it is no fault of theirs. Others who provided essential information are Steven Davis of the American Horticultural Therapy Association, Rick Bonlender and Vernon Bryant of the American Community Gardening Association, Jane Meyer of the Master Gardeners International Corporation, Terry Sharrer of the Association of Living Historical Farms and Museums, Virginia Wall of the Association of Zoological Horticulture, Bernadette Callery of the New York Botanical Garden, Barbara Pitschel of the Strybing Arboretum, Holly Shimizu of the U.S. Botanic Garden, Henry Gilbert of the National Agricultural Library, Trevor Cole of Dominion Arboretum, Mary Walker of the New England Wild Flower Society, Robert Norton of the Agricultural Research Service, Frank Sanchis of the National Trust for Historic Preservation, Scott Kunst, Barbara Barton, Barbara Ingrum, and Nancy Martelli.

Special thanks also to the staff of *American Horticulturist*, Kathy Fisher, Beth Wiesner, and Martha Palermo, Executive Director of AHS Frank Robinson, Director of Programs Joe Keyser, Cathy Gau, the rest of the AHS staff, and most of all to Liisa.

NORTH AMERICAN HORTICULTURE

1

National Organizations

The organizations listed in this chapter share a common interest in horticulture in the broad sense of the term. Most are nonprofit organizations; some are trade associations that function as lobbying groups. Some deal with specific plants (All-America Rose Selections) or types of plants (Herb Society of America, North American Fruit Explorers). Others are concerned with a specific type of gardening (Bonsai Canada, American Rock Garden Society) while still others are professional organizations for a discipline that involves horticulture (Garden Writers Association of America, American Society of Landscape Architects). Eligibility requirements for membership vary—some are expressly for amateur gardeners while others require a professional affiliation.

National grower trade associations are listed at the end of the chapter.

A valuable resource for information on national horticultural organizations is Barbara Barton's *Gardening by Mail* (Boston: Houghton Mifflin Co., 1990). Barton also lists plant and seed sources, and garden supplies and services.

NATIONAL HORTICULTURAL ORGANIZATIONS

UNITED STATES

1.1
Alan Chadwick Society
50 Oak Mountain Dr.
San Rafael, CA 94903

Purpose: Biodynamic organic gardening organization.

1.2
All-America Selections
1311 Butterfield Rd.
Suite 310
Downers Grove, IL 60515
708-963-0770

Purpose: To test new, unsold flowers and vegetables grown from seed.
Garden: 220 display gardens and 62 test gardens across the U.S.
Cross-reference: 20

1.3
All-America Gladiolus Selections
11734 Road 33½

Madera, CA 93638
209-645-5329

Purpose: To provide testing and evaluation of gladiolus seedlings.
Garden: Numerous test gardens in Canada and the U.S.

1.4
All-America Rose Selections
221 N. LaSalle St., #3900
Chicago, IL 60601
312-372-7090; (FAX) 312-372-6160

Purpose: To test new rose varieties and report to the public on any that may have outstanding qualities.
Garden: 24 test gardens and 135 public gardens across the U.S.
Cross-reference: 20

1.5
American Association of Botanical Gardens and Arboreta, Inc.
786 Church Rd.
Wayne, PA 19087
215-688-1120

Purpose: To support public gardens and their professionals.

Publications: *The Public Garden*, quarterly; *AABGA Newsletter*, monthly.
Other: Consulting services, job clearinghouse, workshops.

1.6
American Association of Landscape Contractors
155 Throop St.
North Babylon, NY 11704
516-661-1966

Purpose: To promote the business interests of landscape contractors and related industries and to maintain high standards of ethical practices and cooperation.
Publication: *AALC Journal*, annual.
Scholarships: Scholarship of $750 to third semester horticulture student with ambitions of becoming a landscape contractor.

1.7
American Association of Nurserymen
1250 I St., N.W., Suite 500
Washington, DC 20005
202-789-2900; (FAX) 202-789-1893

Purpose: To represent the interests of the entire nursery business.
Grants: Research grants.
Scholarships

1.8
American Bonsai Society, Inc.
P.O. Box 358
Keene, NH 03431
603-352-9034

Purpose: To promote the art of bonsai and to provide information, advice, and supplies.
Publications: *Bonsai: Journal of the American Bonsai Society*, quarterly; *ABStracts*, 3/year.

1.9
American Botanical Council
P.O. Box 201660
Austin, TX 78720
512-331-8868; (FAX) 512-331-1924

Purpose: Disseminates information on herbs and medicinal plants, promotes research on medicinal plants, fosters plant conservation, and seeks to increase public awareness of the role of medicinal plants in public health care.
Publication: *HerbalGram*, quarterly.
Other: Speakers' bureau, "Classical Botanical Reprints" series of reprints from journals.

1.10
American Chestnut Foundation
West Virginia University
College of Agriculture & Forestry
P.O. Box 6057
Morgantown, WV 26506-6057
507-743-8570

Purpose: To conduct and fund research and applied projects leading to the restoration of the American

chestnut as an important element of forests in the eastern U.S.
Publication: *Journal of the American Chestnut Foundation*, biannual.
Garden: Research farm at Rt. 1, Box 17, Meadowview, VA 24361, with more than 2,000 plant specimens.
Garden not open to public.
Grants: Research grants on all aspects of chestnut biology, particularly those aimed at countering the chestnut blight.

1.11
American Community Gardening Association
325 Walnut St.
Philadelphia, PA 19106
215-625-8280; (FAX) 215-625-8288

Purpose: To support the national movement for public gardening and greening, to create a network for those involved in garden organizing, and to create professionalism among community gardening organizers.
Publications: *Journal of Community Gardening*, quarterly; *Multilogue*, bimonthly.
Cross-reference: 19

1.12
American Floral Marketing Council
1601 Duke St.
Alexandria, VA 22314
703-836-8700; (FAX) 703-836-8705

Purpose: To increase the consumption of floral and plant products.

1.13
American Greenhouse Vegetable Growers
P.O. Box 25058
Colorado Springs, CO 80936
719-531-0505; (FAX) 719-531-0506

Purpose: To distribute technical information to greenhouse vegetable growers.
Publication: *American Greenhouse Vegetable Growers*, bimonthly.

1.14
American Guild of Flower Arrangers
c/o Bob Thomas
Rosewood Plantation
Rt. 1, Box 71
Lamont, FL 32336

Purpose: Professional organization of flower arrangers.

1.15
American Herb Association
P.O. Box 353
Rescue, CA 95672
916-626-5046

Publication: *Newsletter*, quarterly.

1.16
American Horticultural Society
7931 East Boulevard Dr.
Alexandria, VA 22308
703-768-5700; (FAX) 703-765-6032

Purpose: To inspire and educate both amateur and professional horticulturists, and to provide horticultural information to individuals, scientific organizations, educational institutions, and commercial enterprises.

Publications: *American Horticulturist*, bimonthly; *American Horticulturist* News Edition, bimonthly.

Garden: Dahlia test garden, All-America Rose Selections display garden, ivy collection, hosta collection, wildflower meadow, perennials, annuals, spring bulbs, herb garden, orchard, historic boxwoods, and specimen trees.

Library: 14.97

Seed/specimen exchange.

Other: Book Program, Gardeners' Information Service, Annual Seed Program, travel program. The National Backyard Compost Demonstration Park demonstrates a variety of composting methods and provides information and classes on backyard composting. AHS is located at River Farm, once a part of George Washington's estate. The original 1757 house is still used.

Cross-references: 2.19, 2.66, 20.226

1.17

American Horticultural Therapy Association
9220 Wightman Rd., Suite 300
Gaithersburg, MD 20879
301-948-3010

Purpose: To promote and develop horticultural therapy as a therapeutic and rehabilitative medium for persons who are disabled or disadvantaged.

Publications: *People Plant Connection*, 11/year; *Journal of Therapeutic Horticulture*, annual.

Scholarships: $250 scholarship for academic achievement.

Other: Administers Horticulture Hiring the Disabled, a national employment project.

Cross-reference: 13

1.18

American Pomological Society
102 Tyson Building
University Park, PA 16802
814-865-2571

Purpose: To promote fruit variety and rootstock improvement through breeding and evaluation, and to publish current information on fruit variety introductions and performances of existing varieties.

Publication: *Fruit Varieties Journal*, quarterly.

1.19

American Rock Garden Society
15 Fairmead Rd.
Darien, CT 06820

Purpose: To promote the study and growth of alpine, saxatile, and other low-growing perennials.

Publication: *Bulletin of the American Rock Garden Society*, quarterly.

Seed/specimen exchange.

Other: Slide library.

1.20

American Seed Trade Association
1030 15th St., N.W., Suite 964
Washington, DC 20005
202-223-4080; (FAX) 202-293-2617

Purpose: To promote the business interests of persons, firms, and corporations engaged in the seed industry in the U.S., Canada, and Mexico.

1.21

American Society for Enology and Viticulture
P.O. Box 1855
Davis, CA 95617
916-753-3142

Purpose: An educational and research group that supports the study of grapes and wine making.

Publication: *American Journal of Enology and Viticulture*, quarterly.

1.22

American Society of Agricultural Engineers
2950 Niles Rd.
St. Joseph, MI 49085
616-429-0300; (FAX) 616-429-3852

Purpose: A professional and technical organization of those interested in engineering knowledge and technology for agriculture, associated industries, and related resources.

1.23

American Society of Consulting Arborists
700 Canterbury Rd.
Clearwater, FL 34624
813-446-3356

Purpose: To improve consultation service on the use and care of shade and ornamental trees.

Publication: *Arboriculture Consultant*, bimonthly.

Other: Workshops.

1.24

American Society of Golf Course Architects
221 N. LaSalle St.
Chicago, IL 60601
312-372-7090

Purpose: Professional association of golf course architects.

1.25

American Society of Landscape Architects, Inc.
4401 Connecticut Ave., N.W.
5th Floor
Washington, DC 20008-2302
202-686-ASLA

Purpose: A professional organization dedicated to the education and continuing development of landscape architects.

Publications: *Landscape Architecture*, monthly; *Landscape Architecture News Digest*, 10/year.

Other: Employment referral service.

1.26
Associated Landscape Contractors of America
405 N. Washington St.
Falls Church, VA 22046
703-241-4004

Purpose: Professional organization for the landscape contracting industry.
Publication: *Action Letter*, monthly.
Other: Educational programs, trade shows, conferences.

1.27
Association for Women in Landscaping
P.O. Box 22562
Seattle, WA 98122
206-282-0137

Purpose: To promote the participation of women in horticulture.
Publication: *AWL Quarterly*, quarterly.
Other: Hosts the annual Women in Horticulture Conference, seminars.

1.28
Association for Zoological Horticulture
Oklahoma City Zoo
2101 N.E. 50th St.
Oklahoma City, OK 73111
405-424-3344

Purpose: Professional group of zoo horticulturists dedicated to promoting naturalistic landscaping and botanical collections appropriate to animal exhibits.
Publication: *Zoo Horticulture*, quarterly.
Seed/specimen exchange.
Cross-reference: 18

1.29
Association of American Pesticide Control Officials
2004 Le Suer Rd.
Richmond, VA 23229
804-288-8181

Purpose: To promote uniformity in the regulation of pesticides among state officials and the U.S. Environmental Protection Agency.
Publication: *Official Publication*, annual.

1.30
Association of American Plant Food Control Officials
Division of Regulatory Services
University of Kentucky
Lexington, KY 40546
606-257-2668; (FAX) 606-257-7351

Purpose: To promote uniformity of fertilizer regulation in North America.
Publication: *Official Publication*, annual.

1.31
Association of American Seed Control Officials
Oregon DOA, Commodity Inspection
635 Capitol St., N.E.
Salem, OR 97310-0110
503-378-3774

Purpose: A professional association of government-employed seed control officials.

1.32
Association of Consulting Foresters, Inc.
5410 Grosvenor Ln.
Suite 205
Bethesda, MD 20814
301-530-6795

Purpose: A professional organization for consulting foresters that seeks to promote the best professional, economic, and scientific management of forests in the U.S.
Publication: *The Consultant*, quarterly.

1.33
Association of Professional Landscape Designers
221 Morris Rd.
Ambler, PA 19002
215-643-5926; (FAX) 301-933-2916

Purpose: A professional association for landscape designers dedicated to encouraging high ethical standards in the industry.
Publication: *APLD News*, 3/year.

1.34
Association of Specialty Cut Flower Growers, Inc.
155 Elm St.
Oberlin, OH 44074
216-774-2887

Purpose: To unite growers engaged in the production and marketing of field grown and specialty floral crops through education regarding culture, research, and sales of cut flowers.
Publication: *The Cut Flower Quarterly*, quarterly.

1.35
Atlantic Seedsmen's Association
298 E. McCormick Ave.
State College, PA 16801
814-237-0330

Purpose: To disseminate information concerning seeds, crops, markets, and other matters of interest to seedsmen, to foster helpful relations with experiment station, extension, and regulatory personnel, and to cooperate in the advancement of agriculture and horticulture.
Publication: *ASA Newsletter*, quarterly.

1.36
Bio Integral Resource Center
P.O. Box 7414
Berkeley, CA 94707
415-524-2567

Purpose: To provide information on least toxic pest management.
Publications: *Common Sense Pest Control Quarterly*, quarterly; *The IPM Practitioner*, 10/year.

Library: Number of books: 10,000.
Lending policy: Library not open to public.

1.37
Bio-Dynamic Farming and Gardening Association, Inc.
P.O. Box 550
Kimberton, PA 19442
215-935-7797

Purpose: To advance the practices and principles of biodynamic agriculture.
Publications: *Biodynamics*, quarterly; *Bio-Dynamics News & Events*, bimonthly.

1.38
Biological Urban Gardening Services
P.O. Box 76
Citrus Heights, CA 95611
916-726-5377

Purpose: Assists homeowners and landscape managers in reducing or eliminating the use of agricultural chemicals from urban landscapes.
Publication: *BUGS Flyer*, quarterly.
Garden: Test garden for experimenting with products and techniques.
Garden not open to public.
Library: Number of books: 500; number of periodicals: 20.
Lending policy: Library not open to public.

1.39
Bonsai Clubs International
2636 West Mission Rd., #277
Tallahassee, FL 32304
904-575-1442

Purpose: To increase knowledge and skill in the art of bonsai.
Publication: *Bonsai Magazine*, bimonthly.
Library: Slide and book lending library.

1.40
Center for Urban Horticulture
University of Washington
GF-15
Seattle, WA 98195
206-543-8616

Purpose: An academic unit dedicated to excellence in research, teaching, and public service in urban horticulture, the new science dealing with the interactions of people and plants in the urban environment.
Garden: Demonstration and test gardens at Washington Park Arboretum (16.417) and at Union Bay (3501 N.E. 41st St., Seattle) with over 6,000 plant specimens. Emphasis on woody plants.
Library: 14.99
Other: Professional seminars, public classes and lectures, plant information services, slide and photograph lending library, and speakers bureau.
Cross-references: 12.338, 12.461, 15.205, 16.415.

1.41
Chestnut Growers Exchange
P.O. Box 12632
Portland, OR 97212-0632
503-243-1322

Purpose: To promote and develop a commercial chestnut industry in the U.S.
Publication: *Chestnutworks*, biannual.

1.42
Council of Horticulture Association Executives
1250 I St., N.W., Suite 500
Washington, DC 20005
202-789-2900

Purpose: An organization for executives of horticultural organizations.

1.43
Council of Tree & Landscape Appraisers
1250 I St., N.W., Suite 504
Washington, DC 20005
202-789-2592; (FAX) 202-789-1893

Purpose: To establish conformity, expertise, and stature within the tree, landscape, and nursery professions regarding the appraisal of tree and landscape values, and to communicate facts related to that subject to the public.

1.44
Council on Botanical and Horticultural Libraries, Inc.
c/o John F. Reed
The New York Botanical Garden
Bronx, NY 10458
212-220-8728; (FAX) 212-220-6504

Purpose: To improve communication between individuals and institutions concerned with the development, maintenance, and use of libraries of botanical and horticultural literature.
Publications: *Newsletter*, 3-4/year; *CBHL Plant Bibliographies*, occasional.
Other: Circulation of libraries' duplicate lists and want lists.

1.45
Elm Research Institute
Harrisville, NH 03450
800-FOR-ELMS

Purpose: To develop support for research on Dutch Elm Disease.
Publication: *Research Institute News Briefs*, sporadic.
Other: Elm Hot Line information service.

1.46
Eucalyptus Improvement Association
P.O. Box 1963
Diamond Springs, CA 95619

Purpose: To encourage the development and utilization of genetically improved planting stock of eucalyptus for fuelwood, pulp, windbreaks, and other wood products on farm and rangeland in California, and to provide public information on the establishment, management, and harvest of eucalyptus plantations.

1.47
Fertilizer Institute
501 Second St., N.E.
Washington, DC 20002
202-675-8250; (FAX) 202-544-8123

Purpose: A lobbying organization for fertilizer dealers and manufacturers.
Publications: *Action*, monthly, free; *Dealer Progress Magazine*, bimonthly.

1.48
Floral Marketing Division
Produce Marketing Association
1500 Casho Mill Rd.
P.O. Box 6036
Newark, DE 19714-6036
302-738-7100; (FAX) 302-731-2409

Purpose: Floral division of a trade association for the fresh produce and floral industries.
Publications: *PMA Newsletter*, biweekly; *Floral Marketing Directory & Buyer's Guide*.
Other: Floral training programs.

1.49
Florists' Transworld Delivery Association
29200 Northwestern Highway
Southfield, MI 48034
313-355-9300

Purpose: Florist trade association.
Publications: *FTD News*, bimonthly; *Florist*, monthly.

1.50
Forest Farmers Association, Inc.
Box 95385
Atlanta, GA 30347
404-325-2954; (FAX) 404-325-2955

Purpose: To represent the interests of Southern timberland owners.
Publication: *Forest Farmer*, 10/year.
Library.
Lending policy: Reference only.

1.51
Future Farmers of America
5632 Mount Vernon Memorial Highway
Alexandria, VA 22309
703-360-3600; (FAX) 703-360-5524

Purpose: A national organization for young people preparing for careers in agriculture.
Publications: *FFA New Horizons*, bimonthly; *Between Issues*, bimonthly; *Update* monthly.
Scholarships: Scholarships for students studying horticulture and general agriculture.
Library: Number of books: 300; number of periodicals: 15.
Lending policy: Library not open to public.

1.52
FWH Seed Exchange
P.O. Box 651
Pauma Valley, CA 92061
619-742-1720

Purpose: A gardeners' seed exchange that facilitates the trading and selling of seeds through the mail.
Publication: *FWH Seed Exchange Bulletin*, annual.

1.53
Garden Centers of America
1250 I St., N.W.
Suite 500
Washington, DC 20005
202-789-2900; (FAX) 202-789-1893

Purpose: A trade association representing garden center retailers and retail nurseries.
Publication: *GCA Newsletter*, bimonthly.

1.54
Garden Writers Association of America
1218 Overlook Rd.
Eustis, FL 32726
904-589-8888

Purpose: Association of professional garden writers dedicated to improving the standards of horticultural journalism.
Publications: *Quill & Trowel*, bimonthly; *GWAA Membership Directory*, annual.
Scholarships: Three annual scholarships to students studying in fields leading to horticultural communication.

1.55
Gardens for Peace
P.O. Box 13428
Atlanta, GA 30324-9998

Purpose: To identify and link established gardens throughout the world where contemplation and meditation by individuals and communities will foster a climate for peace among all people and a respect for the environment.

1.56
Gay and Lesbian Horticulture Society
P.O. Box 2641
Champaign, IL 61825-2641

Purpose: To provide a professional organization for gays and lesbians in horticulture, botany, landscaping, and all plant sciences.
Publication: *Newsletter*, 2–3/year, free.

1.57
Golf Course Superintendents Association of America
1617 Saint Andrews Dr.
Lawrence, KS 66047-1707
913-841-2240

Purpose: A professional association for golf course superintendents.
Publications: *Golf Course Management*, monthly; *Newsline*, 11/year.

Grants: Research grants.
Scholarships.
Other: Educational seminars, audiovisual library.

1.58
Hardwood Research Council
P.O. Box 34518
Memphis, TN 38184-0518
901-377-1824; (FAX) 901-382-6419

Purpose: A nonprofit scientific and educational
organization that seeks to promote and strengthen
programs of research on critical hardwood issues and to
disseminate related information.

1.59
Herb Research Foundation
P.O. Box 2602
Longmont, CO 80501
303-449-2265

Purpose: A nonprofit research and educational
organization dedicated to raising funds for research on
medicinal plants and herbal products and providing
reliable research data to members, the public, and the
press.
Publication: *HerbalGram*, quarterly.
Grants: Research grants for research on herbal folk
medicines, herbal teas, and other herbal products.

1.60
Herb Society of America
9019 Kirtland Chardon Rd.
Mentor, OH 44060
216-256-0514

Purpose: To further the knowledge and use of herbs.
Publications: *The Herbarist*, annual; *HSA News*, quarterly.
Garden: Demonstration garden at the United States
National Arboretum (16.83) with over 700 specimens.
Grants: Research grants for scientific, academic, or artistic
investigation of herbal plants.
Seed/specimen exchange.

1.61
Hobby Greenhouse Association
8 Glen Terrace
Bedford, MA 01730
617-275-0377

Purpose: To promote greenhouse gardening as a hobby or
avocation, and to disseminate practical information on
greenhouse maintenance and growing techniques.
Publications: *Hobby Greenhouse*, quarterly; *Directory of
Manufacturers of Hobby Greenhouses*, quarterly, free.
Seed/specimen exchange.
Other: Slide, cassette, and videotape library for members.

1.62
Home Orchard Society
P.O. Box 776
Clackamas, OR 97015
503-630-3392

Purpose: To promote home fruit growing.
Publication: *Pome News*, quarterly.
Garden: Home Orchard Society Arboretum at Clackamas
Community College (12.271).
Garden not open to public.
Other: Scion and rootstock exchanges.

1.63
Horticultural Research Institute
1250 I St., N.W., Suite 500
Washington, DC 20005
202-789-2900; (FAX) 202-789-1893

Purpose: The research division of the American
Association of Nurserymen devoted to the advancement
of the landscape and nursery industry through research
and education.
Publication: *Journal of Environmental Horticulture*,
quarterly.
Grants: Awarded over $100,000 for research in 1989–90.

1.64
Hydroponic Society of America
P.O. Box 6067
Concord, CA 94524
415-682-4193

Purpose: To promote hydroponics.
Publication: *Hydroponics/Soilless Growers*, bimonthly.

1.65
Ikebana International
C.P.O. Box 1262
Tokyo, 101-91, Japan

Purpose: To promote the art of ikebana.
Other: Many chapters in North America.

1.66
Indoor Citrus and Rare Fruit Society
490 Spring Grove Rd.
Hollister, CA 95023
Publication: *IC&RF Society Newsletter*, quarterly.
Garden: Test garden. Over 50 plant specimens including
20 citrus varieties. Garden not open to public.
Library: Number of books: 200; number of periodicals:
42.
Lending policy: Library not open to public.
Seed/specimen exchange.
Other: Free seeds to members, plant locator service.

1.67
Indoor Gardening Society of America, Inc.
5305 SW Hamilton St.
Portland, OR 97221
503-292-9785

Purpose: To encourage the growing of plants under
fluorescent lights, in greenhouses, and on windowsills.
Publication: *The Indoor Garden*, bimonthly.
Seed/specimen exchange.

1.68

International Dwarf Fruit Tree Association
2500 Valley Rd.
Winchester, VA 22601
703-667-8330; (FAX) 703-667-5692

Purpose: To develop and disseminate information on the use of rootstocks and orchard culture to growers.
Publications: *Compact News*, sporadic; *Compact Fruit Tree*, annual.

1.69

International Herb Growers & Marketers Association
P.O. Box 281
Silver Spring, PA 17575
717-285-4252

Purpose: Trade association for businesses involved in the growing and marketing of herbs and herb-related products.
Publication: *The Herb Grower and Marketer*, bimonthly.
Library: Number of books 500; number of periodicals: 30.
Lending policy: Library not open to public.

1.70

International Plant Propagators' Society, Inc.
Center for Urban Horticulture
GF-15, University of Washington
Seattle, WA 98195
206-543-8602

Purpose: Dedicated to the science of plant propagation.
Publication: *Annual Proceedings*, annual.

1.71

International Society of Arboriculture
P.O. Box 908
Urbana, IL 61801
217-328-2032

Purpose: A scientific and educational organization devoted to the dissemination of knowledge in the care and preservation of shade and ornamental trees.
Publication: *Journal of Arboriculture*, monthly.
Grants: Research grants. $20,000 awarded in 1990.
Other: Arborbase Information Clearinghouse.

1.72

International Society of Tropical Foresters, Inc.
5400 Grosvenor Ln.
Bethesda, MD 20814
301-897-8720

Purpose: To transfer technology and science to a network of members concerned with the management, protection, and wise use of tropical forests.
Publication: *ISTF News*, quarterly.
Other: Workshops, symposia.

1.73

Interstate Professional Applicators Association, Inc.
P.O. Box 1377
Milton, WA 98354-1377

Purpose: An organization of professional pesticide users.
Publication: *Pest Control Progress*, quarterly.

1.74

Irrigation Association
1911 North Fort Meyer Dr.
Suite 1009
Arlington, VA 22209-1630
703-524-1200; (FAX) 703-524-9544

Purpose: A trade organization representing the irrigation industry at national and international levels.
Publication: *Irrigation News*, monthly.

1.75

Landscape Maintenance Association, Inc.
P.O. Box 728
Largo, FL 34649
813-584-2312

Purpose: To upgrade the landscape maintenance industry through professional seminars, education, certification, and the establishment of standards.
Publication: *Landscape Maintenance Newsletter*, monthly.

1.76

Landscape Materials Information Service
P.O. Box 216
Callicoon, NY 12723
914-887-4401

Purpose: A nonprofit association that acts as a clearinghouse for information for nurserymen, landscape architects, landscape contractors, and related professionals.

1.77

Lawn Institute
P.O. Box 108
Pleasant Hill, TN 38578
615-277-3722

Purpose: To serve the needs of communicators for information on the care of lawns and sports turf.
Publications: *Press Kits*, quarterly; *Harvests Newsletter*, quarterly; *Lawn-O-Gram Newsletter*, quarterly.

1.78

Mailorder Association of Nurseries
8683 Doves Fly Way
Laurel, MD 20723
301-490-9143

Purpose: To promote the general interests of the horticultural industry servicing home gardeners by direct mail.
Publication: *Newsletter*, quarterly.

1.79

Master Gardeners International Corporation (MaGIC)
2904 Cameron Mills Rd.
Alexandria, VA 22302
703-683-6485

Purpose: To promote and improve Extension Master Gardening and home horticulture, and to increase communication among Master Gardeners, Extension personnel, and horticulturists.
Publication: *MaGIC Lantern*, quarterly.
Other: Directory of Master Gardening programs.
Cross-reference: 12

1.80
Municipal Arborists and Urban Foresters Society
P.O. Box 1255
Freehold, NJ 07728-1255
201-431-7903

Purpose: A professional organization representing the municipal branch of the International Society of Arboriculture.
Publication: *Newsletter*, bimonthly.

1.81
Musser International Turfgrass Foundation
P.O. Box 124
Sharon Center, OH 44274
216-239-2383

Purpose: To give fellowship and scholarship grants to graduate students who are pursuing doctoral degrees in the turfgrass sciences.
Scholarships: Scholarships in the turfgrass sciences.

1.82
National Agricultural Chemicals Association
1155 15th St., N.W.
Washington, DC 20005
202-296-1585; (FAX) 202-463-0474

Purpose: Trade organization representing producers of pesticide products.

1.83
National Arbor Day Foundation
100 Arbor Ave.
Nebraska City, NE 68410
402-474-5655

Purpose: A nonprofit educational organization dedicated to tree planting and conservation.
Publications: *Arbor Day*, bimonthly; *Tree City U.S.A. Bulletin*, bimonthly.

1.84
National Arborist Association
The Meeting Place Mall
Rt. 101, P.O. Box 1094
Amherst, NH 03031-1094
693-673-3311

Purpose: To bring together persons and firms engaged in arboriculture, to advance the shade tree preservation industry, and to promote legislation.
Publication: *The Reporter*, monthly.

1.85
National Association of Plant Patent Owners
1250 I St., N.W., Suite 500
Washington, DC 20005
202-789-2900; (FAX) 202-789-1893

Purpose: To foster general environmental enhancement by improving varieties of plants.

1.86
National Association of State Foresters
Hall of the States
444 North Capitol St., N.W.
Washington, DC 20001
202-624-5415; (FAX) 202-624-5407

Purpose: A national nonprofit association representing the directors of state forestry agencies.
Publication: *NASF Washington Update*, monthly, free.

1.87
National Bark and Soil Producers Association
13542 Union Village Circle
Clifton, VA 22024
703-830-5367; (FAX) 703-830-0281

Purpose: To promote product quality and industry standards for bark mulch and soil products.
Publications: *Bark and Soil Producers Report*, quarterly; *National Bark and Soil Product Index*, annual, free.

1.88
National Garden Bureau, Inc.
1311 Butterfield Rd., Suite 310
Downers Grove, IL 60515
708-963-0770

Purpose: To disseminate gardening information on seed to garden communicators in North America.

1.89
National Gardening Association
180 Flynn Ave.
Burlington, VT 05401
802-863-1308

Purpose: To promote gardening activities at home, in institutions, and in the workplace for the betterment of the individual and humanity.
Publication: *National Gardening Magazine*, monthly.
Library: Number of books: 1,000; number of periodicals: 50.
Lending policy: Library not open to public.
Seed/specimen exchange.
Other: Gardening Answer Service, seed search service.
Cross-references: 13.189, 19.95

1.90
National Greenhouse Manufacturers Association
P.O. Box 567
Pana, IL 62557
217-562-2644

Purpose: Trade organization for greenhouse manufacturers.

1.91
National Institute on Park and Grounds Management
P.O. Box 1936
Appleton, WI 54913
414-733-2301

Purpose: To inform and educate professionals in charge of large outdoor areas.
Grants: Research grants to recognized turf research programs relating to sports turf.

1.92
National Junior Horticultural Association
441 E. Pine St.
Fremont, MI 49412
616-924-5237

Purpose: To interest young people in horticulture.
Publication: *Going & Growing*, 3/year.

1.93
National Landscape Association
1250 I St., N.W., Suite 500
Washington, DC 20005
202-789-2900; (FAX) 202-789-1893

Purpose: To educate and promote the interests of
landscape professionals.
Publication: *NLA Landscape News*, bimonthly.

1.94
National Lawn and Garden Distributors Association
1900 Arch St.
Philadelphia, PA 19103
215-564-3484; (FAX) 215-564-2175

Purpose: To further the profitability of all segments of
the lawn and garden industry through the promotion of
harmonious relations between distributor and supplier,
while providing a forum for the exchange of ideas and
industry information.
Publication: *NLGDA Membership Directory*, annual.

1.95
National Pest Control Association
8100 Oak St.
Dunn Loring, VA 22027
703-573-8330

Purpose: Trade organization for the pest control industry.
Publication: *Pest Management*, monthly.

1.96
National Roadside Vegetation Management Association
309 Center Hill Rd.
Wilmington, DE 19807
302-655-9993

Purpose: To exchange ideas and information on safety
and operational functions, roadside beautification, and
ecological soundness, and to establish and administer
standards of recognition and professionalism.
Publication: *Proceeds of Annual Meeting*, annual, free.

1.97
National Wildflower Research Center
2600 FM 973 North
Austin, TX 78725-4201
512-929-3600

Purpose: To promote the conservation and use of native
plants in planned landscapes through research and
education.
Publications: *Wildflower, the Newsletter of the NWRC*,
bimonthly; *Wildflower, Journal of the NWRC*, biannual.
Garden: Research plots with over 100 plant specimens
including the rare penstemon, *Salvia penstemonoides*.
Library: Number of books: 450; number of periodicals:
60.
Lending policy: Library not open to public.
Other: Seminars, conferences, tours.

1.98
National Xeriscape Council
P.O. Box 163172
Austin, TX 78716-3172
512-392-6225

Purpose: To promote, encourage, assist, facilitate, and
establish community programs for water conservation
through sound landscaping practices.
Publication: *Xeriscape News*, bimonthly.

1.99
North American Fruit Explorers
Rt. 1, Box 94
Chapin, IL 62628
217-245-7589

Purpose: Dedicated to amateur and professional fruit
growers who share an interest in the discovery,
cultivation, and appreciation of superior fruit varieties.
Publication: *The Pomona*, quarterly.
Grants: Research grants.
Seed/specimen exchange.
Other: Mail lending library for members, plant cutting
exchange.

1.100
North American Horticultural Supply Association
1900 Arch St.
Philadelphia, PA 19103
215-564-3484; (FAX) 215-564-2175

Purpose: An association of full-service distributors and
manufacturers who are committed to improving the
distribution of horticultural supplies for the betterment
of the entire industry.
Publications: *Industry Events Calendar*, biannual;
Membership Directory, annual.

1.101
North American Mycological Association
3556 Oakwood
Ann Arbor, MI 48104-5213
313-971-2552

Purpose: An organization of individuals, families, and
local clubs interested in mushrooming.
Publications: *McIlvaineae*, annual; *The Mycophile*,
bimonthly.
Other: Compiles the North American Mushroom
Poisoning Case Registry.

1.102
Northern Nut Growers Association
9870 S. Palmer Rd.
New Carlisle, OH 45344
513-878-2610

Purpose: To generate interest in the growing and
propagation of superior nut species for amateurs and
professionals.
Publications: *Nutshell*, quarterly; *Annual Report*, annual.
Grants: Research grants.
Seed/specimen exchange.

1.103

Outdoor Power Equipment Institute, Inc.
341 South Patrick St., Old Town
Alexandria, VA 22314
703-549-7600

Purpose: A trade association composed of manufacturers and their suppliers of consumer and commercial outdoor power equipment, dedicated to promoting the outdoor power equipment industry.
Publications: *Consumer Monthly Report*, monthly; *Monthly Import/Export Report*, monthly; *Industry Outlook*, annual; *Executive Update Newsletter*, monthly.
Other: Audiovisual programs, trade shows.

1.104

Pacific Northwest Fruit Tester's Association
1101 West Orchard Ave.
Selah, WA 98942
509-697-8133

Purpose: To compile evaluations of the fruit testers into an annual report.
Publications: *Testers Annual Report*, annual; *Testers Newsletter*, 3–4/year.
Seed/specimen exchange.
Other: Scion wood exchange program.

1.105

Perennial Plant Association
3383 Schirtzinger Rd.
Columbus, OH 43026
614-771-8431

Purpose: To promote perennials throughout industry and the American gardening public.
Publications: *Perennial Plant Association*, quarterly; *Perennial Plant Proceedings*, annual.
Grants: Grants for production and marketing research.
Other: Symposia, seminars.

1.106

Plant Amnesty
906–NW 87th St.
Seattle, WA 98117
206-783-9813

Purpose: To promote better pruning and responsible garden management practices through a combination of humor, controversy, and sound horticultural advice.
Publication: *Plant Amnesty Newsletter*, quarterly.
Library: Number of books: 20; Number of periodicals: 15.
Specializes in pruning-related information and trees.
Lending policy: Reference only.
Other: Educational literature.

1.107

Pond Society
P.O. Box 449
Acworth, GA 30101
404-975-0277; (FAX) 404-975-0277

Purpose: Dedicated to helping people become successful pond keepers.
Publications: *Pondscapes Journal*, quarterly; *Pondscapes Magazine*, monthly; *Pondscapes National Product and Service Directory*, quarterly.
Seed/specimen exchange.

1.108

Professional Grounds Management Society
10402 Ridgland Road
Suite 4
Cockeysville, MD 21030
301-667-1833

Purpose: To assist managers in developing techniques and management skills to assure outstanding grounds management.
Publication: *Grounds Management Forum*, monthly.
Scholarships: Two $500 or one $1,000 scholarship yearly to college, associate degree, or graduate student studying grounds management or related field.

1.109

Professional Lawn Care Association of America
1000 Johnson Ferry Rd., N.E.
Suite C-135
Marietta, GA 30068-2112
404-977-5222

Purpose: To advance the mutual interests of members' lawn care companies.
Publication: *Pro Source*, bimonthly.

1.110

Professional Plant Growers Association
P.O. Box 27517
Lansing, MI 48909
517-694-7700; (FAX) 517-694-8560

Purpose: To advance the interests of the bedding and container plant industry.
Publications: *PPGA News*, monthly; *PPGA Buyers Guide*, annual, free.
Garden: Demonstration garden at 1980 N. College Rd., Mason, MI, 48854, with All-America Selections and FloraSelect winners.
Scholarships: Awards an average of 15 scholarships, ranging from $750–$2,000, per year to students of horticulture.
Cross-reference: 20.140.

1.111

Rare Fruit Council International
P.O. Box 561914
Miami, FL 33256
305-378-4457

Purpose: To bring tropical fruit to Florida and to encourage people to grow more fruit.
Publication: *Tropical Fruit News*, monthly.

1.112

Rare Pit & Plant Council
251 West 11th St.
New York, NY 10014
212-255-9256

Purpose: To promote the growing of tropical fruits and vegetables from the seeds of the fruits.
Publication: *The Pits*, 10/year.
Seed/specimen exchange.

1.113
Society of American Florists
1601 Duke St.
Alexandria, VA 22314
703-836-8700; (FAX) 703-836-8705

Purpose: A trade association aiming to unify and lead the floriculture industry in addressing governmental and industry issues, and increase the profitability of the industry.
Publications: *SAF—Business News for the Floral Industry*, monthly; *Dateline: Washington*, 24/year.

1.114
Society of American Foresters
5400 Grosvenor Ln.
Bethesda, MD 20814
301-897-8720; (FAX) 301-897-3690

Purpose: To advance the science, education, practice, and teaching of professional forestry and to use the skill of the profession to benefit society.
Publication: *Journal of Forestry*, monthly.
Library: 14.53

1.115
Society of Municipal Arborists
7000 Olive Blvd.
University City, MO 63130
314-862-6767

Purpose: A professional organization that promotes and improves the practice of municipal arboriculture and stimulates public interest in tree planting and preservation.
Publication: *City Trees*, bimonthly.
Grants: Research grants through the Tree Research Trust Fund.

1.116
Southern Fruit Fellowship
Rt. 3, Box 268
Columbus, MS 39701
601-328-2189

Purpose: To exchange information and plants of southern fruits and vegetables.
Publication: *Southern Fruit Fellowship*, quarterly.
Seed/specimen exchange.
Other: Plant exchange program.

1.117
Sports Turf Managers Association
400 North Mountain Ave., Suite 301
Upland, CA 91786
714-981-9199

Purpose: Dedicated to the education of those involved in the management of sports turf facilities using natural grass.

Publication: *The Sports Turf Manager*, quarterly.
Grants: Research grants for research on turf or related areas.
Scholarships: Two annual scholarships for college students enrolled in a four-year program in turf management who plan a career in sports turf management (exclusive of golf).
Other: Regional education seminars.

1.118
Terrarium Association
P.O. Box 276
Newfane, VT 05345
802-365-4721

Purpose: Provides terrarium-related publications to individuals, schools, libraries, and garden clubs. Not a membership organization.
Garden: Test garden at Old Parsonage, Grassy Brook Rd., Brookline, VT with various types of terraria on display. Open by appointment only.
Other: Terraria answer service.

1.119
United Fresh Fruit and Vegetable Association
727 North Washington St.
Alexandria, VA 22314
703-836-3410; (FAX) 703-836-7745

Purpose: The leading organization serving the fresh produce industry, representing growers, shippers, wholesalers, foodservice operators, retailers, and allied industry suppliers.
Publications: *United Newswire*, biweekly; *Potato Letter*, monthly; *Resource Distribution*, quarterly; *Produce Merchandiser*, monthly.

1.120
United States Golf Association
Green Section
P.O. Box 708
Far Hills, NJ 07931
201-234-2300; (FAX) 201-234-9687

Purpose: To collect and distribute information regarding the proper maintenance and upkeep of golf courses.
Publication: *Green Section Record*, bimonthly.
Grants: In 1989 made 21 grants totaling $620,300 to 14 universities.
Library.
Lending policy: Reference only.

1.121
Vinifera Wine Growers Association
P.O. Box P
The Plains, VA 22171
703-754-8564

Purpose: To assist home and commercial vinyardists and vintners to grow the noble European varieties and make superior table wines.
Publication: *Journal of the Vinifera Wine Growers Association*, quarterly.

Garden: Demonstration garden at Highbury Farm, Road 628, The Plains, VA 22171, with about 200 vines of Chardonnay, Riesling, Pinot Noir, Cabernet Sauvignon, Gewurztraminer, and all Old World vinifera varieties.

Library: Number of books: 500; number of periodicals: 40.

1.122

Vocational Agriculture Service
1401 South Maryland Dr.
Urbana, IL 61801
217-333-3871

Purpose: Provides vocational agricultural and horticultural materials for all phases of secondary education programs.

Other: The VAS catalog lists over 1,600 titles in booklet, slidefilm, slideset, and transparency format. It includes items on floriculture, house plants, fruits, landscaping, turfgrass, and vegetables.

1.123

Walnut Council
5603 W. Raymond St.
Suite 0
Indianapolis, IN 46241
317-244-3312; (FAX) 317-244-3386

Purpose: To advance the cultural practices, science, and technology of walnut propagation, encourage walnut tree planting, foster management of existing trees, promote walnut product use, and disseminate information.

Publication: *Walnut Council Bulletin*, quarterly.

1.124

Wholesale Florists & Florist Suppliers of America
5313 Lee Highway
Arlington, VT 22207
703-241-1100; (FAX) 703-237-6438

Purpose: To enhance the economic opportunities of floral wholesale jobbers and those floral suppliers who market through them.

Publication: *LINK Magazine*, 10/year.

1.125

Wholesale Nursery Growers of America
1250 I St., N.W., Suite 500
Washington, DC 20005
202-789-2900; (FAX) 202-789-1893

Purpose: To serve the interests of wholesale nursery growers.

Publication: *The Grower*, bimonthly.

1.126

Woman's National Farm and Garden Association, Inc.
2402 Clearview Dr.
Glenshaw, PA 15116
412-486-7964

Purpose: To encourage conservation of natural resources, establish scholarships in agriculture, horticulture, ocean-

ography and related professions, and to develop opportunities for those so trained.

Scholarships: Scholarships for students of horticulture, horticultural therapy, and landscape design.

CANADA

1.127

Bonsai Canada
12 Beardmore Crescent
Willowdale, ON M2K 2P5

Purpose: To promote and communicate information on the art of bonsai.

Publication: *Bonsai Canada*, annual.

1.128

Canadian Golf Superintendents Association
2000 Weston Rd., Suite 203
Weston, ON M9N 1X3
416-249-7304; (FAX) 416-249-8467

Purpose: Professional association for golf superintendents.

Publication: *Greenmaster*, 8/year.

Other: Employment referral service.

1.129

Canadian Horticultural Council
Danny Dempster, Exec. V-P
1101 Prince of Wales Dr.
Suite 310
Ottawa, ON K2C 3W7
613-226-4187

Purpose: An industry association, dedicated to the improvement of the horticultural and allied industries in Canada.

1.130

Canadian Horticultural Therapy Association
P.O. Box 399
Hamilton, ON L8N 3H8
416-529-7618; (FAX) 416-577-0375

Purpose: To promote the use of horticulture as a therapeutic modality.

Publication: *CHTA Newsletter*, quarterly.

Cross-reference: 13

1.131

Canadian Nursery Trades Association
1293 Matheson Blvd.
Mississauga, ON L4W 1R1
416-629-1367

Purpose: Nursery trade association.

1.132

Canadian Organic Growers
P.O. Box 6408, Station "J"
Ottawa, ON K2A 3Y6

Purpose: To promote and disseminate information on organic growing techniques.

Publication: *COGnition*, quarterly.

Seed/specimen exchange.
Other: Heritage Seed Program, mail lending library.

1.133
Canadian Ornamental Plant Foundation
c/o Lloyd Muray, P.O. Box 725
Durham, ON N0G 1R0
519-369-5382

Purpose: To facilitate the introduction of new and
improved ornamental cultivars.
Publication: *Newsletter*, quarterly.

1.134
Canadian Produce Marketing Association
310—1101 Prince of Wales Dr.
Ottawa, ON K2C 3W7
613-226-4187; (FAX) 613-226-2984

Purpose: A trade association of businesses involved in the
distribution and marketing of fresh fruits and
vegetables.
Publication: *Communique*, monthly.

1.135
Canadian Society for Herbal Research
P.O. Box 82, Station A
Willowdale, ON M2N 5S7
705-432-2418

Purpose: Public education concerning herbalism.
Publication: *The Herbalist*, quarterly.
Garden: Test garden of medicinal herb plants in
Sunderdale, ON. Garden not open to public.
Library: Number of books: 150; number of periodicals: 0.
Lending policy: Library not open to public.
Other: Database of medicinal plants.

1.136
Canadian Society of Landscape Architects
306 Metcalfe St.
Ottawa, ON K2P 1S2
613-232-6342

Publication: *Landscape Architects Review*, bimonthly.

1.137
Canadian Wildflower Society
75 Ternhill Crescent
North York, ON M3C 2E4
416-499-7907

Purpose: To encourage the study, conservation, and
cultivation of the native wild flora of North America.
Publication: *Wildflower*, quarterly.
Seed/specimen exchange.
Other: Annual native plant sales, workshops.

1.138
Flowers Canada
155 Suffolk St. West
Guelph, ON N1H 2J7
519-823-2670; (FAX) 519-823-8920

Purpose: A national trade association whose purpose is to
promote the interests of floriculture and ornamental
horticulture.
Publication: *Flowers Canada News*, bimonthly.

1.139
Garden Research Exchange
536 MacDonnell St.
Kingston, ON K7K 4W7
613-542-6547

Purpose: A home gardeners' research exchange that
publishes research reports received from gardeners in
Canada and the United States.
Publication: *Vegetable Garden Research*, annual.

1.140
La Société d'animation du Jardin et de l'Institut
botaniques
Jardin botanique, local 125
4101 est, rue Sherbrooke
Montreal, QC H1X 2B2
514-872-1493

Purpose: Dedicated to increasing the knowledge of
horticulture and botany, and especially of the flora of
Quebec.
Publications: *Quatre-Temps*, 3/year; *Liaison-SAJIB*, 3/year.

1.141
Society of Commercial Seed Technologists
P.O. Box 1712
Brandon, MB R7A 6S3
(FAX) 204-328-7400

Purpose: To encourage professional standards and
cooperation between official seed analysts and
commercial agencies.
Publication: *SCT Newsletter*.

NATIONAL GROWER TRADE ASSOCIATIONS

The following are trade associations for fruit, nut, vegetable, mushroom,
turf, and Christmas tree grower organizations. They may be able to provide
information about specific plants or references to other helpful sources.

For more information on horticultural and agricultural trade associations, consult the annual *Directory of American Agriculture* (Pierre, South Dakota: Agricultural Resources & Communications, Inc.). This useful directory
lists national and state organizations, governmental offices and programs, and
much more.

UNITED STATES

1.142

American Corn Growers Association
P.O. Box 18157
Washington, DC 20036
202-223-9802

1.143

American Cranberry Growers Association
581 Magnolia
Pemberton, NJ 08068
609-894-9231

1.144

American Dry Pea & Lentil Association
5071 Highway 8 West
Moscow, ID 83843
208-882-3023; (FAX) 208-882-6406

1.145

American Mushroom Institute
907 East Baltimore Pike
Kennett Square, PA 19348
215-388-7807

1.146

American Peanut Research and Education Society
376 Ag. Hall
Oklahoma State University
Stillwater, OK 74078
405-744-6423

1.147

American Sugarbeet Growers Association
1156 15th St., N.W.
#1020
Washington, DC 20005
202-833-2398; (FAX) 202-785-7312

1.148

American Viticultural Association
P.O. Box 1146
Middletown, CA 95461
707-987-2385; (FAX) 707-987-9351

1.149

American Wine Society
3006 Latta Rd.
Rochester, NY 14612
716-225-7613

1.150

Concord Grape Association
5775 Peachtree-Dunwoody Rd.
Suite 500-D
Atlanta, GA 30342
404-252-3663

1.151

Cranberry Institute
1101 Fifteenth St., N.W.
Suite 202
Washington, DC 20005
202-785-3232

1.152

Federated Pecan Growers Association
Knapp Hall
LSU
Room 214
Baton Rouge, LA 70803
504-388-2222

1.153

Fresh Garlic Association
P.O. Box 2410
Sausalito, CA 94966-2410

1.154

Green Gem Bluegrass Growers Association
607 Great Western Bldg.
Spokane, WA 99201
509-624-9263

1.155

International Apple Institute
P.O. Box 1137
McLean, VA 22101
703-442-8850; (FAX) 703-790-0845

1.156

International Pumpkin Association
2155 Union St.
San Francisco, CA 94123
415-346-4446

1.157

International Wild Rice Association
P.O. Box 903
Grand Rapids, MN 55744
213-327-6796

1.158

Jojoba Growers Association
2201 East Camel Back
Suite 130-B
Phoenix, AZ 85016
602-956-8887

1.159

Leafy Greens Council
Box 76067
St. Paul, MN 55175

1.160

Merion Bluegrass Association
12341 25th N.E.
Seattle, WA 98125
206-365-7548

1.161
Mushroom Growers Association
18 S. Water Market
Chicago, IL 60608
312-421-7088

1.162
National Christmas Tree Association
611 East Wells St.
Milwaukee, WI 53202
414-276-6410; (FAX) 414-276-3349

1.163
National Corn Growers Association
1000 Executive Pkwy.
Suite 105
St. Louis, MO 63141
314-275-9915; (FAX) 314-275-7061

1.164
National Dry Bean Council
P.O. Box 6008
Saginaw, MI 48608
517-790-3010; (FAX) 517-790-3747

1.165
National Lime Association
3601 N. Fairfax Dr.
Arlington, VA 22201
703-243-5463

1.166
National Onion Association
One Greeley National Plaza
Suite 510
Greeley, CO 80631
303-353-5895; (FAX) 303-353-5897

1.167
National Peach Council
P.O. Box 122
Vance, SC 29163
803-492-7724

1.168
National Peanut Council of America
1500 King St.
Suite 301
Alexandria, VA 22314
703-838-9500; (FAX) 703-838-9508

1.169
National Sunflower Association
4023 N. State St.
Bismarck, ND 58501
701-224-3019; (FAX) 701-224-2498

1.170
North American Blueberry Council
P.O. Box 166
Marmora, NJ 08223
609-399-1559; (FAX) 609-399-1590

1.171
Popcorn Institute
111 E. Wacker Dr.
Suite 600
Chicago, IL 60601
312-644-6610

1.172
Potato Association of America
Hancock Agricultural Research Stn.
Rt. 1, Box 115
Hancock, WI 54943
715-249-5712

1.173
Southeastern Peanut Association
P.O. Box 70157
Albany, GA 31707
912-888-2508; (FAX) 912-888-5150

1.174
Southwestern Peanut Growers Association
P.O. Box 338
Gorman, TX 76454
817-734-2222; (FAX) 817-734-2288

1.175
Sweet Potato Council of the U.S.
P.O. Box 14
Marsh Hill Rd.
McHenry, MD 21541
301-387-9537

CANADA

1.176
Canadian Christmas Tree Growers Association
66 Pleasant St.
Bridgewater, NS B4N 1N1

1.177
Canadian Mushroom Growers Association
310-1101 Prince of Wales Dr.
Ottawa, ON K2C 3W7
613-226-4187

1.178
Northern Nut Growers Association
R.R. 1
Niagara on the Lake, ON L0S 1J0
416-262-4927

1.179
Wild Blueberry Association of North America
18 Floral Ave.
Fredericton, NB E3A 1K7
506-472-2517

2

Plant Societies

As gardeners develop more knowledge about their hobby or profession, many develop an interest in a single species or a related plant group. This interest in specialization is so widespread that scores of plant societies have been formed in North America solely to promote African violets, camellias, roses, and some less well-known plants such as penstemons and cycads.

Most of these organizations publish newsletters or magazines and sponsor plant and seed exchanges. Many encourage research, promote hybridization of new varieties for the general market, and work toward the standardization of nomenclature. Most also have local chapters, some of which maintain test and demonstration gardens.

Addresses of many of the listed organizations change annually with the election of new officers.

UNITED STATES

2.1
African Violet Society of America, Inc.
P.O. Box 3609
Beaumont, TX 77704-3609
409-839-4725

Publication: *African Violet Magazine*, bimonthly.
Cross-reference: 10.69

All-America Selections
Downers Grove, IL

Cross-reference: 1.2

All-America Gladiolus Selections
Madera, CA

Cross-reference: 1.3

All-America Rose Selections
Chicago, IL

Cross-reference: 1.4

2.2
American Bamboo Society
P.O. Box 640
Springville, CA 93265
209-539-2145

Publications: *American Bamboo Society Newsletter*, bimonthly; *Journal of the American Bamboo Society*, irregular.

Garden: Demonstration and test gardens at Quail Botanical Gardens (16.47).
Seed/specimen exchange.

2.3
American Begonia Society
P.O. Box 56
Bio Dell, CA 95562-0056
619-461-6906

Publication: *The Begonian*, bimonthly.
Cross-reference: 10.8

2.4
American Boxwood Society
Box 85
Boyce, VA 22620
703-939-4646

Publication: *The Boxwood Bulletin*, quarterly.
Garden: Boxwood garden at Orland E. White Arboretum (16.401).
Garden not open to public.
Grants: Research grants for projects on *Buxus*.
Cross-reference: 10.11

2.5
American Calochortus Society
260 Alden Rd.
Hayward, CA 94541
415-276-2414

Publication: *Mariposa,* quarterly.
Garden: Demonstration garden at East Bay Regional Parks Botanic Garden in Berkeley, California; test garden at Chabot College, Livermore, California.

2.6
American Camellia Society
P.O. Box 1217
Fort Valley, GA 31030
912-967-2358

Publications: *The Camellia Journal,* quarterly; *Camellia Yearbook,* annual.
Garden: Massee Lane Gardens: approximately 30 species and 1,500 specimens of camellias; also a rose garden, Japanese garden, and a landscaped greenhouse.
Grants: American Camellia Society Endowment Fund partially supports four researchers in the study of diseases, cold and heat hardiness, and the eradication of camellia and petal blight.
Library: 14.25

American Chestnut Foundation
Morgantown, WV

Cross-reference: 1.10

2.7
American Conifer Society
Box 314
Perry Hall, MD 21128
301-256-5595

Publication: *Bulletin,* quarterly.

2.8
American Daffodil Society, Inc.
1686 Grey Fox Trails
Milford, OH 45150
513-248-9137

Publication: *The Daffodil Journal,* quarterly.
Garden: Several test gardens, the largest, at Whetstone Park (16.316), includes over 800 cultivars and/or species.
Grants: Research grants.
Library: Number of books: 75. Historical and other works on daffodils, including *RHS Yearbook* from 1913 to present.
Lending policy: Members only by special arrangement.

2.9
American Dahlia Society
10 Roland Pl.
Wayne, NJ 07470
201-694-4864

Publication: *Bulletin,* quarterly.
Grants: Research grants.
Library.

2.10
American Fern Society
Dept. of Botany
University of Vermont
Burlington, VT 05401

Publications: *Fiddlehead Forum,* bimonthly; *American Fern Journal,* quarterly; *Pterologia,* irregular.
Seed/specimen exchange.
Other: Spore exchange.

2.11
American Fuchsia Society
San Francisco County Fair Bldg.
9th Ave. & Lincoln Way
San Francisco, CA 94122
408-257-0752

Publication: *The American Fuchsia Society Bulletin,* bimonthly.
Scholarships: Scholarships at local colleges.
Library: Number of books: 100. All volumes pertain to fuchsia.
Lending policy: Reference only.
Cross-reference: 10.28

2.12
American Ginger Society
P.O. Box 100
Archer, FL 32618
904-495-9168

Publication: *Zingiber,* irregular.
Garden: Demonstration garden at Tom Wood's Nursery in Archer, Florida, includes 175 species and varieties of Zingiberaceae.
Garden not open to public.
Seed/specimen exchange.
Other: Slide collection, computer database on species.

2.13
American Gloxinia and Gesneriad Society
Box 5493
Beverly Farms, MA 01915-9997

Publication: *The Gloxinian,* bimonthly.
Seed/specimen exchange.
Other: Lending library of slide programs.
Cross-reference: 10.29

2.14
American Gourd Society
P.O. Box 274
Mount Gilead, OH 43338-0274
419-946-3302

Publication: *The Gourd,* quarterly.
Seed/specimen exchange.

2.15
American Hemerocallis Society
1454 Rebel Dr.
Jackson, MS 39211
601-366-4362

Publication: *Daylily Journal,* quarterly.
Other: Slide library.
Cross-reference: 10.36

American Herb Association
Rescue, CA
Cross-reference: 1.15

2.16
American Hibiscus Society
P.O. Box 321540
Cocoa Beach, FL 32932-1540
407-783-2576

Publication: *The Seed Pod,* quarterly.
Grants: Research grants.
Scholarships: Scholarships by individual chapters.
Seed/specimen exchange.
Other: Slide library, educational forums.

2.17
American Hosta Society
5206 Hawksbury Ln.
Raleigh, NC 27606
919-851-4784

Publication: *The Hosta Journal,* biannual.

2.18
American Iris Society
7414 East 60th St.
Tulsa, OK 74145
918-627-0706

Publication: *Bulletin of the American Iris Society,* quarterly.
Grants: Occasional grants for programs pertaining to the genus *Iris.*
Scholarships: One annual scholarship to a postgraduate student in the plant science field.
Seed/specimen exchange.
Other: Judge accreditation and training program.
Cross-reference: 10.42

2.19
American Ivy Society
P.O. Box 520
West Carrollton, OH 45449-0520
513-434-7069

Publications: *The Ivy Journal,* biannual; *The Ivy Newsletter,* biannual.
Garden: Display and hardiness trial gardens at the American Horticultural Society (1.16), Brookside Gardens (16.184), Chicago Botanic Garden (16.138), College of William and Mary, Longwood Gardens (16.348), Mendocino Coast Botanical Gardens (16.38), Miracosta College, the Smithsonian Institution, the University of British Columbia (16.449), Botanical Gardens of Volusis, Inc., and the American Ivy Society Hardiness Trials Garden at Lebanon, Ohio.
Library: Volumes on *Hedera.*
Lending policy: Reference only.
Other: Identification service.
Cross-reference: 10.35

2.20
American Orchid Society
6000 South Olive Ave.
West Palm Beach, FL 33405
407-585-8666

Publications: *AOS Bulletin,* monthly; *Awards Quarterly,* quarterly; *Lindleyana,* quarterly.
Grants: Research grants.
Scholarships.

2.21
American Penstemon Society
1569 South Holland Ct.
Lakewood, CO 80226
303-986-8096

Publication: *Bulletin of American Penstemon Society,* biannual.
Garden: Demonstration gardens at Cox Arboretum (16.297) and 1410 Eudora St., Denver, Colorado. Garden not open to public.
Grants: Occasional small research grants.
Scholarships: Occasional scholarships.
Library.
Lending policy: Circulating.
Seed/specimen exchange.
Cross-reference: 10.57

2.22
American Peony Society
250 Interlachen Rd.
Hopkins, MN 55343

Publication: *Bulletin,* quarterly.
Seed/specimen exchange.
Cross-reference: 10.55

2.23
American Poinsettia Society
P.O. Box 706
Mission, TX 78572-1256
512-585-1256
Other: Society is only partly active at the present and is not accepting new memberships.

American Pomological Society
University Park, PA
Cross-reference: 1.18

2.24
American Primrose Society
6620 N.W. 271st Ave.
Hillsboro, OR 97124
503-640-4582

Publication: *Primroses,* quarterly.
Seed/specimen exchange.
Other: Slide library.

2.25
American Rhododendron Society
P.O. Box 1380
Gloucester, VA 23061
804-693-4433

Publication: *American Rhododendron Society Journal,* quarterly.

Grants: Research grants handled by the American Rhododendron Society Research Foundation.
Scholarships: Some chapters have scholarships.
Seed/specimen exchange.
Other: Pollen bank, slide library.

American Rock Garden Society
Darien, CT
Cross-reference: 1.19

2.26
American Rose Society
P.O. Box 30000
Shreveport, LA 71130
318-938-5402

Publications: *The American Rose*, monthly; *The American Rose Annual*, annual.
Garden: Test garden of old garden roses at 8877 Jefferson Paige Rd. in Shreveport.
Grants: Research grants for rose-related research.
Library: 14.42
Cross-references: 2.80, 10.68

2.27
American Willow Growers Network
RD 1, Box 124A
South New Berlin, NY 13843
607-847-8264

Publication: *Newsletter*, annual.
Garden: Test garden. Over 60 varieties of *Salix*, including basketry, soil stabilization, and windbreak willows.
Seed/specimen exchange.
Other: Cutting exchange.

2.28
Aril Society International
5500 Constitution, N.E.
Albuquerque, NM 87110
505-255-8207

Publication: *Yearbook*.

2.29
Australian Plant Society of California
P.O. Box 50722
Pasadena, CA 91115

Publication: *The Australian Plant Society Newsletter*, quarterly.
Other: Membership includes a subscription to the quarterly *Australian Plants*, published by the Society for Growing Australian Plants.

2.30
Azalea Society of America, Inc.
P.O. Box 34536
West Bethesda, MD 20827-0536
301-585-5269

Publication: *The Azalean*, quarterly.

2.31
Bromeliad Society, Inc.
2488 East 49th
Tulsa, OK 74105

Publication: *Journal*, bimonthly.
Seed/specimen exchange.
Cross-reference: 10.10

2.32
Bromeliad Study Group of Northern California
1334 S. Van Ness
San Francisco, CA 94110
415-647-8114

2.33
Cactus and Succulent Society of America
1535 Reeves St.
Los Angeles, CA 90035
213-556-1923

Publications: *Cactus and Succulent Journal*, bimonthly; *CSSA Newsletter*, bimonthly.
Grants: Research grants for the study of cactus and succulents in the field and the laboratory.
Library: Number of books: 200; number of periodicals: 3.
Lending policy: Reference only.
Seed/specimen exchange.
Other: Slide library, speakers' bureau.

2.34
California Rare Fruit Growers
The Fullerton Arboretum
California State University, Fullerton
Fullerton, CA 92634

Publications: *The Fruit Gardener*, quarterly; *Journal*, annual.
Garden: Display gardens at the San Diego Zoo (18.7) and at California Polytechnic (12.25).

Chestnut Growers Exchange
Portland, OR
Cross-reference: 1.41

2.35
Colorado Water Garden Society
528 South Alcott
Denver, CO 80219
303-922-9559

Publication: *News—The Newsletter of CWGS*, bimonthly.

2.36
Cryptanthus Society
3629 Bordeaux Court
Beaumont, TX 76016-2809
409-835-0644

Publication: *Journal*, quarterly.
Other: Slide library.

2.37
Cycad Society
1161 Phyllis Ct.
Mountain View, CA 94040
415-964-7898

Publication: *The Cycad Newsletter*, quarterly.
Seed/specimen exchange.
Other: Seed and pollen bank.

2.38
Cymbidium Society of America
6881 Wheeler Ave.
Westminster, CA 92683
714-894-5421

Publication: *The Orchid Advocate*, bimonthly.

2.39
Dwarf Iris Society of America
3167 E. U.S. 224
Ossian, IN 46777
219-597-7403

Publication: *Dwarf Iris Society Newsletter*, 3/year.
Garden: Display gardens across the U.S.

Elm Research Institute Harrisville, NH
Cross-reference: 1.45

2.40
Epiphyllum Society of America
P.O. Box 1395
Monrovia, CA 91017
818-447-9688

Publication: *The Bulletin*, bimonthly.

Eucalyptus Improvement Association
Diamond Springs, CA
Cross-reference: 1.46

2.41
Gardenia Society of America
P.O. Box 879
Atwater, CA 95301
209-358-2231

Publication: *Gardenia Quarterly*, quarterly.

2.42
Gesneriad Hybridizers Association
4115 Pillar Dr., Rt. 1
Whitmore Lake, MI 48189

Publication: *Crosswords*, 3/yr.

2.43
Gesneriad Society International
2119 Pile St.
Clovis, NM 88101-3597

Publication: *Gesneriad Journal*, bimonthly.

2.44
Hardy Fern Foundation
P.O. Box 60034
Richmond Beach, WA 98160-0034

Garden: Future display garden will be located at the
Rhododendron Species Foundation Garden (2.79),
Washington. The first test garden will be at the New
York Botanical Garden (16.286).
Garden not open to public.

2.45
Hardy Plant Society
Mid-Atlantic Group
49 Green Valley Rd.
Wallingford, PA 19086

Publication: *Newsletter*, quarterly.
Other: An affiliate of the Hardy Plant Society of Great
Britain.

2.46
Hardy Plant Society of Oregon
33530 S.E. Bluff Rd.
Boring, OR 97009
503-643-2146

Publication: *HPSO Bulletin*, biannual.
Grants: Annual grants for any horticultural purpose in
Oregon.
Library: Number of books: 25.
Lending policy: Reference only.
Seed/specimen exchange.

2.47
Heliconia Society International
Flamingo Gardens
3750 Flamingo Rd.
Fort Lauderdale, FL 33330
305-473-2955

Publication: *Heliconia Society International Bulletin*,
quarterly.
Garden: Demonstration and test gardens at Flamingo
Gardens (16.88), Harold Lyon Arboretum (16.111), and
the National Tropical Botanical Garden (16.121).
Flamingo Gardens has over 300 specimens.

Herb Research Foundation
Longmont, CO
Cross-reference: 1.59

Herb Society of America
Mentor, OH
Cross-reference: 1.60

Heritage Rose Foundation
Raleigh, NC
Cross-reference: 17.15

Heritage Roses Group—North Central
Monroe, WI
Cross-reference: 17.16

Heritage Roses Group—Northeast
Clinton Corners, WI
Cross-reference: 17.17

Heritage Roses Group—Northwest
Kent, WA
Cross-reference: 17.18

Heritage Roses Group—South Central
Austin, TX
Cross-reference: 17.19

Heritage Roses Group—Southeast
Raleigh, NC
Cross-reference: 17.20

Heritage Roses Group—Southwest
Berkeley, CA
Cross-reference: 17.21

Heritage Roses Group—Southwest
Pacific Grove, CA
Cross-reference: 17.22

Historic Iris Preservation Society
Colorado Springs, CO
Cross-reference: 17.23

2.48
Holly Society of America, Inc.
304 North Wind Rd.
Baltimore, MD 21204
301-825-8133

Publication: *Holly Society Journal*, quarterly.
Garden: Many demonstration and test gardens.
Garden not open to public.
Grants: Research grants.
Cross-reference: 10.41

Home Orchard Society
Clackamas, OR
Cross-reference: 1.62

2.49
Hoya Society International, Inc.
P.O. Box 54271
Atlanta, GA 30308

Publication: *The Hoya*, quarterly.
Garden: Demonstration garden at the Atlanta Botanical
    Garden (16.103) with over 200 specimens including
    H. imbricata and *H. darwinii*.
Grants: Occasional research grants.

Indoor Citrus and Rare Fruit Society
Los Altos, CA
Cross-reference: 1.66

2.50
International Aroid Society, Inc.
Box 43–1853
Miami, FL 33143
305-271-3767

Publications: *Aroideana*, quarterly; *Newsletter*, 11/yr.
Cross-reference: 10.6

2.51
International Camellia Society
P.O. Box 750
Brookhaven, MS 39601-0150
601-833-2718; (FAX) 601-833-6891

Publication: *The International Camellia Journal*, annual.

2.52
International Carnivorous Plant Society
Fullerton Arboretum
California State University
Fullerton, CA 92634

Publication: *Carnivorous Plant Newsletter*, quarterly.
Seed/specimen exchange.

International Dwarf Fruit Tree Association
Winchester, VA ·
Cross-reference: 1.68

2.53
International Geranium Society
4610 Druid St.
Los Angeles, CA 90032-3202
213-222-6809

Publication: *Geraniums Around the World*, quarterly.
Seed/specimen exchange.

International Gladiolus Hall of Fame
Greeley, CO
Cross-reference: 17.24

2.54
International Golden Fossil Tree Society, Inc.
201 West Graham Ave.
Lombard, IL 60148-3333
708-627-5636

Publication: *The Irregular Newsletter of the International
    Golden Fossil Tree Society*, 4–5/year.

2.55
International Lilac Society, Inc.
P.O. Box 315
Rumford, ME 04276
207-562-7453

Publications: *Lilacs*, quarterly; *Proceedings*, annual.
Grants: Grants for research in the genus *Syringa*.
Seed/specimen exchange.

2.56
International Oak Society
14870 Kingsway Dr.
New Berlin, WI 53151
414-786-0383

Seed/specimen exchange.

2.57
International Oleander Society
P.O. Box 3431
Galveston, TX 77552-0431
409-762-9334

Publication: *Nerium News*, quarterly.
Seed/specimen exchange.
Other: Slide program available to members.

2.58
International Ornamental Crabapple Society
Morton Arboretum
Lisle, IL 60532
708-698-0074

Publication: *Malus*, biannual.
Garden: Test garden with 49 cultivars.
Seed/specimen exchange.

2.59
International Palm Society
P.O. Box 368
Lawrence, KS 66044
913-843-1235

Publication: *Principes*, quarterly.
Seed/specimen exchange.

2.60
International Tropical Fern Society
8720 S.W. 34th St.
Miami, FL 33165
305-221-0502

2.61
International Water Lily Society
School of Science, Southwest Texas State University
San Marcos, TX 78666
512-245-2119

Publication: *Water Garden Journal*, quarterly.
Garden: Several demonstration gardens across the
country.
Grants: Research grants for the identification and
registration of *Nymphaea* and *Nelumbo*.
Other: Education slide sets on water gardening.

2.62
Los Angeles International Fern Society
P.O. Box 90943
Pasadena, CA 91109-0943

Publications: *LAIFS Journal; Monthly Fern Lesson*,
monthly.

2.63
Magnolia Society, Inc.
907 South Chestnut St.
Hammond, LA 70403-5102
504-542-9477

Publication: *Magnolia: Journal of the Magnolia Society*,
biannual.
Seed/specimen exchange.
Other: Slide library.
Cross-reference: 10.48

2.64
Marigold Society of America, Inc.
P.O. Box 112
New Britain, PA 18901
215-348-5273

Publication: *Amerigold Newsletter*, quarterly.
Seed/specimen exchange.
Other: Slide set on marigolds.

2.65
Mycological Society of San Francisco
P.O. Box 11321
San Francisco, CA 94101
415-759-0495

Publication: *Mycena News*, 9/year.
Scholarships: One or two scholarships annually ranging to
$1,000 to college students of mycology in northern
California.
Library: Number of books: 300 of mycological interest.
Lending policy: Members only, reference only.
Other: Classes, forays.

2.66
National Chrysanthemum Society, Inc.
10107 Homar Pond Dr.
Fairfax Station, VA 22039
703-978-7981

Publication: *The Chrysanthemum*, quarterly.
Garden: Demonstration garden at the American
Horticultural Society (1.16).

2.67
National Cucumber Conference
c/o Todd Wehner
Dept. of Horticultural Sciences
Raleigh, NC 27695-7609

2.68
National Fuchsia Society
11507 E. 187th St.
Artesia, CA 90701
213-865-1806

Publication: *The Fuchsia Fan*, bimonthly.

2.69
Nebraska Herbal Society
P.O. Box 4493
Lincoln, NE 68504
402-466-6897

Publication: *Nebraska Herbal Society Newsletter*, quarterly.
Garden: Demonstration garden at Folsom Children's Zoo
(18.30) with over 60 plant specimens.
Seed/specimen exchange.

North American Fruit Explorers
Chapin, IL
Cross-reference: 1.99

2.70
North American Gladiolus Council
11102 West Calumet Rd.
Milwaukee, WI 53224-3109
414-354-7859

Publications: *North American Gladiolus Society Bulletin*,
quarterly; *Gladio-Grams*, quarterly.
Garden: Test garden at the Gulf Coast Research and
Education Center (11.108), with thousands of
specimens.
Other: Speakers bureau.
Cross-reference: 10.30

2.71
North American Heather Society
62 Elma-Monte Rd.
Elma, WA 98541
206-482-3258

Publication: *Heather News*, quarterly.
Grants: Grants to establish heather plantings.

2.72
North American Lily Society, Inc.
P.O. Box 272
Owatonna, MN 55060
507-451-2170

Publications: *Quarterly Bulletin*, quarterly; *Lily Yearbook*,
annual.
Library: Books on lilies available for loan by mail to
members.
Seed/specimen exchange.

North American Mycological Association
Ann Arbor, MI
Cross-reference: 1.101

Northern Nut Growers Association
New Carlisle, OH
Cross-reference: 1.102

2.73
Northwest Fuchsia Society
P.O. Box 33071
Seattle, WA 98133-0071
206-364-7735

Publication: *Fuchsia Flash*, 10/year.

2.74
Northwest Perennial Alliance
P.O. Box 45574
University Station
Seattle, WA 98145
206-525-6245

Publication: *The Perennial Post*, quarterly.

Pacific Northwest Fruit Tester's Association
Selah, WA
Cross-reference: 1.104

2.75
Pacific Orchid Society
P.O. Box 1091
Honolulu, HI 96808

Publication: *Hawaiian Orchid Journal*, quarterly.

2.76
Peperomia Society International
5240 W. 20th St.
Vero Beach, FL 32960

Publication: *The Gazette*, quarterly.
Seed/specimen exchange.

Perennial Plant Association
Columbus, OH
Cross-reference: 1.105

2.77
Plumeria Society of America
P.O. Box 22791
Houston, TX 77227-2791
713-780-8326

Publication: *Plumeria Potpourri*, quarterly.
Cross-reference: 10.63

Rare Fruit Council International
Miami, FL
Cross-reference: 1.111

Rare Pit & Plant Council
New York, NY
Cross-reference: 1.112

2.78
Reblooming Iris Society
1146 W. Rialto
Fresno, CA 93705
209-229-6434

Publication: *The Reblooming Iris Recorder*, biannual.

2.79
Rhododendron Species Foundation
P.O. Box 3798
Federal Way, WA 98063
206-661-9377

Publication: *RSF Newsletter*, quarterly.
Garden: Demonstration garden at Weyerhaeuser Way S.,
Federal Way, Washington, with 8,500 rhododendron
species and over 1,000 taxa of other plants.
Library: 14.100
Seed/specimen exchange.
Other: Plant, seed, and pollen distribution.
Cross-reference: 2.44

2.80
Rose Hybridizers Association
3245 Wheaton Rd.
Horseheads, NY 14845
607-562-8592

Publication: *Newsletter*, quarterly.

Garden: Test gardens at the American Rose Society (2.26) and Boerner Botanical Gardens (16.419).
Library: Number of books: 300; number of periodicals: 250.
Lending policy: Books available to members by mail.
Seed/specimen exchange.

2.81
Saintpaulia International
1650 Cherry Hill Rd. South
State College, PA 16803
814-237-7410

Publication: *Saintpaulia International News*, bimonthly.
Seed/specimen exchange.

2.82
Sedum Society
Rt. 2, Box 130
Sedgwick, KS 67135
316-796-0496

Publications: *Sedum Society Newsletter*, quarterly.
Seed/specimen exchange.
Other: Slide library, cuttings exchange.

2.83
Sempervivum Fanciers Association
37 Ox Bow Ln.
Randolph, MA 02368
617-963-6737

Publication: *Sempervivum Fanciers Association Newsletter*, quarterly.
Seed/specimen exchange.

2.84
Society for Japanese Irises
16815 Falls Rd.
Upperco, MD 21155
301-374-4788

Publication: *The Review*, biannual.
Garden: Display gardens across America.
Library.
Lending policy: Books available to members by mail.

2.85
Society for Louisiana Irises
P.O. Box 40175
Lafayette, LA 70504
318-264-6203

Publication: *Newsletter*, quarterly.
Grants: Occasional research grants.

2.86
Society for Pacific Coast Native Iris
4333 Oak Hill Rd.
Oakland, CA 94605
415-638-0658

Publication: *The Almanac*, biannual.
Seed/specimen exchange.

2.87
Society for Siberian Irises
631 G24 Highway
Norwalk, IA 50211

Publication: *The Siberian Iris*, biannual.
Garden: Display gardens in Canada and the United States.
Other: Circulating slide library.

2.88
Solanaceae Enthusiasts
3370 Princeton Ct.
Santa Clara, CA 95051
408-241-9440

Publication: *Solanaceae Enthusiasts*, quarterly.
Seed/specimen exchange.

Southern Fruit Fellowship
Columbia, MS
Cross-reference: 1.116

2.89
Species Iris Group of North America
150 North Main St.
Lombard, IL 60148
312-627-1421

Publication: *SIGNA*, biannual.
Seed/specimen exchange.

2.90
Spuria Iris Society
10521 Bellarose Dr.
Sun City, AZ 85351
602-977-2354

Publication: *Spuria Newsletter*, biannual.

Texas Rose Rustlers
Houston, TX
Cross-reference: 17.37

2.91
Tropical Flowering Tree Society
Fairchild Tropical Garden
10901 Old Cutler Rd.
Miami, FL 33156
305-248-0818

Publication: *Newsletter*, quarterly.
Seed/specimen exchange.
Other: Field trips.

2.92
Tubers
c/o Steve Neal
HCR 4, Box 169-D
Gainesville, MO 65655-9729

Publication: *Tater Talk*, biannual.

Walnut Council
Indianapolis, IN
Cross-reference: 1.123

2.93
Woody Plant Society
c/o Betty Ann Mech
1315–66th Ave. Northeast
Minneapolis, MN 55432
612-574-1197

Publication: *Bulletin*, biannual.

2.94
World Pumpkin Confederation
14050 Gowanda State Rd.
Collins, NY 14034
716-532-5995; (FAX) 716-532-5690

Publication: *World Pumpkin Confederation Newsletter*,
  quarterly.
Garden: Test garden. Giant varieties of pumpkin.
Garden not open to public.
Seed/specimen exchange.

CANADA

2.95
African Violet Society of Canada
1573 Arbordale Ave.
Victoria, BC V8N 5J1
(604) 477-7561

Publication: *Chatter*, quarterly.

2.96
Alpine Garden Club of British Columbia
Box 5161–MPO
Vancouver, BC V6B 4B2

Publication: *Alpine Garden Club Bulletin*, 5/year.
Seed/specimen exchange.
Other: Lectures, field trips.

2.97
British Columbia Begonia and Fuchsia Society
6773 Tyne St.
Vancouver, BC V5S 3M4

2.98
British Columbia Lily Society
5510 239th St.
Langley, BC V3A 7N6
604-534-4729

Publication: *BCLS Newsletter*, 5/year.
Grants: Research grants.
Library: Number of books: 5; number of periodicals: 25.
Lending policy: Reference only.

2.99
Canadian Chrysanthemum and Dahlia Society
140 Centennial Rd.
Scarborough, ON M1C 1Z5
416-286-5798

Publications: *CCDS Bulletin*, monthly; *CCDS Yearbook*,
  annual.

2.100
Canadian Geranium and Pelargonium Society
4040 West 38th Ave.
Vancouver, BC V6N 2Y9

Publication: *Storksbill*, quarterly.

2.101
Canadian Gladiolus Society
1274–129A St.
Ocean Park, BC V4A 3Y4
604-536-8200

Publications: *Bulletin*, annual; *Canadian Gladiolus Annual*,
  annual.

2.102
Canadian Iris Society
199 Florence Ave.
Willowdale, ON M2N 1G5
416-225-1088

Publication: *Canadian Iris Society Newsletter*, quarterly.
Garden: Demonstration garden at the Royal Botanical
  Gardens (16.473) with hundreds of varieties of irises.
Grants: Occasional research grants for research on
  improving growing methods or disease and insect
  control.
Library: Circulating iris library for members only.

2.103
Canadian Orchid Society
128 Adelaide St.
Winnipeg, MB R3A 0W5
204-943-6870

Publication: *The Canadian Orchid Journal*, quarterly.

Canadian Ornamental Plant Foundation
Durham, ON

Cross-reference: 1.133

2.104
Canadian Peony Society
1246 Donlea Crescent
Oakville, ON L6J 1V7
416-845-5380

Seed/specimen exchange.

2.105
Canadian Prairie Lily Society
22 Red River Rd.
Saskatoon, SK S7M 2C2
306-747-3776

Publication: *C.P.L.S. Newsletter*, quarterly.
Seed/specimen exchange.

2.106
Canadian Rose Society
686 Pharmacy Ave.
Scarborough, ON M1L 3H8
519-751-4850

Publications: *The Rosarian*, 3/year; *Canadian Rose Annual*,
  annual.

Garden: Numerous demonstration gardens across Canada.
Library: Circulating rose library for members

Canadian Society for Herbal Research
Willowdale, ON

Cross-reference: 1.135

2.107
Dahlia Society of Nova Scotia
Box 83
Great Village, NS B0M 1L0
902-668-2838

Publication: *The Dahlia Society of Nova Scotia Newsletter*, quarterly.
Seed/specimen exchange.
Other: Dahlia tuber exchange program.

2.108
Newfoundland Alpine & Rock Garden Club
Memorial University Botanical Garden
University of Newfoundland
St. John's, NF A1C 5S7
709-737-8590

2.109
Ontario Herbalists Association
7 Alpine Ave.
Toronto, ON M6P 3R6
416-536-3835

Publication: *The Canadian Journal of Herbalism*, quarterly.
Garden: Herb garden at Balance Life Gardens (16.457), Ontario, with 400 specimens; open by appointment.

2.110
Ontario Rock Garden Society
c/o Andrew Osyany
Box 146
Shelburne, ON L0N 1S0

Publication: *Ontario Rock Garden Society Journal*, 10/year.
Seed/specimen exchange.
Other: A chapter of the American Rock Garden Society.

2.111
Rhododendron Society of Canada
5200 Timothy Crescent
Niagara Falls, ON L2E 5G3
416-357-5981

Publication: *Bulletin*, biannual.
Seed/specimen exchange.

2.112
Royal Saintpaulia Club
c/o Ms. A. Moffett
Box 198
Sussex, NB E0E 1P0

2.113
Saskatchewan Perennial Society
Forestry Farm Park & Zoo
Sutherland P.O.
Saskatoon, SK S7N 2H0
306-242-5329

Publication: *Flora Borealis*, bimonthly.
Library: Number of books: 10; number of periodicals: 20. Emphasis on perennials.
Lending policy: For members only, reference only.

2.114
Toronto Cactus & Succulent Club
P.O. Box 334
Brampton, ON L6V 2L3
416-877-6013

Publication: *Cactus Factus*, 8/yr.

3

Native Plant Societies and Botanical Clubs

Native plant societies and state botanical clubs are involved in the protection and preservation of plants indigenous to a state or region and the study of local and regional flora. Their activities range from conducting plant inventories and plant propagation to restoration projects, plant rescues, and the sponsorship of habitat protection legislation. Most also conduct educational programs, lead field trips, and public a newsletter.

Although there is no national native plant society as such, the Center for Plant Conservation (8.20), the National Wildflower Research Center (1.97), and the Canadian Wildflower Society (1.137) are engaged in many similar concerns as the groups listed in this chapter, and may be able to provide more information on native plant societies in your area. The New England Wild Flower Society (3.22) also functions as a national native plant organization. They publish a very useful pamphlet, *Botanical Clubs and Native Plant Societies of the United States*, which is periodically updated. Another essential source for those interested in native plants is the *1988 Plant Conservation Resource Book* (Center for Plant Conservation: Jamaica Plain, 1988). (See the introduction to Chapter 8 for a description.)

Many of the addresses listed below change on a regular basis when new officers are elected for the organization.

UNITED STATES

AK
3.1
Alaska Native Plant Society
c/o Verna Pratt
P.O. Box 141613
Anchorage, AK 99514
907-333-8212

Publication: *Newsletter,* biannual.

AL
3.2
Alabama Wildflower Society
11120 Ben Clements Rd.
Northport, AL 35476
205-339-2541

Publication: *The Alabama Wildflower Society Newsletter,*
    biannual.
Scholarships: Blanche E. Dean Scholarship to college

students who show the promise of making contributions to Alabama in the field of botany or ornithology.
Recent awards have been of $500.
Other: Field trips, programs.

AR
3.3
Arkansas Native Plant Society
c/o Dr. James Gulden
Dept. of Forest Resources
University of Arkansas, Monticello
Monticello, AR 71655
501-460-1049

AZ
3.4
Arizona Native Plant Society
P.O. Box 41206
Sun Station
Tucson, AZ 85717

Publication: *The Plant Press,* 3/year.

29

CA
3.5
California Botanical Society
Mona Bourell
Dept. of Botany
California Academy of Sciences
San Francisco, CA 94118

Publication: *Madrono*, quarterly.

3.6
California Native Plant Society
909 12th St., Suite 116
Sacramento, CA 95814
916-447-2677

Publications: *Fremontia*, quarterly; *Bulletin*, quarterly.
Grants: Research grants.
Scholarships.
Other: Educational programs, rare plant program,
   publishes the *Inventory of Rare and Endangered Vascular
   Plants of California*.

3.7
California Spring Blossom & Wildflower Association
San Francisco County Fair Bldg.
9th Ave. and Lincoln Way
San Francisco, CA 94122

Scholarships: Two $150 scholarships awarded to students
   in the horticulture program at San Francisco City
   College.
Library: Number of books: 100.
Lending policy: Library not open to public.

3.8
Southern California Botanists
Dept. of Biology
Fullerton State University
Fullerton, CA 92634
714-449-7034

Publication: *Crossosoma*, bimonthly.

CO
3.9
Colorado Native Plant Society
P.O. Box 200
Fort Collins, CO 80522

Publication: *Aquilega*, bimonthly.
Other: Field trips, seminars, workshops, classes.

CT
3.10
Connecticut Botanical Society
Osborn Memorial Laboratory
167 Prospect St.
New Haven, CT 06511

Grants: Supports research of the Nature Conservancy.
Library.
Lending policy: Library not open to public.
Cross-reference: 15.41

FL
3.11
Florida Native Plant Society
5505 N.W. 92 Way
Gainesville, FL 32606
904-378-2054

Publication: *The Palmetto*, quarterly.
Garden: Local chapters have demonstration gardens.
Seed/specimen exchange.
Other: Conferences, workshops.

GA
3.12
Georgia Botanical Society
1645 Kellogg Springs Dr.
Dunwoody, GA 30338
404-396-6858

Publications: *Georgia Botanical Society Newsletter*,
   bimonthly; *Tipularia*, biannual.
Seed/specimen exchange.
Other: Field trips, special programs.

HI
3.13
Hawaiian Botanical Society
3190 Maile Way
Dept. of Botany
University of Hawaii
Honolulu, HI 96822

Publication: *Newsletter of the Hawaiian Botanical Society*,
   irregular.
Grants: For projects relating to Hawaiian botany.

ID
3.14
Idaho Native Plant Society
P.O. Box 9451
Boise, ID 83707

Publication: *Sage Notes*, bimonthly.
Other: Slide collection of Idaho's rare and sensitive
   species; annual rare plant meeting.

IL
3.15
Illinois Native Plant Society
R.R. 1, Box 495A
Westville, IL 61883
217-662-2142

Publication: *The Harbinger*, quarterly.
Library.
Lending policy: Library not open to public.
Other: Field trips, maintains prairie plot with Illinois
   endangered species in it.

3.16
Southern Illinois Native Plant Society
c/o Dr. Robert Mohlenbrock
Botany Dept., Southern Illinois University
Carbondale, IL 52901

KS
3.17
Kansas Wildflower Society
Mulvane Art Center
Washburn University
Topeka, KS 66621
913-295-6324

Publication: *Kansas Wildflower Society Newsletter*, quarterly.
Seed/specimen exchange.
Other: Education programs, field trips.

KY
3.18
Kentucky Native Plant Society
Dept. of Biological Sciences
Eastern Kentucky University
Richmond, KY 40475

Publication: *KNPS Newsletter*, quarterly.
Seed/specimen exchange.
Other: Field trips, workshops.

LA
3.19
Louisiana Native Plant Society
717 Giuffrias
Metairie, LA 70001
504-831-2342

Publication: *LNPS Newsletter*, quarterly.
Seed/specimen exchange.

3.20
Louisiana Project Wildflower
Lafayette Natural History Museum
637 Girard Park Dr.
Lafayette, LA 70503
318-235-6181

Publication: *Louisiana Project Wildflower Newsletter*, quarterly.

MA
3.21
New England Botanical Club, Inc.
Harvard University Herbaria
22 Divinity Ave.
Cambridge, MA 02138
617-495-2365

Publication: *Rhodora*, quarterly.
Grants: Annual award of up to $1,000 for graduate
students to pursue field studies in systematic botany or
plant ecology in connection with the flora of New
England.
Library: Number of books: 200; number of periodicals: 2.
Lending policy: Reference only.
Other: Herbaria of vascular and nonvascular plants of
New England.

3.22
New England Wild Flower Society, Inc.
Garden in the Woods
Hemenway Rd.
Framingham, MA 01701
508-877-6574

Publication: *Wild Flower Notes*, annual.
Garden: Includes 50 nationally endangered species
including double trillium, double bloodroot, showy
lady's-slipper, yellow lady's-slipper, pitcher plants,
western skunk cabbage, Braun's holly fern,
turkeybeard, double marsh marigold, white Virginia
bluebells, white bleeding heart, and pixie moss. One of
the best wildflower gardens in North America. Also 15
acres of landscaped gardens including a lily pond,
rhododendron grove, lady's-slipper paths, and a bog.
Grants: Funds research concerning cultivation and
propagation of temperate North American native plants.
Average $1,000 per grant.
Library: 14.48
Seed/specimen exchange.
Other: Slide library with 25,000 color slides of native
plants; many publications; public programs.
Cross-reference: 12.167

ME
3.23
Josselyn Botanical Society
c/o Marilyn Dwelley
P.O. Box 41
China, ME 04926

Publication: *Christmas Newsletter*, annual.
Scholarships: Scholarships to botany students.

MI
3.24
Michigan Botanical Club
Matthaei Botanical Gardens
1800 Dixboro Rd.
Ann Arbor, MI 48105
313-998-7061

Publication: *Michigan Botanist*, quarterly.
Other: Big tree survey; field trips.

MN
3.25
Minnesota Native Plant Society
220 Biological Sciences Center
1445 Gortner Ave.
St. Paul, MN 55108
612-625-1234

Publication: *The Plant Press*, 3/yr.
Seed/specimen exchange.
Other: Field trips.

MO
3.26
Missouri Native Plant Society
c/o John Darel
P.O. Box 176
Dept. of Natural Resources
Jefferson City, MO 65102

MS
3.27
Mississippi Native Plant Society
P.O. Box 2151
Starkville, MS 39759
601-325-7570

Publication: *Mississippi Native Plant Society Newsletter*,
  quarterly.
Garden: 3 gardens at various locations.
Other: Field trips, lectures.

MT
3.28
Montana Native Plant Society
P.O. Box 992
Bozeman, MT 59771-0992
406-587-0120

Publication: *Kelseya*, quarterly.
Other: Field trips.

NC
3.29
North Carolina Wild Flower Preservation Society
c/o North Carolina Botanical Garden
Totten Center,
CB #3375, UNC-CH
Chapel Hill, NC 27599-3375

Publication: *North Carolina Wild Flower Preservation Society
  Newsletter*, biannual.
Grants: See below.
Scholarships: Tom and Bruce Shinn scholarship and
  grant fund is awarded to undergraduate students for
  research on the biology or horticulture of native
  plants.

3.30
Southern Appalachian Botanical Club, Inc.
Dept. of Biology, Catawba College
Salisbury, NC 28144
704-637-4442

Publication: *Castanea*, quarterly.
Other: Co-sponsor of the Wildflower Pilgrimage in the
  Great Smoky Mountains National Park.

NJ
3.31
Native Plant Society of New Jersey
P.O. Box 1295
Morristown, NJ 07962-1295

Publication: *Newsletter*, quarterly.
Grants: Funds research into propagation of terrestrial
  orchids.
Seed/specimen exchange.

NM
3.32
Native Plant Society of New Mexico
443 Live Oak Loop N.W.
Albuquerque, NM 87122
505-356-3942

Publication: *Newsletter*, bimonthly.
Seed/specimen exchange.

NV
3.33
Northern Nevada Native Plant Society
P.O. Box 8965
Reno, NV 89507
702-358-7759

Publication: *Newsletter*, 9/year.
Seed/specimen exchange.

NY
Center for Plant Conservation
Jamaica Plain, NY
Cross-reference: 8.20

3.34
Niagara Frontier Botanical Society
Buffalo Museum of Science
1020 Humboldt Pkwy.
Buffalo, NY 14211
716-896-5200

Publication: *Clintonia*, quarterly.
Other: Monthly lectures (September through May).

3.35
Syracuse Botanical Club
101 Ambergate Rd.
Dewitt, NY 13214
315-445-0985

3.36
Torrey Botanical Club
New York Botanical Garden
200th St. and Southern Blvd.
Bronx, NY 10458
212-220-8987; (FAX) 212-220-6504

Publications: *Bulletin of the Torrey Botanical Club*,
  quarterly; *Memoirs of the Torrey Botanical Club*, irregular.
Grants: $250 research grants for graduate student
  research.
Other: Field trips, lectures.

OH
3.37
Ohio Native Plant Society
6 Louise Dr.
Chagrin Falls, OH 44022
216-338-6622

Publication: *On the Fringe*, bimonthly.
Grants: $500/year for work in the area of native plants.
Seed/specimen exchange.
Other: Field classes.

OR
3.38
Native Plant Society of Oregon
1920 Engel Ct., N.W.
Salem, OR 97304
503-585-9419

Publication: *NPSO Bulletin*, monthly.

PA
3.39
Botanical Society of Western Pennsylvania
401 Clearview Ave.
Pittsburgh, PA 15205
412-921-1797

Publication: *Wildflowers*, monthly.
Garden: Demonstration garden in Titus Bog Natural Area.
Library: Number of books: 500.
Lending policy: Library not open to public.
Other: Weekly field trips.

3.40
Muhlenberg Botanical Society
North Museum
Franklin & Marshall College
College Ave.
Lancaster, PA 17604
717-393-6529

3.41
Pennsylvania Native Plant Society
1806 Commonwealth Bldg.
316 Fourth Ave.
Pittsburgh, PA 15222

RI
3.42
Rhode Island Wild Plant Society
12 Sanderson Rd.
Smithfield, RI 02917
401-789-6405

Publication: *Newsletter*, quarterly.

SD
3.43
Great Plains Botanical Society
P.O. Box 461
Hot Springs, SD 57747
605-745-3397

TN
3.44
Tennessee Native Plant Society
Dept. of Botany
University of Tennessee
Knoxville, TN 37996-1100
615-974-2256

Publication: *TNP*, bimonthly.
Seed/specimen exchange.
Other: Field trips.

TX
3.45
El Paso Native Plant Society
7760 Maya
El Paso, TX 79912
915-584-8690

Publication: *El Paso Native Plant Society Newsletter*,
   monthly.
Seed/specimen exchange.

National Wildflower Research Center
Austin, TX
Cross-reference: 1.97

3.46
Native Plant Project
P.O. Box 1433
Edinburg, TX 78540-1433
512-380-0310

Publication: *The Sabal*, 8/year.

3.47
Native Plant Society of Texas
210 W. 8th St., Suite A
P.O. Box 891
Georgetown, TX 78627
512-863-7794

Publication: *NPSOT News*, bimonthly.
Grants: Occasional.
Seed/specimen exchange.
Other: Field days.

3.48
Native Prairies Association of Texas
P.O. Box 331376
Fort Worth, TX 76163
817-292-5588

Publication: *The Native Prairies Association of Texas
   Newsletter*, quarterly.
Library: Number of books: 50; number of periodicals: 6.
   All of the proceedings of the North American Prairie
   Conferences.
Lending policy: Reference only.
Seed/specimen exchange.
Other: Active role in management of 40-acre Penn Prairie
   in Southwest Dallas.

UT
3.49
Utah Native Plant Society
c/o Red Butte Gardens and Arboretum
Building 436, University of Utah
Salt Lake City, UT 84112
801-581-5322

Publication: *The Sego Lily*, bimonthly.
Grants: Funding for graduate students of Utah institutions
   to study and/or prepare recovery plans for threatened,
   endangered, or sensitive native plants of Utah.
Seed/specimen exchange.
Other: Field trips, lectures, educational programs.

VA
3.50
Virginia Native Plant Society
P.O. Box 844
Annandale, VA 22003
703-368-3943

Publication: *Bulletin*, quarterly.

Library: Number of books: 50.
Lending policy: Library not open to public.
Other: Workshops, field trips, garden tours.

VT
3.51
Vermont Botanical and Bird Club, Inc.
HCR #65, River Rd.
Killington, VT 05751
801-442-3646

Publication: *Vermont Botanical and Bird Clubs Newsletter*,
  sporadic.

WA
3.52
Washington Native Plant Society
Dept. of Botany, KB-15
University of Washington
Seattle, WA 98195
206-543-1976

Publication: *Douglasia*, quarterly.
Grants: For plant inventories.
Other: Field trips, workshops, programs.

WI
3.53
Botanical Club of Wisconsin

c/o Wisconsin Academy of Sciences, Arts and Letters
1922 University Ave.
Madison, WI 53705
608-263-1692

Publications: *Bulletin of the Botanical Club of Wisconsin*,
  quarterly; *Wisconsin Flora*, irregular.
Other: Field trips, workshops, spring wildflower fair.

WY
3.54
Wyoming Native Plant Society
P.O. Box 1471
Cheyenne, WY 82003

Publication: *Newsletter*, 3/year.
Scholarships: For research on the native flora of
  Wyoming.

CANADA

ON
Canadian Wildflower Society
North York, ON

Cross-reference: 1.137

4

Scholarly and Scientific Organizations

Membership in scholarly organizations is usually restricted to professional horticulturists, agronomists, or botanists, rather than amateurs. Most members are likely to be research scientists. These organizations usually publish scholarly journals. Many have regional or local branches.

One reference that may be of use to those seeking more information on scholarly and scientific organizations is Elisabeth B. Davis's *Guide to Information Sources in the Botanical Sciences* (Littleton, Colorado: Libraries Unlimited, Inc., 1987).

UNITED STATES

4.1
American Bryological and Lichenological Society
Dept. of Biology
University of Nebraska at Omaha
Omaha, NE 68182-0072

Purpose: devoted to the scientific study of all aspects of the biology of bryophytes and lichen-forming fungi.
Publication: *The Bryologist*, quarterly.
Seed/specimen exchange.
Cross-reference: 15.66

American Chestnut Foundation
Morgantown, WV

Cross-reference: 1.10

4.2
American Phytopathological Society
3340 Pilot Knob Rd.
St. Paul, MN 55121
612-454-7250; (FAX) 612-454-0766

Purpose: Devoted to the study of plant diseases and their control.
Publications: *Phytopathology*, monthly; *Plant Disease*, monthly; *Molecular Plant-Microbe Interactions*, bimonthly.

4.3
American Plant Life Society
P.O. Box 985
National City, CA 92050
619-474-3530

Purpose: Dedicated to research and education, particularly in the *Amaryllis*, *Iris*, and *Lilium* families.
Publications: *Herbertia*, annual; *Newsletter*, quarterly.
Library: 14.12
Seed/specimen exchange.
Cross-reference: 10.4

American Pomological Society
University Park, PA

Cross-reference: 1.18

American Society for Enology and Viticulture
Davis, CA

Cross-reference: 1.21

4.4
American Society for Horticultural Science
113 South West St.
Alexandria, VA 22314-2824
703-836-4606; (FAX) 703-836-2024

Purpose: To promote and encourage national and international interest in scientific research and education in horticulture.
Publications: *Journal of the American Society for Horticultural Science*, bimonthly; *HortScience*, monthly; *ASHS Newsletter*, monthly.
Scholarships: Scholarship fund established, but no awards yet.
Library.
Lending policy: Reference only.
Other: Placement service for jobs in horticultural science.

American Society of Agricultural Engineers
St. Joseph, MI

Cross-reference: 1.22

4.5
American Society of Pharmacognosy
P.O. Box 13145
Pittsburgh, PA 15243
412-648-8579; (FAX) 412-648-1086

Purpose: Dedicated to the promotion, growth, and development of pharmacognosy and all sciences related to or dealing in natural products.
Publication: *Journal of Natural Products,* bimonthly.
Grants: Research grants in pharmacognosy and natural products chemistry.

4.6
American Society of Plant Taxonomists
c/o Neil A. Harriman, Biology Dept.
University of Wisconsin-Oshkosh
Oshkosh, WI 54901
414-424-1002

Purpose: To promote teaching and research in plant taxonomy.
Publication: *Systematic Botany,* quarterly.
Grants: Grants of $500 to members to conduct research in plant systematics.

4.7
American Society of Plant Physiologists
15501 Monona Dr.
Rockville, MD 20855
301-251-0560

Purpose: Professional society of plant physiologists, plant biochemists, and other plant scientists.
Publication: *ASPP Newsletter,* bimonthly.

4.8
American Type Culture Collection
Plant Germplasm Collection
12301 Parklawn Dr.
Rockville, MD 20852
301-881-2600; (FAX) 301-770-2587

Purpose: An independent, nonprofit service institution dedicated to the collection, preservation, and distribution of authentic cultures of living organisms, including plant tissue culture.
Other: Scientific databases, training programs, and workshops.
Cross-reference: 15.98

4.9
Association of Systematics Collections
730 11th St., N.W., 2nd Floor
Washington, DC 20001
202-347-2850; (FAX) 202-347-0072

Purpose: To support and represent natural history museums, herbaria, universities, and botanical gardens that have biological collections used for research.
Publications: *ASC Newsletter,* bimonthly; *Washington Initiative,* monthly.
Other: Information on funding opportunities, information on botanical collections and researchers, and workshops.

4.10
Botanical Society of America
Dept. of Ecology and Evolutionary Biology
University of Connecticut
Storrs, CT 06269-3043
203-486-4322

Purpose: Supports botanical projects, provides opportunity for the presentation of research, and provides direction for the publication of studies.
Publications: *American Journal of Botany,* monthly; *Plant Science Bulletin,* quarterly.

Center for Urban Horticulture
Seattle, WA
Cross-reference: 1.40

Chihuahuan Desert Research Institute
Alpine, TX
Cross-reference: 8.21

Council on Botanical and Horticultural Libraries, Inc.
Bronx, NY
Cross-reference: 1.44

4.11
Desert Legume Program
University of Arizona
P.O. Box 3607
College Station
Tucson, AZ 85722
602-621-9492

Purpose: To develop a seed bank of legumes, native or adapted to arid regions around the world, and to evaluate arid-climate legumes for potential uses and disseminate this information.
Publication: *ARIDUS,* quarterly.
Garden: Test garden at the Campus Agricultural Center, 2120 East Allen Rd., Tucson, Arizona 85719 with 85 plant specimens. Open by appointment.
Seed/specimen exchange.
Other: Seminars.
Cross-references: 12.19, 12.398

Elm Research Institute
Harrisville, NH
Cross-reference: 1.45

4.12
Flora of North America
Missouri Botanical Garden
P.O. Box 299
St. Louis, MO 63166
314-577-9550; (FAX) 314-577-9558

Purpose: A project to prepare, edit, and publish a synoptic flora of the 1,700–2,000 species of vascular plants of North America north of Mexico, and to develop a computerized database from this information.
Publication: *Flora of North America Newsletter,* bimonthly, free.

Hardwood Research Council
Memphis, TN

Cross-reference: 1.58

Horticultural Research Institute
Washington, DC

Cross-reference: 1.63

4.13

Hunt Institute for Botanical Documentation
Carnegie Mellon University
Pittsburgh, PA 15213
412-268-2434

Purpose: A research institute studying all aspects of the
history of plant science, including bibliography,
biography, iconography, and art.
Publications: *Huntia*, irregular; *Bulletin*, biannual.
Library: 14.86

4.14

Interamerican Society for Tropical Horticulture
18905 S.W. 280 St.
Homestead, FL 33031
305-246-6340

Purpose: To foster and encourage research on
horticultural crops, particularly those grown in the
tropics.
Publication: *Proceedings*, annual.

International Erosion Control Association
Steamboat Springs, CO

Cross-reference: 8.39

International Plant Propagators' Society, Inc.
Seattle, WA

Cross-reference: 1.70

4.15

Laboratory of Tree-Ring Research
University of Arizona
Building 58
Tucson, AZ 85721
602-621-2191; (FAX) 602-621-8229

Purpose: To conduct dendrochronological research
addressing aspects of environmental impacts on tree
growth, climatic studies, archeological studies, and
other information stored in three ring time series.
Publication: *Tree-Ring Bulletin*, annual.
Other: Dates wood with clear annual growth rings.

4.16

Landscape Architecture Foundation
1733 Connecticut Ave., N.W.
Washington, DC 20009
202-223-6229

Purpose: An education and research vehicle for the
landscape architecture profession in the U.S.
Publication: *News Journal of the Landscape Architecture
Foundation*, biannual, free.
Library: 14.19

Musser International Turfgrass Foundation
Sharon Center, OH

Cross-reference: 1.81

4.17

Mycological Society of America
c/o Donald Pfister, Secretary
Farlow Herbarium
Harvard University, 20 Divinity Ave.
Cambridge, MA 02138

Purpose: To promote and advance the science of
mycology and its numerous subdisciplines.
Publications: *Mycologia*, bimonthly; *Mycological Society of
America Newsletter*, biannual.

National Association of Plant Patent Owners
Washington, DC

Cross-reference: 1.85

4.18

O. J. Noer Research Foundation, Inc.
301 South 61st St.
Milwaukee, WI 53214
414-258-7337

Purpose: To promote scientific research in turfgrass and
related fields.
Publication: *Annual Report*.
Grants: Research grants for research on turfgrass and
related fields.
Library.

4.19

Organization for Flora Neotropica
The New York Botanical Garden
Bronx, NY 10458-5126
212-220-8628

Purpose: To prepare and publish a complete flora of
tropical America.
Publication: *Flora Neotropica*, irregular.

4.20

Phycological Society of America
Dept. of Botany
Louisiana State University
Baton Rouge, LA 70803
504-388-8558

Purpose: An organization for educators, researchers, and
others interested in the pure, applied, or avocational
study and utilization of algae.
Publication: *Journal of Phycology*, quarterly.

4.21

Phytochemical Society of North America
Dept. of Biological Sciences
Goucher College
Towson, MD 21204
301-337-6303

Purpose: To promote phytochemical research and
communication.
Publication: *Newsletter*, quarterly.

Smithsonian Institution, Dept. of Botany
Washington, DC

Cross-reference: 8.71

4.22
Society for Economic Botany
Center for American Archeology
P.O. Box 22
Kampsville, IL 62053
618-653-4432

Purpose: To develop interdisciplinary channels of
   communication among groups concerned with the past,
   present, and future use of plants.
Publication: *Economic Botany*, quarterly.

4.23
Soil Science Society of America
677 S. Segoe Rd.
Madison, WI 53711
608-273-8080; (FAX) 608-273-2021

Purpose: Dedicated to the advancement of the discipline
   and practice of soil science through the acquisition and
   dissemination of scientific information on soils in
   relation to crop production, environmental quality, and
   wise land use.
Publications: *Soil Science Society of America Journal*,
   bimonthly; *Journal of Production Agriculture*, quarterly;
   Journal of Environmental Quality, quarterly.
Library: Number of books: 500; number of periodicals: 50.
Lending policy: Library not open to public.

4.24
Weed Science Society of America
309 West Clark St.
Champaign, IL 61820
217-356-3182

Purpose: To promote the development of knowledge
   concerning weeds and their control.
Publications: *Weed Science*, quarterly; *Weed Technology*,
   quarterly; *WSSA Newsletter*, quarterly.

CANADA

4.25
Canadian Botanical Association
Institut botanique
Université de Montreal
4010, rue Sherbrooke est
Montreal, PQ H1X 2B2

Publication: *Canadian Botanical Association Bulletin*,
   quarterly.

4.26
Canadian Phytopathological Society
Dept. of Environmental Biology
University of Guelph
Guelph, ON N1G 2W1
519-824-4120

Publications: *Canadian Journal of Plant Pathology*,
   quarterly; *CPS News*, quarterly.

4.27
Canadian Society for Horticultural Science
907–151 Slater St.
Ottawa, ON K1P 5H4

Purpose: A professional organization dedicated to
   promoting and fostering the science of horticulture.
Publication: *Newsletter*, sporadic.

5

National Garden Club Associations

National garden club associations are involved in a variety of horticultural activities including gardening, civic beautification, conservation, landscape preservation, and education. All of the listed organizations have regional, state, or provincial chapters, and numerous local chapters. Many of these chapters sponsor horticultural scholarships, maintain local gardens, and provide other services to the community. For more information on these, contact the national offices.

UNITED STATES

5.1
Garden Club of America
598 Madison Ave.
New York, NY 10022
212-753-8287; (FAX) 212-753-0134

Purpose: To stimulate the knowledge and love of gardening.
Publications: *GCA Bulletin*, biannual; *Newsletter*, bimonthly.
Grants: Awards in tropical botany.
Scholarships: Scholarships in landscape architecture, agriculture, horticulture, conservation, ecology, and related fields.
Library: 14.66
Other: Lectures, seminars, symposia.

5.2
Men's Garden Club of America
5560 Merle Hay Rd.
Box 241
Johnston, IA 50131
515-278-0295

Purpose: To advance gardening and excellence in horticulture throughout the nation.
Publications: *The Gardener*, bimonthly; *MGCA Newsletter*, bimonthly.

Scholarships.
Library: 14.36

5.3
National Council of State Garden Clubs
4401 Magnolia Ave.
St. Louis, MO 63110
314-776-7574

Purpose: Umbrella organization for state garden clubs.
Publication: *The National Gardener*, bimonthly.
Garden: Demonstration gardens throughout the country.
Grants: Institutional research grants.
Scholarships: 28 annual scholarships of $3,500 to students in horticulture or related subjects.
Library.
Lending policy: Reference only.

CANADA

5.4
Garden Clubs Canada
6 Compton Place
London, ON N6C 4G4
519-681-7089

Purpose: An umbrella organization to facilitate communication between Canadian garden clubs.
Publication: *Garden Clubs Canada Newsletter*, 3/year.

6

State and Provincial Horticultural Societies

State and provincial horticultural societies have been established throughout North America to promote and encourage the development of horticulture. The nature of these organizations varies from state to state, province to province. Some are membership organizations directed toward the amateur gardener and horticulture in the broadest sense. Others are oriented toward commercial growers and are concerned with a small range of plants.

Many state and provincial horticultural societies publish a newsletter with pertinent information on local horticulture. A few have horticultural libraries.

Many of the addresses listed below change on a regular basis as new officers are elected for the society.

UNITED STATES

AK
6.1
Alaska Horticultural Association
P.O. Box 1909
Palmer, AK 99645

Purpose: To promote interest and education in every phase of horticulture where plants may be used for ornamental or economic purposes.

AR
6.2
Arkansas State Horticultural Society
Room 306, Plant Science Bldg.
University of Arkansas
Fayetteville, AR 72701
501-575-2603

Purpose: To promote horticultural interests in general and those of the state of Arkansas in particular.
Publication: *Proceedings of the Annual Meeting*, annual.
Grants: Grants to University of Arkansas Agricultural Experiment Station.
Scholarships: Six to eight $500–$1,000 scholarships to university students of horticulture or related plant sciences.
Other: Giant pumpkin and watermelon growing contests.

CA
6.3
California Horticultural Society
California Academy of Sciences
San Francisco, CA 94118
415-566-5222

Purpose: To advance the knowledge and appreciation of ornamental horticulture in California.
Publication: *Newsletter*, monthly.
Grants: Annual grants.
Scholarships: Horticultural scholarships for special projects.
Library: Number of books: 350.
Lending policy: Circulating to members.
Seed/specimen exchange.
Other: Field trips, members receive *Pacific Horticulture*, quarterly.

6.4
Saratoga Horticultural Foundation
15185 Murphy Ave.
San Martin, CA 95046

Purpose: To introduce better plants for California landscapes.
Publication: *Quarterly Report*.
Garden: Test garden devoted to new and unusual species, 200 specimens.
Library: Number of books: 270; number of periodicals: 19.

Lending policy: Reference only.
Seed/specimen exchange.
Other: Speakers bureau.

6.5
Southern California Horticultural Institute
P.O. Box 49798
Barrington Station
Los Angeles, CA 90049

Purpose: Dedicated to the improvement of the Southern
  California environment by providing horticultural
  information and encouraging a greater interest in
  horticulture.
Publication: *Bulletin*, monthly.
Other: Members also receive *Pacific Horticulture*.

6.6
Western Horticultural Society
P.O. Box 60507
Palo Alto, CA 94306
415-948-3410

Purpose: The advancement of and education about
  horticulture.
Publication: *WHS Newsletter and Plant Notes*, 10/year.
Seed/specimen exchange.
Other: Field trips, lectures, members receive *Pacific
  Horticulture*, quarterly.

CT
6.7
Connecticut Horticultural Society
150 Main St.
Wethersfield, CT 06109
203-529-8713

Purpose: To promote the interest of horticulture through
  flower shows, lectures, workshops, garden visits, and
  scholarships.
Publication: *The Connecticut Horticultural Society Newsletter*,
  10/year.
Scholarships: Two college scholarships annually to
  students of plant science at the University of
  Connecticut.
Library: Number of books: 2,000.
Lending policy: Circulating.

FL
6.8
Florida Foliage Association
57 E. Third St.
Apopka, FL 32703
407-886-1036; (FAX) 407-886-9585

Purpose: To promote the production, marketing, and
  sales of foliage plants.
Publications: *Florida Foliage Magazine*, monthly; *Florida
  Foliage Locator*, annual; *Foliage News Newsletter*,
  monthly.
Scholarships: Scholarships available through regional
  chapters.

6.9
Florida State Horticultural Society
700 Experiment Station Rd.
Lake Alfred, FL 33850
813-956-1151; (FAX) 813-956-5318

Purpose: Dedicated to the advancement of horticulture in
  Florida by serving as an open forum for growers,
  processors, researchers, and others.
Publications: *Proceedings of the Florida State Horticultural
  Society*, annual; *Index of the Florida State Horticultural
  Society*, 10/year.

IA
6.10
Iowa State Horticultural Society
Wallace State Office Building
Des Moines, IA 50319
515-281-5402; (FAX) 515-281-6236

Purpose: To promote horticultural interests within the
  state of Iowa.
Publication: *Horticulturist*, quarterly.
Garden: Demonstration and test gardens at Iowa State
  University Horticultural Research Farms (12.118).

ID
6.11
Idaho State Horticultural Society
29603 U of I Ln.
Parma, ID 83660
208-722-6701; (FAX) 208-722-6708

Purpose: To promote the Idaho tree fruit and wine grape
  industries through educational meetings and tours.

IL
6.12
Illinois State Horticultural Society
R.R. 1, Box 26
Sidney, IL 61877
217-688-2590

Purpose: To advance the science of pomology and the art
  of horticulture.
Publication: *Transactions of the Illinois State Horticultural
  Society*, annual.
Other: Home of the Illinois Horticultural Hall of Fame.

IN
6.13
Indiana Horticultural Society
Dept. of Horticulture
Purdue University
West Lafayette, IN 47907
317-494-1349; (FAX) 317-494-0391

Purpose: An educational organization for commercial fruit
  producers and advanced amateurs.
Publication: *Hoosier Horticultural Newsletter*, irregular.
Grants: Grants to university staff and organizations for
  work on fruit crops.

KS
6.14
Kansas State Horticultural Society
Dept. of Horticulture
Waters Hall
Kansas State University
Manhattan, KS 66506
913-532-6170; (FAX) 913-532-6949

Purpose: To promote the horticultural industries of
  Kansas.
Publication: *Horticulture Society Newsletter*, quarterly.
Grants: Awards for student research at Kansas State
  University.
Scholarships: Annual scholarship awards.

KY
6.15
Kentucky State Horticultural Society
P.O. Box 469
Princeton, KY 42445
502-365-7541; (FAX) 502-365-2667

Purpose: Educational and public service activities relating
  to growing fruit.
Publication: *Proceedings of Annual Meeting*, annual.
Grants: Small grants to the University of Kentucky.
Other: Educational meetings.

MA
6.16
Massachusetts Horticultural Society
300 Massachusetts Ave.
Boston, MA 02115
617-536-9280

Purpose: To promote the art and science of horticulture.
Publication: *The Leaflet*, bimonthly.
Library: 14.47
Other: Sponsors the New England Flower Show.

6.17
Worcester County Horticultural Society
30 Tower Hill Rd.
Boylstown, MA 01505
508-869-6111

Purpose: An educational organization for advancing
  horticultural science, and encouraging and improving
  the practice of horticulture.
Publication: *Grow with Us*, bimonthly.
Scholarships: $500–$2,000 to a university student (must
  be a resident of New England or attending a New
  England college or university) majoring in horticulture
  or a related field.
Library: 14.49
Other: Flower shows, classes, workshops, trips, apple
  scion program.

MD
6.18
Horticultural Society of Maryland
1563 Sherwood Ave.
Baltimore, MD 21239
301-352-7863

Purpose: Education, support of horticulture in the
  community, and fellowship among those interested in
  horticulture.
Scholarships: $1,000 awarded annually to an ornamental
  horticultural student from Maryland.
Other: Members receive *Green Scene*, the magazine of the
  Pennsylvania Horticultural Society.

MI
6.19
Michigan State Horticultural Society
338 Plant and Soil Sciences Bldg.
MSU
East Lansing, MI 48824
517-355-5194; (FAX) 517-353-0890

Purpose: To sponsor an annual meeting and trade show
  for commercial fruit growers and regional meetings on
  fruit culture and marketing.
Publication: *Annual Report of the Michigan State
  Horticultural Society*, annual.
Grants: Grants for research on fruit production,
  harvesting, processing, and marketing.
Other: Educational seminars and trade shows.

MN
6.20
Minnesota State Horticultural Society
1970 Folwell Ave.
#161 Alderman Hall
St. Paul, MN 55108
612-624-7752

Purpose: To promote the art and science of horticulture
  in Minnesota.
Publications: *Minnesota Horticulturist*, 9/year; *Garden Club
  Dispatch*, monthly.
Library: Number of books: 1,500; number of periodicals:
  300; 110 horticultural videotapes, 75 horticultural slide
  programs.
Lending policy: Circulating for members, reference only
  for nonmembers.
Other: Educational programs, Minnesota Green land
  stewardship program, tours.
Cross-reference: 19.46

MO
6.21
Missouri State Horticultural Society
P.O. Box 417
Columbia, MO 65205
314-442-3207

Purpose: An organization of commercial fruit growers
  established for the purpose of education,
  communication, and fraternization.
Publication: *Missouri State Horticultural Society Newsletter*,
  bimonthly.

ND
6.22
North Dakota State Horticultural Society
P.O. Box 5658
North Dakota State University
Fargo, ND 58105
701-237-8161

Purpose: To promote horticulture throughout the state via the dissemination of information.

Publication: *North Dakota Horticulture*, monthly.

Scholarships: The Harry M. Graves Scholarship to a North Dakota State University student.

NJ

6.23

New Jersey State Horticultural Society
P.O. Box 116
Clayton, NJ 08312
609-863-0110; (FAX) 609-881-4191

Purpose: To promote horticulture in New Jersey.

NY

6.24

Horticultural Alliance of the Hamptons
P.O. Box 202
Bridgehampton, NY 11932
516-537-2223

Purpose: To encourage awareness and excellence in horticulture on the South Fork of Long Island.

Publication: *Newsletter*, quarterly, free.

Library: Number of books: 1,000 including 200 rare books; number of periodicals: 15.

Lending policy: Circulating and noncirculating reference books.

Seed/specimen exchange.

Other: Lectures, garden tours.

6.25

Horticultural Society of New York
128 West 58th St.
New York, NY 10019
212-757-0915

Purpose: To promote the knowledge and love of horticulture, and to beautify and enhance the urban environment.

Publication: *Newsletter*, quarterly.

Library: 14.68

Other: Community gardening program, "Project Greenworks" horticultural therapy program for prisoners, classes, gardening information line; sponsors the New York Flower Show.

6.26

Long Island Horticultural Society
c/o Donald Brodman
44 North Kings Ave.
Lindenhurst, NY 11757
516-884-1679

Publication: *Newsletter*, 11/yr.

6.27

Metro Hort Group, Inc.
P.O. Box 1113
New York, NY 10185-0010
212-799-0711; (FAX) 212-288-1207

Purpose: To provide a forum for professionals of the New York metropolitan horticultural community to share information and discuss current topics of interest.

Other: Referral service for horticultural professionals, job listings.

6.28

New York State Horticultural Society
2981 Ridgeway Ave.
Rochester, NY 14606

Purpose: Trade association dedicated to promoting the horticulture industry in New York.

Publication: *Newsletter*, bimonthly.

Other: Trade shows, educational programs, insurance program.

OK

6.29

Oklahoma Horticulture Society
OSU Technical Branch
900 N. Portland
Oklahoma City, OK 73107
405-945-3358

Purpose: To foster and stimulate interest in horticulture in Oklahoma.

Publication: *Horizons*, quarterly.

OR

6.30

Oregon Horticultural Society
P.O. Box 1246
McMinnville, OR 97128
503-472-7910

Purpose: To disseminate horticulture production and marketing information.

Publication: *Oregon Horticultural Society Annual Meeting Proceedings*, annual.

Scholarships: Two or three full tuition scholarships to horticulture students at Oregon State University.

PA

6.31

Horticultural Society of Western Pennsylvania
P.O. Box 5126
Pittsburgh, PA 15206
412-392-8540

Purpose: To develop an appreciation for the value of horticulture, botany, and the conservation of natural resources through education, research, and the establishment of a botanical garden.

Publications: *Newsletter; Journal*, annual.

6.32

Pennsylvania Horticultural Society
325 Walnut St.
Philadelphia, PA 19106-2777
215-625-8250

Purpose: To inspire residents of the Delaware Valley and beyond to practice the art and science of horticulture.

Publication: *The Green Scene*, bimonthly.

Library: 14.90

Other: Sponsors the Philadelphia Flower Show;

Horticultural Hotline (plant information service): Philadelphia Green Program (cummunity gardening).

Cross-reference: 19.86

6.33
State Horticultural Association of Pennsylvania
General Delivery
Loganville, PA 17342
717-428-2070

Purpose: An educational group of tree and small fruit
growers.
Publication: *Pennsylvania Fruit News*, monthly.
Grants: Research grants.
Other: Summer orchard tour.

PR
6.34
Puerto Rico Horticultural Society, Inc.
Box 4104
San Juan, PR 00936
809-753-6911

SD
6.35
South Dakota State Horticultural Society
R.R. 2, Box 50
Brookings, SD 57006
605-688-5965

Purpose: Dedicated to the advancement of the art and
science of horticulture throughout South Dakota.
Publication: *SDSHS Newsletter*, 3/year.

TN
6.36
East Tennessee Horticultural and Landscape Association
P.O. Box 51112
Knoxville, TN 37950-1112
615-974-7324

Purpose: To circulate useful information to members, to
develop and maintain integrity, goodwill, and good
business practices within the industry, and to promote a
professional image and advance landscape horticulture
in the public esteem.
Publication: *ETHLA Newsletter*, bimonthly.
Grants: Research grants to the Development Trust Fund
at the University of Tennessee's Dept. of Ornamental
Horticulture and Landscape Design.
Scholarships: $300 annually to university, 4-H, or FFA
horticulture students to travel to national horticulture
competitions.

6.37
Tennessee Fruit and Vegetable Horticultural Society
Morgan Hall
University of Tennessee
Knoxville, TN 37901
615-974-7271

Purpose: To improve the profitability of the fruit and
vegetable enterprises and the welfare of horticultural
producers, marketers, and consumers.

TX
6.38
Texas State Horticultural Society
4348 Carter Creek Pkwy.
Suite 101

Bryan, TX 77802
409-846-1752; (FAX) 409-846-1752

Purpose: To promote and encourage all phases of
horticulture in Texas.
Publication: *The Texas Horticulturist*, monthly.

VA
6.39
Virginia State Horticultural Society
1219 Stoneburner St.
P.O. Box 718
Staunton, VA 24401
703-332-7790; (FAX) 703-332-7792

Purpose: To promote and foster all the horticultural
interests of its members within Virginia. Mainly a
professional fruit growers organization.
Publication: *Virginia Fruit*, bimonthly.
Library.
Lending policy: Reference only.

WA
6.40
Northwest Horticultural Council
P.O. Box 570
Yakima, WA 98907
509-453-3193; (FAX) 509-457-7615

Purpose: A trade organization representing the deciduous
fruit industry of Oregon and Washington.

6.41
Northwest Horticultural Society
Isaacson Hall
University of Washington, GF-15
Seattle, WA 98195
206-527-1794

Purpose: To help establish the coastal Northwest as the
horticultural center of the globe and to increase
horticultural knowledge.
Grants: Various grants—recently awarded $5,000 to the
Hardy Fern Society to support their research and
display garden.
Scholarships: $1,000 annual scholarship to a student in
the graduate program at the University of Washington
Center for Urban Horticulture.
Library: 14.99
Lending policy: Reference only.
Seed/specimen exchange.
Other: Lectures, workshops, tours.

6.42
Washington State Horticultural Association
Box 136
Wenatchee, WA 98807
509-662-2067

Purpose: To represent the interests of tree-fruit growers
before the state government.
Publications: *Annual Meeting Proceedings*, annual; *Annual
Warehouse Seminar Proceedings*, annual.
Grants: Travel and study funds.

CANADA

AB

6.43

Alberta Horticultural Association
Box 223
Lacombe, AB T0C 1S0
403-782-3053

6.44

Calgary Horticultural Society
2405 9th Ave., S.E.
Calgary, AB T2G 4T4
403-262-5609

Purpose: To promote horticultural excellence in Calgary.
Publication: *CHS Newsletter*, 8/year.
Library: Number of books: 400; number of periodicals: 10.
Lending policy: Reference only; open to members only.
Other: Plant exchanges, lectures.

6.45

Western Canadian Society for Horticulture, Inc.
c/o Roger Vick
University of Alberta Devonian Botanic Garden
Edmonton, AB T6G 2E1
403-987-3054; (FAX) 403-492-7219

Purpose: To foster and coordinate horticulture on the prairies and Northwest Territories.
Publication: *W.C.S.H. Grapevine*, quarterly.

6.46

Agassiz Agricultural and Horticultural Association
Box 78
Agassiz, BC V0M 1A0
604-796-2534

MB

6.47

Manitoba Horticultural Association
908 Norquay Building
Winnipeg, MB R3C 0P8
204-945-3856

NB

6.48

New Brunswick Horticultural Society
New Brunswick Dept. of Agriculture
Plant Industry Branch
P.O. Box 6000
Fredericton, NB E3B 5H1
506-453-2108

NF

6.49

Newfoundland Horticultural Society
P.O. Box 10099
St. Johns, NF A1A 4L5
709-712-4604; (FAX) 709-772-6064

Purpose: To promote gardening through meetings, flower shows, garden visits, and special interest groups.
Publication: *Down to Earth*, 11/year.
Library: Number of books: 200.
Lending policy: Open to members only.

ON

6.50

Horticultural Society of Parkdale & Toronto
P.O. Box 455, Station C
Toronto, ON M6J 3P5
416-530-4269

Purpose: To encourage interest in, the practice of, and learning about horticulture.
Publication: *Newsletter*, 9/year.
Seed/specimen exchange.
Other: Community greenhouse.

6.51

Ontario Horticultural Association
R.R. 3
Englehart, ON P0J 1H0
705-544-2474

Purpose: To provide leadership and assist in the promotion of education and interest in all areas of horticulture.
Publication: *Annual Report*.
Scholarships: Two scholarships for students of horticulture at the University of Guelph.

6.52

Ottawa Horticultural Society
P.O. Box 8921
Ottawa, ON K1G 3J2

PE

6.53

Prince Edward Island Rural Beautification Society
Box 1194
Charlottetown, PE C1A 7M8

PQ

6.54

Federation des societes d'horticulture et d'ecologie
4545, ave. Pierre-de-Coubertin
C.P. 1000, Succ. M.
Montreal, PQ H1V 3R2

6.55

Ornamental Horticultural Society
SIHOQ
Jardin van den Hende
Université Laval
Ste.-Foy, PQ T5J 1H3

SK

6.56

Saskatchewan Horticultural Association
P.O. Box 152
Balacarres, SK S0G 0C0
306-334-2201

7

Garden Centers

Garden centers have been established in many cities for the purpose of promoting horticulture and educating citizens about the merits of gardening. These garden centers are often supported by membership groups, by local government funds, or by private endowment. They provide a particularly valuable function in cities that do not have botanical gardens or arboreta.

Services of garden centers include workshops, classes, conservation and beautification programs, outreach programs, flower judging workshops, gardening information, horticultural consultation, and community gardening and horticultural therapy programs. Many garden centers also have sizable libraries and extensive demonstration gardens.

If your city is not represented below, call your local extension officer to see if your area has a garden center.

UNITED STATES

AL
7.1
Hillcrest Garden Center
1632 South Court St.
Montgomery, AL 36101

7.2
Red Mountain Garden Center
3808 Old Leeds Rd.
Birmingham, AL 35213

AZ
Valley Garden Center
Phoenix, AZ
Cross-reference: 16.14

7.3
Yuma Garden Center
Century House
248 Madison Ave.
Yuma, AZ 85364

CA
7.4
Cottonwood Meadows Garden Center
Rt. 2, Box 1017
Oakdale, CA 95361

7.5
Heather Farms Garden Center
1540 Marchbanks Dr.
Walnut Creek, CA 94596
415-947-1679
Cross-reference: 13.60

7.6
Lakeside Park Trial and Show Gardens and Garden
  Center
666 Belleview Ave.
Oakland, CA 94612
415-273-3208

Martin Art and Garden Center
Ross, CA
Cross-reference: 16.37

7.7
Orange County Memorial Garden Center
Orange County Fair Grounds
Costa Mesa, CA 92626

7.8
Sacramento Garden Center
McKinley Park
Sacramento, CA 95801

7.9
San Diego Floral Association and Garden Center
Casa del Prado
Balboa Park
San Diego, CA 92101

7.10
Stockton Garden Center
1000 North Hunter St.
Stockton, CA 95200

7.11
Woodside-Atherton Garden Center
1475 Portola Rd.
Woodside, CA 94062

CT
7.12
Greenwich Garden Center
Montgomery Pinetum
Bible St.
Cos Cob, CT 06807
203-869-9242

Cross-reference: 14.16

7.13
Knox Parks Horticultural Center
150 Walbridge Rd.
West Hartford, CT 06119
203-523-4276

7.14
New Canaan Garden Center
Box 4
New Canaan, CT 06840

7.15
Roxbury-Bridgewater Garden Center
Grand Building
Roxbury, CT 06897

7.16
Wilton Garden Center
Ridgefield Rd.
Historic Town Hall
Wilton, CT 06897

DE
7.17
Wilmington Garden Center
503 Market St. Mall
Wilmington, DE 19801
302-658-1913

Cross-references: 14.22, 19.20

FL
7.18
Key West Garden Center
White St.
Key West, FL 33014

7.19
Miami Beach Garden Center and Conservatory
2000 Garden Center Dr.
Miami Beach, FL 33139
305-673-7256

GA
7.20
Augusta Garden Center
Sixth and Telfair Sts.
Augusta, GA 30900

7.21
Macon Garden Center
730 College St.
Macon, GA 31200

7.22
Marietta Garden Center
505 Kenshaw Ave.
Marietta, GA 30060

7.23
Thomasville Garden Center
1002 East Broad St.
Thomasville, GA 31792

IA
Des Moines Botanical Center
Des Moines, IA
Cross-reference: 16.126

7.24
Marshalltown Garden Center
709 Center St.
Marshalltown, IA 50158

7.25
Pearson Memorial Garden Center
Maquoketa, IA 52060

IL
7.26
Moline Garden Center
3300 Fifth Ave.
Riverside Park
Moline, IL 61265

7.27
Washington Park Horticultural Center
Fayette & Chatham Rds.
Springfield, IL 62704

KS
7.28
Independence Garden Center
Riverside Park
Independence, KS 67301

7.29
Manhattan Horticulture Center
1600 Laramie St.
Manhattan, KS 66502

7.30
Merritt Horticultural Center
411 North 61st St.
Kansas City, KS 66102
913-299-9254

Cross-references: 13.26, 13.93

LA
R.S. Barnwell Memorial Garden and Art Center
Shreveport, LA

Cross-reference: 17.126

LA
7.31
While Away Hours Garden Center
5919 Magazine St.
New Orleans, LA 70100

MA
Berkshire Garden Center
Stockbridge, MA

Cross-reference: 16.176

MD
Cylburn Wild Flower Preserve and Garden Center
Baltimore, MD

Cross-reference: 16.185

MI
7.32
Chadwick Garden Center
Public Museum East Bldg.
233 Washington, S.E.
Grand Rapids, MI 49503

Detroit Garden Center
Detroit, MI

Cross-reference: 17.154

7.33
Grosse Pointe Garden Center
32 Lake Shore Dr.
Grosse Point Farms, MI 48236
313-881-4594

MO
7.34
Kansas City Garden Center
5300 Pennsylvania Ave.
Kansas City, MO 64112

7.35
Powell Horticultural & Natural Resource Center
Rt. 1
Kingsville, MO 64061

MS
7.36
Crosby Garden Center
Shamrock Inn
Crosby, MS 39633

7.37
Hattiesburg Garden Center
Hutchinson Ave.
Hattiesburg, MS 39401

7.38
Jackson Garden Center
1912 West Capital St.
Jackson, MS 39200

7.39
Planters Hall Garden Center
906 Monroe St.
Vicksburg, MS 39180

NC
7.40
Winston-Salem Garden Center
801 West Fourth St.
Winston-Salem, NC 27100

NH
7.41
New Hampshire Federation Garden Center
State Library
Concord, NH 03301

NJ
Deep Cut Park Horticultural Center
Middletown, NJ

Cross-reference: 16.254

NM
7.42
Alamogordo Garden Center
Tenth and Orange Sts.
Alamogordo, NM 88310

NY
7.43
Garden Center of Rochester
5 Castle Park
Rochester, NY 14620

Cross-reference: 14.65

7.44
Robertson Garden Center
30 Front St.
Binghamton, NY 13905

OH
7.45
Canton Garden Center
1615 Stadium Park, N.W.
Canton, OH 44718

7.46
Civic Garden Center of Greater Cincinnati
2715 Reading Rd.
Cincinnati, OH 45206
513-221-0981

Cross-references: 14.79, 19.78

7.47
Dayton Council Garden Center
1820 Brown St.
Dayton, OH 45409

Franklin Park Conservatory and Garden Center
Columbus, OH

Cross-reference: 16.301

Garden Center of Greater Cleveland
Cleveland, OH

Cross-reference: 16.302

7.48
Wegerzyn Horticultural Center
1301 East Siebenthaler Ave.
Dayton, OH 45414
513-277-6545

Cross-references: 14.82, 19.80, 20.193

OK
7.49
Ardmore Garden Center
502 Stanley Blvd.
Ardmore, OK 73401

Tulsa Garden Center
Tulsa, OK

Cross-reference: 16.320

PA
Fairmount Park Horticultural Center
Philadelphia, PA

Cross-reference: 16.341

7.50
Greensburg Garden & Civic Center
951 Old Salem Rd.
Greensburg, PA 15601
412-836-1123

7.51
Harrisburg Area Civic Garden Center
5810 Linglestown Rd.
Harrisburg, PA 17112

7.52
Pittsburgh Civic Garden Center
1059 Shady Ave.
Pittsburgh, PA 15232
412-441-4442

Cross-references: 13.45, 13.171, 14.91

7.53
River Crest Horticultural Center
Rt. 39
Mont Clare, PA 19453
215-935-9738

Cross-reference: 13.174

SC
7.54
Sumter Garden Center
West Liberty St.
Sumter, SC 29150

TN
7.55
Goldsmith Civic Garden Center
750 Cherry Rd.
Memphis, TN 38117
901-685-1566

TX
7.56
Amarillo Garden Center
1400 Streit Dr.
Amarillo, TX 79106
806-352-6573

Austin Area Garden Center
Austin, TX

Cross-reference: 16.377

7.57
Corpus Christi Garden Center
5325 Greely Dr.
Corpus Christi, TX 78400

Dallas Civic Garden Center
Dallas, TX

Cross-reference: 16.381

7.58
El Paso Garden Center
3105 Grant Ave.
El Paso, TX 79900

7.59
Garden Center in Tyrrell Park
Beaumont Council of Garden Clubs
P.O. Box 7962
Beaumont, TX 77726-7962
409-842-3135

Cross-reference: 14.96

Houston Garden Center
Houston, TX

Cross-reference: 16.385

7.60
Irving Garden Arts Center
906 Senter Rd.
Irving, TX 75060

7.61
Lubbock Garden-Arts Center
4215 University
Lubbock, TX 79413

7.62
Nell Pape Garden Center
1805 North Fifth St.
Waco, TX 76700

7.63
Old River Terrace Garden Center
523 Crockett
Channelville, TX 77530

7.64
Pasadena Garden Center
610 West Shaw
Pasadena, TX 75501

UT
7.65
Utah Garden Center
1596 East 21st St.
Sugar House Park
Salt Lake City, UT 84100

VA
Green Spring Farm Park
Alexandria, VA

Cross-reference: 16.396

7.66
Hampton Roads Garden Center
Museum Dr.
Newport News, VA 23601

7.67
Norfolk Garden Center
Educational Building
749 East 29th St.
Norfolk, VA 23500

7.68
Richmond Garden Center
Hermitage Rd.
Richmond, VA 23200

7.69
Roanoke Garden Center
2713 Avenham Ave., S.W.
Roanoke, VA 24001

WV
7.70
Sunrise Garden Center
746 Myrtle Rd.
Charleston, WV 25314
304-344-8035

7.71
Wheeling Garden Center
Oglebay Park
Wheeling, WV 26003
304-242-0665

Cross-reference: 14.102

CANADA

ON
7.72
Civic Garden Centre
777 Lawrence Ave., East
North York, ON M3C 1P2
416-445-1551

Cross-reference: 14.105

~ 8 ~

Conservation

This chapter is divided into three sections. The first lists by region the offices of the U.S. Fish and Wildlife Service—the agency responsible for implementing the Endangered Species Act. The second section lists a number of conservation groups that deal with preserving soil, seeds, plants, and natural areas. The third lists arboreta and botanical gardens that are a part of the Center for Plant Conservation network.

There are a number of very useful directories that provide more comprehensive information on conservation organizations in North America. The Center for Plant Conservation's *1988 Plant Conservation Resource Book* (Jamaica Plain: Center for Plant Conservation, 1988) lists state rare plant laws, endangered plant lists, federal and state government contacts, botanists, individuals in state heritage programs, native plant societies, state contacts with the Nature Conservancy, and national organizations dealing with plant conservation. The annual *Conservation Directory* (Washington, D.C.: National Wildlife Federation) details organizations, agencies, and officials in the United States and Canada concerned with natural resource use and management at the national, state, and provincial levels. *Healthy Harvest III* (Washington, D.C.: Potomac Valley Press, 1989) is a directory of sustainable agricultural and horticultural organizations.

U.S. FISH AND WILDLIFE SERVICE ENDANGERED SPECIES PROGRAM

The goal of the U.S. Fish and Wildlife Endangered Species Program is to stop endemic plant and animal endangerments and extinctions caused by man's influence on wild ecosystems and to restore the species to the point where they are no longer threatened or endangered. The program involves numerous phases, including the listing and delisting of Endangered and Threatened species, the procurement of current population data on these species, the appointment of species recovery teams, and the preparation of environmental assessments or impact statements.

The threatened and endangered status of individual species is constantly changing. The U.S. Fish and Wildlife Office of Endangered Species periodically issues new lists indicating the status of species being considered for federal protection as well as those already protected. To obtain a copy of the list of plant and animal species already under federal protection, write for *Endangered & Threatened Wildlife and Plants*. To obtain a listing of plants under consideration for federal protection, write for the current issue of the *Federal Register, Endangered and Threatened Wildlife and Plants; Review of Plant Taxa for Listing as Endangered or Threatened Species*. Both publications are available, free of charge, from the Publication Unit, U.S. Fish and Wildlife Service, Washington, D.C. 20240. Also, the University of Michigan School of Natural Resources publishes the *Endangered Species Update* (with the USFWS *Endangered Species Technical Bulletin* inserted) ten times a year. For a subscription, write the School of Natural Resources, University of Michigan, Ann Arbor, MI 48109-1115.

For more information on the Endangered Species Program, write to the Office of Endangered Species, U.S. Fish and Wildlife Service, Department of the Interior, Washington, DC 20240.

U.S. FISH AND WILDLIFE SERVICE REGIONAL OFFICES

REGION 1 (CALIFORNIA, HAWAII, IDAHO, NEVADA, OREGON, WASHINGTON, AMERICAN SAMOA, COMMONWEALTH OF THE NORTHERN MARIANA ISLANDS, GUAM, AND THE PACIFIC TRUST TERRITORIES)

8.1
Regional Director
U.S. Fish and Wildlife Service
Lloyd 500 Bldg., Suite 1692
Portland, OR 97232
503-231-6118

REGION 2 (ARIZONA, NEW MEXICO, OKLAHOMA, AND TEXAS)

8.2
Regional Director
U.S. Fish and Wildlife Service
P.O. Box 1306
Albuquerque, NM 87103
505-766-2321

REGION 3 (ILLINOIS, INDIANA, IOWA, MICHIGAN, MINNESOTA, MISSOURI, OHIO, AND WISCONSIN)

8.3
Regional Director
U.S. Fish and Wildlife Service
Federal Bldg., Fort Snelling
Twin Cities, MN 55111
612-725-3500

REGION 4 (ALABAMA, ARKANSAS, FLORIDA, GEORGIA, KENTUCKY, LOUISIANA, MISSISSIPPI, NORTH CAROLINA, SOUTH CAROLINA, TENNESSEE, PUERTO RICO, AND THE U.S. VIRGIN ISLANDS)

8.4
Regional Director
U.S. Fish and Wildlife Service
Richard B. Russell Federal Bldg.
75 Spring St., S.W.
Atlanta, GA 30303
404-331-3580

REGION 5 (Connecticut, Delaware, District of Columbia, Maine, Maryland, Massachusetts, New Hampshire, New Jersey, New York, Pennsylvania, Rhode Island, Vermont, Virginia, and West Virginia)

8.5
Regional Director
U.S. Fish and Wildlife Service
One Gateway Center, Suite 700
Newton Corner, MA 02158
617-965-5100

REGION 6 (COLORADO, KANSAS, MONTANA, NEBRASKA, NORTH DAKOTA, SOUTH DAKOTA, UTAH, AND WYOMING)

8.6
Regional Director
U.S. Fish and Wildlife Service
P.O. Box 25486
Denver Federal Center
Denver, CO 80225
303-236-7920

REGION 7 (ALASKA)

8.7
Regional Director
U.S. Fish and Wildlife Service
1011 E. Tudor Rd.
Anchorage, AK 99503
907-786-3542

REGION 8 (RESEARCH AND DEVELOPMENT)

8.8
U.S. Fish and Wildlife Service
Research and Development
Washington, DC 20240
703-358-1710

CONSERVATION ORGANIZATIONS

UNITED STATES

8.9

Abundant Life Seed Foundation
Box 772 (1029 Lawrence)
Port Townsend, WA 98368
206-385-5660; (FAX) 206-385-9288

Purpose: To acquire, propagate, preserve, and distribute the seeds of native plants on the North Pacific Rim through catalogs, workshops, and publications.
Publication: *Seed Midden*, 2–4/year.
Other: Maintains seed production gardens; has conducted a tree survey of Port Townsend, WA, and a wildflower survey of Highway 20 in Washington state.

Alan Chadwick Society
San Rafael, CA
Cross-reference: 1.1

Alberene Seed Foundation
Keene, VA
Cross-reference: 17.2

8.10

American Forest Council
1250 Connecticut Ave., N.W.
Washington, DC 20036
202-463-2459

Purpose: To promote the development and productive management of the nation's commercial forest lands.
Publication: *American Tree Farmer*, quarterly.

8.11

American Forestry Association
P.O. Box 2000
Washington, DC 20013
202-667-3300; (FAX) 202-667-7751

Purpose: To maintain and improve the health and value of trees and forests, and to attract and cultivate the interests of citizens, industry, and government in trees and forest resources.
Publications: *American Forests*, bimonthly; *AFA Resource Hotline*, biweekly; *Urban Forest Forum*, bimonthly, free.
Other: Global Releaf campaign aimed at expanding forest area and reducing deforestation as a means of combatting global warming.

American Horticultural Society
Alexandria, VA
Cross-reference: 1.16

8.12

American Nature Study Society
5881 Cold Brook Rd.
Homer, NY 13077

Purpose: To promote excellence in nature education and writing through publications, field trips, workshops, and awards.
Publication: *Nature Study*, quarterly.
Other: Many programs are plant-related; co-sponsored Children's Gardening Symposium held at Brooklyn Botanical Garden in 1989.

American Society for Environmental History
Newark, NJ
Cross-reference: 17.5

American Type Culture Collection
Rockville, MD
Cross-reference: 4.8

8.13

Aprovecho Institute
80574 Hazelton Rd.
Cottage Grove, OR 97424
503-942-9434

Purpose: To demonstrate alternatives to consumerism by linking people with technologies that make the best use of their skills and resources.
Publication: *News from Aprovecho*, 5/year.
Other: Fava Bean Project to make information on favas widely available, to test different varieties, breed new varieties, and increase U.S. seed stock.

8.14

Association of Natural Bio-Control Producers
2108 Park Marina Dr.
Redding, CA 96001

Purpose: A new, nonprofit trade organization of producers and distributors of beneficial insects and predatory organisms.

8.15

Atlantic Center for the Environment
39 South Main St.
Ipswich, MA 01938
508-356-0038; (FAX) 508-356-7322

Purpose: To promote public involvement in natural resource management through education, policy, and research programs in Atlantic Canada, eastern Quebec, and northern New England.
Publication: *Nexus*, quarterly.
Other: Habitat education and assessment programs, plant conservation.

8.16

Audubon Naturalist Society of the Central Atlantic States, Inc.
8940 Jones Mill Rd.
Chevy Chase, MD 20815
301-652-9188

Purpose: Public education, conservation, and natural science.
Publications: *Naturalist News*, 10/year; *Atlantic Naturalist*, annual; *Naturalist Review*, 10/year.

Bio Integral Resource Center
Berkeley, CA
Cross-reference: 1.36

Bio-Dynamic Farming and Gardening Association, Inc.
Kimberton, PA

Cross-reference: 1.37

Biological Urban Gardening Services
Citrus Heights, CA

Cross-reference: 1.38

8.17
Botanical Dimensions
P.O. Box 807
Occidental, CA 95465
707-874-1531

Purpose: Dedicated to collecting living plants, seeds, and surviving plant lore from cultures practicing folk medicine in the tropics worldwide.
Publication: *Plant Wise*, quarterly.
Other: Maintains an extensive living collection (not open to the public) on the "Big Island" of Hawaii; offers research grants to collection efforts in targeted areas; maintains a seed exchange program for private collectors.

8.18
Botanical Peace Corps
P.O. Box 1368
Sebastopol, CA 95473

Purpose: To identify, collect, preserve, trans-ship, and propagate seeds and living specimens of rare and potentially valuable endangered plants.

Butterbrooke Farm
Oxford, CT

Cross-reference: 17.8

8.19
California Urban Forests Council
c/o Forestry Division
1320 N. Eastern Ave.
Los Angeles, CA 90063
213-267-2481

Purpose: To educate, promote, and implement urban forestry programs.
Publication: *Urban Forestry—The Quarterly*, quarterly.
Other: Promotes tree planting activities.

8.20
Center for Plant Conservation
Missouri Botanical Garden
St. Louis, MO 63166
314-664-1200

Purpose: Dedicated to conserving rare and endangered plants of the U.S. through cultivation and research.
Publications: *Plant Conservation*, quarterly; *Bulletin Board*, quarterly.
Other: Maintains a National Collection of Endangered Plants through a network of 20 botanical gardens and arboreta. (See following section of chapter.)

Central Prairie Seed Exchange
Topeka, KS

Cross-reference: 17.9

8.21
Chihuahuan Desert Research Institute
P.O. Box 1334
Alpine, TX 79831
915-837-8370

Purpose: To promote the understanding of the Chihuahuan Desert through scientific research and education.
Publications: *Chihuahuan Desert Discovery*, biannual; *Chihuahuan Desert Newsbriefs*, biannual.
Other: 40-acre arboretum with extensive collection of Chihuahuan cacti and succulents; involved in revegetation on Big Bend National Park in Texas, and rescue of threatened plants.

8.22
Children of the Green Earth
307 N. 48th
Seattle, WA 98103
206-781-0852

Purpose: To inspire and assist people in helping children and youngsters to plant trees and connect with others around the world.
Publication: *Children of the Green Earth Newsletter*, biannual.
Other: Tree planting partnerships with deforested countries.

8.23
Committee for National Arbor Day
P.O. Box 333
West Orange, NJ 07052
201-731-0840; (FAX) 201-731-6020

Purpose: To secure passage of Congressional legislation calling for the President to proclaim the last Friday in April as National Arbor Day.
Publication: *National Arbor Day Review*, quarterly, free.
Other: Annual distribution of coniferous seedlings to local school.

Corns
Turpin, OK

Cross-reference: 17.10

8.24
Council on the Environment of New York City
51 Chambers St., Room 228
New York, NY 10007
212-566-0990

Purpose: To promote environmental awareness among New Yorkers and develop solutions to environmental problems.
Publication: *New York City Environmental Bulletin*, bimonthly, free.
Other: Open Space Greening Program provides material and technical assistance to community-sponsored open

space projects; Plant-A-Lot helps neighborhood groups create and maintain green open spaces; Green Bank is a 50/50 matching fund available to groups with existing parks and gardens; Grow Truck lends gardening tools and provides on-site technical assistance; School Greening Project helps classes in elementary schools beautify their grounds.
Cross-reference: 19.66

8.25
Delaware Nature Society
P.O. Box 700
Hockessin, DE 19707
Purpose: The study and conservation of local flora and fauna.
Other: Annual tree, shrub, and perennials sale.

Desert Legume Program
Tucson, AZ
Cross-reference: 4.11

8.26
Earth First!
P.O. Box 5871
Tucson, AZ 85703
602-622-1371
Purpose: To preserve the biological diversity of Earth by taking an uncompromising stand in its defense.
Publication: *Earth First!* Journal, 8/year.
Other: Grazing Task Force, Rainforest Action Group, and Redwood Action Team; Earth First! members have engaged in demonstrations and other actions to prevent logging of America's old growth forests and the destruction of western public lands by cattle.

8.27
Ecology Action
5798 Ridgewood Rd.
Willits, CA 95490
707-459-0150
Purpose: Research and education in small-scale, highly productive agriculture based on biodynamic and French intensive horticulture.
Publication: *Ecology Action Newsletter*, quarterly.
Other: Research garden; sale of vegetable, herb, and flower seeds; book publishing.

8.28
Educational Communications
P.O. Box 35473
Los Angeles, CA 90035-0473
213-559-9160
Purpose: To improve the quality of the environment on Earth through a speaker's bureau, radio and television documentaries, public service announcements, and other forms of public education.
Publication: *The Compendium Newsletter*, bimonthly.
Other: Produced television shows on water conservation gardens, and many radio shows on ethnobotany.

8.29
Environmental Defense Fund, Inc.
257 Park Ave. South
New York, NY 10010
212-505-2100
Purpose: Dedicated to improving and protecting environmental quality and public health.
Publication: *EDF Letter*, bimonthly.

8.30
Foliage for Clean Air Council
405 N. Washington St.
Falls Church, VA 22046
703-534-5268
Purpose: A communications clearinghouse on the use of foliage and blooming plants and flowers to improve indoor air quality.

8.31
Forest Resource Center
Rt. 2, Box 156A
Lanesboro, MN 55949
507-467-2437
Purpose: To promote the wise economic, educational, and recreational use of our natural resources.
Publication: *Shiitake News*, 3/year.
Other: Has the largest hardwood log shiitake mushroom research facility in the United States.

8.32
Friends of the Earth
218 D St., S.E.
Washington, DC 20003
202-544-2600
Purpose: Committed to the preservation, restoration, and rational use of the earth.
Publication: *Friends of the Earth*, monthly.

8.33
Friends of the Trees Society
P.O. Box 1466
Chelan, WA 98816
Purpose: A network of individuals, local groups, and organizations working to preserve forests and plant trees.
Publication: *International Green Front Report*, biennial.

Garden Conservancy
Cold Spring, NY
Cross-reference: 17.13

Grain Exchange
Salinas, KS
Cross-reference: 17.14

8.34
Grand Prairie Friends of Illinois
P.O. Box 36
Urbana, IL 61801
217-244-2177

Purpose: A nonprofit conservation group involved in education, management, and preservation of native tall grass prairie, specifically remnants in east-central Illinois.

Publication: *Grand Prairie Friends Notes*, quarterly.

Other: Management of prairie remnants and seed collections for prairie reconstructions.

8.35
Greater Yellowstone Coalition
P.O. Box 1874
Bozeman, MT 59715
406-586-1593; (FAX) 406-586-0851

Purpose: Dedicated to the preservation and protection of the Greater Yellowstone ecosystem.

Publication: *The Greater Yellowstone Report*, quarterly.

Other: Compiled an ecosystem-wide inventory of rare plants.

8.36
Greensward Foundation, Inc.
104 Prospect Park West
Brooklyn, NY 11215
212-269-0927

Purpose: To work for the improvement of America's natural landscape parks.

Publication: *A Little News*, 3/year, free.

Other: The Camperdown Fund works towards saving old trees.

8.37
Harmonious Earth Research
45 Kahele St.
Kihei
Maui, HI 96753
808-879-0930

Purpose: To research and demonstrate sustainable agricultural and residential edible landscapes using permaculture principles.

Other: Researches tropical, perennial vegetables suitable for home gardeners.

8.38
Harmony Research Farm
P.O. Box 1028
Cave Junction, OR 97523
503-592-4371

Purpose: To engage in research focused on preparation for upcoming climatic changes due to cyclic movements, ozone depletion, and the greenhouse effect.

Other: Studies self-sufficient gardening with year-round production and genetic integrity of seed selection. Test garden (not open to the public) with over 70 medicinal herbs including immune system, anticancer, and antiviral plants.

8.39
International Erosion Control Association
P.O. Box 4904
Steamboat Springs, CO 80477
303-879-3010; (FAX) 303-879-8563

Purpose: To promote the worldwide exchange of information concerning effective and economic methods of erosion control.

Publication: *IECA Report*, quarterly.

8.40
International Tree Crops Institute USA, Inc.
P.O. Box 4460
Davis, CA 95617
916-753-4535

Purpose: Applied research and public education on multipurpose trees for windbreak, fuelwood, erosion control, land reclamation, fodder, and wildlife habitat.

Other: Cooperative tree improvement program to select fast-growing trees for agroforestry in California and similar climate zones.

8.41
Izaak Walton League of America
1401 Wilson Blvd., Level B
Arlington, VA 22209
703-528-1818; (FAX) 703-528-1836

Purpose: One of the oldest conservation organizations in America, the League works to defend the nation's soil, air, woods, waters, and wildlife.

Publications: *Outdoor America*, quarterly; *Outdoor Ethics*, quarterly, free; *SPLASH!*, quarterly, free.

KUSA Society, The
Ojai, CA

Cross-reference: 17.25

8.42
Michigan Nature Association
7981 Beard Rd., Box 102
Avoca, MI 48006
313-324-2626

Purpose: To carry on a program of natural history study and education and to acquire, maintain, and protect nature sanctuaries and preserves, natural areas, and examples of Michigan's flora and fauna.

Other: The MNA's program protects samples of 29 of Michigan's 30 known natural habitats containing 82% of its native plant species. Virtually all of the state's native tree species are found on MNA lands.

8.43
Missouri Prairie Foundation
P.O. Box 200
Columbia, MO 65205
816-361-1700

Purpose: To further the scientific study, esthetic appreciation, historical understanding, educational involvement, and preservation of Missouri's native grassland heritage.

Publication: *Missouri Prairie Journal*, quarterly.

Other: Research on native prairie plants.

National Arbor Day Foundation
Nebraska City, NE

Cross-reference: 1.83

8.44

National Association of Conservation Districts
408 East Main
League City, TX 77573
713-332-3402; (FAX) 713-332-5259

Purpose: NACD represents nearly 3,000 local soil and
water conservation districts and their state associations.
Publications: *The District Leader*, monthly, free; *Tuesday
Letter*, monthly.
Other: Committees on cropland conservation, forestry,
Great Plains and public lands, pasture, and range.

8.45

National Audubon Society
950 Third Ave.
New York, NY 10022
212-832-3200; (FAX) 212-593-6254

Purpose: Devoted to the protection of plants, animals,
and their habitats, the conservation of resources, and
sound environmental policy through education,
research, sanctuaries, and activism.
Publications: *Audubon*, bimonthly; *Audubon Activist*,
bimonthly.

8.46

National Coalition Against the Misuse of Pesticides
530 7th St., S.E.
Washington, DC 20003
202-543-5450

Purpose: To assist individuals, organizations, and
communities with information on pesticides and their
alternatives.
Publication: *Pesticides and You Newsletter*, 5/year.

National Heirloom Flower Seed Exchange
Cambridge, MA

Cross-reference: 17.27

8.47

National Tree Society
P.O. Box 10808
Bakersfield, CA 93389
805-589-6912

Purpose: To preserve the delicate balance of Earth's
biosphere by planting and encouraging others to plant
and care for trees.
Other: Various states have local chapters and treebanks
with indigenous species for planting.

National Wildflower Research Center
Austin, TX

Cross-reference: 1.97

8.48

National Wildlife Federation
1400 Sixteenth St., N.W.
Washington, DC 20036-2266
202-797-6800

Purpose: Conservation and educational organization
dedicated to arousing public awareness of the need for
wise use, proper management, and conservation of the
natural resources upon which all life depends: air,
water, soils, minerals, forests, plant life, and wildlife.
Publications: *National Wildlife*, bimonthly; *International
Wildlife*, bimonthly; *Ranger Rick*, monthly; *Your Big
Backyard*, monthly.

8.49

National Woodland Owners Association
374 Maple Ave. E., Suite 210
Vienna, VA 22180
702-255-2700

Purpose: A nationwide organization of nonindustrial
private woodland owners that works to promote forestry
and the best interests of woodland owners.
Publications: *Woodland Report*, 8/year; *National Woodlands
Magazine*, quarterly.

National Xeriscape Council
Austin, TX

Cross-reference: 1.98

Native Seeds/ SEARCH
Tucson, AZ

Cross-reference: 17.29

8.50

Natural Resources Defense Council
1350 New York Ave., N.W.
Washington, DC 20005
202-783-7800; (FAX) 202-783-5917

Purpose: To promote wise (sustainable) use of natural
resources and the protection of the environment.
Publication: *Memo on Endangered Plants*, every 6 weeks,
free.
Other: Plant Conservation Project promotes programs to
protect plant species and communities.

8.51

Nature Conservancy
1815 North Lynn St.
Arlington, VA 22209
703-841-5300; (FAX) 703-841-1283

Purpose: To preserve biological diversity—rare plants,
animals, and natural communities—worldwide by
protecting their habitats.
Publication: *The Nature Conservancy Magazine*, bimonthly.
Other: The Nature Conservancy maintains botanists in
offices throughout the country. It has numerous plant
protection projects.

8.52

New Alchemy Institute
237 Hatchville Rd.
East Falmouth, MA 02536
508-564-6301; (FAX) 508-457-9680

Purpose: New Alchemy Institute serves students,
teachers, households, and small-scale farmers with
ecological research and education projects on food,
energy, water, and waste treatment systems.

Publication: *The New Alchemy Quarterly*, quarterly.

Other: Consulting in integrated pest management, greenhouse and interior plantscape management, green manure, and much else.

8.53

New England Forestry Foundation, Inc.
85 Newbury St.
Boston, MA 02116
617-437-1441

Purpose: To promote wise and effective forest management.

Publication: *The Timberline*, quarterly.

8.54

New Forests Project
731 Eighth St., S.W.
Washington, DC 20003
202-547-3800; (FAX) 202-546-4784

Purpose: To help stop the destruction of the Earth's forests while assisting the poorest people of developing countries to improve the quality of their lives by planting trees.

Publication: *The New Forests Project Newsletter*, irregular, free.

Other: Land reform project in Guatemala and the World Seed Project which distributes seed and supporting materials.

8.55

Nitrogen Fixing Tree Association
P.O. Box 680
Waimanalo, HI 96795
808-259-8555; (FAX) 808-262-4688

Purpose: To promote the use of nitrogen-fixing trees for fuel, fodder, timber, and fertilizer in developing countries.

Publications: *Nitrogen Fixing Tree Research Reports*, annual; *Lencaena Research Reports*, annual; *NFTA News & Highlight*, sporadic.

Other: NFTA sponsors a cooperative planting program and maintains a seed bank of nitrogen-fixing tree species.

8.56

North Dakota Natural Science Society
P.O. Box 8238
Grand Forks, ND 58202-8238
701-777-3650

Purpose: Encourages the recording of observations of nature in the Great Plains and the communication of research on the North American grasslands and their biota.

Publication: *The Prairie Naturalist*, quarterly.

Other: Annual wildflower field trips; Natural Areas Committee documents natural areas in the state.

8.57

Permaculture
7781 Lenox Ave.
Jacksonville, FL 32221

Purpose: Training, information, and consulting for environmentally integrated habitation.

Publications: *Robin Newsletter*, sporadic; *The International Permaculture Solutions Journal*, sporadic.

Other: Tree bank for threatened species; Forest Ecosystem Rescue Network; lectures and workshops.

8.58

Planet Drum Foundation
P.O. Box 31251
San Francisco, CA 94131
415-285-6556

Purpose: To provide an effective grassroots approach to ecology that emphasizes sustainability, regional self-reliance, and community self-determination.

Publication: *Raise the Stakes*, biannual.

Other: Native plant garden; published *A Green City Program for San Francisco Bay Area Cities and Towns*.

8.59

Prairie/Plains Resource Institute
1307 L St.
Aurora, NE 68818
402-694-5535

Purpose: Dedicated to the preservation and restoration of prairie and other unique plains habitats and to education regarding the natural and cultural history of the region.

Publication: *Prairie/Plains Journal*, biannual.

Other: Prairie restoration work.

8.60

Rachel Carson Council, Inc.
8940 Jones Mill Rd.
Chevy Chase, MD 20815
301-652-1877

Purpose: To provide information in environmental and human health effects of toxic pesticide contamination.

8.61

Rainforest Action Network
301 Broadway, Suite A
San Francisco, CA 94133
415-398-4404; (FAX) 415-398-2732

Purpose: A nonprofit activist organization working to save the world's rainforests.

Publications: *World Rainforest Report*, quarterly; *Action Alerts*, monthly.

Other: International campaign to stop the import of hardwoods from primeval tropical rainforests and promote ecologically sound plantations on already degraded land.

8.62

Rare Seed Locators, Ltd.
2140 Shattuck Ave., Drawer 2479
Berkeley, CA 94704

Purpose: A seed finder's service and seed exchange with the purpose of bringing selected rare and endangered species under cultivation.

Other: Those wishing to participate should send a list of wants or a list of seeds available (alphabetically listed by Latin name) and a self-addressed stamped envelope.

8.63

San Francisco Friends of the Urban Forest
512 Second St., 4th Floor
San Francisco, CA 94107
415-543-5000

Purpose: To promote and facilitate the planting of street trees through education, community organizing, and fundraising.
Publication: *Treescapes*, quarterly.
Other: Tree plantings.

8.64

Save America's Forests
1742 18th St., N.W.
Room 201
Washington, DC 20009
202-667-5150

Purpose: A coalition of forest reform groups and individuals around the country working to make management for ecological values the top priority for America's National Forests.

8.65

Save the Prairie Society
10327 Elizabeth
Westchester, IL 60154
708-865-8736

Purpose: To preserve and manage the 80-acre Wold Road Prairie.
Publication: *Prairie Pointer*, irregular.

8.66

Save-the-Redwoods League
114 Sansome St., Room 605
San Francisco, CA 94104
415-362-2352

Purpose: To purchase redwood and giant sequoia forest for protection in state and national parks.
Publication: *Save-the-Redwoods League Bulletin*, biannual.
Other: Besides ongoing purchase of redwoods, the League is involved in reforestation projects.

Scatterseed Project
Farmington, ME
Cross-reference: 17.31

8.67

Scenic America
216 7th St., S.E.
Washington, DC 20003
202-546-1100

Purpose: To preserve America's scenic landscapes by providing information, technical assistance, and legal advice on all forms of landscape preservation including tree preservation and billboard control.
Publication: *Scenic America Newsletter*, bimonthly.

8.68

Seattle Tilth Association
4649 Sunnyside Ave N.
Seattle, WA 98103
206-633-0451

Purpose: To support and promote organic gardening and farming, urban ecology, composting, recycling, and environmental issues.
Publication: *Seattle Tilth Newsletter*, monthly.
Other: Demonstration gardens showing organic gardening techniques, Xeriscaping, and space-saving techniques.

Seed Savers Exchange
Decorah, IA
Cross-reference: 17.32

Seed Saving Project
Davis, CA
Cross-reference: 17.33

8.69

Sierra Club
730 Polk St.
San Francisco, CA 94109
415-776-2211

Purpose: To explore, enjoy, and protect the wild places of the Earth; to practice and promote the responsible use of the Earth's ecosystems and resources; to educate and enlist humanity to protect and restore the quality of the natural and human environment.
Publication: *Sierra*, bimonthly.

8.70

Siskiyou Permaculture Resources Group
P.O. Box 874
Ashland, OR 97520
503-488-0311

Purpose: To educate the population of our region on sustainable landscaping, permaculture design, and ecological principles.
Publication: *The Grapevine*, quarterly.
Other: Regular courses on topics like edible landscaping and ecological landscaping.

8.71

Smithsonian Institution, Dept. of Botany, Plant Conservation Unit
National Museum of Natural History
NHB 166, 10th & Constitution Aves.
Washington, DC 20560
202-357-2027; (FAX) 202-786-2563

Purpose: Contributes to efforts to identify, protect, and conserve endangered and threatened plants within the United States and other regions of the world.
Publication: *Biological Conservation Newsletter*, monthly, free.
Other: Centers of Plant Endemism Project aims to identify all principal areas in the United States where plants grow in a restricted habitat; IUCN/SI Latin America Plant Project is an international collaboration to compile information on endangered plants and habitats in Latin America and make recommendations for their protection.

Society of American Foresters
Bethesda, MD

Cross-reference: 1.114

Society of Municipal Arborists
University City, MO

Cross-reference: 1.115

8.72
Soil and Water Conservation Society
7515 NE Ankeny Rd.
Ankeny, IA 50021-9764
515-289-2331

Purpose: To advance the science and art of good land and
water use.

Talavaya Center
Espanola, NM

Cross-reference: 17.36

8.73
Tallgrass Prairie Alliance
P.O. Box 379
Baldwin City, KS 66006
913-594-3172

Purpose: To promote the preservation of the North
American prairie.

8.74
Tennessee Conservation League
11 Music Circle South, Suite 5
Nashville, TN 37203
615-254-7364

Purpose: Conservation education and advocacy which
promotes the wise use and management of Tennessee's
natural resources to benefit animals, forests, and
people.
Publication: *Tennessee Out-of-Doors*, bimonthly.
Other: Releaf Tennessee encourages school children to
plant and nurture trees.

8.75
Texas Organization for Endangered Species
P.O. Box 12773
Austin, TX 78711-2773
512-329-9627; (FAX) 512-327-2453

Purpose: A nonprofit organization for scientists, resource
managers, and others interested in the conservation of
rare species and natural communities.
Publication: *News and Notes*, biannual.
Other: Plant Committee.

Thomas Jefferson Center for Historic Plants
Charlottesville, VA

Cross-reference: 17.38

8.76
Traffic (U.S.A.)
World Wildlife Fund
1250 24th St., N.W.
Washington, DC 20037
202-293-4800; (FAX) 202-293-9211

Purpose: A program of the World Wildlife Fund; part of
an international network that monitors global trade in
wildlife and wildlife products.
Publication: *TRAFFIC (USA)*, quarterly, free.
Other: TRAFFIC investigates the international trade of
numerous species of plants, including a major project
on the international bulb trade and trade in endangered
species of plants found in Chinese medicinals.

8.77
Treepeople
12601 Mulholland Dr.
Beverly Hills, CA 90210
818-753-4600

Purpose: An environmental problem-solving organization
that promotes personal involvement, community action,
and global awareness.
Publication: *Seedling News*, bimonthly.
Other: School program, public education, tree plantings,
educational vegetable gardens, nursery.

8.78
Trees for Life, Inc.
1103 Jefferson
Wichita, KS 67203
316-263-7294

Purpose: The primary aim of Trees for Life is to end
hunger in the developing countries by planting fruit
and fuel trees.
Publications: *LifeLines*, 3/year, free; *Update!*, 3/year, free.
Other: Provides a free tree growing kit (seeds,
instructions, growing containers, and teacher
workbooks) to elementary school children in the U.S.

8.79
Trees for Tomorrow
Natural Resources Education Center
P.O. Box 609, 611 Sheridan St.
Eagle River, WI 54521
715-479-6456

Purpose: Natural resources education.
Publication: *NORTHBOUND*, quarterly.
Other: Tree seedling sales program.

8.80
Vermont Natural Resources Council
9 Bailey Ave.
Montpelier, VT 05602
802-223-2328; (FAX) 802-223-0600

Purpose: To foster a land ethic in Vermont, including
better policies for land use, forestry, the protection of
air and water quality, and wildlife.
Publication: *Vermont Environmental Report*, quarterly.

8.81
Western Earth Support Co-op
P.O. Box 269
Westcliffe, CO 81252
719-746-2275

Purpose: To promote bioregionalism, deep ecology, nature religions, and environmental employment in western North America.
Publications: *Western Environmental Jobletter*, monthly; *West Wind*, quarterly.
Other: Involved in saving old growth forest vegetation in Colorado.

8.82
Wilderness Society
1400 I St., N.W.
10th Floor
Washington, DC 20005
202-842-3400

Purpose: Devoted to preserving wilderness and wildlife, protecting America's prime forests, parks, rivers, and shorelands, and fostering an American land ethic.
Publication: *Wilderness*, quarterly.

8.83
Wildlife Information Center, Inc.
629 Green St.
Allentown, PA 18102
215-434-1637; (FAX) 215-434-2706

Purpose: Disseminates information on wildlife conservation, education, and research, and advocates nonkilling wildlife uses such as observation, photography, drawing, painting, and ecotourism.
Publications; *Wildlife Activist*, quarterly; *Wildlife Issues and News*, quarterly.
Other: Promotes wetlands protection, tropical and temperate rainforest protection, and the protection of native wildflowers.

8.84
World Forestry Center
4033 SW Canyon Rd.
Portland, OR 97221
503-228-1367; (FAX) 503-228-3624

Purpose: To improve understanding of the importance of well-managed forests and related resources.
Publication: *Forest World*, quarterly.
Other: Demonstration tree farm.

8.85
World Green
Rt. 5, Box 142
Rockwall, TX 75087
214-475-3662

Purpose: To supply trial packages of seeds from food plants developed for the southwestern United States to nurserymen in other arid regions of the world.

8.86
World Wildlife Fund–U.S.
1250 24th St., N.W.
Washington, DC 20037
202-293-4800

Purpose: Dedicated to protecting endangered wildlife and wildlands.
Publication: *Focus*, bimonthly.

8.87
Xerces Society
10 S.W. Ash St.
Portland, OR 97204
503-222-2788

Purpose: To preserve biodiversity through the conservation of rare and endangered invertebrates and their habitats.
Publications: *Wings, Essays on Invertebrate Conservation*, 3/year; *Butterfly Count*, annual.

CANADA

8.88
Canadian Forestry Association
185 Somerst St. West
Suite 203
Ottawa, ON K2P 0J2
613-232-1815; (FAX) 613-232-4210

Purpose: To promote the integrated use, sustainable development, and stewardship of Canada's forests and related land, water, and wildlife resources.
Publication: *Forestry on the Hill*, bimonthly.

8.89
Canadian Institute of Forestry
151 Slater St.
Suite 1005
Ottawa, ON K1P 5H3
613-234-2242; (FAX) 613-234-6181

Purpose: To advance the stewardship of Canada's forest resources, provide national leadership in forestry, promote competence among forestry professionals, and foster public awareness of Canadian and international forestry issues.
Publication: *The Forestry Chronicle*, bimonthly.

8.90
Canadian Nature Federation
453 Sussex Dr.
Ottawa, ON K1N 6Z4

Purpose: To promote the understanding, appreciation, and enjoyment of nature, and the conservation of the environment.
Publication: *Nature Canada*, quarterly.

Canadian Organic Growers
Ottawa, ON

Cross-reference: 1.132

8.91
Canadian Plant Conservation Programme
The Aboretum
University of Guelph
Guelph, ON N1G 2W1
519-824-4120, ext. 3093; (FAX) 519-763-9598

Purpose: To conserve genetic diversity of species and populations of rare native plants and heritage ornamentals in Canada, and to enhance communications and cooperation among concerned institutions maintaining living collections.
Publication: *CPCP Newsletter*, biannual, free to appropriate institutions.

Canadian Wildflower Society
North York, ON

Cross-reference: 1.137

8.92
Canadian Wildlife Federation
1673 Carling Ave.
Ottawa, ON K2A 3Z1
613-725-2191

Purpose: To foster understanding of natural processes so that people may live in harmony with the land and its resources.
Publication: *International Wildlife*, bimonthly.

8.93
Friends of the Earth
251 Laurier Ave. West
Suite 701
Ottawa, ON K1P 5J6
613-230-3352; (FAX) 613-232-4354

Purpose: A national and international environmental research and advocacy organization, working with 35 sister organizations throughout the world. Its main work in Canada is on ozone depletion, global warming, energy policy, pesticides, and the monitoring of the federal government's environmental record.
Publications: *Earth Words*, quarterly; *Atmosphere*, quarterly.
Other: Involved in reforming the Pest Controls Products Act; published "How to Get Your Lawn & Garden Off Drugs"; other urban antipesticides projects.

8.94
Hamilton Naturalists' Club
Postal Station E
Box 5182
Hamilton, ON L8S 4L3

Purpose: A field naturalists' club studying and helping to preserve the flora and fauna of the region.
Publication: *The Wood Duck*, 8/year.
Other: Wildflower meadow and woodland restoration projects.

Heritage Seed Program
Uxbridge, ON

Cross-reference: 17.40

8.95
Manitoba Naturalists' Society
302-128 James Ave.
Winnipeg, MB R3B 0N8
204-943-9029

8.96
Nature Conservancy of Canada
794A Broadview Ave.
Toronto, ON M4K 2P7
416-469-1701

Purpose: Dedicated to the acquisition and preservation of ecologically significant natural land areas throughout Canada.

8.97
Toronto Field Naturalists
20 College St., Suite 4
Toronto, ON M5G 1K2
416-968-6255

Purpose: To encourage the appreciation and preservation of Ontario's natural heritage of plants, animals, and landforms.
Publication: *Toronto Field Naturalist*, 8/year.
Other: Botany group with regular meetings and outings; plant inventories; owns Jim Baillie Nature Reserve (forested wetland).

THE CENTER FOR PLANT CONSERVATION NETWORK OF ARBORETA AND BOTANICAL GARDENS

The following gardens help to maintain seed banks and living collections of rare and endangered plants of North America.

AZ

8.98
Arboretum at Flagstaff, The
P.O. Box 670
S. Woody Mountain Rd.
Flagstaff, AZ 86002
602-774-1441

Cross-reference: 16.7

8.99
Desert Botanical Garden
1201 North Galvin Pkwy.
Phoenix, AZ 85008
602-941-1225

Cross-reference: 16.10

CA
8.100
Rancho Santa Ana Botanic Garden
1500 North College Ave.
Claremont, CA 91711-3101
714-625-8767; (FAX) 714-626-7670

Cross-reference: 16.48

8.101
University of California Botanical Garden at Berkeley
Centennial Dr.
University of California
Berkeley, CA 94720
415-642-3343

Cross-reference: 16.60

CO
8.102
Denver Botanic Garden
909 York St.
Denver, CO 80206
303-331-4000; (FAX) 303-331-4013

Cross-reference: 16.65

FL
8.103
Bok Tower Gardens
P.O. Box 3810
Lake Wales, FL 33859-3810
813-676-1408

Cross-reference: 16.86

8.104
Fairchild Tropical Garden
10901 Old Cutler Rd.
Miami, FL 33156
305-667-1651

Cross-reference: 16.87

HI
8.105
National Tropical Botanical Garden
P.O. Box 340
Lawai, HI 96765
808-332-7324; (FAX) 808-332-9765

Cross-reference: 16.121

8.106
Waimea Falls Park Arboretum & Botanical Gardens
56-864 Kamehameha Hwy.
Haleiwa
Oahu, HI 96712
808-638-8511

Cross-reference: 16.124

MA
8.107
Arnold Arboretum of Harvard University
125 Arborway
Jamaica Plain, MA 02130
617-524-1715; (FAX) 617-524-1418

Cross-reference: 16.173

MO
8.108
Missouri Botanical Garden
P.O. Box 299
St. Louis, MO 63166
314-577-5100

Cross-reference: 16.214

NC
8.109
North Carolina Botanical Garden
3375 Totten Center
Univ. of North Carolina—Chapel Hill
Chapel Hill, NC 27599-3375
919-962-0522

Cross-reference: 16.232

NE
8.110
Nebraska Statewide Arboretum
112 Forestry Sciences Laboratory
University of Nebraska, Lincoln
Lincoln, NE 68583-0823
402-472-2971

Cross-reference: 16.247

NY
8.111
New York Botanical Garden
200th and Southern Blvd.
Bronx, NY 10458-5126
212-220-8700; (FAX) 212-220-6504

Cross-reference: 16.286

OH
8.112
Holden Arboretum
9500 Sperry Rd.
Mentor, OH 44060
216-256-1110; (FAX) 216-256-1655

Cross-reference: 16.304

OR
8.113
Berry Botanic Garden
11505 S.W. Summerville Ave.
Portland, OR 97219
503-636-4112

Cross-reference: 16.322

TX
8.114
Mercer Arboretum and Botanic Gardens
22306 Aldine Westfield Rd.
Humble, TX 77338
713-443-8731

Cross-reference: 16.387

8.115
San Antonio Botanical Gardens
555 Funston Pl.
San Antonio, TX 78209
512-821-5115

Cross-reference: 16.389

UT
8.116
State Arboretum of Utah
Building 436
University of Utah
Salt Lake City, UT 84112
801-581-5322

Cross-reference: 16.392

9

State, Provincial, and Local Horticultural Organizations

This chapter lists state, provincial, regional, and local professional associations, trade associations, garden club organizations, and other horticultural groups. The addresses of these organizations change regularly when new officers are elected.

For more information on state and local horticultural organizations, see the annual *Directory of American Agriculture* (Pierre, South Dakota: Agricultural Resources & Communications, Inc.).

UNITED STATES

AK
9.1
Alaska Federation of Garden Clubs
2801 Bennett Dr.
Anchorage, AK 99517

Contact: Mrs. Albert J. Silk

9.2
Alaska Seed Growers Association
SR B, Box 7634
Palmer, AK 99645

Contact: Ted Pyrah, President

9.3
Alaska Society of Professional Soil Scientists
Soil Conservation Service
Suite 300, 200 E 9th Ave.
Anchorage, AK 99501-3687

Contact: Joseph Moore, President

AL
9.4
Alabama Blueberry Association
Rt. 1, Box 331A
Ashland, AL 36251
205-396-5385

Contact: Mark Walker, President

9.5
Alabama Christmas Tree Association
124 St. Clair Ln.
Huntsville, AL 35811
205-851-9375

Contact: George Brown

9.6
Alabama Forestry Association
555 Alabama St.
Montgomery, AL 36104
205-265-8733

Contact: Gene Davenport, President

9.7
Alabama Fruit Growers Association
1021 Crestwood Dr.
Auburn, AL 36830
205-887-3930

9.8
Alabama Nurserymen's Association
P.O. Box 9
Auburn, AL 36831-0009
205-821-5148

Contact: Judy P. Copeland, Executive Secretary

9.9
Alabama Pecan Growers Association
P.O. Box 68
Elmore, AL 36025
205-567-6434

9.10
Alabama Seedmen's Association
P.O. Box 2546
Auburn, AL 36831
205-821-7400

9.11
Alabama Soil Fertility Society
P.O. Box 6101
Montgomery, AL 36106
205-832-4280

Contact: Ken Carr, President

9.12
Alabama State Florists Association
1727 6th Ave., S.E.
Decatur, AL 35601
205-355-6991

Contact: John McBride, President

9.13
Alabama State Soil & Water Conservation Committee
P.O. Box 3336
Montgomery, AL 36193
205-242-2620

Contact: James Plaster, Executive Secretary

9.14
Alabama Sweet Potato Association
Rt. 14, Box 870
Cullman, AL 35055
205-734-4898

Contact: W. C. Kress, President

9.15
East Alabama Florists Association
Rt. 1, Box 31
Wetumpka, AL 36902

9.16
Garden Club of Alabama
124 Highland Pl.
Sheffield, AL 35660

Contact: Mrs. Gene Castleberry

9.17
Huntsville Herb Society
4747 Bob Wallace Ave.
Huntsville, AL 35805
205-830-4447

9.18
Huntsville/Madison County Botanical Garden Society
P.O. Box 281
Huntsville, AL 35804
205-534-3270

9.19
Professional Soil Classifiers Association of Alabama
2241 Johns Circle
Auburn, AL 36830
205-826-7156

Contact: James H. Brown, Executive Officer

9.20
South Alabama Botanical and Horticultural Society
c/o John Bowen
P.O. Box 8382
Mobile, AL 36608

9.21
South Alabama Florists Association
3928 Airport Blvd.
Mobile, AL 36608

Contact: Carolyn Hendrix

AR
9.22
Arkansas Federation of Garden Clubs, Inc.
Box 609
Cherokee Village, AR 72525

Contact: Mrs. L. Morgan Yost

9.23
Arkansas Florist Association
P.O. Box 243
Morrilton, AR 72110

9.24
Arkansas Forestry Association
620 Plaza West
McKinley & Lee Sts.
Little Rock, AR 72205
501-663-8260

Contact: Merf Pavlovich, Executive Vice-President

9.25
Arkansas Greenhouse Association
11740 Maumelle
North Little Rock, AR 72116

9.26
Arkansas Nursery Association
P.O. Box 55295
Little Rock, AR 72225
501-225-0029

Contact: Faith Walsham, Executive Secretary

9.27
Arkansas Pest Control Association
1724 Charlotte Ct.
Little Rock, AR 72204
501-664-5608

Contact: Gordon Barnes, Executive Director

9.28
Arkansas Plant Food Educational Society
P.O. Box 5504
North Little Rock, AR 72119
501-673-2954

Contact: Glen Stroh, President

9.29
Arkansas Seed Dealers Association
Stratton Seed Co.
P.O. Box 32
Stuttgart, AR 72160
501-673-4433

Contact: Jim Craig

9.30
Arkansas Seed Growers Association
Loewer Oaks and Grain Seed Inc.
Rt. 1, Box 233
Wynne, AR 72396
501-697-3416

Contact: Carl Loewer, President

9.31
Arkansas Soil and Water Conservation Commission
One Capitol Mall, Suite 2D
Little Rock, AR 72201
501-682-1611

Contact: J. Randy Young, Executive Director

9.32
Arkansas Turf & Grass Association
P.O. Box 206
Searcy, AR 72143

9.33
Central Arkansas Fruit Growers
2901 W. Roosevelt
Little Rock, AR 72204

9.34
Christmas Tree Association
Rt. 8, Box 940
Benton, AR 72015

9.35
State Plant Board
P.O. Box 1069
Little Rock, AR 72203
501-225-1598

Contact: Jim Shatsar, Manager

AZ
9.36
Arizona Apple Growers Association
Star Rt. 1, Box 114
Wilcox, AZ 85643
602-384-2414

Contact: Alan Seitz, President

9.37
Arizona Christmas Tree Association
3111 E. Marilyn
Phoenix, AZ 85032
602-971-6513

Contact: Steve Sutherland, President

9.38
Arizona Citrus Institute
P.O. Box 1388
Mesa, AZ 85211
602-964-8615

Contact: Jim Mast, President

9.39
Arizona Federation of Garden Clubs
508 Highland Dr.
Prescott, AZ 86303

Contact: Mrs. Harry Wagner

9.40
Arizona Grain & Seed Association
P.O. Box T
1901 N. Trekell
Casa Grande, AZ 85222
602-836-8228

Contact: John Skelley, President

9.41
Arizona Grape Growers Association
P.O. Box 115
Dateland, AZ 85333
602-454-2772

Contact: Bob Rush, President

9.42
Arizona Landscape Contractors Association
2720 E. Thomas Rd.
Phoenix, AZ 85016
602-956-4252

Contact: Sharon Dewey, Executive Director

9.43
Arizona Nursery Association
444 W. Camelback Rd., No. 302
Phoenix, AZ 85013
602-241-0317

Contact: Deborah Flowers, Executive Vice-President

9.44
Arizona State Florists Association
15215 N. Cave Creek Rd.
Phoenix, AZ 85032
602-744-9069

Contact: Fordyce Steinhour

9.45
Jojoba Growers Association
2201 E. Camelback Rd., Suite 130-B
Phoenix, AZ 85016
602-956-8893

Contact: Vicky Hubbard

9.46
Phoenix Vegetable District
P.O. Box 7
Tolleson, AZ 85353
602-936-1873

Contact: Don McGaffee, President

9.47
Southern Arizona Pistachio Growers Association
601 N. Wilment, Suite 53
Tucson, AZ 85711
602-745-1611

Contact: Gary Auerbach, President

9.48
Vegetable Growers Association
3121 N. 19th Ave., No. 190
Phoenix, AZ 85015
602-266-6225

Contact: Shelly Tunis, Executive Director

9.49
Western Growers Association
3121 N. 19th Ave., No. 190
Phoenix, AZ 85015
602-266-6149

Contact: Shelly Tunis, Legal Consultant

CA
9.50
Apricot Producers of California
1064 Woodland Ave.
Modesto, CA 93351-1349
209-524-0801

Contact: Les Rose, Operations Vice-President

Australian Plant Society of California
Pasadena, CA

Cross-reference: 2.29

Bromeliad Study Group of Northern California
San Francisco, CA

Cross-reference: 2.32

9.51
Calavo Growers of California
15661 Red Hill Ave.
Tustin, CA 92680-7321
714-259-1166

Contact: Al Nangelos, President

9.52
California Apricot Advisory Board
1280 Boulevard Way, Suite 107
Walnut Creek, CA 94595
415-937-3660

Contact: Gene Stokes, General Manager

9.53
California Arboretum Foundation
301 N. Baldwin Ave.
Arcadia, CA 91006
818-447-8207

9.54
California Artichoke Advisory Board
P.O. Box 747
Castorville, CA 95012
408-633-4411

Contact: Patricia Boman, Manager

9.55
California Association of Nurserymen
1419 21st St.
Sacramento, CA 95814
916-448-2881

9.56
California Avocado Commission
1251 East Dyer Rd.
Santa Anna, CA 92705
714-558-6761

Contact: Mark Affleck, President

9.57
California Avocado Society
P.O. Box 4816
Saticay, CA 93004
805-644-1184

Contact: Victor S. Pankey, President

9.58
California Beet Growers, Ltd.
P.O. Box 222
Brawley, CA 92227-0222
619-344-1366

Contact: Jack Flemming, Director

9.59
California Cactus Growers Association
28136 Brodiaea
Moreno Valley, CA 92388
714-242-7860

9.60
California Celery Research Advisory Board
531-D North Alta Ave.
Dinuba, CA 93618
209-591-0434

Contact: Dana Dickey, Manager

9.61
California Christmas Tree Association
1451 Danville Blvd., #200
Alamo, CA 94507
415-837-7463

Contact: Sharon Burke

9.62
California Chrysanthemum Growers Association
788 San Antonio Rd.
Palo Alto, CA 94303
415-494-1451

9.63
California Citrus Research Board
117 West Ninth St., Room 913
Los Angeles, CA 90015
213-627-3041

Contact: Billy J. Peightal, Manager

9.64
California Cling Peach Advisory Board
160 Spear St., Suite 1330
Box 7111
San Francisco, CA 94105
415-541-0100

Contact: Thomas Krugman, General Manager

9.65
California Council of Landscape Architects
2451 Potomac St.
Oakland, CA 94602

9.66
California Date Administrative Committee
45701 Monroe, Suite H
Indio, CA 92201
619-347-4510

Contact: Anne Ezell, Manager

9.67
California Dry Bean Advisory Board
531-D North Alta Ave.
Dinuba, CA 93618
209-591-4866

Contact: Jerry Munson, Manager

9.68
California Fertilizer Association
1700 I St., Suite 130
Sacramento, CA 95814
916-441-1584

Contact: Steven R. Beckley

9.69
California Fig Advisory Board
3425 N. First
Fresno, CA 93726
209-224-3447

Contact: Ron Klamm, Manager

9.70
California Flower Council
P.O. Box 1365
Davis, CA 95617-1365
415-543-3790

Contact: Stephen Oku, President

9.71
California Fresh Market Tomato Advisory Board
531-D North Alta Ave.
Dinuba, CA 93618
209-591-0437

Contact: Ed Beckman, Manager

9.72
California Garden Clubs, Inc.
7540 Granite Ave.
Orangeville, CA 95662

Contact: Mrs. Allan Nielsen, President

9.73
California Grape & Tree Fruit League
1540 E. Shaw Ave., Suite 120
Fresno, CA 93710
209-226-6330

9.74
California Iceberg Lettuce Advisory Board
512 Pajaro St.
Salinas, CA 93901
408-443-3205

Contact: Harold Bradshaw, Manager

9.75
California Iceberg Lettuce Commission
2801 Monterey-Salinas Hwy.
Box 3354
Monterey, CA 93942
408-375-8277

9.76
California Kiwifruit Commission
1540 River Park Dr., Suite 110
Sacramento, CA 95815
916-929-3740

Contact: Mark Houston, President

9.77
California Landscape Contractors Association, Inc.
2226 K St.
Sacramento, CA 95816
916-448-2522

9.78
California Landscape Industry Association, Inc.
2226 K St.
Sacramento, CA 95816
916-448-2522

9.79
California Landscape and Irrigation Council
14408 E. Whittier Blvd., Suite A5
Whittier, CA 90605
213-941-4900

9.80
California Macadamia Society
P.O. Box 1290
Fallbrook, CA 92028
619-743-8081

Contact: Jeannine T. Howell, Secretary-Treasurer

9.81
California Melon Research Board
531-D North Alta Ave.
Dinuba, CA 93618
209-591-0435

9.82
California Pistachio Commission
1915 N. Fine Ave.
Fresno, CA 93727
209-252-3345

Contact: Ron Sether, President

9.83
California Potato Research Advisory Board
531-D North Alta Ave.
Dinuba, CA 93618
209-591-0436

Contact: Jim Melban, Manager

9.84
California Prune Board
World Trade Center, Room 103
San Francisco, CA 94111
415-986-1452

Contact: F.W. Davis, Executive Director

9.85
California Raisin Advisory Board
3455 North First, Box 5335
Fresno, CA 93755
209-224-7010; (FAX) 209-224-7016

Contact: Clyde Nef, Manager

California Rare Fruit Growers
Fullerton, CA
Cross-reference: 2.34

9.86
California Redwood Association
405 Enfrente Dr., Suite 200
Novato, CA 94949
415-382-0662

9.87
California State Florist Association
650 Fifth St.
San Francisco, CA 94107
415-495-6780

9.88
California Strawberry Advisory Board
41 Hanger Way
P.O. Box 269
Watsonville, CA 95077-0269
408-724-1301

Contact: Dave Riggs, President

9.89
California Table Grape Commission
P.O. Box 5498
Fresno, CA 93755
209-224-4997

Contact: Bruce Obbink, President

9.90
California Tree Fruit Agreement
701 Fulton Ave.
P.O. Box 255383
Sacramento, CA 95865
916-483-9261

Contact: Jonathan W. Field, Manager

California Urban Forests Council
Los Angeles, CA
Cross-reference: 8.19

9.91
California Weed Conference
P.O. Box 609
Fremont, CA 94537-0609
415-790-1252

Contact: Wanda M. Graves, Treasurer

9.92
Design Associates Working with Nature
1442a Walnut St., Box 101
Berkeley, CA 94709
415-664-1315

9.93
Diamond Walnut Growers
P.O. Box 1727
1050 S. Diamond St.
Stockton, CA 95201
209-467-6000

Contact: Gerald Barton, President

9.94
Fresh Garlic Association
P.O. Box 2410
Sausalito, CA 94966-2410
408-383-5057; (FAX) 415-381-8185

Contact: David Martin, Executive Officer

9.95
International Pumpkin Association
2155 Union St.
San Francisco, CA 94123
415-346-4446

Los Angeles International Fern Society
Pasadena, CA

Cross-reference: 262

Mycological Society of San Francisco
San Francisco, CA

Cross-reference: 2.65

9.96
Northern California Flower Growers and Shippers
  Association
1150 Elko Dr.
Sunnyvale, CA 94089
408-734-2883

9.97
Northern California Turfgrass Council
P.O. Box 268
Lafayette, CA 94549
415-283-6162

9.98
Orange County Floral Association
120 North El Camino Real
San Clemente, CA 92672
714-498-3454

Contact: Linda Thompson

9.99
Oxnard Pest Control Association
P.O. Box 1187
Oxnard, CA 93032
805-483-1024

Contact: Ed Frost, Senior Executive Officer

9.100
Professional Interior Plantscapers' Association of Southern
  California
4000 Birch St., #180
Newport Beach, CA 92660

9.101
Redwood Empire Florists Association
Claudia's Florist
P.O. Box 466
Sonoma, CA 95476

9.102
Sacramento Retail Florists Association
1901 Del Paso Rd.
Sacramento, CA 95815
916-885-7692

Contact: Ed Thompkins

9.103
Sacramento Tree Foundation
1540 River Parks #201
Sacramento, CA 95815
916-924-TREE

9.104
San Diego County Flower Association
P.O. Box 87
Encinitas, CA 92024
714-753-5727

Contact: Marilu Johnson

9.105
San Diego Floral Association
Caseo del Prado
Balboa Park
San Diego, CA 92101-1619
619-232-5762

Cross-references: 14.6, 21.14

9.106
San Fernando Valley Florists Association
14840 Nordhoff St. #3
Panorama City, CA 91402
818-781-7109

9.107
San Francisco Flower Growers Association
644 Brannan St.
San Francisco, CA 94107
415-781-8410

Contact: Angelo Stagnaro, Jr.

9.108
San Francisco Flower Show, Inc.
San Francisco County Fair Building
Ninth Ave. at Lincoln Way
San Francisco, CA 94122
415-221-5724

San Francisco Friends of the Urban Forest
San Francisco, CA

Cross-reference: 8.63

9.109
San Joaquin Florist Association
9898 North Jackstone Rd.
Stockton, CA 95212
209-466-9509

Contact: Dorothy Plotz

9.110
Santa Barbara County Nursery and Flower Growers
  Association
P.O. Box 4038
Santa Barbara, CA 92103
805-963-1793

9.111
Southern California Floral Association
756 Wall St.
Los Angeles, CA 90014

9.112
Southern California Heritage Roses
3637 Empire Dr.
Los Angeles, CA 90034

9.113
Strybing Arboretum Society
Ninth Ave. at Lincoln Way
San Francisco, CA 94122
415-661-1514

9.114
Teleflora
12233 West Olympic Blvd.
Los Angeles, CA 90064
213-826-5253

9.115
Theodore Payne Foundation
10459 Tuxford St.
Sun Valley, CA 91352

9.116
Western Growers Association
17620 Fitch St.
Irvine, CA 92714
714-863-1000

Contact: David L. Moore, President

CO
9.117
Colorado Apple Administrative Committee
2267 R 25 Rd.
Cedaredge, CO 81413
303-856-3873

Contact: Dan Williams, Chairman

9.118
Colorado Certified Potato Growers Association
7148 N. County Rd. 2E
Monte Vista, CO 81144
719-852-4659

Contact: Robert Mattive, President

9.119
Colorado Corn Growers Association
1121 N. Park Ave.
Johnstown, CO 80534
303-732-4470

Contact: Bruce Kauffman, President

9.120
Colorado Federation of Garden Clubs, Inc.
Box 186
Idaho Springs, CO 80452

Contact: Mrs. Dode Mehrer

9.121
Colorado Foundation Seed Project
Dept. of Agronomy
Plant Science Bldg.
Colorado State University
Fort Collins, CO 80523
303-491-6202

Contact: Jim Stanelle, Manager

9.122
Colorado Greenhouse Growers Association
2785 N. Speer Blvd., Suite 230
Denver, CO 80211
303-433-6423

Contact: Robert A. Briggs, Executive Vice-President

9.123
Colorado Nurserymen's Association
746 Riverside Dr.
Rural Route, Box 2676
Lyons, CO 80540
303-747-2662

Contact: Cary G. Hall, Executive Director

9.124
Colorado Potato Growers Exchange
2401 Larimer St.
Denver, CO 80205
303-292-0159

Contact: Rebecca Derry, General Manager

9.125
Colorado Seed Growers Association
Dept. of Agronomy
Plant Science Bldg.
Colorado State University
Fort Collins, CO 80523
303-491-6202

Contact: James Stanelle, Manager

9.126
Colorado Seedmen's Association
24 Baylor Dr.
Longmont, CO 80501
303-776-0013

Contact: Doris Engstrom, Secretary-Treasurer

9.127
Colorado Sugarbeet Growers Association
Greeley National Plaza
Suite 620
Greeley, CO 80631
303-352-6875

Colorado Water Garden Society
Denver, CO
Cross-reference: 2.35

9.128
Rocky Mountain Christmas Tree Association
3807 Capitol Dr.
Fort Collins, CO 80526
303-284-6351

Contact: Ray Mehaffey

9.129
Rocky Mountain Plant Food & Agricultural Chemicals
    Association
2150 S. Bellaire St., Room 204
Denver, CO 80222
303-753-9067

Contact: W. L. Gordon

9.130
United Floral Industry of Colorado
2785 North Speer Blvd., Suite 230
Denver, CO 80211
303-433-6423

Contact: Evelyn Reichert, Executive Secretary

CT
9.131
Allied Florists Association of Central Connecticut, Inc.
421 Campbell Ave.
West Haven, CT 06516
203-934-2653

Contact: Charles Barr

9.132
Connecticut Association for Living Historical and
    Agricultural Museums
P.O. Box 134
Redding, CT 06875
203-938-8048

Contact: Nancy Campbell, Regional Representative

9.133
Connecticut Christmas Tree Association
820 West St.
Guilford, CT 06437
203-457-0664

Contact: Michael Pochan, Secretary

9.134
Connecticut Council on Soil and Water Conservation
State Office Bldg.
165 Capitol Ave.
Hartford, CT 06106
203-566-7234

Contact: Allen Bennett, Executive Director

9.135
Connecticut Extension Council
132 Daly Rd.
Hartford, CT 06248
203-486-4125

Contact: Lorna Marcisenuk, President

9.136
Connecticut Florists Association
590 Main St.
Monroe, CT 06468
203-268-9000

Contact: Robert V. Heffernan, Executive Vice-President

9.137
Connecticut Forest and Park Association
16 Meriden Rd., Rt. 66
Middletown, CT 06457
203-346-2372

Contact: George M. Milne, President

9.138
Connecticut Gladiolus Society
Canton, CT 06019
203-693-4219

Contact: Norman Adams, President

9.139
Connecticut Nurserymen's Association
Horticultural Associates, Inc.
24 West Rd., Suite 53
Vernon, CT 06066
203-872-2095

Contact: Larry Carville, Executive Secretary

9.140
Connecticut Nut Growers Association
409 Elm St.
Windsor Locks, CT 06096
203-623-0703

Contact: Edward Straughn, Treasurer

9.141
Connecticut Orchid Society
432 Undermountain Rd.
Salisbury, CT 06068
203-435-2263

Contact: Judy Becker, President

9.142
Connecticut Pomological Society
46 High View Terrace
Enfield, CT 06082
203-749-7516

Contact: Tom Moriarty, Secretary

9.143
Connecticut Rhododendron Society
14 Roundhill Rd.
Granby, CT 06035

Contact: Frans Krot, President

9.144
Connecticut Tree Protective Association
P.O. Box 344
New Haven, CT 06513-0344
203-772-3662

9.145
Connecticut Vegetable and Bedding Plant Growers
  Association
Rt. 6, Box 165A
Bethel, CT 06801
203-797-4176

Contact: Joseph Maisano, Jr., Secretary

9.146
Federated Garden Clubs of Connecticut
P.O. Box 672
Wallingford, CT 06492
203-265-2101

Contact: Lee Bauerfield, President

DC
9.147
Mid-Atlantic Florists Association
1817 Columbia Rd., N.W.
Washington, DC 20009
202-667-7800

DE
9.148
Delaware Agricultural Museum Association
866 N. DuPont Hwy.
Dover, DE 19901
302-734-1618

9.149
Delaware Association of Nurserymen
R.R. 1, Box 233
Felton, DE 19943
302-398-4118

Contact: Pete Gerardi, President

9.150
Delaware Council on Soil and Water Conservation
Dept. of Natural Resources and Environmental Control
Division of Soil and Water Conservation
P.O. Box 1401
Dover, DE 19903
302-736-4411

Contact: John A. Hughes, Director

9.151
Delaware Federation of Garden Clubs, Inc.
2306 Jamaica Dr.
Wilmington, DE 19810

Contact: Mrs. Robert Weeks

9.152
Delaware Fruit Growers Association
c/o T.S. Smith and Sons
P.O. Box 275
Bridgeville, DE 19933
302-337-8271

Contact: Charles Smith, President

Delaware Nature Society
P.O. Box 700
Hockessin, DE 19707

Cross-reference: 8.25

9.153
Delaware Potato Growers Association
665 Shellcross Lake Rd.
Middletown, DE 19709
302-378-2084

Contact: Ed Baker, President

9.154
Delaware Vegetable Growers Association
R.D. 1, Box 144
Magnolia, DE 19962
302-697-7681

Contact: Joe Jackewicz, Jr.

9.155
Mardel Watermelon Growers Association
R.D. 3, Box 234
Laurel, DE 19956
302-875-3091

Contact: George J. Collins, President

9.156
Maryland, Delaware Watermelon Association
Rt. 3, Box 234
Laurel, DE 19956
302-875-3091

Contact: George Collins, President

9.157
Maryland-Delaware Plant Food and Crop Protection
  Association
103 Linden Ave.
Georgetown, DE 19947
302-856-2155

Contact: William W. Henderson, Secretary

FL
9.158
Allied Florists of N.E. Florida
2330 Oak St.
Jacksonville, FL 32204
904-388-1038

Contact: John Mullis

9.159
Broward County Florists Association
4461 Sheridan St.
Hollywood, FL 33021
305-966-8262

Contact: Harvey Pearson

9.160
Central Florida Fern Co-op
P.O. Box 588
Pierson, FL 32080
904-749-4911

Contact: Robert Zahra

9.161
Central Florida Florists Association
2425 Sandy Ln.
Orlando, FL 32818
305-295-5692

Contact: Pat Vobornik, President

9.162
Citrus Administrative Committee
P.O. Box 24508
Lakeland, FL 33802-3103
813-682-3103

Contact: Arthur B. Chadwell, Manager

9.163
Florida Fruit & Vegetable Association
P.O. Box 140155
Orlando, FL 32814-0155
407-894-1351

9.164
Florida Avocado Administrative Committee
P.O. Box 188
Homestead, FL 33090-0188
305-247-0848

Contact: Shirley J. Manchester, Administrator

9.165
Florida Christmas Tree Association
208 Newins-Ziegler Hall
University of Florida
Gainesville, FL 32611
904-392-4826

Contact: Roger Webb

9.166
Florida Citrus Commission
P.O. Box 148
Lakeland, FL 33802-0148
813-682-0171

Contact: Dan L. Gunter, Executive Director

9.167
Florida Citrus Mutual
P.O. Box 89
Lakeland, FL 33802
813-682-1111

9.168
Florida Citrus Nurserymen's Association
P.O. Box 145
Winter Haven, FL 33882
813-294-4763

9.169
Florida Federation of Garden Clubs
P.O. Box 1604
Winter Park, FL 32790
407-647-7016

Contact: Mrs. Kenton H. Haymans

9.170
Florida Fern Growers Association
P.O. Box 3426
DeLand, FL 32723
904-736-3376

Contact: Helen Shields

9.171
Florida Fertilizer and Agrichemical Association
P.O. Box 9326
Winter Haven, FL 33883-9326
813-293-4827

Contact: John King, Executive Director

9.172
Florida Flower Association, Inc.
P.O. Box 1569
Fort Myers, FL 33901
813-332-1771

Contact: Jane Restum

Florida Foliage Association
Apopka, FL
Cross-reference: 6.8

9.173
Florida Forestry Association
P.O. Box 1696
Tallahassee, FL 32302
904-222-5646

Contact: William K. Cook, President

9.174
Florida Grape Growers Association
1255 Farrell Dr.
DeLeon Springs, FL 32130
904-985-1821

Contact: John Halloway

9.175
Florida Irrigation Society
808 N. John St.
Orlando, FL 32808
407-291-9074

9.176
Florida Lime Administrative Committee
P.O. Box 188
Homestead, FL 33090-0188
305-247-0848

Contact: Shirley J. Manchester, Administrator

9.177
Florida Lychee Growers Association
1404 N. Garfield Ave.
Deland, FL 32724

Contact: Ms. R. B. Hall

9.178
Florida Mango Forum
20900 S.W. 376 St.
Homestead, FL 33034

Contact: Stephen F. Vandas, Director

9.179
Florida Nurserymen & Growers Association
Box 16796
Temple Terrace, FL 33687

9.180
Florida Ornamental Growers Association
P.O. Box 942
Alva, FL 33920-0942
813-728-2535

Contact: Mike Hackman

9.181
Florida Peanut Producers Association
1204 Cliff St.
P.O. Box 447
Graceville, FL 32440
904-263-6130; (FAX) 904-263-6210

Contact: Buster Smith, Executive Director

9.182
Florida Pecan Growers Association
1143 Fifield Hall
University of Florida
Gainesville, FL 32611
904-392-1996

Contact: Tim Crocker

9.183
Florida Seedsman & Garden Supply Association
P.O. Box 368
Yalaha, FL 34797
904-728-3600

Contact: William Weyand, President

9.184
Florida State Florists Association
1612 S. Dixie Hwy.
Lake Worth, FL 33460
407-585-9491

Contact: Charlotte D. Garrison

9.185
Florida Strawberry Growers Association
P.O. Box 2631
Plant City, FL 33566
813-752-6822

Contact: William McClelland, President

9.186
Florida Sweet Corn Exchange
P.O. Box 140155
Orlando, FL 32814-0155
407-894-1351

Contact: Wayne Crain

9.187
Florida Tomato Committee
P.O. Box 140635
Orlando, FL 32814-3071
407-894-3071

Contact: David Neill, Chairman

9.188
Florida Tomato Exchange
P.O. Box 140635
Orlando, FL 32814-0635
407-894-3071

Contact: Jeff Gargiulo, President

9.189
Florida Turfgrass Association
302 S. Graham Ave.
Orlando, FL 32803-6399
407-898-6721

Contact: M. J. McLaughlin, President

9.190
Florida Vegetable Exchange
P.O. Box 140155
Orlando, FL 32814-0155
407-894-1351

Contact: Joe Obucina, President

9.191
Florida Watermelon Association
P.O. Drawer 1604
Marianna, FL 32446
904-482-4758

Contact: W. J. Peacock III, President

9.192
Florists Association of Greater Miami
14700 N.W. 7th Ave.
Miami, FL 33168
305-769-3843

Contact: Gus de Laflor

9.193
Gulf Citrus Growers Association, Inc.
P.O. Box 1319
LaBelle, FL 33935
813-675-2180

9.194
Indian River Citrus League
P.O. Box 519
Vero Beach, FL 32961
407-562-2728

9.195
Retail Florists Association of Northeast Florida
1942 Rogero Rd.
Jacksonville, FL 32211
904-744-7411

Contact: Steve Cox

9.196
Southern Nurserymen's Association
2126 Palm Vista Dr.
Apopka, FL 32712
407-889-3808; (FAX) 407-889-5004

9.197
Southwest Florida Florists Association
1305-A Homestead Rd.
Lehigh Acres, FL 33936
813-369-1311

Contact: Jim Phebus

9.198
Tropical Fruit Growers of South Florida, Inc.
Dade-IFAS Cooperative Extension Service
18710 S.W. 288 St.
Homestead, FL 33030-2309

GA
9.199
Forest Farmers Association
P.O. Box 95385
Atlanta, GA 30347
404-325-2954

Contact: B. Jack Warren, Executive Vice-President

9.200
Garden Club of Georgia, Inc.
7 Woodland Dr.
Cartersville, GA 30120

Contact: Mrs. E. Carl White

9.201
Georgia Apple Council
Extension Horticultural Dept.
University of Georgia
Athens, GA 30602
404-542-2861

Contact: Dr. Stephen C. Myers

9.202
Georgia Blueberry Association
P.O. Box 1982
Alma, GA 31510
912-632-8667

Contact: M. E. Smith, Manager

9.203
Georgia Christmas Tree Association
P.O. Box 187
Riverdale, GA 30274
404-361-2823

Contact: Rick Peel

9.204
Georgia Christmas Tree Growers Association
Rural Development Center
P.O. Box 1209
Tifton, GA 31793
912-386-3418

Contact: David Moorehead

9.205
Georgia Commercial Flower Growers Association
P.O. Box 6784
Athens, GA 30604
404-542-2340

Contact: Carolyn E. Cox, Secretary

9.206
Georgia Corn Growers Association
P.O. Box 27
Girard, GA 30426
912-569-4814

Contact: Jimmy Dixon, President

9.207
Georgia Forestry Association
40 Marietta St., N.W., Suite 1020
Atlanta, GA 30303-2806
404-522-0951

Contact: Robert L. Izlar, Executive Director

9.208
Georgia Forestry Commission
P.O. Box 819
Macon, GA 31298-4599
912-744-3237

Contact: John Mixon, Director

9.209
Georgia Grape Growers Association
Fox Vineyards
225 Highway 11
Covington, GA 30209
404-787-5402

Contact: John Fuchs

9.210
Georgia Landscape Professionals
Medical College of Georgia
RA-100
Augusta, GA 30910-1903

Contact: Larry Ward

9.211
Georgia Mountain Apple Growers Association
R.R. 2, Box 184
Ellijay, GA 30540
404-276-1144

Contact: Jerry Smith, President

9.212
Georgia Muscadine Growers Association
R.R. 1
Wray, GA 31798
912-468-7878

Contact: Gary Paulk, President

9.213
Georgia Peach Council
126 Cochran Dr.
Byron, GA 31008
912-956-4155

Contact: Joseph H. Cleveland, First Vice-President

9.214
Georgia Peanut Commission
P.O. Box 967
Tifton, GA 31793
912-386-3470; (FAX) 912-386-3501

Contact: Don Koehler, Executive Director

9.215
Georgia Peanut Producers Association
P.O. Box 390
Blakely, GA 31723
912-723-8221

Contact: Walt Mitchell

9.216
Georgia Pecan Growers Association
18 Quail St.
Leesburg, GA 31763
912-759-2187

Contact: Jane Crocker, Executive Secretary

9.217
Georgia Pest Control Association
One Executive Concourse
Suite 103
Duluth, GA 30136
404-476-0827

Contact: Valera Jessee, Director

9.218
Georgia Plant Food Educational Society
15 Chatuachee Crossing
Savannah, GA 31411
912-598-0654

Contact: Robert Farrow, Secretary-Treasurer

9.219
Georgia Seed Development Commission
2420 South Milledge Ave.
Athens, GA 30605
404-548-7688

Contact: Dr. J. Earl Elsner

9.220
Georgia Seedmen's Association
147 South Stratford Dr.
Athens, GA 30605
404-548-7688

Contact: Dr. Harold Loden, Executive Secretary

9.221
Georgia State Florists Association
Tucker Flower Shop, Inc.
2249 Idlewood Rd.
Tucker, GA 30084
404-938-7158

Contact: Monroe W. Brown

9.222
Georgia State Nurserymen's Association
190 Spring Tree Rd.
Athens, GA 30605
404-542-2471

Contact: Jake Tinga, Executive Secretary

9.223
Georgia Sweet Potato Growers Association
P.O. Box 314
Ocilla, GA 31774
912-468-7848

Contact: Joey Veal

9.224
Georgia Sweet Potato Improvement Association
Horticulture Dept.
Coastal Plain Experiment Station
P.O. Box 748
Tifton, GA 31793-5401
912-386-3357

Contact: Dr. Melvin R. Hall, Secretary

9.225
Georgia Turfgrass Foundation
Soil Testing Laboratory
2400 College Station Rd.
Athens, GA 30605
404-542-5350

Contact: Gill Landry, Secretary

9.226
Georgia Vegetable Growers Association
P.O. Box 748
Tifton, GA 31793
912-386-3904

Contact: Smittle Doyle, Secretary-Treasurer

9.227
Georgia Watermelon Association
P.O. Box 38
Morven, GA 31638
912-775-2580

Contact: H. R. Lawson, Secretary-Treasurer

9.228
Georgia Wholesale Florists Association
471 Glen Iris Dr., N.E.
Atlanta, GA 30308
404-523-7736

Contact: Robert Blei

9.229
Greater Augusta Florists Association
934 Baker Ave.
Augusta, GA 30904
404-736-3938

Contact: Nancy W. Whittle, President

HI
9.230
Aloha Arborist Association
c/o Dudley Hulbert
Trees of the Tropics
3265 Paumaka Pl.
Honolulu, HI 96822
808-988-3761

9.231
Anthurium Association of Hawaii
Hawaiian Greenhouses
P.O. Box 1
Pahoa, HI 96778
808-968-6228

Contact: Alan Kuahara, President

9.232
Big Island Association of Nurserymen
P.O. Box 4365
Hilo, HI 96720
808-959-5888

Contact: Roy Shigenaga, President

9.233
Dendrobium Orchid Growers Association of Hawaii
P.O. Box 17341
Honolulu, HI 96817
808-259-8311

Contact: Tosh Sugita, President

9.234
Florist Association of Hawaii
P.O. Box 2893
Honolulu, HI 96802
808-947-7511

Contact: Ann Nakata, Secretary

9.235
Hawaii Anthurium Growers Cooperative
170 Wiwoole St.
Hilo, HI 96720
808-935-6681

Contact: William J. Mori, Manager

9.236
Hawaii Anthurium Industry Association
875 Komohana
Hilo, HI 96720
808-968-6239

Contact: Dan Hata

9.237
Hawaii Association of Nurserymen
c/o Robert DeNeve
Hawaiian Phoenix
P.O. Box 1046
Keaau, HI 96749
808-965-7137

9.238
Hawaii Federation of Garden Clubs, Inc.
P.O. Box 25401
Honolulu, HI 96825

Contact: Mrs. John W. Johnson

9.239
Hawaii Guild of Professional Gardeners
c/o Kevin Mulkern
P.O. Box 25491
Honolulu, HI 96825
808-396-6595

9.240
Hawaii Landscape and Irrigation Contractors Association
c/o Gregory Culver, Pua Lani
Landscape Design, P.O. Box 1769
Kilua, HI 96734
808-261-7517

9.241
Hawaii Macadamia Nut Association
Mack Farms
F.R. 25
Captain Cook, HI 96704
808-328-2435

Contact: Charles Young

9.242
Hawaii Orchid Promotion Council
73-4399 Laui St. No. 10
Kailu-Kona, HI 96740
808-965-9570

Contact: Greg Braun, President

9.243
Hawaii Protea Association
P.O. Box 68
Kula, HI 96790
808-878-6273

Contact: Paul Winiarski, Manager

9.244
Hawaii State Guava Association
1209 C Kaumana St.
Hilo, HI 96720
808-935-9842

Contact: Doug Felix, President

9.245
Hawaii Tropical Flower and Foliage Association
P.O. Box 817
Kaneohe, HI 96744
808-237-8515

Contact: John Morgan, President

9.246
Hawaii Turfgrass Association
c/o John Gillis III
Oahu Country Club
150 Country Club Rd.
Honolulu, HI 96817
808-595-3256

9.247
Kawai Nui Heritage Foundation
c/o Martha McDaniel
Box 1101
Kailua, HI 96734

9.248
Landscape Industry Council of Hawaii
The Tree People
3620 Waialae Ave., Room 3
Honolulu, HI 96816
808-734-5963

Pacific Orchid Society
Honolulu, HI

Cross-reference: 2.75

9.249
Pineapple Growers Association of Hawaii
P.O. Box 3380
Honolulu, HI 96801
808-544-5415

Contact: Joseph W. Hartley, Jr., President

IA
9.250
Allied Florists of Des Moines
1546 45th St.
Des Moines, IA 50311
515-255-6191

Contact: Mary Scot

9.251
Cedar Rapids Professional Florists
1800 Ellis Blvd., N.W.
Cedar Rapids, IA 52405
319-366-1826

Contact: Robert Pierson

9.252
Federated Garden Clubs of Iowa
Windsong
Rt. 4, Box 223
Marshalltown, IA 50158

Contact: Mrs. G. B. Cox

9.253
Iowa Christmas Tree Growers Association
c/o Mark Howell, President
Rt. 1
Cumming, IA 50061
515-981-4762

9.254
Iowa Corn Growers Association
360 West Towers
1200 35th St.
West Des Moines, IA 50265
515-225-9242

Contact: Evan Stadlman, Executive Director

9.255
Iowa Fertilizer and Chemical Association
323 University Ave.
Des Moines, IA 50314
515-282-9659

Contact: Dan Frieberg, Executive Vice-President

9.256
Iowa Fruit and Vegetable Growers Association
c/o Max Hagen, President
3333 33rd Ave., S.W.
Cedar Rapids, IA 52404
319-396-8573

9.257
Iowa Nurserymen's Association
c/o Ken Peckosh, President
3990 Blairs Ferry Rd.
Cedar Rapids, IA 52402
319-393-5946

9.258
Iowa Nut Growers Association
c/o W.J. Christiansen, President
Box 629
Newell, IA 50568
712-272-4404

9.259
Iowa Turfgrass Institute
c/o Mike Agnew, President
Iowa State University
Ames, IA 50011
515-294-0027

9.260
Society of Iowa Florists
c/o Joni Tietz, President
2275 Independence
Waterloo, IA 50707
319-234-6883

ID
9.261
Elwyee Beet Growers Association
P.O. Box 9
Mountain Home, ID 83647
208-587-8182

Contact: Jack Post, President

9.262
Idaho Apple Commission
P.O. Box 909
Parma, ID 83660
208-722-5111

Contact: Larry Link, Executive Director

9.263
Idaho Association of Pea and Lentil Producers
5071 Highway 8
Moscow, ID 83843
208-882-3023

Contact: Harold Blain

9.264
Idaho Bean Commission
P.O. Box 9433
Boise, ID 83707
208-334-3520

Contact: Harold West

9.265
Idaho Cherry Commission
P.O. Box 909
Parma, ID 83660
208-722-5111

Contact: Larry Link, Executive Director

9.266
Idaho Mint Commission
P.O. Box 576
Caldwell, ID 83606
208-459-4470

Contact: Lars Aaland, Secretary

9.267
Idaho Mint Growers Association
18638 Prescott Ln.
Nampa, ID 83687
208-466-6031

Contact: Maurice Woodard, President

9.268
Idaho Nursery Association
c/o Carla Nakano, Executive Director
Hillside Nursery
2350 Hill Rd.
Boise, ID 83702
208-343-2545

9.269
Idaho Onion Growers Association
120 Third St., Box 430
Parma, ID 83660
208-722-5044

Contact: Jerry Stone, Secretary

9.270
Idaho Potato Commission
P.O. Box 1068
Boise, ID 83701
208-334-2350

9.271
Idaho Seed Council
Grassland West Co.
P.O. Box A
Culdesac, ID 83524
208-843-2219

Contact: Roger Styner

9.272
Idaho Soil Improvement Committee/Far West Fertilizer
   Association
29603 U of I Ln.
Parma, ID 83660
208-722-6701

Contact: Brad Brown

9.273
Idaho State Federation of Garden Clubs, Inc.
1218 N. 25th St.
Boise, ID 83702

Contact: F. Ruth Thacker

9.274
Idaho State Florists Association
121 Third Ave. West
Gooding, ID 83330
208-934-4358

Contact: Dolly Goicoechea

9.275
Idaho Sugar Beet Association
R.R. 5, Box 69
Rupert, ID 83350
208-532-4365

Contact: George Grant, President

9.276
Idaho Water Garden Society
2570 S. Swallowtail Ln.
Boise, ID 83706

9.277
Idaho Weed Control Association
1330 Flier Ave. East
Twin Falls, ID 83301
208-734-3600

Contact: Gail Malbert, Secretary

9.278
Idaho-Eastern Oregon Onion Committee
P.O. Box 909
Parma, ID 83660
208-722-5111

Contact: Larry Link, Manager

9.279
Idaho-Eastern Oregon Potato Committee
P.O. Box 2192
Idaho Falls, ID 83403
208-529-8057

9.280
Idaho-Eastern Oregon Seed Association
P.O. Box 399
Parma, ID 83651
208-466-0764

Contact: Lavern Hansen

9.281
Idaho-Oregon Fruit and Vegetable Association
P.O. Box 909
Parma, ID 83660
208-722-5111

Contact: Larry Link, Manager

9.282
Inland Empire Christmas Tree Association
P.O. Box 200
Santa, ID 83866

9.283
Intermountain Grass Growers Association
E. 2375 Mullen Ave.
Post Falls, ID 83854
208-773-5862

Contact: Dennis L. Carlson, Executive Secretary

9.284
Kootenai Valley Nursery Growers Association
Levig Nursery
H.C.R. 60, Box 31
Bonners Ferry, ID 83805
208-267-2136

Contact: Les Levig

9.285
North Idaho Berry Association
6370 Kaniksu Shores Circle
Sandpoint, ID 83864
208-263-1245

Contact: Harry Menser

9.286
Potato Growers of Idaho
P.O. Box 949
Blackfoot, ID 83221
208-785-1110

Contact: John Rooney, Executive Director

9.287
Southeast Idaho Weed Control Association
5500 South 5th
Pocatello, ID 83204
208-233-9591

Contact: Susie Going, Secretary

9.288
Washington Dry Pea and Lentil Commission
P.O. Box 8566
Moscow, ID 83843
208-882-3023

Contact: Harold Blain, Administrator

IL
9.289
Chicagoland Home & Flower Show
Trade Expositions & Associated Management, Ltd.
Tyler Creek Plaza
Box 7338
Elgin, IL 60121-7338
708-888-TEAM

9.290
Garden Clubs of Illinois, Inc.
23W163 Blackberry Ln.
Glen Ellyn, IL 60137

Contact: Mrs. William H. Laycock

Grand Prairie Friends of Illinois
Urbana, IL

Cross-reference: 8.34

9.291
Illinois Christmas Tree Association
Rt. 1, Box 315
Ashmore, IL 61912
217-349-8688

Contact: Marsha Blair

9.292
Illinois Corn Growers Association
2415 E. Washington
P.O. Box 1623
Bloomington, IL 61702-1623
309-557-3257

Contact: Kent Chidley, President

9.293
Illinois Fertilizer and Chemical Association
Box 186
St. Anne, IL 60964
815-427-6644

Contact: Lloyd Burling, President

9.294
Illinois Forage and Grassland Council
c/o Don Graffis
W301 Turner Hall, Univ. Of Illinois
1102 South Goodwin Ave.
Urbana, IL 61801
217-333-4424

9.295
Illinois Fruit and Vegetable Growers Foundation
P.O. Box 218
Cissna Park, IL 60924
815-457-2037

Contact: Dan Hinkle, Chairman

9.296
Illinois Ginseng Growers Association
P.O. Box 462
Wateska, IL 60970
815-432-4022

Contact: Kathy Judge, Secretary-Treasurer

9.297
Illinois Nurserymen's Association
c/o Randy Vogel, Executive Director
Suite 1702
Hilton Hotel
Springfield, IL 62701
217-525-6222

9.298
Illinois Seed Dealers Association
508 South Broadway
Urbana, IL 61801
217-367-4053

Contact: Steve Barwick, President

9.299
Illinois Specialty Crops Association
1701 Towanda Ave.
Bloomington, IL 61701

Contact: Don Naylor, Manager

9.300
Illinois State Florist Association
c/o Dan Iron, Executive Director
505 South 23rd St.
Mattoon, IL 61938

9.301
Illinois Vegetable Growers Association
17510 Garden Valley Rd.
Woodstock, IL 60098
815-568-7023

Contact: Henry Boi, Executive Secretary

9.302
Ornamental Growers Association of Northern Illinois
P.O. Box 67
Batavia, IL 60510
708-879-0520

Contact: Bonnie Zaruba, Executive Secretary

Save the Prairie Society
Westchester, IL
Cross-reference: 8.65

9.303
Tri-City Florist Club
2513 30th Avenue Ct.
Moline, IL 61265
319-764-3330

Contact: Harry Johnson

IN
9.304
Garden Club of Indiana
Rt. 1, Box 34
Geneva, IN 46740
219-334-5453

Contact: Barbara Yoder

9.305
Indiana Herb Association
1701 Towanda Ave.
Bloomington, IN 61701

9.306
Indiana Arborists Association, Inc.
c/o Nathan Matthews, Secretary
Public Service Indiana
1000 East Main St.
Plainfield, IN 46168
317-838-1441

9.307
Indiana Association of Consulting Foresters
Forest Management Services
735 Maple Ave.
Terre Haute, IN 47804
812-235-2025

Contact: Larry J. Owen, Secretary

9.308
Indiana Association of Nurserymen, Inc.
202 East 650 North
West Lafayette, IN 47906
317-497-1100

Contact: Phillip Carpenter, Executive Secretary

9.309
Indiana Association of Professional Soil Classifiers
1217 S. Elm St.
Crawfordsville, IN 47993
219-244-7953

Contact: Douglass Wolf, President

9.310
Indiana Christmas Tree Growers Association
S.E. Purdue Ag. Center, Box 155
Butlerville, IN 47223
812-485-6977

Contact: John Seifert, Secretary

9.311
Indiana Corn Growers Association
8770 Guion Rd., Suite A
Indianapolis, IN 46268
317-876-9311

Contact: Rowe Sargent, President

9.312
Indiana Extension Agents Association
Courthouse
Brownstone, IN 47220
812-358-6101

Contact: Carolyn Gordon, President

9.313
Indiana Flower Growers Association
c/o Allen Hammer
Horticultural Dept.
HORT Building, Purdue University
West Lafayette, IN 47907
317-494-1335

9.314
Indiana Forestry and Woodland Owners Association
2505 Radcliffe Ave.
Indianapolis, IN 46227
317-787-6508

Contact: M. Gene Hassler, Executive Secretary

9.315
Indiana Fruit Growers Cooperative Association, Inc.
c/o Jerry Chandler, President
P.O. Box 224
Stilesville, IN 46180
317-539-6255

9.316
Indiana Nut Growers Association
c/o Jerry W. Lehman, President
7280 Shady Ln.
Terre Haute, IN 47802
812-232-2024

9.317
Indiana Pest Control Association
451 E. 38th St.
Indianapolis, IN 46205
317-925-9292

Contact: Dan Everts, President

9.318
Indiana Plant Food and Agricultural Chemicals
  Association
101 West Washington St.
Suite 1313 East Tower
Indianapolis, IN 46204
317-632-4028

Contact: T. Jeffrey Boese, President

9.319
Indiana Seed Trade Association
c/o Steve Houghton, President
Central Indiana Supply Co., Inc.
3610 Shelby St.
Indianapolis, IN 46227
317-788-7013

9.320
Indiana Vegetable Growers Association
c/o James E. Simon, Horticulture Dept.
HORT Building, Purdue University
West Lafayette, IN 47907
317-494-1328

9.321
Indiana Winegrowers Guild
Butler Winery
1022 North College
Bloomington, IN 47401
812-339-7233

Contact: Jim Butler, President

9.322
Indianapolis Landscape Association
A.M. Rust Landscape & Equipment Supply
12001 Michigan Rd., N.W.
Zionsville, IN 46077
317-873-2050

Contact: William Brown, Executive Secretary

9.323
Indianapolis Museum of Art—Horticultural Society
610 College Ln.
Indianapolis, IN 46240
317-253-0925

Contact: Mrs. Robert H. Brunner, Secretary

9.324
Northeast Indiana Flower Growers Association
North Manchester Greenhouse
R.R. 4, Box 11
North Manchester, IN 46962
219-982-2493

Contact: Gary Eberly

9.325
Northwestern Indiana Nurserymen's Association
Serviscape, Inc.
P.O. Box 8658
Michigan City, IN 46360
219-872-9412

Contact: Pete Sinnott

9.326
State Florists Association of Indiana
P.O. Box 20189
Indianapolis, IN 46220
317-253-0500

Contact: E. A. Schoenberger, Executive Director

9.327
Tri-State Azalea Society
8220 Schissler Rd.
Evansville, IN 47712
812-985-9388

Contact: Robin Hahn, President

9.328
West Central Indiana Flower Growers Association
c/o Jess Vandergraff
Vandergraff's Greenhouse
3617 Morehouse Rd.
West Lafayette, IN 47906
317-463-1239

KS
9.329
County Weed Directors Association of Kansas
Butler County Weed Dept.
Rt. 3, Box 19A
El Dorado, KS 67042
316-321-5190

Contact: Riley Waters

9.330
Greater Wichita Florists Association
6125 East 13th St.
Wichita, KS 67218
316-686-0275

Contact: Glendora Henson

9.331
Horticulture Alumni & Friends
Rt. 2, Box 535
St. George, KS 66535
913-776-0397

Contact: Carl Meyers, President

9.332
Kansas Arborists Association
P.O. Box 708
Lawrence, KS 66044
913-841-1246

Contact: George Osborn, President

9.333
Kansas Associated Garden Clubs
811 Sunset Dr.
Lawrence, KS 66044

Contact: Mrs. Vernon Carlsen

9.334
Kansas Associated Nurserymen
Johnson's Garden Center
2707 W. 18th St.
Wichita, KS 67203

Contact: Marty Johnson, President

9.335
Kansas Christmas Tree Association
2610 Clafin Rd.
Manhattan, KS 66502
913-537-7050

Contact: William L. Loucks

9.336
Kansas Christmas Tree Growers
c/o Ruth Bramham, President
Rt. 6, Box 367
Lawrence, KS 66044
913-842-7055

9.337
Kansas Corn Growers Association
Rt. 4, Box 282
Lawrence, KS 66044
913-843-6949

Contact: Roger Pine, President

9.338
Kansas Extension Agents Association
County Extension Agent 4-H
2110 Harper
Lawrence, KS 66046
913-843-7058

Contact: Lindy Lindquist

9.339
Kansas Fertilizer and Chemical Association
816 S.W. Tyler
Topeka, KS 66612
913-234-0463

Contact: Mike VanCampen, President

9.340
Kansas Flower, Lawn, and Garden Show
c/o Bob Heifner, Chairman
4707 W. 6th St.
Topeka, KS 66608
913-272-1487

9.341
Kansas Fruit Growers
Rt. 1
Courtland, KS 66939

Contact: Dan Kuhn, President

9.342
Kansas Grape Growers & Wine Makers Association
2602 Larken Dr.
Wichita, KS 67216

Contact: Lee Beadles, President

9.343
Kansas Greenhouse Growers
Jackson's Greenhouse
1933 Lower Silver
Topeka, KS 66608
913-232-3416

Contact: Annette Jackson, President

9.344
Kansas Nurserymen's Association
5530 S.W. 19th
Topeka, KS 66604

Contact: John W. Tonkin, Executive Secretary

9.345

Kansas Nut Growers Association
c/o Ronald Curtis, President
Rt. 1
Parsons, KS 67357

9.346

Kansas Seed Industry Association
2000 Kimball Ave.
Manhattan, KS 66502
913-532-6118

Contact: Gary Drussell, President

9.347

Kansas State Florists
c/o Lynne Moss, President
The Flower Shop
390 E. 4th St.
Pratt, KS 67124
316-672-2948

9.348

Kansas Turfgrass Association
Hidden Lakes G.C.
6020 S. Greenwich Rd.
Derby, KS 67037

Contact: John Wright, President

9.349

Kansas Vegetable Growers
c/o Ron Meier, Vice-President
J.C. Meier & Sons
4827 NW 25th
Topeka, KS 66618

KY

9.350

Allied Florists of Metro Louisville
3812 Saint Germain Ct.
Louisville, KY 40207
502-896-1616

Contact: Jack Worland

9.351

Appalachian Alternative Crops
414 S. Wenzel St.
Louisville, KY 40204-1049
502-589-0975

9.352

Garden Club of Kentucky
P.O. Box 1653
Middleboro, KY 40965

Contact: Mrs. Charles D. Huddleston

9.353

Kentucky Arborists' Association
10105 Afton Rd.
Louisville, KY 40223
502-245-4036

Contact: Larry Lose, Treasurer

9.354

Kentucky Corn Growers Association
Dept. of Agronomy
N-122 Ag-Science Bldg., North
University of Kentucky
Lexington, KY 40546-0091
606-257-3975

Contact: Dr. Morris Bitzer

9.355

Kentucky Fertilizer and Agricultural Chemical Association
P.O. Box 21743
Lexington, KY 40522
606-226-2288; (FAX) 606-266-2240

Contact: J. Tracy Spencer, Executive Director

9.356

Kentucky Florist Association
Parkway Florist
3954 Cane Run Rd.
Louisville, KY 40211
502-778-1666

Contact: Harriet Miller

9.357

Kentucky Greenhouse Association
Sunshine Greenhouse
1105 Moser Rd.
Louisville, KY 40223
502-244-0707

Contact: Tom Welsh, President

9.358

Kentucky Nurserymen's Association
Korfhage Florist & Nursery
9611 Taylorsville Rd.
Louisville, KY 40299
502-447-3641

Contact: Bob Korfhage President

9.359

Kentucky Seed Dealers Association
Caudill Seed Co.
1201 Story Ave.
Louisville, KY 40206
502-583-4402

Contact: Pat Caudill, Manager

9.360

Kentucky Seed Improvement Association
P.O. Box 12008
Lexington, KY 40579-2008
606-257-2971

Contact: Billy Huffines, President

9.361

Kentucky Soil and Water Conservation Committee
691 Teton Trail
Frankfort, KY 40601
502-564-3080

Contact: Stanley Head, Director

9.362
Kentucky Vegetable Growers Association
8545 River Rd.
Hebron, KY 41048
606-689-4622

Contact: Howard Hempfling, President

LA
9.363
Allied Florists of S.W. Louisiana
314 Helen St.
Lake Charles, LA 70601

Contact: Annie Funk

9.364
Baton Rouge Area Retail Florists
8255 Florida Blvd.
Baton Rouge, LA 70806

Contact: W. Irwin Pugh

9.365
Federated Council of New Orleans Garden Clubs, Inc.
787 Jewell St.
New Orleans, LA 70124
504-282-5077

Contact: Mrs. Donald Miester

9.366
Louisiana Association of Nurserymen
c/o Warren Meadows, Secretary
4560 Essen Ln.
Baton Rouge, LA 70809
504-766-3471

9.367
Louisiana Association of Plant Pathologists and
  Nematologists
Rice Research Station, LSU
P.O. Box 1429
Crowley, LA 70527

Contact: Donald E. Groth, President

9.368
Louisiana Forestry Association
P. O. Drawer 5067
Alexandria, LA 71307-5067
318-443-2558

Contact: Buck Vandersteen, Executive Director

9.369
Louisiana Garden Club Federation
139 Citrus Rd.
River Ridge, LA 70123

Contact: Mrs. E. D. Massa

9.370
Louisiana Greenhouse Growers Association
University of Southwestern Louisiana
P.O. Box 44433
Lafayette, LA 70504
318-231-5348

Contact: Jackie Dunlop, President

9.371
Louisiana Nurserymen's Association
770 Robert Rd.
Slidell, LA 70458

Contact: Louis Parr, President

9.372
Louisiana Pecan Growers Association
610 Texas St., Suite 400
Shreveport, LA 71101
318-424-3185

Contact: Brent Young, President

9.373
Louisiana Seedmen's Association
P.O. Box 12360
Alexandria, LA 71315-2360
318-473-0505

Contact: Ragan K. Nelson, Executive Secretary

9.374
Louisiana State Florists Association
12760 Triple B Rd.
Greenwell Spring, LA 70739

Contact: Dean Marcotte

9.375
Louisiana Sweet Potato Association
P.O. Box 1442
Bastrop, LA 71220
318-281-8660

Contact: Jerry Self

9.376
Louisiana Turfgrass Association
P.O. Box 531
Abita Springs, LA 70420-0531
504-892-1892

Contact: Edward F. Bonner, President

9.377
Louisiana Vegetable Growers Association
Rt. 1, Box 588-D
Belle Chasse, LA 70037

Contact: C. J. Bechnel, President

9.378
Louisiana-Mississippi Christmas Tree Association
Rt. 2, Box 325
Angie, LA 70426
504-848-5133

Contact: Kevin M. Steele

9.379
New Orleans Garden Society
3914 Prytania St.
New Orleans, LA 70115

9.380
New Orleans Retail Florists Association
1015 Veterans Blvd.
Metairie, LA 70005
504-837-5338

Contact: Janet Schnidler

MA
9.381
Garden Club Federation of Massachusetts, Inc.
547 Central Ave.
Needham, MA 02194

Contact: Mrs. George A. Dennett

9.382
Green Industry Council
P.O. Box 171
Sutton, MA 01590
508-476-3007

Contact: Phyllis Gillespie, Executive Director

9.383
Massachusetts Christmas Tree Association
P.O. Box 733
Millbury, MA 01527
617-865-0457

9.384
Massachusetts Flower Growers
105 Everett St.
Concord, MA 01742

Contact: Robert T. Luczai

9.385
Massachusetts Forestry Association
P.O. Box 202
Princeton, MA 01541
508-464-2759

Contact: Robert M. Ricard, Executive Director

9.386
Massachusetts Tree Wardens' and Foresters' Association
81 Charter Rd.
Acton, MA 01720
508-264-9629

Contact: Dean A. Charter, President

MD
9.387
Allied Florists of Greater Baltimore
113 West Franklin St.
Baltimore, MD 21201
301-752-3332

Contact: Tom Shaner

9.388
Federated Garden Clubs of Maryland, Inc.
128 Round Bay Rd.
Severna Park, MD 21146

Contact: Mrs. Hal D. Tray

9.389
Landscape Contractor Association of Maryland, D.C. and
  Virginia
c/o Beth Palys
13305 Oakwood Dr.
Rockville, MD 20850

9.390
Maryland Allied Florists Association of Greater Baltimore
17 Warren Rd., Suite 22-A
Pikeville, MD 21208
301-486-2979

Contact: Gary Okshansky, Executive Director

9.391
Maryland Arborist Association
c/o Ed Hogarth, Jr.
3861 Old Federal Hill Rd.
Jarrettsville, MD 21084

9.392
Maryland Christmas Tree Association
P.O. Box 314
Perry Hall, MD 21128

9.393
Maryland Extension Specialists Association
Ag. Engineering
University of Maryland
College Park, MD 20742
301-454-3901

Contact: Dr. David Doss

9.394
Maryland Forests Association
623 Shore Acres Rd.
Arnold, MD 21012
301-757-3581

9.395
Maryland Grape Growers Association
18517 Kingshill Rd.
Germantown, MD 20874
301-972-1325

Contact: James Russell

9.396
Maryland Greenhouse Growers Association
P.O. Box 432
Perry Hall, MD 21128

9.397
Maryland Nurserymen's Association
2 Troon Ct.
Baltimore, MD 21236

9.398
Maryland Peninsula Horticultural Society
P.O. Box 1836
Salisbury, MD 21801
301-749-6141

Contact: Wayne V. Shaff, Secretary

9.399
Maryland Seeding Association
1409 Spencerville Rd.
Spencerville, MD 20868
301-384-6300

Contact: Dianna Patton, Secretary

9.400
Maryland State Apple Commission
P.O. Box 917
Hagerstown, MD 21740-0917
301-733-8777

Contact: Lois J. Swaim, Secretary

9.401
Maryland State Soil Conservation Committee
Maryland Dept. of Agriculture
50 Harry S. Truman Pkwy.
Annapolis, MD 21401
301-841-5863

Contact: Louise Lawrence, Executive Secretary

9.402
Maryland Turfgrass Association
P.O. Box 283
Burtonsville, MD 20866

Contact: Dianna Lee Patton, Secretary

9.403
Maryland Turfgrass Council
P.O. Box 223
White Marsh, MD 21162

Contact: Cheryl Gaultney, Secretary

9.404
Maryland Vegetable Growers Association
1122 Holzapfel Hall
College Park, MD 20742
301-454-3007

Contact: Dr. Charles McClurg

9.405
Maryland Winery and Grape Growers Advisory Board
5120 Wissioming Rd.
Bethesda, MD 20816
202-872-1097

Contact: Ann Milne, Chairperson

9.406
Mid-Atlantic Nurserymen's Trade Show
P.O. Box 314
Perry Hall, MD 21128

9.407
National Capital Area Federation of Garden Clubs, Inc.
8916 Wooden Bridge Rd.
Potomac, MD 20854

Contact: Mrs. Frederick Caldwell

ME
9.408
Allied Florists of Greater Portland
781 Roosevelt Trail
Windham, ME 04062
207-892-4627

9.409
Garden Club Federation of Maine, Inc.
Box 56
Salisbury Cove, ME 04672

Contact: Mrs. Steen L. Meryweather

9.410
Maine Arborist Association
c/o Terry Traver
Box 199A, Rt. 100
Gray, ME 04039
207-657-3256

9.411
Maine Blueberry Commission
32 Coburn Hall
University of Maine Campus
Orono, ME 04469
207-581-1475

Contact: Ed McLaughlin, Executive Director

9.412
Maine Christmas Tree Association
Whitten Rd.
R.R. 1, Box 696
Kennebunk, ME 04043
207-985-3778

Contact: Diane Holmes

9.413
Maine Dry Bean Growers Association
West Ridge Rd. RFD
Skowhegan, ME 04976
207-474-8865

Contact: Kenneth Hogate

9.414
Maine Federated Garden Club
RFD 1, Box 1040
Corinna, ME 04928
207-278-5994

Contact: Mrs. Phillip Burril

9.415
Maine Nurserymen's Association
PST-SMVTI
South Portland, ME 04106

Contact: Richard Churchill, Executive Secretary

9.416
Maine Organic Farmers and Gardeners Association
P.O. Box 2176
Augusta, ME 04338
207-622-3118

Contact: Nancy Ross

9.417
Maine Pomological Society
R.R. 2, Box 621
Bridgton, ME 04009-9546
207-647-2869

Contact: Tom Gyger, Corresponding Secretary

9.418
Maine Potato Quality Control Board
744 Maine St., Room 1
Presque Isle, ME 04769
207-769-5061

Contact: Jane Fowler

9.419
Maine State Florists' Association
Butler Twins Florist
Maine Ave.
Gardiner, ME 04345

Contact: Richard Butler, President

9.420
Maine Tree Crop Alliance
c/o Environmental Science Bldg.
Unity College
Unity, ME 04988
207-948-6161

Contact: Jack Kertesz, Director

9.421
Maine Tree Farm Committee
Scott Paper Co.
Woodlands Dept.
R.R. 1, Box 400
Fairfield, ME 04937
207-453-2527

Contact: Paul Memmer, Chairman

9.422
Maine Vegetable and Small Fruit Growers Association
Rt. 2, Box 910
Litchfield, ME 04350
207-724-3881

Contact: Steve Goodwin, President

9.423
Mid-Coast Greenhouse Growers Association
Knox-Lincoln County Extension Office
375 Main St.
Rockland, ME 04841

Contact: Herb Annis

9.424
Mid-Maine Greenhouse Growers Association
Everlasting Farm
2106 Essex St.
Bangor, ME 04401

Contact: Michael Zuck

9.425
Southern Maine Greenhouse Growers Association
Androscoggin-Sagadahoc County
Extension Office, 277 Minot Ave.
Auburn, ME 04210

Contact: Vivian Holmes

MI
9.426
Allied Florists Association of Greater Detroit
1515 East 11 Mile Rd.
Royal Oak, MI 48067
313-542-8866

Contact: Robert Heron

9.427
Detroit Allied Florists
1515 East Eleven Mile Rd.
Royal Oak, MI 48067
313-542-8866

9.428
Federated Garden Club of Michigan
1509 Orchard Ln.
Niles, MI 49120
616-683-3030

Contact: Mrs. Peter H. Butus

9.429
Garden Club of Michigan
7113 Greer Rd.
Howell, MI 48843
517-546-2649

Contact: Mrs. Douglas Roby, President

9.430
Great Lakes Sugar Beet Growers Association
320 Plaza North
Saginaw, MI 48604
517-792-1531

Contact: Robert Young, Executive Vice-President

9.431
Great Lakes Vegetable Growers Convention
440 Plant and Soil Science Bldg.
East Lansing, MI 48824
517-353-7156

Contact: Hugh Price

9.432
Green Industries Council
1200 North Telegraph
Pontiac, MI 48053
313-681-9435

Contact: Greg Patchan, Chairman

9.433
Lawn Sprayers Association of Michigan
305 Livernois
Ferndale, MI 48220

Contact: Bill Olsen, Executive Secretary

9.434
Michigan Agrichemical and Fertilizer Association
18470 Sharon
Oakley, MI 48649
517-845-7301

Contact: Carol Schmiege

9.435
Michigan Apple Committee
7692 Ashley Ave.
Belding, MI 48809
616-794-3839

Contact: Thomas Heffron

9.436
Michigan Asparagus and Plum Advisory Boards
P.O. Box 23218
Lansing, MI 48909
517-323-7000

Contact: Harry Foster, Executive Director

9.437
Michigan Association of Cherry Producers
2220 University Park Dr., Suite 200
Okemos, MI 48864
517-347-0010

Contact: Richard Johnson, Executive Director

9.438
Michigan Association of Consulting Foresters
Rt. 1, Box 562
Lake Ann, MI 49650

Contact: Lee Eckstrom, President

9.439
Michigan Association of Nurserymen
819 N. Washington Ave.
Suite 2
Lansing, MI 48906
517-487-1282

Contact: Richard Seely, President

9.440
Michigan Bean Commission
100 A North Main St.
Leslie, MI 49251
517-694-0581

Contact: Jim Byrun, Executive Secretary

9.441
Michigan Cactus and Succulent Society
Louis Kilbert
5601 Coomer Rd.
W. Bloomfield, MI 48033
313-681-4791

9.442
Michigan Cherry Committee
3640 Kinnley Rd.
Ludington, MI 49431
616-843-2256

Contact: Art Lister, Jr., Chairman

9.443
Michigan Christmas Tree Association
343 S. Union St.
P.O. Box 128
Sparta, MI 49345
616-887-9008

Contact: Barry Brand, Executive Secretary

9.444
Michigan Corn Growers Association
Rt. L, Moore Rd.
Ekton, MI 48731
517-375-4067

Contact: Joe Sweeney

9.445
Michigan Floral Association
2420 Science Pkwy.
P.O. Box 590
Okemos, MI 48864
517-349-5754

Contact: Larry E. Andrick, Executive Vice-President

9.446
Michigan Forest Association
1558 Barrington
Ann Arbor, MI 48103
313-665-8279

Contact: McClain B. Smith, Executive Director

9.447
Michigan Foundation Seed Association
P.O. Box 22155
Lansing, MI 48909
517-332-3546

Contact: Richard Isles, Manager

9.448
Michigan Grape Society
8191 U.S. 31-33
Berrien Springs, MI 49103
616-471-7486

Contact: Edwin Kerlikowski

Michigan Nature Association
Avoca, MI

Cross-reference: 8.42

9.449
Michigan Nut Growers Association
4975 Grand River Rd.
Owosso, MI 48867
517-651-5278

Contact: William Nash, President

9.450
Michigan Onion Committee
P.O. Box 306
Hudsonville, MI 49426
616-669-1250

Contact: Byron Carpenter, Executive Director

9.451
Michigan Pear Research Association
455 State St.
Fremont, MI 49412
616-928-2516

Contact: Pat Chase, Secretary-Treasurer

9.452
Michigan Red Tart Cherry Growers
Box 30960
Lansing, MI 48909
517-323-7000

Contact: Rick Olson, Executive Director

9.453
Michigan State Florists Association
2420 Science Pkwy.
Okemos, MI 48864
517-349-5754

9.454
Michigan Tree Farm Committee
P.O. Box 299
Escanaba, MI 49829
906-265-6109

Contact: Jim Dunn, Chairman

9.455
Michigan Turfgrass Association
P.O. Box 995
Novi, MI 48050
313-348-8110

Contact: Gordon LaFontaine, Executive Director

9.456
Michigan Vegetable Council
440 Plant and Soil Science Bldg.
Michigan State University
East Lansing, MI 48824
517-353-7156

Contact: Hugh Price

MN
9.457
Central Minnesota Vegetable Growers Association
8562 Bower Ct.
Inver Grove Court, MN 55075

Contact: Larry Cermak

9.458
Federated Garden Clubs of Minnesota
1295 Dodd Rd., W.
St. Paul, MN 55118

Contact: Mrs. Herbert Larson

Forest Resource Center
Lanesboro, MN
Cross-reference: 8.31

9.459
Irrigators Association of Minnesota
Box 387
Brooten, MN 56316
612-346-2605

Contact: Jack Anderson, Executive Director

9.460
Minnesota Apple Growers Association
P.O. Box 23
LaCrescent, MN 55947
507-895-4750

Contact: Gordon Yates, Treasurer

9.461
Minnesota Berry Growers Association
North Star Gardens
19050 Manning Trail North
Marine on St. Croix, MN 55047
612-433-5850

Contact: Paul N. Otten
Cross-reference: 21.80

9.462
Minnesota Christmas Tree Association
10304 94th Ave. North
Maple Grove, MN 55369
612-425-3742

Contact: Carl E. Vogt

9.463
Minnesota Commercial Flower Growers
P.O. Box 130307
St. Paul, MN 55113
612-633-4986

Contact: Jen McCarthy, Executive Director

9.464
Minnesota Corn Growers Association
402 South Matorville Ave.
Kasson, MN 55944
507-634-4252

Contact: Alan Larsen, President

9.465
Minnesota Dry Edible Bean Council
Rt. 3, Box 520
Frazee, MN 56544
218-334-6351

Contact: Timothy Courneya, Executive Vice-President

9.466
Minnesota Forestry Association
220 First Ave., N.W.
Room 210
Grand Rapids, MN 55744
218-326-4200

Contact: John R. Suffron, Executive Director

9.467
Minnesota Fruit and Vegetable Growers Association
15709 Round Lake Blvd., N.W.
Anoka, MN 55303
612-421-4668

Contact: James Knoll, President

9.468
Minnesota Herb Society
1728 Logan Ave. South
Minneapolis, MN 55403
612-374-5135

Contact: Aly Sayae, President

9.469
Minnesota Nursery and Landscape Association
P.O. Box 130307
St. Paul, MN 55113
612-633-4987

Contact: Jen McCarthy, Executive Director

9.470
Minnesota Pesticide Information and Education
2916 South Shore Dr.
Prior Lake, MN 55372
612-447-1187

Contact: Terry Ambroz

9.471
Minnesota Plant Food and Chemical Association
1821 University Ave., Rm. 253
St. Paul, MN 55140
612-644-6235

Contact: Craig Sallstrom

9.472
Minnesota Turf Association
2537 Northdale Blvd.
Coon Rapids, MN 55433
612-755-4244

Contact: Curt Cling, Secretary

9.473
North Central Florists Association
641 Old Highway 8, S.W.
New Brighton, MN 55112
612-633-6666

Contact: Brian Pletscher, President

9.474
North Dakota Edible Bean Council
R.R. 3, Box 520
Frazee, MN 56544
218-334-6351

Contact: Duane Mergner, Chairman

9.475
North Star Lily Society, Inc.
16100 14th Ave., North
Plymouth, MN 55447
612-473-0001

9.476
Northern Minnesota Blue Grass Growers Association
R.R. 1
Roseau, MN 56751
218-463-1251

Contact: Ed Baumgartner, President

9.477
Southern Minnesota Sugarbeet Association
Renville, MN 56284
612-329-8305

Contact: Arlo Gordon

9.478
Southern Vegetable Growers Association
Hollandale, MN 56045
507-889-6271

Contact: Larry Reynan

MO
9.479
Federated Garden Clubs of Missouri, Inc.
10 Chipper Rd.
St. Louis, MO 63131

Contact: Mrs. J. Herman Belz

9.480
Floral Industry of Greater St. Louis
14 Media Dr.
St. Louis, MO 63146
314-872-3396

Contact: Thelma Schlobohm

9.481
Garden Club of St. Louis
33 Granada Way
St. Louis, MO 63124
314-997-5185

Contact: Peggy Jones, Vice-President

9.482
Greater St. Louis Flower Growers
9201 Highway BB
Hillsboro, MO 63050
816-279-6339

Contact: Douglas Schmidt

9.483
Midland Empire Florists Association
624 North 6th
St. Joseph, MO 64501
816-279-2277

9.484
Missouri Association of Nurserymen
7911 Spring Valley Rd.
Raytown, MO 64138
816-353-1203

Contact: Pat Klapis

9.485
Missouri Blueberry Council
R.R. 6, Box 436
Rolla, MO 65401
314-364-2837

Contact: Charles Croom

9.486
Missouri Christmas Tree Association
Rt. 2, Box 250
Harrisonville, MO 64701
816-887-2433

Contact: Dick Sylvester

9.487
Missouri Corn Growers Association
3702 West Truman Blvd., Suite 120
Jefferson City, MO 65109
314-893-4181

Contact: Phil Aylward, President

9.488
Missouri Grape Growers Association
P.O. Box 630
Jefferson City, MO 65102
314-751-3374

Contact: James Ashby

9.489
Missouri Nut Growers Association
421 N. West Briarcliff Rd.
Kansas City, MO 64116
816-453-6842

Contact: Gordon Kempf, President

Missouri Prairie Foundation
Columbia, MO

Cross-reference: 8.43

9.490
Missouri Seed Improvement Association
3211 La Mone
Industrial Blvd.
Columbia, MO 65201-8245
314-449-0586

Contact: Jerry Schuerenberg, Executive Secretary

9.491
Missouri State Florists Association
412 West 60th Terrace
Kansas City, MO 64113
816-333-7187

Contact: Louise Benson

9.492
Ozark Florist Association
1221 West 7th St.
Joplin, MO 64801
417-781-3310

9.493
Ozarks Regional Herb Growers and Marketers Association
c/o Leanna Potts
717 Glenview
Joplin, MO 64840

9.494
Professional Florists of Mississippi Valley, Inc.
P.O. Box 112
Arnold, MO 63010

Contact: Everst Wanner

9.495
Springfield Florists Association
310 West Walnut
Springfield, MO 65801

Contact: Ron Reynolds

9.496
St. Louis Horticulture Society
711 Auber Ridge Ct.
Manchester, MO 63011
314-227-4897

Contact: Sheldon Korklan

MS
9.497
Garden Clubs of Mississippi
P.O. Box 265
Drew, MS 38737

Contact: Mrs. H. T. Miller, Jr.

9.498
Mississippi Association of Plant Pathologists and
    Nematologists
Drawer PG
Mississippi State, MS 39762
601-325-3138

Contact: Dr. Gary W. Lawrence

9.499
Mississippi Forestry Association
620 North State St.
Jackson, MS 39202-3398
601-354-4936

Contact: Steve Corbitt, Executive Vice-President

9.500
Mississippi Nurserymen's Association
P.O. Box 5446
Mississippi State, MS 39762
601-325-1682

Contact: Dr. David Tatum

9.501
Mississippi Peach Growers Association
P.O. Box 5446
Mississippi State, MS 39762
601-325-3223

Contact: Freddie Rasberry, Secretary-Treasurer

9.502
Mississippi Pecan Growers Association
P.O. Box 5446
Mississippi State, MS 39762
601-325-3223

Contact: Freddie Rasberry, Secretary-Treasurer

9.503
Mississippi Seed Improvement Association
P.O. Drawer MS
Mississippi State MS 39762
601-325-3211

Contact: R. C. Milner, Executive Secretary

9.504
Mississippi Seedsmen's Association
427 Springwood Circle
Terry, MS 39170
601-373-9006

Contact: Rush Reed, President

9.505
Mississippi Soil and Water Conservation Commission
P.O. Box 23005
Jackson, MS 39225-3005
601-359-1281

Contact: Benny Golf, Chairman

9.506
Mississippi Sweet Potato Council
P.O. Box 5207
Mississippi State, MS 39762
601-325-7773

Contact: Benny Graves, Secretary-Treasurer

9.507
Mississippi Turfgrass Association
MCES, Box 5446
Mississippi State, MS 39762
601-325-3935

Contact: David Nagel

9.508
Mississippi-Louisiana Blueberry Association
P.O. Box 193
Poplarville, MS 39470
601-795-4525

Contact: John Braswell, Executive Secretary

9.509
North Mississippi Allied Florists
P.O. Box 1005
Corinth, MS 38834
601-286-5292

Contact: Milford Browder

MT
9.510
Association of Montana Turf and Ornamental Professionals
P.O. Box 375
Milltown, MT 59851
406-549-6929

9.511
Big Sky Bio-Dynamic Group
398 Sweathouse Rd.
Victor, MT 59875

9.512
Council on Urban Forestry
126 West Spruce St.
Missoula, MT 59802
406-721-4095

9.513
Great Northern Botanicals Association
P.O. Box 362
Helena, MT 59624

9.514
Montana Association of Nurserymen
P.O. Box 1871
Bozeman, MT 59771-1871
406-586-6042

9.515
Montana Christmas Tree Association
385 Lake Blaine Rd.
Kalispell, MT 59901
406-755-2783

Contact: Linda McHenry

9.516
Montana Federation of Garden Clubs
P.O. Box 356
Ashland, MT 59003

Contact: Mrs. Anthony Hanic

9.517
Montana Potato Improvement Association
1411 Willow Rd.
Deer Lodge, MT 59722
406-693-2307

Contact: Leo Nichols, President

9.518
Montana Seed Growers Association
Leon Johnson Hall
Montana State University
Bozeman, MT 59717
406-994-3516

Contact: Howard F. Bowman, Manager

9.519
Montana Seed Trade Association
P.O. Box 1338
Townsend, MT 59644
406-266-5244

Contact: Harry Johnson, Secretary

9.520
Montana State Florists Association
1134 Utah Ave.
Butte, MT 59701

Contact: Jack Brown

9.521
Montana Weed Control Association
521 1st Ave., N.W.
Great Falls, MT 59404
406-727-2804

Contact: Jim Freeman, President

NC
9.522
Corn Growers Association of North Carolina
1213 Ridge Dr.
Raleigh, NC 27607
919-828-8188

Contact: Jan Crenshaw, Executive Director

9.523
Garden Club of North Carolina, Inc.
848 Shoreland Rd.
Winston-Salem, NC 27106

Contact: Mrs. E. L. Swaim

9.524
Gourd Village Garden Club
4008 Green Level Rd. West
Apex, NC 27502
919-362-4357

Contact: Mary Ann Rood, President

9.525
North Carolina Apple Growers Association
P.O. Box 58
Edneyville, NC 28727
704-685-7768

Contact: Evelyn Hill, Executive Secretary

9.526
North Carolina Association of Nurserymen
P.O. Box 400
Knightdale, NC 27545

9.527
North Carolina Christmas Tree Growers Association, Inc.
617 Fountain Pl.
Burlington, NC 27215

9.528
North Carolina Christmas Tree Growers Association
P.O. Box 524
Sparta, NC 28675
919-372-4355

Contact: Richard M. Woodie

9.529
North Carolina Co-op Bulb Growers Association
Rt. 1, Box 2
Rocky Point, NC 28457
919-675-2497

9.530
North Carolina Commercial Flower Growers Association,
    Inc.
Rt. 1, Box 5
Eden, NC 27288

9.531
North Carolina Fresh Vegetable Growers Association, Inc.
Rt. 6, Box 400
Dunn, NC 28334

9.532
North Carolina Grape Growers Association
Rt. 3, Box 3810
Burgaw, NC 28425
919-259-5474

Contact: Cliff Bannerman, President

9.533
North Carolina Greenhouse Growers Association, Inc.
Dept. of Horticultural Science
Box 7609, North Carolina State U.
Raleigh, NC 27695-7609

9.534
North Carolina Herb Association
N.C. Dept. of Agriculture
Marketing Division
P.O. Box 27647
Raleigh, NC 27611
919-733-7136

Contact: Ross Williams

9.535
North Carolina Landscape Contractors Association, Inc.
P.O. Box 400
Knightdale, NC 27545

9.536
North Carolina Peach Growers Society, Inc.
P.O. Box 519
Candor, NC 27229

9.537
North Carolina Peanut Growers Association
P.O. Box 1709
Rocky Mount, NC 27802
919-446-8060

Contact: Norfleet Sugg, Executive Secretary

9.538
North Carolina Potato Association
P.O. Box 138
Shawboro, NC 27973
919-232-3188

Contact: Jimmy Ferebee, President

9.539
North Carolina Primrose Growers Association
Rt. 3
Angier, NC 27501
919-639-6370

Contact: Ronnie Wheller, President

9.540
North Carolina Trellised Tomato Growers Association
570 Brevard Rd.
Asheville, NC 28806
704-253-1691

Contact: Mike Ferguson

9.541
North Carolina Watermelon Association
Murfreesboro Farms, Inc.
P.O. Box 216
Murfreesboro, NC 27855
919-398-5830

Contact: Frances Bunch

9.542
North Carolina Yam Commission
108 West Main St.
Benson, NC 27504-1344
919-894-2166

Contact: Billy Corbett, President

9.543
Southeastern Blueberry Council
Rt. 2, Box 610
Castle Hayne, NC 28429
919-675-2314

Contact: Dr. Mike Mainland

9.544
Southern Retail Florist Association
P.O. Box 9058
Charlotte, NC 28299-9058
704-322-6405

Contact: Avanelle Finger

9.545
Turfgrass Council of North Carolina, Inc.
P.O. Box 5395
Cary, NC 27511

ND
9.546
North Dakota Corn Growers Association
R.R. 1, Box 106
Page, ND 58064
701-668-2760

Contact: Robert Thompson, President

9.547
North Dakota Dry Bean Seed Growers Association
Hastings Hall, NDSU
Fargo, ND 58105
701-237-7931

Contact: William Ongstad, President

9.548
North Dakota Federation of Garden Clubs
R.R. 3
Jamestown, ND 58401

Contact: Lois Forrest

9.549
North Dakota Natural Science Society
P.O. Box 8238
Grand Forks, ND 58202-8238
701-777-3650

Paul B. Kannowski, Secretary-Treasurer

9.550
North Dakota Nursery and Greenhouse Association
1107 Airport Rd.
P.O. Box 2601
Bismarck, ND 58502
701-258-0165

Contact: Lee W. Hinds

9.551
North Dakota State Potato Council
P.O. Box 1744
Grand Forks, ND 58206
218-773-3633

Contact: Lloyd A. Schmidt, Executive Vice-President

9.552
North Dakota State Seed Commission
State Seed Dept.
University Station, Box 5257
Fargo, ND 58105
701-237-7927

Contact: Doug Johansen

9.553
North Dakota State Soil Conservation Committee
State Highway Bldg.
608 E. Boulevard
Bismarck, ND 58505-0790

Contact: Robert Senechal, Jr.

9.554
North Dakota Weed Control
600 West Main Riverside
West Fargo, ND 58078
701-282-5487

Contact: George Bishoff, President

NE
9.555
Federated Garden Clubs of Nebraska
Rt. 1, Box 70
Palisade, NE 69040

Contact: Mrs. Robert Keating

9.556
Nebraska Potato Development
P.O. Box 339
Alliance, NE 69301
308-762-1674

Contact: Gary Leever

9.557
Nebraska Arborists Association
8015 West Center
Omaha, NE 68124

9.558
Nebraska Association of Nurserymen
P.O. Box 80177
Lincoln, NE 68501
402-476-3852

Contact: David S. McBride

9.559
Nebraska Beet Growers Association
R.R. 1, Box 260
Minatare, NE 69356
308-783-1087

Contact: John Masser, President

9.560
Nebraska Christmas Tree Association
7020 Holdrege
Lincoln, NE 68505
402-472-2167

Contact: Lorrie Adams

9.561
Nebraska Corn Growers Association
301 Centennial Mall South
4th Floor, Box 95107
Lincoln, NE 68509-5107
402-471-2676

Contact: Randy Klein

9.562
Nebraska Dry Bean Growers Association
R.R. 2
Mitchell, NE 69357
308-632-8614

Contact: Cliff Walker, President

9.563
Nebraska Fertilizer and Agricultural Chemical Institute
1111 Lincoln Mall, Suite 308
Lincoln, NE 68508-1528
402-476-1528

Contact: Robert Anderson, President

9.564
Nebraska Florists Society
1024 East 14th St.
Fremont, NE 68025
402-721-0984

Contact: Mel Schwanke

Nebraska Herbal Society
Lincoln, NE

Cross-reference: 2.69

9.565
Nebraska Nutgrowers
c/o William Gustafson
5219 Leighton Ave.
Lincoln, NE 68504

9.566
Nebraska Popcorn Growers Association
R.R. 2, Box 286
Elgin, NE 68636
402-843-2411

Contact: Steve Koinzan, President

9.567
Nebraska Potato Council
Western Potatoes
P.O. Box 755
Alliance, NE 69301
308-762-4917

Contact: Gene Kerschner, President

9.568
Nebraska Turf Grass Foundation
R.R. 13
Lincoln, NE 68505
402-786-3444

Contact: Maury Spence, President

9.569
Nemokan Floral Association
Greens Greenhouses
Bell St. at 14th
Fremont, NE 68025
402-721-0984

Contact: Mel Schwanke, Executive Vice-President

9.570
Omaha Council of Garden Clubs
4014 South 14 St.
Omaha, NE 68107

Prairie/Plains Resource Institute
Aurora, NE

Cross-reference: 8.59

9.571
South Omaha Horticultural Society
6064 Grant St.
Omaha, NE 68104

9.572
United Florists
5602 Read St.
Omaha, NE 68152
402-571-7800

Contact: Jim Spitalnick

NH
9.573
Granite State Garden and Flower Show
c/o NH Landscaping Inc.
6 Oak Hill Rd.
Hooksett, NH 03106

9.574
National Potato Council for Vermont-New Hampshire
Hudson, NH 03051
Contact: William Rodonis, Sr.

9.575
New England Fruit Growers Council on the Environment
Poverty Lane Orchards
R.R. 1, Box 610
West Lebanon, NH 03784
603-448-1511
Contact: Steven M. Wood, Executive Director

9.576
New Hampshire Arborists Association
45 Elwyn Rd.
Portsmouth, NH 03801

9.577
New Hampshire Blueberry Growers Association
Wolfeboro, NH 03894
Contact: Peter MacKenzie, President

9.578
New Hampshire Federation of Garden Clubs
Box 10
Windham, NH 03087
Contact: Mrs. William Murdock

9.579
New Hampshire Fruit Growers Association
Poverty Lane Orchard
Poverty Ln.
West Lebanon, NH 03784

9.580
New Hampshire Grape Growers Association
Mayhill Orchard
P.O. Box 218
Goffstown, NH 03045
603-529-2404
Contact: F. Cameron Ludwig, President

9.581
New Hampshire Landscape Association
45 Elwyn Rd.
Portsmouth, NH 03801

9.582
New Hampshire Plant Growers Association
8 Meserve Rd.
Durham, NH 03824

9.583
New Hampshire Potato Growers Association
63 Water St.
Lancaster, NH 03496
603-788-2131
Contact: Leighton C. Pratt, Secretary

9.584
New Hampshire Small Fruit Growers Association
c/o Robert Harrington, President
White Oaks Rd.
Laconia, NH 03246

9.585
New Hampshire Vegetable Growers Association
c/o Phil Ferdinando
Rt. 102
Derry, NH 03038

9.586
Seacoast Growers Association
P.O. Box 385
Stratham, NH 03885
603-778-1039
Contact: Edie Barker, President

NJ
9.587
Allied Florists of Delaware Valley, Inc.
Delaware Valley Wholesale Florist
520 Mantua Blvd. North
Sewell, NJ 08080
215-271-7270

9.588
Cultivated Sod Association of New Jersey, Inc.
Soil & Crops Dept.
Cook College, P.O. Box 231
New Brunswick, NJ 08903
201-932-9453

9.589
Farmers and Gardeners Association of New Jersey, Inc.
R.D. 1, Box 152-X
Holmdel, NJ 07733
201-264-0256

9.590
Garden Club of New Jersey
Box 622
West Milford, NJ 07480
Contact: Mrs. Carmine Grossi

9.591
Jersey Fruit Cooperative Association, Inc.
101 West Main St.
Moorestown, NJ 08057
609-235-5330

9.592
Jersey Shore Professional Florists Association
P.O. Box 253
Avon By The Sea, NJ 07717
201-774-6626

Contact: Patricia Gann

9.593
New Jersey Apple Institute
P.O. Box 413
Belvidere, NJ 07823
201-475-6506

9.594
New Jersey Board of Tree Experts
Dept. of Environmental Protection
CN 404
Trenton, NJ 08625
609-292-2520

9.595
New Jersey Christmas Tree Growers Association, Inc.
Box 29, River Rd.
Mays Landing, NJ 08330

9.596
New Jersey Federation of Shade Tree Commissions
Blake Hall, Cook College
P.O. Box 231
New Brunswick, NJ 08903
201-246-3210

9.597
New Jersey Flower and Garden Show, Inc.
141 Union Ave., Suite 4
Middlesex, NJ 08846
201-560-9020

9.598
New Jersey Forestry Association
P.O. Box 51
Chatham, NJ 07928-0051
609-771-8301

Contact: Ronald Sheay, Secretary

9.599
New Jersey Nursery & Landscape Association, Inc.
Pennington Professional Center
Building A, Suite 3, 65 S. Main St.
Pennington, NJ 08534
609-737-0890

9.600
New Jersey Peach Council, Inc.
152 Ohio Ave.
Clementon, NJ 08021
609-784-1001

9.601
New Jersey Plant & Flower Growers Association, Inc.
7 Toucan Ct.
Wayne, NJ 07470
201-696-4087

9.602
New Jersey Shade Tree Federation
P.O. Box 231
Blake Hall, Cook College
Rutgers University
New Brunswick, NJ 08903
201-246-3210

Contact: William J. Porter, Executive Secretary

9.603
New Jersey Small Fruits Council, Inc.
152 Ohio Ave.
Clementon, NJ 08021
609-784-1001

9.604
New Jersey State Florists' Association, Inc.
7 Toucan Ct.
Wayne, NJ 07470
201-696-4087

9.605
New Jersey State Soil Conservation Committee
New Jersey Dept. of Agriculture
CN 330
Trenton, NJ 08625
609-292-5540

Contact: Samuel Race, Executive Secretary

9.606
New Jersey State Sweet Potato Industry Association
5449 Genoa Ave.
Vineland, NJ 08360
609-691-8210

Contact: James Bertonazzi, President

9.607
New Jersey Tree Farm Committee
Bureau of Forest Management
CN 404
Trenton, NJ 08625

9.608
New Jersey Turfgrass Association
Soils and Crops Dept., Cook College
P.O. Box 231
New Brunswick, NJ 08903

9.609
Pesticide Association of New Jersey
P.O. Box 463
Clayton, NJ 08312

9.610
Plant Food Educational Society
Dept. of Soil and Crops
Cook College, P.O. Box 231
New Brunswick, NJ 08903
201-932-9872

9.611
Vegetable Growers' Association of New Jersey, Inc.
Blake Hall, Room 212, Cooke College
P.O. Box 231
New Brunswick, NJ 08903
201-932-9395

NM
9.612
Friends of the Rio Grande Botanic Garden, Inc.
208 Carlisle Blvd., S.E.
Albuquerque, NM 87106
505-268-7738

Contact: Patricia Sussmann, President

9.613
New Mexico Agricultural Chemical and Plant Food
 Association
NAPI
P.O. Drawer 1318
Farmington, NM 87499
505-763-3428

Contact: Charlie Higgins, President

9.614
New Mexico Association of Nurserymen
P.O. Box 667
Estancia, NM 87016
505-384-2726

Contact: Roger DeAngelis, President

9.615
New Mexico Garden Clubs, Inc.
7000 Seminole Rd., N.E.
Albuquerque, NM 87110

Contact: Mrs. Donald Wood

9.616
New Mexico Organic Growers Association
1312 Lobo Pl., N.E.
Albuquerque, NM 87106
505-268-5504

9.617
New Mexico Peanut Commission
220 North Mike
Portales, NM 80130
505-478-2316

Contact: Wayne Baker, Chairman

9.618
New Mexico Peanut Growers Association
East Star Rt., Box 141
Portales, NM 88130
505-276-8368

Contact: David Sanpors, President

9.619
New Mexico Pistachio Association
R.R. 1, Box 257D
Alamogordo, NM 88310
505-437-0602

Contact: Thomas M. McGinn, President

9.620
New Mexico Soil and Water Conservation Commission
Box 2362
Silver City, NM 88062
505-538-9495

Contact: James W. Harrison, Chairman

9.621
New Mexico Vine and Wine Society
LaVina Vineyard
P.O. Box 121
Chamberino, NM 88027
505-882-2092

Contact: Clarence Cooper, President

9.622
Permaculture Drylands
P.O. Box 1812
Santa Fe, NM 87504-1812
505-982-2063

9.623
Southwest Turfgrass Association
New Mexico State University
P.O. Box 3Q
Las Cruces, NM 88003
915-584-9752

Contact: Kurt Desiderio, President

9.624
Western Irrigated Pecan Growers Association
6135 Shalem County Trail
Las Cruces, NM 88005
505-523-5062

Contact: Les Fletcher, President

9.625
Western Pecan Growers Association
P.O. Box 315
Mesilla Park, NM 88047
505-526-6165

Contact: Gary V. Arnold, President

NV

9.626
Desert Turf/Landscape Conference and Show
c/o Southwest Lawn and Landscape
1458 E. Tropicana, Suite 390
Las Vegas, NV 89119

9.627
Nevada Garden Clubs, Inc.
P.O. Box 27624
Las Vegas, NV 89126

Contact: Mrs. Charles Gorley

9.628
Nevada Nurserymen's Association
4850 Kilda Circle
Las Vegas, NV 89112
702-361-3501

Contact: Bobbie Pennock

9.629
Nevada Seed Council
P.O. Box 700
Lovelock, NV 89419
702-273-2901

Contact: Jim Robertson

9.630
North Nevada Florists Association
P.O. Box 5955
Reno, NV 89513

Contact: Judith A. Mitchell

9.631
Sierra Nevada Landscape Architects
1150 Corporate Blvd.
Sparks, NV 89502

Contact: Steve Williams, President

NY
9.632
Associated Florists of Greater Rochester
3480 Brockport Rd.
Spencerport, NY 14559

Contact: Evelyn Sorce

9.633
Cayuga Biodynamic Association
Britton Rd.
Aurora, NY 13026

Council on the Environment of New York City
New York, NY

Cross-reference: 8.24

9.634
Eastern Long Island Biodynamic Group
c/o Williams
P.O. Box 1582
Sag Harbor, NY 11963

9.635
Empire State Soil Fertility Association
Philps Supply Inc.
Box 190
Philps, NY 14532
315-548-3525

Contact: Greg Bodine

9.636
Federated Garden Clubs of New York State, Inc.
Box 25, Minor Rd.
Brewster, NY 10509

Contact: Mrs. Vincent De Somma

9.637
Finger Lakes Wine Growers Association
Canadaigua Wine Co.
116 Buffalo St.
Canadaigua, NY 14424
716-394-7900

Contact: Stafford Krause

9.638
Garden Society
New York Botanical Garden
Bronx, NY 10458

9.639
Long Island Cauliflower Association
300 Midwood Rd.
Cutchogue, NY 11935
516-734-6586

Contact: Andre Cybulski

9.640
Long Island Flower Growers Association
40 Moriches Ave.
East Moriches, NY 11940
516-878-2828

Contact: Ute Leuthardt

9.641
Long Island Nurserymen's Association
1281 Old Country Rd.
Riverhead, NY 11901
516-727-7080

Contact: Penny Bone

9.642
Long Island Retail Florists
69 Rt. 111
Dix Hills, NY 11746

Contact: H. Vezeris

9.643
Metro Retail Florists Association
95 Lexington Ave.
New York, NY 10016

Contact: Rita D'Alessandro

9.644
New York Apple Research Association
P.O. Box 350
Fishers, NY 14453
716-924-2171

Contact: Kenneth Pollard, Secretary

9.645
New York Cherry Growers Association
P.O. Box 350
Fishers, NY 14453
715-924-2171

Contact: Kenneth Pollard, Secretary

9.646
New York Christmas Tree Association
2947 East Bayard St. Ext.
Seneca Falls, NY 13148
315-568-8173

Contact: John B. Webb

9.647
New York Corn Growers Association
2719 High St.
Clyde, NY 14433
315-923-5361

Contact: Dick Magade, President

9.648
New York Florists Club
15 East 28th St.
New York, NY 10016

Contact: Joseph Valenty, Secretary

9.649
New York Seed Improvement Cooperative
P.O. Box 218
Ithaca, NY 14851
607-257-2233

Contact: Don K. Shardlow, Executive Vice-President

9.650
New York State Flower Industries Research & Education
  Fund, Inc.
R.D. 5
Ballston Spa, NY 12020

9.651
New York State Forest Owners Association
Box 123
Boonville, NY 13309
315-942-4593

Contact: Ruth J. Thoden, Executive Secretary

9.652
New York State Green Council
P.O. Box 612
Latham, NY 12110

Contact: Elizabeth Seme

9.653
New York State Nurserymen's Association
Gary Gasparini Landscaping
4005 Split Rock Rd.
Camillus, NY 13031
315-488-4261

Contact: Gary Gasparini

9.654
New York State Seed Association
P.O. Box 462
Geneva, NY 14456
315-787-2217

Contact: M. T. Vittum, Secretary-Treasurer

9.655
New York State Small Fruit Growers Association
R.D. 2, Box 238A
Locke, NY 13092
315-497-1120

Contact: Mary Ann Grisamore

9.656
New York State Soil and Water Conservation Commission
1 Winners Circle, Capital Plaza
Albany, NY 12235
518-457-3738

Contact: David Pendergast, Executive Director

9.657
New York State Turfgrass Association
P.O. Box 612
Latham, NY 12110
800-873-TURF

Contact: Stephen Smith, President

9.658
New York State Vegetable Growers Association
P.O. Box 356
Ithaca, NY 14851-0356
607-539-7648

Contact: Jean Warholic, Secretary-Treasurer

9.659
New York State Wine Grape Growers
350 Elm St.
Penn Yan, NY 14527
315-536-2853

Contact: John Martini, President

9.660
New York Turf & Landscape Association, Inc.
P.O. Box 307
Scarsdale, NY 10583

9.661
Niagara Frontier Retail Florists
431 Delaware Ave.
Buffalo, NY 14202
716-856-9000

Contact: Marc Kotarski

9.662
Pesticide Association of New York State
4561 Frank Gay Rd.
Marcellus, NY 13108
315-673-1303

Contact: Dr. David H. Marsden

9.663
Spring Valley Regional Biodynamic Group
241 Hungry Hollow Rd.
Spring Valley, NY 10977

9.664
United Florists of Western New York
118 South Forest Rd.
Williamsville, NY 14221
716-632-1290

Contact: Frank Mischler

OH
9.665
Allied Florists of Toledo Area
3450 W. Central Ave.
Suite 336
Toledo, OH 43606
419-531-0125

Contact: Greg Kuehnle

9.666
Central Ohio Allied Florists Association
1359 Darcann Dr.
Columbus, OH 43220
614-457-4320

Contact: Barbara Robertson

9.667
Cincinnati Flower Growers Association
3150 Compton Rd.
Cincinnati, OH 45239
513-729-1134

Contact: Jim Krismer

9.668
Florists Association of Greater Cleveland
226 Hanna Bldg.
Cleveland, OH 44115
216-523-1341

Contact: Hary Gard Shaffer, Jr.

9.669
Garden Club of Ohio, Inc.
883 Bexley Dr.
Perrysburg, OH 43551

Contact: Mrs. Donald De Cessna

9.670
Greater Akron Florists Association
1034 W. Main St.
Revenna, OH 44266

Contact: Leland D. Deaner

9.671
Ohio Association of Garden Clubs
402 Craggy Creek Dr.
Chippewa Lake, OH 44215
216-769-2210

Contact: Jan Harman

9.672
Ohio Christmas Tree Association
P.O. Box 373
Burton, OH 44021
216-834-4206

Contact: Ture L. Johnson

9.673
Ohio Christmas Tree Growers
1155 Brockway Rd.
Rome, OH 44085
216-294-2292

Contact: John Henson

9.674
Ohio Corn Growers Association
1100 East Center St.
Marion, OH 43302
614-383-2676

Contact: Mike Wagner, Executive Director

9.675
Ohio Fertilizer and Pesticide Association
P.O. Box 151
Worthington, OH 43085
614-885-1067

Contact: John C. Foltz, Executive Director

9.676
Ohio Floriculture Foundation
2001 Fyffe Ct.
Columbus, OH 43210

9.677
Ohio Florists' Association
1301 Worthington Woods Blvd.
Worthington, OH 43085

Contact: Ronald Cornell

9.678
Ohio Forestry Association
1301 Worthington Woods Blvd.
Worthington, OH 43085
614-846-9456; (FAX) 614-846-9457

Contact: Ronald C. Cornell, Executive Director

9.679
Ohio Fruit Growers Society
Two Nationwide Plaza
P.O. Box 479
Columbus, OH 43216
614-249-2424; (FAX) 614-249-2200

Contact: Mike Pullins

9.680
Ohio Greenhouse Cooperative Association
430 Bagley Rd.
Berea, OH 44017
216-243-5600

Contact: Joan Stemmer, Secretary

9.681
Ohio Horticulture Council
4680 Indianola Ave.
Columbus, OH 43214
614-261-6834

Contact: David Kelly

9.682
Ohio Junior Horticulture Association
5759 Sandalwood, N.E.
North Canton, OH 44721
216-492-3252

Contact: Dennis Waldman, Executive Officer

9.683
Ohio Land Improvement Contractor Association
P.O. Box 116
Dublin, OH 43017
614-889-2314

Contact: Larry Trask, Executive Secretary

9.684
Ohio Nurserymen's Association
2021 E. Dublin-Granville Rd.
Columbus, OH 43229

Contact: William Statler

9.685
Ohio Nut Growers Association
1807 Linbergh, N.E.
Massillon, OH 44646
216-833-7373

Contact: Raymond E. Silvis, Secretary-Treasurer

9.686
Ohio Potato Growers Association
4680 Indianola Ave.
Columbus, OH 43214
614-261-6834

Contact: David Kelly, Manager

9.687
Ohio Seed Dealers Association
5055 Cambrian Dr.
Columbus, OH 43220
614-459-1018

Contact: Dick Moore, Executive Secretary

9.688
Ohio Seed Improvement Association
6150 Avery Rd., Box 477
Dublin, OH 43017-0477
614-889-1136

Contact: John Armstrong, Manager

9.689
Ohio Soil and Water Conservation Commission
Fountain Square, Bldg. E-2
Columbus, OH 43224
614-265-6610

Contact: Lawrence G. Vance, Executive Secretary

9.690
Ohio Vegetable and Potato Growers Association
Two Nationwide Plaza
P.O. Box 479
Columbus, OH 43216
614-249-2424; (FAX) 614-249-2200

Contact: Mike Pullins, Executive Director

OK
9.691
Horticulture Industries Show
Estes Chemical
9410 E. 55th St.
Tulsa, OK 74145

Contact: Larry Lindsey

9.692
Oklahoma Christmas Tree Association
Box 7, Echo Canyon
Claremore, OK 74017

Contact: Loren Loomis

9.693
Oklahoma Fertilizer and Chemical Association
P.O. Box 1747, 2309 North 10th
Enid, OK 73702
405-233-9516

Contact: Joe N. Hampton, Executive Vice-President

9.694
Oklahoma Fruit Growers' Association
Rohm Hass Co.
404 Grandview Blvd.
Muskogee, OK 74403

Contact: Gary Plunkett

9.695
Oklahoma Garden Clubs, Inc.
115 W. First
Atoka, OK 74525

Contact: Mrs. W. A. Williams

9.696
Oklahoma Greenhouse Growers Association
400 North Portland
Oklahoma City, OK 73107

Contact: Diane Satterlee

9.697
Oklahoma Herb Growers' Association
Rt. 1, Box 847
Ramona, OK 74061

Contact: Doris Bankes

9.698
Oklahoma Horticulture Industries Council
400 North Portland
Oklahoma City, OK 73107
405-942-5276

Contact: Diane Satterlee

9.699
Oklahoma Nurserymen's Association
400 North Portland
Oklahoma City, OK 73107

Contact: Diane Satterlee

9.700
Oklahoma Peanut Commission
Box D
Madill, OK 73446
405-795-3622

Contact: Bill Flanagan, Executive Secretary

9.701
Oklahoma Pecan Commission
P.O. Box 434
Sand Springs, OK 74063
918-245-8854

Contact: Mike Spradling, Chairman

9.702
Oklahoma Pecan Growers' Association
Noble Foundation
P.O. Box 2180
Ardmore, OK 73402

Contact: Scott Landgraf

9.703
Oklahoma Plant Food Education Society
1216 West Willow, Suite C
Enid, OK 73701
405-242-1211

Contact: Kent Prickett, Secretary

9.704
Oklahoma Sod Producers Association
5638 Rockford
Tulsa, OK 74105

Contact: Ray Volentine

9.705
Oklahoma State Florists' Association
Stumpf Flowers
1212 E. Frank Phillips
Bartlesville, OK 74003

Contact: Ron Stumpf

9.706
Oklahoma Turfgrass Research Foundation
Dept. of Transportation
10206 Stoneham
Oklahoma City, OK 73120

Contact: Micky Dolan

9.707
Oklahoma Vegetable Association
Rt. 2, Box 1788
Hydro, OK 73048

Contact: Dean Smith

9.708
Oklahoma Vegetable Growers
Rt. 1, Box 441
Bixby, OK 74008-9801
918-366-3289

Contact: Naomi Shanks, Secretary-Treasurer

9.709
Tulsa Florists Association
514 Plaza Ct.
Sand Spring, OK 74063
918-245-5856

Contact: Norma Coble

OR
9.710
Central Oregon Potato Growers Association
5287 S.W. Macy Ln.
Culver, OR 97734
503-546-4424

Contact: Richard Macy, President

Hardy Plant Society of Oregon
Boring, OR

Cross-reference: 2.46

Harmony Research Farm
Cave Junction, OR

Cross-reference: 8.38

9.711
Idaho-Eastern Oregon Onion Committee
2641 Highway 201
Nyssa, OR 97918
503-372-2889

Contact: Nobel Marinaka, Jr.

9.712
Northwest Christmas Tree Association
P.O. Box 3366
Salem, OR 97302

Contact: Bryan Ostlund, Executive Secretary

9.713
Nut Growers Society of Oregon, Washington, and British
  Columbia
P.O. Box 23126
Tigard, OR 97223
502-639-3118

9.714
Oregon Association of Nurserymen
2780 S.E. Harrison Ave.
Suite 102
Milwaukee, OR 97222
503-653-8733

9.715
Oregon Bartlett Pear Commission
813 S.W. Alder, Suite 601
Portland, OR 97205
503-223-8139

Contact: Herb Diede, Executive Secretary

9.716
Oregon Blueberry Commission
247 Commercial N.E.
Salem, OR 97301
503-399-8456

Contact: Jan Marie Schroeder

9.717
Oregon Blueberry Growers Association
Rt. 4, Box 338
Cornelius, OR 97113
503-357-5789

Contact: Lloyd Duyck, President

9.718
Oregon Caneberry Commission
247 Commercial N.E.
Salem, OR 97301
503-399-8456

Contact: Jan Marie Schroeder

9.719
Oregon Christmas Tree Association
P.O. Box 3366
Salem, OR 97302
503-364-2942

Contact: Lisa Ostlund

9.720
Oregon Filbert Commission
P.O. Box 23126
8845 S.W. Center Ct.
Tigard, OR 97223
503-639-3118

Contact: Robert Gelhar, Executive Secretary

9.721
Oregon Florists Association
8605 S.W. Terwilliger
Portland, OR 97219
503-246-1311

Contact: Janet Patella

9.722
Oregon Flower Growers Association
1112 S.E. Grand Ave.
Portland, OR 97214
503-232-3212

9.723
Oregon Mint Commission
P.O. Box 3366
Salem, OR 97302
503-364-3346

Contact: Bryan Ostlund

9.724
Oregon Orchardgrass Growers Association
26872 Peoria Rd.
Halsey, OR 97348
503-369-2774

Contact: Larry Warfel, President

9.725
Oregon Orchardgrass Seed Producers Commission
1270 Chemeketa N.E.
Salem, OR 97301
502-370-7019

Contact: John McCulley, Executive Secretary

9.726
Oregon Potato Commission
700 N.E. Multnomah, Suite 460
Portland, OR 97232-4104
503-238-7500

Contact: William Wise

9.727
Oregon Regional Sweet Cherry Commission
Dept. of Agriculture
635 Capitol St. N.E.
Salem, OR 97310-0110
503-378-3787

Contact: Becky Rada

9.728
Oregon Seed Council
866 Lancaster S.E.
Salem, OR 97301
503-585-1157

Contact: David S. Nelson, Executive Secretary

9.729
Oregon Seed League
30742 Vennell
Corvales, OR 97333
503-752-2146

Contact: Rosseta Vennell, President

9.730
Oregon Seed Trade Association
Pickseed West, Inc.
P.O. Box 888
Tangent, OR 97389
503-926-8886

Contact: Kent Wiley

9.731
Oregon Soil and Water Conservation Commission
635 Capitol St., N.E.
Salem, OR 97310
503-378-3810

Contact: George C. Stubbert

9.732
Oregon State Federation of Garden Clubs, Inc.
820 N.W. Elizabeth Dr.
Corvallis, OR 97330

Contact: Mrs. Sam Roller

9.733
Oregon Strawberry Commission
247 Commercial N.E.
Salem, OR 97301
503-399-8456

Contact: Jan Schroeder, Administrator

9.734
Oregon Tall Fescue Commission
866 Lancaster Dr., S.E.
Salem, OR 97301
503-585-1157

Contact: David S. Nelson

9.735
Oregon Tilth
P.O. Box 218
Tualatin, OR 97062

9.736
Oregon Wine Growers Association
P.O. Box 399
Dundee, OR 97115
503-864-2282

Contact: Bill Blosser, President

Siskiyou Permaculture Resources Group
Ashland, OR

Cross-reference: 8.70

9.737
Southwestern Oregon Cranberry Club
Rt. 1, Box 1427
Bandon, OR 97411
503-347-9594

Contact: Ted Frietag, President

9.738
United Horticulture Industry Board
2155 Wells Dr.
Hood River, OR 97383
503-663-5383

Contact: John Wells, Chairman

9.739
Western Oregon Onion Growers Association
P.O. Box 46
Beavercreek, OR 97004
503-632-3122

Contact: Terry Bibby, President

PA
9.740
Allied Florists of Delaware Valley, Inc.
12 Cavalier Dr.
Ambler, PA 19002
215-646-8050

Contact: Robert Cullers

9.741
Burholme Horticultural Society
c/o Mrs. John A. McCarthy
829 Knorr St.
Philadelphia, PA 19111

9.742
Delaware Turfgrass Association
Radley Run Country Club
1100 Country Club Rd.
West Chester, PA 19380
215-793-1660

Contact: Steve Segui

9.743
Garden Club Federation of Pennsylvania
445 N. Front St.
Milton, PA 17847

Contact: Mrs. Henry Hermani

9.744
Mushroom Growers Cooperative Association of
    Pennsylvania
P.O. Box 375
Kennett Square, PA 19348-0375
215-444-3910

Contact: Robert DiMarco, Secretary

9.745
Pennsylvania Christmas Tree Growers' Association
255 Zeigler Rd.
Dover, PA 17315
717-292-5683

9.746
Pennsylvania Co-op Potato Growers
R.D. 2, Box 133
West Finley, PA 15377
412-222-6858

Contact: Nance L. Dorsey, President

9.747
Pennsylvania Florists' Association
2102 Gring Dr.
Wyomissing, PA 19610
215-678-9626

Contact: Barbara Davis

9.748
Pennsylvania Flower Growers
12 Cavalier Dr.
Ambler, PA 19002
215-646-7550

Contact: Robert Cullers

9.749
Pennsylvania Forestry Association
410 East Main St.
Mechanicsburg, PA 17055
717-766-5371

Contact: W. R. Rossman, President

9.750
Pennsylvania Foundation Seed Cooperative
P.O. Box 513
Manheim, PA 17545-0513
717-655-5443

9.751
Pennsylvania Fruit and Vegetable Inspection Association
2301 North Cameron St., Room 312
Harrisburg, PA 17110-9408
717-787-5108

Contact: Sterman Masser, President

9.752
Pennsylvania Grape Industry Association
P.O. Box 35
Brogue, PA 17309-0035
717-927-6192

Contact: Lynn Hunter, Council Chairman

9.753
Pennsylvania Inland Fertilizer Association
115 Washington Ave.
Souderton, PA 18964
215-721-0222

Contact: Arthur McAlaster, President

9.754
Pennsylvania Lime, Fertilizer and Pesticide Society
c/o Robert Thomas
1821 East Branch Rd.
State College, PA 16801
814-238-1508

9.755
Pennsylvania Master Corn Growers Association
Walter C. Johnson & Co.
R.R. 1, Box 52
Julian, PA 16844
814-692-4655

9.756
Pennsylvania Nurserymen's Association
1924 North Second St.
Harrisburg, PA 17102
717-238-1673

9.757
Pennsylvania Nut Growers Association
c/o Glenn L. Helms
R.D. 1, Box 242
Hellertown, PA 18055
215-252-7037

9.758
Pennsylvania Red Cherry Growers Association
Adams County Extension Service
1135 Chambersburg Rd.
Gettysburg, PA 17325
717-334-6271

Contact: Chris Baugher, President

9.759
Pennsylvania Turfgrass Council, Inc.
P.O. Box 417
Bellefonte, PA 16823-0417
814-355-8010

9.760
Pennsylvania Vegetable Growers Association
R.D. 1, Box 392
Northumberland, PA 17857
717-473-8468

Contact: William Troxell, Executive Secretary

9.761
Southeast Grape Industry Association of Pennsylvania
P.O. Box 35
Brogue, PA 17309-0035
717-927-6192

Contact: Arthur Freas, President

9.762
Western Pennsylvania Flower Growers
358 Babcock Blvd.
Gibsonia, PA 15044
412-935-1480

Contact: Ted Eck

RI
9.763
Rhode Island Agricultural Council—Nursery
54 Bristol Ferry Rd.
Portsmouth, RI 02871
401-683-0538

Contact: Peter Vanhof

9.764
Rhode Island Agricultural Council—Turf
SODCO Inc.
P.O. Box 85
Slocum, RI 02877
401-294-3100

Contact: Linda Tucker

9.765
Rhode Island Agricultural Council—Vegetables
South County Trail
Exeter, RI 02822
401-294-2044

Contact: Richard Schartner

9.766
Rhode Island Christmas Tree Growers' Association
Mann School House Rd.
Smithfield, RI 02917

Contact: Dorothy Hyde, Secretary

9.767
Rhode Island Dahlia Society
750 Old Baptist Rd.
North Kingston, RI 02852
401-295-8009

Contact: Walter Taylor, President

9.768
Rhode Island Federation of Garden Clubs
130 Waterway
Saunderstown, RI 02874

Contact: Mrs. G. Dickson Kenney

9.769
Rhode Island Florists Association, Inc.
295A Scituate Ave.
Cranston, RI 02920
401-943-6445

Contact: Diane Carley

9.770
Rhode Island Fruit Growers' Association
R.R. 2, Box 298
North Scituate, RI 02857

Contact: Gilbert Barden, Secretary

9.771
Rhode Island Nurserymen's Association
P.O. Box 515
North Scituate, RI 02857

Contact: Faith Schacht, Secretary

SC
9.772
Carolinas Sod Producers
Patten Seed & Turfgrass Co.
Rt. 2, Box 170
Orangeburg, SC 29115

Contact: Jim Roquemore

9.773
Garden Club of South Carolina
912 Santee Dr.
Florence, SC 29501

Contact: Mrs. John F. C. Hunter

9.774
Greater Charlotte Area Florists Association
1931 East Seventh St.
Charlotte, SC 28204
704-374-1331

Contact: Edward White

9.775
Midland Florists Association
Johnny's Florist
2106 Platt Springs Rd.
West Columbia, SC 29169
803-794-9315

Contact: Johnny Beck

9.776
South Carolina Christmas Tree Growers Association
1433 Firetower Rd., N.E.
Orangeburg, SC 29115
803-533-8133

Contact: Marvin Gaffney, Executive Director

9.777
South Carolina Florists Association
408 Woodland Dr.
Florence, SC 29501
803-669-8191

Contact: Betty R. James

9.778
South Carolina Forestry Association
4811 Broad River Rd.
P.O. Box 21303
Columbia, SC 29221
803-798-4170

9.779
South Carolina Greenhouse Growers' Association
Rt. 1, Box 317
Blythewood, SC 29016

9.780
South Carolina Landscape & Turfgrass Association
City of Spartanburg Parks & Rec. Dept.
P.O. Box 1749
Spartanburg, SC 29304

Contact: David Price

9.781
South Carolina Nurserymen's Association
809 Sunset Dr.
Greenwood, SC 29646

9.782
South Carolina Peach Board
P.O. Box 13413
Columbia, SC 29201
802-253-4036

Contact: Charles Walker, Manager

9.783
South Carolina Peanut Board
P.O. Box 11280
Columbia, SC 29211
803-734-2200

Contact: Norwood McLeod

9.784
South Carolina Tomato Association
52 Shrewsburg Rd.
Charleston, SC 29407
803-571-0056

Contact: Barrett Lawrimore, Executive Director

9.785
South Carolina Watermelon Board
P.O. Box 11280
Columbia, SC 29211
803-734-2200; (FAX) 803-734-2192

Contact: E. Wayne Mack

SD
9.786
South Dakota Association of State Weed and Pest Board
Anderson Bldg.
Pierre, SD 57501
605-773-3796

Contact: Dr. Dennis Clarke, Executive Secretary

9.787
South Dakota Corn Growers Association
4001 Valhalla Blvd., Suite 108
Sioux Falls, SD 57106

Contact: Don Iseminger, Executive Director

9.789
South Dakota Federation of Garden Clubs
P.O. Box 540
Yankton, SD 57078

Contact: Catherine Hladky

9.790
South Dakota Fertilizer and Agricultural Chemical
  Association
121 North Grand
Pierre, SD 57501
605-224-2445

Contact: West Broer, Executive Secretary

9.791
South Dakota Irrigators Association
R.R. 2, Box 79
Conde, SD 57434
605-897-6642

Contact: Tom Huber

9.792
South Dakota Nurserymen's Association
3309 East 10th St.
Sioux Falls, SD 57103
605-336-7743

Contact: Bruce R. Ellingson

9.793
South Dakota Potato Growers Association
P.O. Box 127
Clark, SD 57225
605-532-3311

Contact: Donnally Fjelland, Executive Secretary

9.794
South Dakota Seed Trade Association
410 5th St.
Rapid City, SD 57701
605-342-8414

Contact: Doug Nahrgang, President

TN
9.795
Blue Ridge Florists Club
7112 Oakland Ave.
Johnson City, TN 37501

9.796
Central Tennessee Professional Florists Association
720 North Blythe St.
Gallatin, TN 37066
615-589-2042

Contact: Audry McGee

9.797
Christmas Tree and Shrubbery Growers Association
Rt. 6, Box 520
Johnson City, TN 37601

9.798
Mid-South Christmas Tree Association
P.O. Box 149
Sewanee, TN 37375

Contact: Richard Winslow

9.799
Middle Tennessee Nurserymen's Association
P.O. Box 822
McNinnville, TN 37110
615-473-2993

9.800
Professional Florists Association of Greater Memphis
1295 Jefferson
Memphis, TN 38104

Contact: Linda Floyd

9.801
Professional Florists of the Chattanooga Area
2112 Dayton Blvd.
Chattanooga, TN 27415
615-622-3143

Contact: Harris Downey

9.802
Smoky Mountain Herbal Society
c/o Ornamental Horticulture and Landscape Design Dept.
University of Tennessee
P.O. Box 1071
252 Ellington Plant Science Bldg.
Knoxville, TN 37901

9.803
Tennessee Christmas Tree Association
10219 Bob Gray Rd.
Knoxville, TN 37932

Tennessee Conservation League
Nashville, TN

Cross-reference: 8.74

9.804
Tennessee Corn Growers Association
Rt. 1
Henry, TN 38231
901-243-2231

Contact: Jimmy Tosh, President

9.805
Tennessee Federation of Garden Clubs
958 Brownlee Rd.
Memphis, TN 38116

Contact: Mrs. D. V. Pennington

9.806
Tennessee Flower Growers Association
Box 1071
Knoxville, TN 37901
615-974-1840

9.807
Tennessee Forestry Association
P.O. Box 290693
Nashville, TN 37229
615-883-3832

9.808
Tennessee Foundation Seeds
412 Murfreesboro Rd.
Nashville, TN 37210
615-242-1753

Contact: James Rader, Manager

9.809
Tennessee Fruit and Vegetable Horticultural Association
Rt. 1, Box 129
Bybee, TN 37713

Contact: Patsy Chambers, Executive Secretary

9.810
Tennessee Nurserymen's Association
P.O. Box 57
115 Lyon St.
McMinnville, TN 37110
615-473-3951

9.811
Tennessee State Florists Association
211 Old Hickory Blvd.
Madison, TN 37115

Contact: Jerry Hankin

9.812
Tennessee Turfgrass Association
533 Hagan St.
Nashville, TN 37203

9.813
Tennessee Vegetable Growers Association
Rt. 2, Box 278
Henderson, TN 37075

Contact: Dee Ann Walker, Executive Director

9.814
Tri State Herb Society
c/o Dot Quinn
1322 Orlando Ave.
East Ridge, TN 37412

9.815
West Tennessee Nurserymen's and Landscape
    Association
P.O. Box 241689
Memphis, TN 38124

TX
9.816
Alamo Area Allied Florists' Association
7540 Louis Pasteur Dr.
San Antonio, TX 78229

Contact: Robbie Borden

9.817
Allied Florists of Houston
5002 Morningside
Houston, TX 77005
713-522-7671

Contact: Robert Flagg

9.818
Capitol Area Florists Association
4509 Rim Rock Path
Austin, TX 78745
512-441-5047

Contact: Sharon Boyko

9.819
Central Texas Biodynamic Association
HC 01
Box 24
Dripping Spring, TX 78620

9.820
Dallas Area Historical Rose Group
c/o Joe M. Woodard
8636 Sans Souci Dr.
Dallas, TX 75238

9.821
Fort Worth Cactus & Succulent Society
Fort Worth Botanical Garden Center
Box 181
Decatur, TX 76234

Contact: Ed Maddox

9.822
Gulf Coast Fruit Tree Study Group
1310 Avenue J
South Houston, TX 77587

9.823
Rio Grande Valley Horticultural Society
Box 107
Weslaco, TX 78596
512-968-5000

9.824
Southern Seedsman Association
P.O. Box 569
Tyler, TX 75710
214-597-6637

Contact: John Mass, President

9.825
Texas Association of Nurserymen
Sunbelt Nursery Group
One Ridgemar Centre
6500 W. Freeway, Suite 600
Ft. Worth, TX 76116
817-624-7253

9.826
Texas Bedding Plant Growers
c/o Bob Schmidt
Rt. 3, Box 54
Lubbock, TX 79401
806-746-5858

9.827
Texas Botanical Garden Society
603 W. 9th St.
Austin, TX 78701
512-476-2008; (FAX) 512-476-6436

9.828
Texas Christmas Tree Association
424 American Bank Plaza
Corpus Christi, TX 78475
512-881-8031

Contact: Alton L. Buehring

9.829
Texas Citrus & Vegetable Association
c/o W.E. Weeks
6912 W. Expressway 83
Harlingen, TX 78552-3701
512-423-0340

9.830
Texas Corn Producers Association
218 East Bedford
Dimmitt, TX 79027
806-647-4224

Contact: Carl King, President

9.831
Texas Forestry Association
P.O. Box 1488
Lufkin, TX 75902-1488
490-632-TREE

9.832
Texas Fruit Growers Association
P.O. Drawer CC
College Station, TX 77841
409-846-1752; (FAX) 409-846-1752

Contact: Norman Winter, Executive Secretary

9.833
Texas Garden Clubs, Inc.
7173 Kendallwood
Dallas, TX 75240

Contact: Mrs. Ben P. Denman

9.834
Texas Greenhouse Growers Council
Rt. 3, Box 141
Rockdale, TX 76567

Texas Organization for Endangered Species
Austin, TX

Cross-reference: 8.75

9.835
Texas Peanut Producers Board
P.O. Box 398
Gorman, TX 76454
817-734-2853

Contact: Mary Webb, Executive Director

9.836
Texas Pecan Growers Association
4348 Carter Creek Pkwy.
Suite 401
Bryan, TX 77802
409-846-3285; (FAX) 409-846-1752

Contact: Norman Winter, Executive Director

9.837

Texas Seed Trade Association
6448 Highway 290 East, Suite D-100
Austin, TX 78723
512-371-7185

Contact: Don Ator, Executive Vice-President

9.838

Texas State Florists
c/o Jim Ellison
2107 E. Stone
Brenham, TX 77833
409-836-6011; (FAX) 409-830-1455

9.839

Texas State Soil and Water Conservation Board
P.O. Box 658
Temple, TX 76503
817-773-2250

Contact: Robert G. Buckley, Executive Director

9.840

Texas Sugar Beet Growers Association
Box 1818
Hereford, TX 79045
806-364-1100

Contact: James W. Witherspoon

9.841

Texas Sweet Potato Growers Association
Rt. 1, Box 151A
Yantis, TX 75497
214-383-2870

Contact: Sheilah Rushing, President

9.842

Texas Valley Citrus Committee
P.O. Box 2587
McAllen TX, 78502
512-686-9538

Contact: Leslie Whitlock, Manager

9.843

Texas Vegetable Association
c/o Mike Kirby
Valley Onions, Inc.
P.O. Box 2345
McAllen, TX 78502
512-631-3311

9.844

Texas Watermelon Association
P.O. Box 903
Weatherford, TX 76806
817-594-1045

Contact: Wanda Letson, Secretary-Treasurer

9.845

WesTexas New Mexico Florist Association, Inc.
3404 54th St.
Lubbock, TX 79413
806-797-7582

UT
9.846

Utah Agricultural Chemical and Fertilizer Association
142 East 1900 North
Centerville, UT 84014
801-298-5254

Contact: Matt Swain, Chairman

9.847

Utah Associated Garden Clubs, Inc.
4060 South 1500 East
Salt Lake City, UT 84124

Contact: Mrs. Arlan Headman

9.848

Utah Association of Nurserymen and Landscape
    Contractors
3500 South 900 East
Salt Lake City, UT 84106
801-487-4131

Contact: George S. Hoar, Executive Director

9.849

Utah Horticulture Association
P.O. Box 567
Santaquin, UT 84655
801-754-5601

Contact: Chad Rowley, President

9.850

Utah Onions Association
850 South 2000 West
Syracuse, UT 84041
801-773-0630

Contact: Jerry Hartley, President

9.851

Utah Pistachio Association
P.O. Box 2057
St. George, UT 84771
801-673-1991

Contact: Alan Ragozzine, Executive Director

9.852

Utah Seed Council
Seed Laboratory
Utah Dept. of Agriculture
350 North Redwood Rd.
Salt Lake City, UT 84116
801-538-7182

Contact: Terry Sue Freeman

9.853
Utah Weed Control Association
1190 West 820 North
Provo, UT 84601
801-378-2760

Contact: Larry S. Jeffery, President

VA
9.854
Allied Florists Association of Southeastern Virginia
6301 Ardsley Square, Suite 103-L
Virginia Beach, VA 23464
804-420-5624

9.855
Appalachian Mushroom Growers
Rt. 1, Box 30 BYY
Haywood, VA 22722
703-923-4774

Contact: Maryellen Lombardi

9.856
Association of Virginia Potato and Vegetable Growers
P.O. Box 26
Olney, VA 23418
804-787-3171

Contact: Harvey Belote, President

9.857
Greater Richmond Florists Association
3313 Mechanicsville Pike
Richmond, VA 23222
804-321-2000

Contact: William Gouldin

9.858
Jeffersonian Wine Grapes Grower Society
No. 2 Boar's Head Ln.
Charlottesville, VA 22901
804-296-4188

Contact: Mrs. John B. Rogan, President

9.859
Northern Virginia Nurserymen's Association
5108 N. 26th Rd.
Arlington, VA 22207

Contact: Mary Nugent, President

9.860
State Certified Seed Commission
Virginia Polytechnic Institute
  and State University
Blacksburg, VA 24061
703-231-9801

Contact: Dr. Robert L. Harrison

9.861
Tidewater Turfgrass Association
Parks, City of Virginia Beach
2150 Lynhaven Pkwy.
Virginia Beach, VA 23456
804-471-2027

Contact: Richard Branich

9.862
Virginia Agricultural Chemicals and Soil Fertility
  Association
4800 Broad Meadows Ct.
Glen Allen, VA 23060
804-270-2176

Contact: Mary Jane Neiman, Executive Secretary

9.863
Virginia Christmas Tree Association
Star Rt., 41-T
Deerfield, VA 24432
703-939-4646

Contact: Katherine Ward

9.864
Virginia Christmas Tree Growers Association
Box 33
Riner, VA 24149
703-382-4668

Contact: William Larsen, Secretary-Treasurer

9.865
Virginia Corn Board
P.O. Box 26
Warsaw, VA 22572
804-333-3710

Contact: Paul Rogers, Jr., Chairman

9.866
Virginia Corn Growers Association
10806 Trade Rd.
Richmond, VA 23236
804-379-2099

Contact: David Ottaway, Executive Director

9.867
Virginia Cultivated Turfgrass Association
P.O. Box 213
Chantilly, VA 22021-0213

Contact: W. R. Weeldy

9.868
Virginia Federation of Garden Clubs, Inc.
205 Culpepper Rd.
Richmond, VA 23229

Contact: Mrs. Dewitt B. Casler, Jr.

9.869
Virginia Flora Committee
c/o L. M. Hall
Virginia Academy of Science
Biology Dept., Bridgewater College
Bridgewater, VA 22812

9.870
Virginia Forestry Association
1205 East Main St.
Richmond, VA 23219
804-644-8462; (FAX) 804-788-0734

Contact: Charles F. Finley, Jr.

9.871
Virginia Greenhouse Growers Association
P.O. Box 16108
Chesapeake, VA 23320
804-547-2424

Contact: Janice Stewart, Secretary-Treasurer

9.872
Virginia Herb Growers and Marketers Association
P.O. Box 1176
Chesterfield, VA 23832
703-628-1436

9.873
Virginia Horticulture Council, Inc.
383 Coal Hollow Rd.
Christiansburg, VA 24073-9211
703-382-0904; (FAX) 703-382-2716

9.874
Virginia Nurserymen's Association
P.O. Box 718
Staunton, VA 24401

9.875
Virginia Peanut Board
Rt. 2, Box 137
Dinwiddie, VA 23841
804-469-3118

Contact: Wayne C. Barnes

9.876
Virginia Peanut Growers Association
P.O. Box 149
Capron, VA 23829
804-658-4573

Contact: Russell C. Schools, Executive Secretary

9.877
Virginia Professional Horticulture Conference and Trade
  Show
P.O. Box 6291
Virginia Beach, VA 23456

9.878
Virginia Seedsmen's Association
Smith Seed Inc.
315 Lynn St.
Danville, VA 24511

Contact: Wayne Smith, President

9.879
Virginia Soil Fertility Association
4800 Broad Meadows Ct.
Glen Allen, VA 23060
804-270-2176

Contact: Mary Jane Neiman, Secretary-Treasurer

9.880
Virginia Soil and Water Conservation Board
203 Governor St., Suite 206
Richmond, VA 23219
804-786-2064

Contact: Roland B. Geddes, Director

9.881
Virginia State Apple Board
P.O. Box 718
Staunton, VA 22401
703-332-7790

Contact: Clayton O. Griffin, Director

9.882
Virginia Sweet Potato Association
P.O. Box 26
Olney, VA 23418
804-787-3720

Contact: W. C. Davis, President

9.883
Virginia Turfgrass Association
Waynesboro Country Club
Meadowbrook Rd.
Waynesboro, VA 22980
703-943-2797

Contact: Dave Geiger

9.884
Virginia Turfgrass Council
P.O. Box 9528
Virginia Beach, VA 23450

Contact: Randeen Tharp, Executive Secretary

9.885
Virginia Vineyards Association
P.O. Box 394
Washington, VA 22747
804-296-2604

Contact: Chris Hill, President

9.886
Virginia Wine and Grape Advisory Board
P.O. Box 1163
Richmond, VA 23209
804-786-0481

Contact: Annett Ringwood

9.887
Virginia Wineries Association
P.O. Box 347
Middleburg, VA 22117
703-687-6277

Contact: Archie Smith III, President

9.888
Virginia-Carolina Peanut Association
P.O. Box 499
Suffolk, VA 23434
804-539-2100

Contact: W. Randolph Carter, Executive Secretary

VT
9.889
Federated Garden Clubs of Vermont
Overlake #15
545 S. Prospect St.
Burlington, VT 05401

Contact: Mrs. Howard A. Allen, Jr.

9.890
Hardy Plant Club
c/o Phil Cook
Botany Dept., University of Vermont
Burlington, VT 05405

9.891
Hardy Plant Club of Northwest Vermont
c/o Phil Cook, Dept. of Botany
University of Vermont
Burlington, VT 05405
802-656-0430

Contact: Dr. Phillip W. Cook, Secretary

9.892
New Hampshire-Vermont Christmas Tree Association
R.R. 1, Box 470
Wolcott, VT 05680
802-888-2783

Contact: Pam Dwyer, Executive Secretary

9.893
Vermont Herb Society
Box 151
Hinesburg, VT 05461

Vermont Natural Resources Council
Montpelier, VT
Cross-reference: 8.80

9.894
Vermont Plantsmens Association
Town Rd. 28
Bradford, VT 05033
802-222-9354

Contact: Earl Welch, President

9.895
Vermont Small Fruit and Vegetable Growers Association
Cross Road Farm
Fairlee, VT 05045
802-333-4455

Contact: Tim Taylor, President

9.896
Vermont Tree Fruit Growers Association
Apple Hill Rd.
Bennington, VT 05201
802-447-0144

Contact: Paul Bohn, Jr., President

9.897
Vermont-New Hampshire Potato Growers Association
Williamstown, VT 05679
802-433-5930

Contact: Robert Chappele, President

WA
9.898
Allied Florists of Greater Seattle
Nienaber Advertising
808 106th Ave., N.E., Suite 203
Bellevue, WA 98004
206-455-9881

9.899
Bellevue Botanical Garden Society
P.O. Box 6091
Bellevue, WA 98007

9.900
Far West Fertilizer Association
Pure Gro Co.
Box 8
Toppenish, WA 98948
509-865-2045

Contact: Bob Bell, President

9.901
Green Gem Bluegrass Growers Association
607 Great Western Bldg.
Spokane, WA 99201-1069
509-624-9263

Contact: Dale Severson, Executive Secretary

9.902
Hop Growers of Washington
504 North Naches Ave., Suite 5
Yakima, WA 98901
509-453-4749

Contact: Ann George, Manager

9.903
Inland Empire Flower Growers Club
Jones Wholesaler
P.O. Box 975
Seattle, WA 98155
509-838-2944

Contact: David Montekacco, President

9.904
Northwest Bulb Growers Association
P.O. Box 303
Mt. Vernon, WA 98273
206-424-1375

Contact: Richard Nowadnick, Secretary-Treasurer

9.905
Northwest Christmas Tree Association
1300 Scaring Hawk Ln.
Woodland, WA 98674
206-225-5008

Contact: John Thelen, President

9.906
Northwest Florists Association
406 Main St.
Suite 116
Edmonds, WA 98020
206-778-6162

Contact: Donna Cameron

Northwest Fuchsia Society
Seattle, WA

Cross-reference: 2.73

Northwest Perennial Alliance
Seattle, WA

Cross-reference: 2.74

9.907
Northwest Turfgrass Association
P.O. Box 1367
Olympia, WA 98507
206-754-0825

Contact: Blair Patrick, Executive Director

9.908
Oregon-Washington Pea Growers Association
756 Wauna Vista Dr.
Walla Walla, WA 99362
509-525-7459

Contact: James C. Ferrel, President

9.909
Puget Sound Dahlia Association
P.O. Box 5602
Bellvue, WA 98006

Contact: Roger L. Walker

Seattle Tilth Association
Seattle, WA

Cross-reference: 8.68

9.910
Spokane Area Master Florists
P.O. Box 53
Spokane, WA 92210
509-489-2110

Contact: Patricia Mattson

9.911
Washington Apple Commission
P.O. Box 18
Wenatchee, WA 98801-0018
509-663-9600

Contact: Thomas Hale, President

9.912
Washington Asparagus Growers Association
2810 West Clearwater Ave.
Suite 202
Kennewick, WA 99336
509-783-3094

Contact: Michale Harker, Manager

9.913
Washington Blueberry Commission
1360 Bow Hill Rd.
Bow, WA 98232
206-766-6173

Contact: Dorothy Anderson, Secretary

9.914
Washington Blueberry Growers Association
2462 Zell Rd.
Ferndale, WA 98248
206-366-5311

Contact: Tony Moore, President

9.915
Washington Bulb Commission
P.O. Box 303
Mt. Vernon, WA 98273
206-424-1375

Contact: Richard Nowadnick, Secretary-Treasurer

9.916
Washington Certified Grape Nurserymens Association
Rt. 1, Box 1315
Benton City, WA 99320
509-588-3405

Contact: Tom C. Judkins, Secretary-Treasurer

9.917
Washington Certified Strawberry Plant Growers
   Association
10816 SR 162
Puyallup, WA 98374
206-845-5519

Contact: Ken Spooner, President

9.918
Washington Cranberry Commission
Rt. 1, Box 145
Grayland, WA 98547
206-267-4113

Contact: Martin Paulson, Secretary-Treasurer

9.919
Washington Farm Forestry Association
P.O. Box 7663
Olympia, WA 98501
206-459-0984

Contact: Nels Hanson, Executive Director

9.920
Washington Floricultural Association
12602 145th St. East
Puyallup, WA 98374
206-841-4273

Contact: Dr. Bernard Wesenberg

9.921
Washington Fruit Commission
1005 Tieton Dr.
Yakima, WA 98902
509-453-4837

Contact: Don Severn, President

9.922
Washington Hop Commission
504 North Naches Ave., No. 5
Yakima, WA 98901
509-547-5538

Contact: Ann George, Manager

9.923
Washington Mint Commission
P.O. Box 2111
Pasco, WA 99302
509-547-5538

Contact: Ken Maurer, Executive Secretary

9.924
Washington Mint Growers Association
P.O. Box 2061
Tri Cities, WA 99302
509-547-5538

Contact: Dale Miller, President

9.925
Washington Pesticide Consultants Association
Wilbur-Ellis Co.
7 East Washington Ave.
Yakima, WA 98902
509-248-6171

Contact: Doug Whitener, President

9.926
Washington Potato Growers Association
P.O. Box 377
Othello, WA 99344
509-488-5678

Contact: Larry Jorgenson, Manager

9.927
Washington Potato and Onion Association
108 Interlake Rd.
Moses Lake, WA 98837
509-765-8845

Contact: Henry Michaels, Executive Director

9.928
Washington Red Raspberry Growers Association
816 Loomis Trail Rd.
Lynden, WA 98264
206-595-2481

Contact: Marvin Jarmin, Manager

9.929
Washington Rhubarb Growers Association
P.O. Box 887
Sumner, WA 98390
206-863-7333

Contact: Paul A. Hammack

9.930
Washington Seed Potato Commission
P.O. Box 286
Lynden, WA 98264
206-354-4670

Contact: Doris Roosma, Secretary-Treasurer

9.931
Washington State Federation of Garden Clubs, Inc.
2314 108 S.E.
Bellevue, WA 98004

Contact: Mrs. Ralph Swenson

9.932
Washington State Grape Society
Rt. 1, Box 112
Grandview, WA 98930
509-837-4481

Contact: Henry Charret, Chairman

9.933
Washington State Nursery and Landscape Association
P.O. Box 670
Sumner, WA 98390
206-863-4482

Contact: Steve McGonigal, Executive Director

9.934
Washington State Weed Association
P.O. Box 3056
Tri Cities, WA 99302

Contact: Tim Smith, President

9.935
Washington Strawberry Commission
4430 John Luhr Rd.
Olympia, WA 98506
206-491-1010

Contact: Norval Johanson, Manager

9.936
Washington Tree Fruit Research Commission
Rt. 1, Box 319E
Wapato, WA 98951
509-877-2065

Contact: David Allen, Secretary-Treasurer

9.937
Washington Wine Commission
P.O. Box 61217
Seattle, WA 98121
206-728-2252

Contact: Simon Siegl

9.938
Washington Wine Institute
1932 1st Ave., Room 510
Seattle, WA 98101
206-441-1892

Contact: Simon Siegl, Executive Director

9.939
Washington-North Idaho Seed Association
Valley Grain, Inc.
Quichy, WA 98848
509-787-1561

Contact: Dale Guthrie, President

9.940
Western Cascade Fruit Society
60 Tala Shore Dr.
Port Ludlow, WA 98365
206-293-3484

Contact: John Parker, President

9.941
Western Cascade Tree Fruit Association
9210 131st St. N.E.
Lake Stevens, WA 98258

WI
9.942
Association of Wisconsin Wineries
Highway 188, P.O. Box 87
Prairie du Sac, WI 53578
608-643-6515

Contact: Bob Wollersheim, President

9.943
Ginseng Board of Wisconsin
500 3rd St., Suite 208-2
Wausau, WI 54401
715-845-7300

Contact: Ron Rambadt

9.944
Grape Growers Association
Rt. 1, Box 67
River Falls, WI 54022
715-425-6824

Contact: Bob Tomesh, President

9.945
Greater Milwaukee Florists Association
10855 West Potter Rd.
Wauwatosa, WI 53226
414-774-1498

Contact: Fred Poulsen

9.946
Milwaukee Horticultural Society
c/o Boerner Gardens
5879 S. 92 St.
Hales Corners, WI 53130

9.947
Minnesota Grape Growers Association
Rt. 1, Box 67
River Falls, WI 54022
715-425-6824

Contact: Bob Tomesh, President

9.948
Olbrich Botanical Society
3300 Atwood Ave.
Madison, WI 53704
608-266-4148

9.949
Shiitake Growers Association of Wisconsin
Box 99
Birchwood, WI 54817
715-354-3803

Contact: Carolyn Burnett, President

Trees for Tomorrow
Eagle River, WI
Cross-reference: 8.79

9.950
Wisconsin Arborist Association
7300 Chestnut St.
Wauwatosa, WI 53213
414-258-3000

Contact: Richard Hass, Secretary

9.951
Wisconsin Apple Growers Association
7300 Chestnut St.
Wauwatosa, WI 53213
414-258-3000

Contact: Richard Hass, Secretary

9.952
Wisconsin Apple and Horticultural Council, Inc.
850 Tipperary Rd.
Oregon, WI 53575

Contact: Myra Hann, Executive Secretary

9.953
Wisconsin Berry Growers Association
850 Tipperary Rd.
Oregon, WI 53575
608-835-5464

Contact: Myra Hann, Executive Secretary

9.954
Wisconsin Carrot Growers Association
R.R. 2, Box 70
Randolph, WI 53956
414-326-3534

Contact: Steve Slinger, Secretary-Treasurer

9.955
Wisconsin Central Vegetable Growers Association
R.R. #2
Colloma, WI 54930
715-787-2283

Contact: Hod Chilewski, President

9.956
Wisconsin Cherry Board
521 North Hudson Ave.
Sturgeon Bay, WI 54235
414-743-5447

Contact: Joseph G. Peterson, President

9.957
Wisconsin Christmas Tree Producers Association
P.O. Box 105, 213 Pierce St.
Arlington, WI 53911
608-635-7734

Contact: Virginia Mountford, Executive Secretary

9.958
Wisconsin Corn Growers Association
2976 Triverton Pike
Madison, WI 53711
608-274-7266

Contact: Criss Davis, President

9.959
Wisconsin Cranberry Board
1698 Cranberry Ln.
Wisconsin Rapids, WI 54494
715-886-4159

Contact: Guy A. Gattschalk

9.960
Wisconsin Cranberry Growers Association
P.O. Box 365
Wisconsin Rapids, WI 54495
715-423-2070

Contact: Tom Lochner

9.961
Wisconsin Fertilizer and Chemical Association
1414 East Washington Ave.
Suite 186
Madison, WI 53703
608-255-4001

Contact: Thomas E. O'Connell, Executive Secretary

9.962
Wisconsin Fresh Market Vegetable Growers Association
1375 Pitt Rd.
Brussels, WI 54204
414-825-7023

Contact: Eugene Garbowski, President

9.963
Wisconsin Garden Club Federation, Inc.
75 E. Water St.
Markesan, WI 53946

Contact: Mrs. Walter Seeliger

9.964
Wisconsin Gardens
7777 North 76th St.
Milwaukee, WI 53223-3911
414-354-4830

Contact: Bill Minor, President

9.965
Wisconsin Ginseng Growers Association
500 3rd St., Suite 208-2
Wausau, WI 54401
715-842-3401

Contact: Jay Schmidt

9.966
Wisconsin Land Conservation Association
208 Agricultural Hall
1450 Linden Dr.
Madison, WI 53706
608-262-4583

9.967
Wisconsin Landscape Contractors
5645 South 108th St.
Hales Corner, WI 53130

Contact: Joe Phillips, Executive Secretary

9.968
Wisconsin Landscape Federation
11801 W. Janesville Rd.
Hales Corners, WI 53130

Contact: Joe Phillips, Executive Secretary

9.969
Wisconsin Mint Board
Rt. 3, 124 Siesta Dr.
Montello, WI 53949
608-297-7752

Contact: James Shestock, President

9.970
Wisconsin Nurserymen's Association
11801 West Janesville Rd.
Hales Corners, WI 53130

Contact: Joe Phillips, Executive Secretary

9.971
Wisconsin Potato Growers Auxiliary
211 Lakeview Rd.
Rosholt, WI 54473
715-677-4565

Contact: Jane Zdroik, President

9.972
Wisconsin Potato and Vegetable Growers Association
P.O. Box 327
Antigo, WI 54409

Contact: John R. Martens, Executive Vice-President

9.973
Wisconsin Red Cherry Growers Association
857 City Highway U
Sturgeon Bay, WI 54235
414-743-2984

Contact: Kathleen Petrina, President

9.974
Wisconsin Sod Producers Association
Kellogg Supplies, Inc.
P.O. Box 684
Milwaukee, WI 53201

Contact: Egon Herrman, Secretary

9.975
Wisconsin Turfgrass Association
Blackhawk Country Club
Box 5129
Madison, WI 53705

Contact: Monroe Miller, Secretary

9.976
Wisconsin-Upper Michigan Florists Association
N40-W27928 Glacier Rd.
Pewaukee, WI 53072
414-691-0982

Contact: Stanley C. Foll, Secretary

WV
9.977
Huntington Tri-State Florists
1123-25 Fourth Ave.
Huntington, WV 25701

9.978
Tri-State Growers Association
Box 756
Crab Orchard, WV 25827

Contact: R. L. Webb, Jr.

9.979
West Virginia Allied Florists and Growers Association
Tapham Nursery Garden Center
250 Greeb Bad Rd.
Morgantown, WV 26505
304-292-8733

Contact: Christopher Tapham

9.980
West Virginia Christmas Tree Growers Association
102 Sutton Ave.
Princeton, WV 24740
304-425-5928

Contact: Gene Bailey

9.981
West Virginia Forest and Growers Association
West Virginia University
P.O. Box 6108, Division of Plant Sciences
Evansdale Campus
Morgantown, WV 26506-6108
304-293-6023

9.982
West Virginia Foresters Association
P.O. Box 724
Ripley, WV 25721
304-273-8164

Contact: Richard Waybright, Executive Director

9.983
West Virginia Garden Club, Inc.
1978 Smith Rd.
Charleston, WV 25314

Contact: Mrs. Robert L. Swoope

9.984
West Virginia Grape Growers Association
Pitera Vineyards
Purgitsville, WV 26852
304-289-3493

Contact: Robert Pliska, Executive Secretary

9.985
West Virginia Nurserymen's Association
1601 5th St.
Huntington, WV 25701
304-523-8491

Contact: Mark Springer, President

9.986
West Virginia State Soil Conservation Committee
Gutherie Agricultural Center
Charleston, WV 25305
304-348-2204

Contact: Cleve Benedict, Chairman

9.987
West Virginia Vegetation Management Association
Appalachian Power Co.
P.O. Box 1986
Charleston, WV 25327
304-348-5729

Contact: Lynn Grayston, President

WY
9.988
Wyoming Federation of Garden Clubs, Inc.
P.O. Box 1208
Pinedale, WY 82941

Contact: Mrs. Harry Pelliccione

9.989
Wyoming Weed and Pest Council
2219 Carey Ave.
Cheyenne, WY 82002-0100
307-777-6585

Contact: George Hittle, Coordinator

CANADA

AB
9.990
Alberta Forestry Association
311–10526 Jasper Ave.
Edmonton, AB T5J 1Z7
403-428-7582

9.991
Alberta Regional Lily Society
c/o J. Annett
10922–80th Ave.
Edmonton, AB T6G 0R1

9.992
Bonsai Society of Calgary
c/o G. Hiett
225 22nd Ave., N.E.
Calgary, AB T2E 1T6

9.993
Edmonton Cactus and Succulent Society
c/o Ms. H. Huver
P.O. Box 69
Spruce Grove, AB T0W 2C0

9.994
Edmonton Horticultural Society
11707–150th Ave.
Edmonton, AB T5X 1C1
403-456-7986

Contact: Arlene Smith, Director

9.995
Floral Arts Society of St. Albert
P.O. Box 34
St. Albert, AB T8N 1N2

9.996
Friends of the Devonian Botanic Garden
University of Alberta
Edmonton, AB T6G 2E1

9.997
Landscape Alberta Nursery Trades Association
10215 176th St.
Edmonton, AB T5J 3H1
403-489-1991

Contact: Garry L. Johnson

9.998
Lethbridge and District Horticultural Society
74 Eagle Road North
Lethbridge, AB T1H 4S5

Contact: D. L. Weightman, Secretary

9.999
Orchid Society of Alberta
c/o Jan Weijer
Box 3015
Sherwood Park, AB T8A 2A6

9.1000
St. Albert and District Garden Club
25 Lombard Crescent
St. Albert, AB T8N 3N1

BC
Alpine Garden Club of British Columbia
Vancouver, BC

Cross-reference: 2.96

9.1001
Association of British Columbia Foresters
744 W. Hastings St., No. 510
Vancouver, BC V6C 1A5
604-687-8027

9.1002
British Columbia Blueberry Co-op Association
31852 Marshall Rd.
Abbotsford, BC V2S 4N5

9.1003
British Columbia Certified Seed Potato Growers
   Association
4119–40th St.
Delta, BC V4K 3N2

9.1004
British Columbia Forestry Association
1430–1100 Melville St.
Vancouver, BC V6E 4A6
604-683-7591

9.1005
British Columbia Fruit Growers' Association
Box 160
1473 Water St.
Kelowna, BC V1Y 7N6

9.1006
British Columbia Guild of Flower Arrangers
2268 Kensington Ave.
Burnaby, BC V5B 4E2

British Columbia Lily Society
Langley, BC

Cross-reference: 2.98

9.1007
British Columbia Nursery Trades Association
#107–14914 104th Ave.
Surrey, BC V3R 1M7
604-585-2225

9.1008
British Columbia Raspberry Growers' Association
#204–2464 Clearbrook Rd.
Clearbrook, BC V2T 2X8
604-853-1312

9.1009
Desert Plant Society of Vancouver
c/o Ms. M. Randle
2941 Parker St.
Vancouver, BC V5K 2T9
604-255-0606

9.1010
Garden Club of Vancouver
3185 West 45th Ave.
Vancouver, BC V6N 3L9

9.1011
Interior Greenhouse Growers
c/o 7X Greenhouses
R.R. 1
Keremos, BC V0X 1N0
604-499-5730

9.1012
Okanagan Valley Pollination Society
P.O. Box 186
Vernon, BC V1T 6M2
604-832-2732

9.1013
United Flower Growers' Co-op Association
4085 Marine Way
Burnaby, BC V5J 5E2
604-430-2211

9.1014
Vancouver Ikebana Association
c/o Ms. E. Harman
2293 W. 33rd Ave.
Vancouver, BC V6M 1C1

9.1015
Vancouver Island Rock & Alpine Garden Society
P.O. Box 6507, Station "C"
Victoria, BC V8P 5M4

9.1016
Victoria Flower Arrangers' Guild
1596 Judoba Pl.
Victoria, BC V8N 3K9

9.1017
Victoria Geranium & Fuchsia Society
Box 5266, Station "B"
Victoria, BC V8R 6N4

9.1018
Victoria Gladiolus & Dahlia Society
c/o Johnson Iverson
3314 Fircrest Pl.
Victoria, BC V8P 4B5

9.1019
Victoria Horticultural Society
P.O. Box 5081
Station B
Victoria, BC V8R 6N3

9.1020
Victoria Orchid Society
P.O. Box 337
Victoria, BC V8W 2N2
604-656-3094

Contact: Rona Chalmers

9.1021
Western Greenhouse Growers' Co-op Association
P.O. Box 1236, Station A
15350 No. 10 Hwy.
Surrey, BC V3S 2R3
604-576-8525

MB
9.1022
Landscape Manitoba Nursery Trades Association
104 Parkside Dr.
Winnipeg, MB R3J 3P8
204-889-5981

Contact: Barbara Craig

9.1023
Manitoba Forestry Association, Inc.
900 Corydon Ave.
Winnipeg, MB R3M 0Y4
204-453-3182

Manitoba Naturalists' Society
Winnipeg, MB

Cross-reference: 8.95

9.1024
Manitoba Regional Lily Society
c/o B. Strohman
Box 846
Neepawa, MB R0J 1H0

9.1025
Orchid Society of Manitoba
c/o Ms. D. Jensen
2700 Henderson Hwy.
Winnipeg, MB R2E 0C3

NB
9.1026
Canadian Forestry Association of New Brunswick, Inc.
65 Brunswick St.
Fredericton, NB E3B 1G5
506-455-8372

Contact: David Folster, Secretary

9.1027
Fredericton Botanic Garden Association
P.O. Box 1085
Fredericton, NB E3B 5C2
506-453-4584

Contact: Harold R. Hinds, President

9.1028
Fredericton Garden Club
107 Summer St.
Fredericton, NB E3A 1X7

9.1029
Kennebecasis Garden Club
P.O. Box 702
Rothesay, NB E0G 2W0

9.1030
Landscape New Brunswick Association
New Brunswick Dept. of Agriculture
P.O. Box 6000
Fredericton, NB E3B 5H1
506-453-2108

Contact: Garth Nickerson, Secretary

9.1031
New Brunswick Blueberry Growers Association
Dorchester, NB E0A 1M0
506-379-2248

Contact: Monique Knockwood, Secretary-Treasurer

9.1032
New Brunswick Botanic Garden Association
c/o Daniel Belanger
165 Hebert Blvd.
Edmundston, NB E3V 2S8
506-735-8804

9.1033
New Brunswick Fruit Growers Association, Inc.
1115 Regent St.
Fredericton, NB E3B 3Z2
506-452-8100

9.1034
New Brunswick Potato Agency
P.O. Box 238
Florenceville, NB E0J 1K0
506-392-6022

Contact: Willem Schrage

9.1035
Orchid Society of New Brunswick
c/o Dr. F. H. Predas
University of New Brunswick
P.O. Box 5050
St. John, NB E2L 4L5

NF
Newfoundland Alpine & Rock Garden Club
St. John's, NF

Cross-reference: 2.108

9.1036
Newfoundland Orchid Society
c/o Dr. John Allen
8 Howlett Ave.
St. John's, NF A1B 1K9

NS
9.1037
Atlantic Provinces Nursery Trades Association
130 Bluewater Rd.
Terra Nova Landscaping
Bedford, NS B4B 1G7
902-835-7387

Contact: Tanya Morrison

9.1038
Bio-Dynamic East
Small Farm
R.R. 1
Port Williams, NS B0P 1T0

Contact: Neil Van Nostrand

Dahlia Society of Nova Scotia
Great Village, NS

Cross-reference: 2.107

9.1039
Landscape Atlantic Association
Turf Masters Landscaping Ltd.
P.O. Box 5276
Armdale, NS B3L 4S7
902-443-5321

Contact: Gary Paul, President

9.1040
Landscape Nova Scotia Association
Evergreen Landscaping Co.
15 Purcells Cove Rd.
Halifax, NS B3N 1R2
902-455-8055

Contact: David Pace, Secretary

9.1041
Nova Scotia Association of Garden Clubs
P.O. Box 550
Truro, NS B2N 5E3

9.1042
Nova Scotia Forestry Association
64 Inglis Pl.
Suite 202
Truro, NS B2N 4B4
902-893-4653

9.1043
Orchid Society of Nova Scotia
c/o Dr. J. H. Vandermeullen
Maple Brook House
Frenchmans Rd.
Enfield, NS B0N 1N0

ON
9.1044
Decorative and Horticultural Judges Association
306 Wilson Ave.
Burlington, ON L7L 2M9

9.1045
Garden Club of Ancaster
50 Academy St.
Ancaster, ON L9G 2Y1

9.1046
Garden Club of Burlington
c/o Mrs. R. D. Shoots
5712 Blind Line, R.R. 3
Campbelville, ON L0P L8O

9.1047
Garden Club of Hamilton
2070 Watson Dr.
Burlington, ON L7R 3X4

9.1048
Garden Club of Kitchener-Waterloo
284 Shakespeare Dr.
Waterloo, ON N2L 2T6

9.1049
Garden Club of London
34 Bromleigh Ave.
London, ON N6G 1T9

9.1050
Garden Club of Ontario
c/o Mrs. W. J. E. Spence
228 Dunvegan Rd.
Toronto, ON M5P 2P2

9.1051
Garden Club of Toronto
777 Lawrence Ave. E.
Don Mills, ON M3C 1P2

Hamilton Naturalists' Club
Hamilton, ON
Cross-reference: 8.94

9.1052
Landscape Ontario Horticultural Trades Association
1293 Matheson Blvd.
Mississauga, ON L4W 1R1
416-629-1184
Contact: Jo-Anne Willetts

9.1053
Men's Garden Club
173 Joycey Blvd.
Toronto, ON M5M 2V3

9.1054
Milne House Garden Club
c/o Mrs. Doreen Martindale
64 Colin Ave.
Toronto, ON M5P 2B9

9.1055
Ontario Blueberry Association
R.R. 1
Harrow, ON N0R 1G0

9.1056
Ontario Forestry Association
Suite 209–150 Consumers Rd.
Willowdale, ON M2J 1P9
416-493-4565
Contact: James D. Coates, Executive Vice-President

9.1057
Ontario Fruit & Vegetable Growers' Association
301 Ontario Food Terminal
165 The Queensway
Toronto, ON M8Y 1H8
416-255-4473

9.1058
Ontario Ginseng Growers Association
Box 1062
Waterford, ON N0E 1Y0
Contact: Paul Lucas, President

9.1059
Ontario Herbalists Association
7 Alpine Ave.
Toronto, ON M6P 3R6
416-536-3835
Contact: Anne Bridgman, President

9.1060
Ontario Highbush Blueberry Growers
R.R. 2
Langton, ON N0E 1G0
Contact: Mr. D. Zamecnik

9.1061
Ontario Regional Lily Society
c/o Ms. G. Brown
R.R. 1
Harley, ON N0E 1E0

Ontario Rock Garden Society
Shelburne, ON
Cross-reference: 2.110

9.1062
Ontario Seed Potato Growers Association
Box 109
Everett, ON L0M 1J0
Contact: Rick Bailer, Chairman

9.1063
Ottawa Bonsai Society
c/o Michael Piva
P.O. Box 3126, Station "D"
Ottawa, ON K1P 6H7

9.1064
Society for Biodynamic Farming & Gardening in Ontario
R.R. 3
Acton, ON L7J 2L9

9.1065
Society of Ontario Nut Growers
R.R. 2, Con. 6 Rd.
Niagara-on-the-Lake, ON L0S 1J0
Contact: G. R. Hambleton

9.1066
Toronto Bonsai Society
190 McAllister Rd.
Downsville, ON M3H 2N9
Contact: Eva Davidson

9.1067
Toronto Cactus & Succulent Club
P.O. Box 334
Brampton, ON L6V 2L3
416-767-6433
Contact: Betty Naylor

Toronto Field Naturalists
Toronto, ON
Cross-reference: 8.97

9.1068
Toronto Gesneriad Society
70 Enfield Rd.
Etabicoke, ON M8W 1T9

9.1069
Toronto Hobby Greenhouse Group
c/o Barbara Mitchell
39 Dexter Blvd.
Willowdale, ON M2H 1Z3

9.1070
Toronto Japanese Garden Club
6 Forest Laneway
Suite 105
Willowdale, ON M2N 5Y9
Contact: Mr. M. Nishi

9.1071
Woodstock Horticultural Society
R.R. 2
Burgessville, ON N0J 1C0
Contact: Mary Yeoman

PQ
9.1072
Association Forestière Quebecoise
915 St. Cyrille Ouest, No. 110
Quebec, PQ G1S 1T8
418-681-3588

9.1073
Brome Lake Garden Club
178 Woodard Rd.
R.R. 4
Sutton, PQ J0E 2K0

9.1074
Diggers and Weeders Garden Club
20 Thornhill Ave.
Westmount, PQ K3Y 2E2

9.1075
Federation des Producteurs Maraichers du Quebec
555, boulevard Roland Therrien
Longueuil, PQ J4H 3Y9
418-679-0530
Contact: Jena-Pierre Girary, Vice-Secretary

9.1076
Federation des Producteurs de Fruits et Legumes du
    Quebec
555, boulevard Roland-Therrien
Longueuil, PQ J4H 3Y9
514-679-0530; (FAX) 514-679-5436
Contact: Gilles McDuff

9.1077
Federation des Producteurs de Pommes du Quebec
555, boulevard Roland-Therrien
Longueuil, PQ J4H 3Y9
514-679-0530; (FAX) 514-679-5436

9.1078
Federation des Producteurs de Pommes de Terre du
    Quebec
555, boulevard Roland-Therrien
Longueuil, PA J4H 3Y9
514-679-0530; (FAX) 514-679-5436
Contact: Jacques Mailhot

9.1079
Gesneri-Quebec
L. Hodgson, Pavillon des services
Chambre 2602, Université Laval
2450, boulevard Hochelaga
Ste.-Foy, PQ G1K 7P4

9.1080
Groupe de Bonsai de Quebec
Pavillon des services, Chambre 2601
Université Laval
2450, boulevard Hochelaga
Ste.-Foy, PQ G1K 7P4

9.1081
Institut Québécois du Développement de l'horticulture
    Ornementale
3230, rue Sicotte
Bureau, D-111
St.-Hyacinthe, PQ J2S 2M2
514-774-2181; (FAX) 514-771-6890

9.1082
L'Association des Architectes-Paysagistes du Quebec
407, boulevard Saint-Laurent
Bureau 500
Montreal, PQ H2Y 2Y5
514-395-8566

9.1083
La Federation Interdisciplinaire de l'Horticulture
    Ornementale du Quebec
Jardin Van Den Hende
Université Laval
Ste.-Foy, PQ G1K 7P4
418-659-3561; (FAX) 418-651-7439

9.1084
Les Orchidophiles de Quebec Inc.
C.P. 8857
Ste.-Foy, PQ G1V 4N7

9.1085
Province of Quebec Gladiolus Society
c/o Ms. R. McPhail
11970 Laviene St., Cartierville
Montreal, PQ H4J 1X8

9.1086
Quebec Gladiolus Society
c/o Denis Poirier
505 St. Piere Sud, Box 88
St. Constant, PQ J5A 2E7

9.1088
Societe Internationale d'Arboriculture Quebec, Inc.
Jardin van den Hende
University Laval
Ste.-Foy, PQ G1K 7P4
418-659-3561

Contact: Michel Carrier, President

9.1089
Société de Bonsai de Montreal
4101 est, rue Sherbrooke
Montreal, PQ H1X 2B2
514-872-1782

Contact: M. Claude Gagne, President

9.1090
Syndicat des Producers de Bleuets du Quebec
698, rue Melangon
Saint-Bruno
Lac-St.-Jean, PQ G0W 2L0
418-342-2206

Contact: Jeannot Coté

9.1091
Syndicat des Producteurs d'Oignons du Quebec
6, rue Dumoulin
St.-Remi, PQ J0L 2L0
514-454-3996

Contact: Gerald Pinsonneault

9.1092
Syndicat des Producteurs en Serre du Quebec
555, boulevard Roland-Therrien
Longueuil, PQ J4H 3Y9
514-679-0530

QC
9.1093
Fédération des Sociétés d'horticulture et d'écologie du
    Québec
4545, av. Pierre-de-Coubertin
C.P. 1000, Succ. M.
Montreal, QC H1V 3R2
514-252-3010

9.1094
Garden Club of Montreal
66 St. Sulpice Rd.
Montreal, QC M3Y 2B7

9.1095
L'Association de Biodynamic du Quebec
89 Aqueduc
St. Francois de Montmagny, QC G0R 3A0

9.1096
Quebec Forestry Association, Inc.
Association Forestiere Quebecoise
110
915 St. Cyrille Blvd., West
Quebec City, QC G1S 1T8
418-681-3588

SK
9.1097
Carefree Gesneriad Society
c/o W. Morton
79 Selby Crescent
Regina, SK S4T 6V9

9.1098
Evergreen Garden Club
13 Kootenay Dr.
Saskatoon, SK S7K 1T2

9.1099
Prairie Flower Club
Box 35
White City, SK S9G 5B0

9.1100
Prairie Garden Guild
Box 211
Saskatoon, SK S7K 3K4

Contact: Monty Zary, Treasurer

9.1101
Saskatchewan Christmas Tree Growers Association
Weyerhauser Canada Ltd.
Pulpwood Division, Box 1720
Prince Albert, SK S6V 5T3

Contact: Bob Sutton

9.1102
Saskatchewan Farm Woodlot Association
Box 68
Medstead, SK S0M 1N0

Contact: Hector Shiell

9.1103
Saskatchewan Forestry Association
P.O. Box 400
Prince Albert, SK S6V 5R7
306-763-2189

9.1104
Saskatchewan Fruit Growers Association
Sub P.O. Box 11
Saskatoon, SK S7M 1X0

Contact: Ben Epp

9.1105
Saskatchewan Greenhouse Growers Association
Park Greenhouses
Box 833
Outlook, SK S0L 2N0

Contact: Shirley Park

9.1106
Saskatchewan Nursery Trades Association
Lakeshore Tree Farms Ltd., R.R. 3
Saskatoon, SK S7K 3J6

Contact: Vic Krahn

9.1107
Saskatchewan Orchid Society
P.O. Box 411
Saskatoon, SK S4K 3L3
306-374-2978

Contact: Paul Junk

Saskatchewan Perennial Society
Saskatoon, SK

Cross-reference: 2.113

9.1108
Saskatchewan Vegetable Growers Association
Box 37
Lumsden, SK S0G 3C0

Contact: Bert Wills

9.1109
Saskatoon Horticultural Society
P.O. Box 161
Saskatoon, SK S7K 3K4

9.1110
Unique Flower Arranging Club
836 Shannon Rd.
Regina, SK S4S 5K2

~ 10 ~

Nomenclature and International Registration Authorities

We communicate by words, and the names of plants are all important if we are to know what plants we are discussing. With hundreds of thousands of plant species in the world, an unambiguous and internationally accepted standard of reference is necessary. This exists in the form of the *International Code of Botanical Nomenclature*.

Until the middle of the eighteenth century, plant names varied from country to country and even from scientist to scientist. Some of the early scientific names for plants were merely long descriptions of the plants themselves. In 1753, after a great organizing effort in which he analyzed centuries of previous botanical efforts, Carolus Linnaeus published *Species Plantarum*, a single volume containing a list and brief description of all the plants known to science in his time. In this reference, Linnaeus laid the groundwork for organizing all of the botanical knowledge of his day and developed an orderly system for assigning plant names—binomial nomenclature. The merit of the Linnaean system was internationally acknowledged toward the end of the nineteenth century when the *International Code* was established; today the scientific names assigned to all plants must comply with the well-defined set of regulations.

The Linnaean system of binomial nomenclature classifies a plant as well as names it, so binomials provide a great deal of information about a plant, its characteristics, and its kinship. Each name consists of a generic or genus name and a specific epithet. The generic name assigned to a plant classifies it as a member of a closely related group of species, called a genus (pl. genera), much as a surname ties members of a family together. The specific epithet then designates the individual species within that group much as first names do the individual members of the family. A comparison of the common and botanical names for several small, often-grown trees provides an excellent example. Their common names—peach, cherry laurel, Japanese flowering cherry, apricot, and wild plum—do not imply any botanical relationship, but the botanical names for these plants—*Prunus persica*, *Prunus caroliniana*, *Prunus serrulata*, *Prunus armeniaca*, and *Prunus americana*—indicate that they are, in fact, closely related botanically, since they have all been placed into the genus *Prunus*. Genera are, in turn, grouped into botanically related plant families. The genus *Prunus*, for example, is a member of the rose family, Rosaceae. Other genera in this family include *Malus* (apples), *Rosa* (roses), and *Fragaria* (strawberries).

It is often necessary to make distinctions between forms of plants below the level of species. These are called infraspecific taxa, and the names of these plants are formed by combining the genus, the specific epithet, and a third epithet connected to the name by a term, usually abbreviated, denoting whether it is a subspecies (subsp.), a variety (var.), or a form (forma). Returning to the genus *Prunus* as an example, *Prunus persica* var. *nucipersica* is commonly known as nectarine. In catalogs and other horticultural publications these infraspecific taxa are often misnamed because the specific epithet is eliminated altogether. That is, *Prunus persica* var. *nucipersica* is incorrectly listed as *Prunus nucipersica,* indicating that it is a species when it is actually only a variety of the species *Prunus persica.*

The names of hybrid plants are indicated by the presence of a multiplication sign, ×. For example, *Prunus* × *effusus,* the Duke cherry, is a cross between *Prunus avium* and *Prunus cerasus.*

Even with these guidelines, circumstances often dictate that a plant's accepted name be changed. Under the terms of the *International Code,* the order of publication of a plant name determines which scientific name should correctly be applied to it. *Species Plantarum* has been adopted as the initial reference work for determining these names. The first scientific name applied to a plant, however, is the correct name even if a later name has become the more popular term of reference. In such cases, the plant name must be changed to conform to the earliest reference.

Liberty Hyde Bailey provides an excellent example of this rule in *How Plants Get Their Names* (New York: Macmillan, 1933, pp. 72–73). "A binomial long applied to a plant and appearing continuously in the literature is subject to displacement if an older adequately published scientifically documented name is found. Example is the common greenhouse heliotrope. This is always known in horticulture as *Heliotropium peruvianum,* so named by Linnaeus in the second edition of Species Plantarum, 1762. It turns out, however, that Linnaeus had founded a species *H. arborescens* as early as 1759 in the tenth edition of his Systema Naturae. The two plants are the same, and *Heliotropium arborescens* comes up and *H. peruvianum* goes down in synonymy."

A name change may also be in order when more recent research concludes that a plant is related to a different genus or indeed a different plant family. Various factors may then require a new name for a well-known plant.

Today it has become necessary to provide a system of unique names for cultivated varieties (cultivars) of plants as well. A cultivar is "a horticulturally or agriculturally derived variety of a plant, as distinguished from a natural variety [culti(vated) + var(iety)]" (*The American Heritage Dictionary,* Second College Edition). The procedures for selecting names for these plants are detailed in the *International Code of Nomenclature for Cultivated Plants,* and are administered through a system of registrars who check and record the names chosen for cultivars in particular plant groups. The names of cultivars are either enclosed in single quotes or are preceded by the abbreviation *cv.* (for example, *Prunus persica* 'Alboplena' or *P. persica* cv. Prairie Dawn). An infraspecific taxon may also have cultivated varieties, such as the nectarine *Prunus persica* var. *nucipersica* 'Early Flame'.

If a new form of a plant is developed, the name the originator wishes to apply to his or her creation must first be checked with the registration authority for that genus to assure there is no duplication and then be recorded by the authority for future reference. The following list gives the names and addresses of the registration authorities for many of the more popular plant groups.

For more information on nomenclature and the naming of plants see Liberty Hyde Bailey, *How Plants Get Their Names* (New York: Macmillan, 1933) and *Hortus Third* (New York: Macmillan, 1976); C. Chicheley Plowden, *A Manual of Plant Names* (New York: Philosophical Library, 1969); Allen J. Coombes, *Dictionary of Plant Names* (Portland: Timber Press, 1985);

and William T. Stearn, *Botanical Latin*, 3d ed. (North Pomfret, Vermont: David & Charles Inc., 1983). Also, the Florists' Publishing Company puts out a very handy, periodically updated guide to the pronunciation of botanical names, the *New Pronouncing Dictionary of Plant Names* (Florists' Publishing Co., 111 North Canal St., Chicago, IL 60606-7276).

10.1
Acacia
Geoff Butler
Australian National Botanic Gardens
Black Mountain, ACT 2601
Australia

10.2
Acer
Dr. Thomas M. Antonio
Chicago Botanic Garden
Glencoe, IL
Cross-reference: 16.138

10.3
Aloe
A.J. Bezuidenhout
South African Aloe Breeders Association
P.O. Box 59904
Karen Park, 0118
Pretoria
Republic of South Africa

10.4
Amaryllidaceae including Amaryllis, Brunsvigia, Clivia, ×
Crinodonna, Crinum, Hymenocallis, Lycoris, and Nerine;
excluding Hemerocallis and Narcissus
James M. Weinstock
The American Plant Life Society
10311 Independence
Chatworth, CA 91311
Cross-reference: 4.3

10.5
Amelanchier
Dr. Robert J. Hilton
Arboretum of the University of Guelph
61 Mary St.
Guelph, ON N1G 2A9
Canada
Cross-reference: 16.456

10.6
Araceae
John Banta
International Aroid Society
Rt. 2, Box 144
Alva, FL 33920
Cross-reference: 2.50

10.7
Australian plant genera excluding Leptospermum,
Rhododendron, and Orchidaceae
Geoff Butler
Australian Cultivar Registration Authority (ACRA)
Australian National Botanic Gardens
Black Mountain, ACT 2601
Australia

10.8
Begonia
Carrie Karegeannes
American Begonia Society
3916 Lake Blvd.
Annandale, VA 22003
Cross-reference: 2.3

10.9
Bougainvillea
Dr. Brijendra Singh
Division of Floriculture and Landscaping
Indian Agricultural Research Institute
New Delhi, 110012
India

10.10
Bromeliaceae
Joseph F. Carrone, Jr.
The Bromeliad Society, Inc.
305 North Woodlawn Ave.
Metairi, LA 70001
Cross-reference: 2.31

10.11
Buxus
Lynn R. Batdorf
American Boxwood Society
1409 Elm Grove Circle
Colesville, MD 20904
Cross-reference: 2.4

10.12
Callistephus
Gerda Nolting and Dr. K. Zimmer
Institut für Zierpflanzenbau
Herrenhauser Strasse 2
3000 Hannover-Herrenhausen
Federal Republic of Germany

10.13
Camellia
Tom Savage
International Camellia Society
Hawksview Rd.
Wirlinga 2640, N.S.W.
Australia

135

10.14

Carissa
Dr. Stephen A. Spongberg
Arnold Arboretum
Jamaica Plain, MA

Cross-reference: 16.173

10.15

Chaenomeles
Dr. Stephen A. Spongberg
Arnold Arboretum
Jamaica Plain, MA

Cross-reference: 16.173

10.16

Chrysanthemum, perennials only
H.B. Locke
National Chrysanthemum Society
2 Lucas House
Craven Rd., Rugby
Warwicks, CV21 3JQ
England

10.17

Clematis
Ir. W.A. Brandenburg
Government Institute for Research on Varieties of
   Cultivated Plants
P.O. Box 32
6700
AA Wageningen
Netherlands

10.18

Conifers, dwarf and other garden
John Lewis
Royal Horticultural Society
Orchard Cottage
Hurst, Martock
Somerset
England

10.19

Coprosma
L.J. Metcalf
Royal New Zealand Institute of Horticulture, Inc.
Parks and Recreation Dept.
P.O. Box 58
Invercargill
New Zealand

10.20

Cornus
Dr. Stephen A. Spongberg
Arnold Arboretum
Jamaica Plain, MA

Cross-reference: 16.173

10.21

Cotoneaster
Dr. Jindrich Chmelar and I.A. Nohel
Botanic Gardens and Arboretum
University of Agriculture
662 65
Brno
Czechoslovakia

10.22

Dahlia
David Pycraft
Royal Horticultural Society's Garden
Wisley
Woking
Surrey, GU23 6QB
England

10.23

Delphinium, perennials only
Dr. Alan C. Leslie
Royal Horticultural Society's Garden
Wisley
Woking
Surrey, GU23 6QB
England

10.24

Dianthus
Dr. Alan C. Leslie
Royal Horticultural Society's Garden
Wisley
Woking
Surrey, GU23 6QB
England

10.25

Escallonia
Dr. Elizabeth McClintock
1335 Union St.
San Francisco, CA 94109

10.26

Fagus
Dr. Stephen A. Spongberg
Arnold Arboretum
Jamaica Plain, MA

Cross-reference: 16.173

10.27

Forsythia
Dr. Stephen A. Spongberg
Arnold Arboretum
Jamaica Plain, MA

Cross-reference: 16.173

10.28

Fuchsia
Delight A. Logan
American Fuchsia Society
8710 S. Sheridan Ave.
Reedley, CA 93654

Cross-reference: 2.11

10.29
Gesneriaceae, excluding Saintpaulia
Jimmy D. Dates
American Gloxinia and Gesneriad Society, Inc.
R.R. 1, Box 206A
Galesburg, IL 61401
Cross-reference: 2.13

10.30
Gladiolus
Samuel N. Fisher
North American Gladiolus Council
11345 Moreno Ave.
Lakeside, CA 92040
Cross-reference: 2.70

10.31
Gleditsia
Dr. Stephen A. Spongberg
Arnold Arboretum
Jamaica Plain, MA
Cross-reference: 16.173

10.32
Hardy herbaceous perennials, apart from those genera for which
*   IRAs have been appointed*
Dr. Josef Sieber
Internationale Stauden-Union (ISU)
(International Hardy Plant Union)
Sichtungsgarten
Fachhochschule Weihenstephan
D-8050 Freising
Federal Republic of Germany

10.33
Heather, including Andromeda, Bruckenthalia, Calluna,
*   Daboecia, and Erica*
David McClintock
Heather Society of Great Britain
Bracken Hill
Sevenoaks
Kent, TN15 8JH
England

10.34
Hebe
L.J. Metcalf
Royal New Zealand Institute of Horticulture, Inc.
P.O. Box 58
Invercargill
New Zealand

10.35
Hedera
Dr. Sabina M. Sulgrove
American Ivy Society
5512 Woodbridge
Dayton, OH 45429
Cross-reference: 2.19

10.36
Hemerocallis
W.E. Monroe
American Hemerocallis Society
2244 Cloverdale Ave.
Baton Rouge, LA 70808
Cross-reference: 2.15

10.37
Hibiscus rosa-sinensis
Geoff Harvey
Australian Hibiscus Society
Buderim Garden Centre
P.O. Box 46
Buderim
Queensland, 4556
Australia

10.38
Hosta
Mervin C. Eisel
Minnesota Landscape Arboretum
Chanhassen, MN
Cross-reference: 16.208

10.39
Hyacinthus and other bulbous and tuberous-rooted plants,
*   excluding Tulipa, Dahlia, Lilium, and Narcissus*
Dr. Johan van Scheepen
Royal General Bulbgrowers' Association
P.O. Box 715
2180
Ad Hillegom
Netherlands

10.40
Hydrangea
Dr. Elizabeth McClintock
1335 Union St.
San Francisco, CA

10.41
Ilex
Gene K. Eisenbeiss
Holly Society of America
U.S. National Arboretum
3501 New York Ave., N.E.
Washington, DC 20002
Cross-reference: 2.48

10.42
Iris, excluding bulbous Iris
Kathleen Kay Nelson
American Iris Society
P.O. Box 37613
Omaha, NE 68137
Cross-reference: 2.18

10.43
Kalmia
Dr. Richard A. Jaynes
Broken Arrow Nursery
13 Broken Arrow Rd.
Hamden, CT 06518

10.44
Lagerstroemia
Dr. Donald R. Egolf
United States National Arboretum
Washington, DC

Cross-reference: 16.83

10.45
Lantana
Dr. Stephen A. Spongberg
Arnold Arboretum
Jamaica Plain, MA

Cross-reference: 16.173

10.46
Leptospermum
L.J. Metcalf
Royal New Zealand Institute of Horticulture, Inc.
Parks and Recreation Dept.
P.O. Box 58
Invercargill
New Zealand

10.47
Lilium
Dr. Alan C. Leslie
Royal Horticultural Society's Garden
Wisley
Woking
Surrey, GU23 6QB
England

10.48
Magnolia
Peter J. Del Tredici
American Magnolia Society
c/o Dana Greenhouses
Arnold Arboretum
Jamaica Plain, MA 02130

Cross-reference: 2.63

10.49
Malus, ornamental cultivars
Dr. Stephen A. Spongberg
Arnold Arboretum
Jamaica Plain, MA

Cross-reference: 16.173

10.50
Mango (Mangifera indica L.)
S.N. Pandey
Division of Fruits & Horticulture Technology
Indian Agricultural Research Institute
New Delhi, 110012
India

10.51
Narcissus
Sally Kington
Royal Horticultural Society
P.O. Box 313
Vincent Square
London, SW1P 2PE
England

10.52
Nelumbo and Nymphaea
Phillip R. Swindells
International Water Lily Society
Wycliffe Hall Botanical Gardens
Wycliffe Hall
Barnard Castle
Durham, DL12 9TS
England

10.53
Orchidaceae
John Greatwood
Royal Horticultural Society
13 Courtlands
Haywards Heath
Sussex, RH16 4JD
England

10.54
Other woody plant genera apart from those genera for which IRAs have been appointed
Dr. Donald G. Huttleston
Longwood Gardens
Kennett Square, PA

Cross-reference: 16.348

10.55
Paeonia
Greta Kessenich
American Peony Society
250 Interlachen Rd.
Hopkins, MN 55343

Cross-reference: 2.22

10.56
Pelargonium
J.D. Llewellyn
Australian Geranium Society
'Nyndee' 56 Torokina Ave.
St. Ives, N.S.W., 2075
Australia

10.57
Penstemon
Mark McDonough
American Penstemon Society
4725–119th Ave., S.E.
Bellevue, WA 98006

Cross-reference: 2.21

10.58
Petunia
Gerda Nolting and Dr. K. Zimmer
Institut für Zierpflanzenbau
Herrenhauser Strasse 2
3000 Hannover-Herrenhausen
Federal Republic of Germany

10.59
Philadelphus
Dr. Stephen A. Spongberg
Arnold Arboretum
Jamaica Plain, MA
Cross-reference: 16.173

10.60
Phormium
L.J. Metcalf
Royal New Zealand Institute of Horticulture
Parks and Recreation Dept.
P.O. Box 58
Invercargill
New Zealand

10.61
Pieris
Dr. Stephen A. Spongberg
Arnold Arboretum
Jamaica Plain, MA
Cross-reference: 16.173

10.62
Pittosporum
L.J. Metcalf
Royal New Zealand Institute of Horticulture, Inc.
P.O. Box 58
Invercargill
New Zealand

10.63
Plumeria
John P. Oliver
Plumeria Society of America, Inc.
P.O. Box 22791
Houston, TX 77227-2791
Cross-reference: 2.77

10.64
Populus, forestry cultivars
International Poplar Commission
Via della Terme di Caracalla
00100
Rome
Italy

10.65
*Proteaceae, South African genera only, including Aulax,
Leucadendron, Leucospermum, Mimetes, Orothamnus,
Paranomus, Protea, and Serruria*

E.P. Jordan
Directorate of Plant & Seed Control
Dept. of Agriculture
Private Bag X179
Pretoria 0001
Republic of South Africa

10.66
Pyracantha
Dr. Donald R. Egolf
United States National Arboretum
Washington, DC
Cross-reference: 16.83

10.67
Rhododendron, including Azalea
Dr. Alan C. Leslie
Royal Horticultural Society's Garden
Wisley
Woking
Surrey, GU23 6QB
England

10.68
Rosa
Harold S. Goldstein
American Rose Society
Shreveport, LA
Cross-reference: 2.26

10.69
Saintpaulia
Janet L. Nichols
African Violet Society of America
9 Clover Hill Rd.
Poughkeepsie, NY 12603
Cross-reference: 2.1

10.70
Sempervivum, including Jovibarba and Rosularia
Peter J. Mitchell
The Sempervivum Society
11 Wingle Tye Rd.
Burgess Hill
West Sussex, RH15 9HR
England

10.71
Syringa
Freek Vrugtman
Royal Botanical Gardens
Hamilton, ON
Cross-reference: 16.473

10.72
Tagetes
Gerda Nolting and Dr. K. Zimmer
Institut für Zierpflanzenbau
Herrenhauser Strasse 2
3000 Hannover-Herrenhausen
Federal Republic of Germany

10.73
Tulipa
Dr. Johan van Scheepen
Royal General Bulbgrowers' Association
P.O. Box 175
2180 Ad Hillegom
Netherlands

10.74
Ulmus
Dr. Stephen A. Spongberg
Arnold Arboretum
Jamaica Plain, MA
Cross-reference: 16.173

10.75
Viburnum
Dr. Donald R. Egolf
U.S. National Arboretum
Washington, DC
Cross-reference: 16.83

10.76
Weigela
Dr. Stephen A. Spongberg
Arnold Arboretum
Jamaica Plain, MA
Cross-reference: 16.173

11

United States and Canadian Governmental Programs

U.S. DEPARTMENT OF AGRICULTURE PROGRAMS

The U.S. Department of Agriculture maintains federal and regional research centers and plant introduction stations throughout the country. Listed below are various horticultural research locations and facilities and other agencies that might be of interest to horticulturists, gardeners, and others interested in plants.

11.1
Office of Information
U.S. Department of Agriculture
Washington, DC 20250
202-447-8005

Answers public inquiries regarding the Dept. of
  Agriculture.

COOPERATIVE EXTENSION SERVICE

The Cooperative Extension Service is the most extensive and readily available source of information on horticultural subjects in the United States. The service is a joint effort of the county government, the state land-grant university, and the U.S. Department of Agriculture. Its general mandate is to link research, science, and technology to the needs of people, where they live and work, by providing practical education and information. The C.E.S. communicates results of experiment station research, and provides a variety of services such as soil testing and analysis, diagnosis of plant problems, and plant identification.

Local extension agents may be found in most counties and many major municipalities of each state. Their telephone numbers and addresses are usually found under the county government listing in the phone book as Agricultural Agent or Extension Service. The horticultural knowledge of each agent varies, often depending on the importance of commercial horticulture in the county. All local agents are supported by the resources of the land-grant university

and the U.S.D.A., including fact sheets, circulars, and bulletins dealing with frequently asked questions. Many Extension offices also publish a newsletter that addresses topical horticultural questions. A list of all titles available may be obtained by writing the state contacts listed below.

At the state level, the Extension Service usually has a staff of several horticulturists with different specialties, including enology (the study of wines), fruits, floriculture, forestry, greenhouse horticulture, home horticulture, integrated pest management, landscaping, the nursery trade, ornamentals, soils, turf, vegetables, water, and weeds.

AK
11.2
Wayne G. Vandre
Extension Horticulturist
2221 E. Northern Lights Blvd.
240
University of Alaska
Anchorage, AK 99508
907-279-6575

Cross-references: 12.481, 13.56, 19.1

AL
11.3
Dr. Ronald L. Shumack
Extension Horticulture Dept.
Auburn University
Auburn, AL 36849
205-826-4985

AR
11.4
Kenneth R. Scott
Extension Horticulturist
P.O. Box 391
Little Rock, AR 72203
501-376-6301

Cross-reference: 12.483

11.5
Dr. Gerald Klingaman
Extension Horticulturist
316 Plant Science Bldg.
University of Arkansas
Fayetteville, AR 72701
501-575-2603

AZ
11.6
Dr. Michael Kilby
Extension Horticulturist
University of Arizona
Tucson, AZ 85721
602-621-1400

CA
11.7
William B. Davis
Extension Environmental Horticulturist
University of California
Davis, CA 95616
916-752-0412

11.8
Dr. Tokuji Furuta
Extension Environmental Horticulturist
4114 Batchelor Hall
University of California
Riverside, CA 92521
714-787-3318

Cross-reference: 12.485

CO
11.9
Dr. Kenneth W. Knutson
Extension Associate Professor
Dept. of Horticulture
Fort Collins, CO 80523
303-491-7068

CT
11.10
Edmond L. Marotte
Consumer Horticulturist
Dept. of Plant Science
University of Connecticut
Storrs, CT 06268
203-486-3435

Cross-reference: 12.487

DC
11.11
Pamela Marshall
Extension Horticulturist
1351 Nicholson St., N.W.
Washington, DC 20011
202-282-7410

Cross-reference: 12.488

DE
11.12
Susan Barton
Extension Horticulturist
Townsend Hall
University of Delaware
Newark, DE 19711
302-451-2532

Cross-reference: 12.489

FL
11.13
Dr. Robert J. Black
Extension Urban Horticulture Specialist
Ornamental Horticulture Dept.
University of Florida
Gainesville, FL 32611
904-392-1835

Cross-reference: 12.490

GA
11.14
Dr. A. Jefferson Lewis
Extension Horticulturist
University of Georgia
Athens, GA 30602
404-542-2340

Cross-reference: 12.491

GU
11.15
Dr. Claron D. Bjork
College of Agriculture
University of Guam
UOG Station
Mangilao, GU 96923
617-734-2575

HI
11.16
Dr. Harry C. Bittenbender
Extension Horticulturist
University of Hawaii
Honolulu, HI 96822
808-948-6043

Cross-reference: 12.492

IA
11.17
Dr. Michael L. Agnew
Extension Horticulturist
Dept. of Horticulture
Iowa State University
Ames, IA 50011
515-294-0027

Cross-reference: 12.493

ID
11.18
Larry O'Keeffe
Head
Dept. of Plant, Soil & Entomological Sciences
University of Idaho
Moscow, ID 83843-4196
208-885-6277

IL
11.19
James Schmidt
Extension Specialist
104 Ornamental Horticulture Bldg.
1107 W. Dorner Dr.
University of Illinois
Urbana, IL 61801
217-333-2125

Cross-reference: 12.495

IN
11.20
B. R. Lerner
Extension Horticulturist
Dept. of Horticulture
Purdue University
West Lafayette, IN 47907
317-494-1311

Cross-reference: 12.496

KS
11.21
Dr. Frank D. Morrison
Extension Program Leader
Waters Hall 227
Kansas State University
Manhattan, KS 66506
913-532-6173

Cross-reference: 12.497

KY
11.22
Dr. John Strang
Extension Horticulturist
Agriculture Science Center North
Room N-318
University of Kentucky
Lexington, KY 40546
606-257-5685

Cross-reference: 12.498

LA
11.23
Dr. Thomas E. Pope
Horticulture Specialist
Room 155
J. C. Miller Horticulture Bldg.
Louisiana State University
Baton Rouge, LA 70803
504-388-2222

Cross-reference: 12.499

MA
11.24
Kathleen M. Carool
Extension Specialist, Horticulture
French Hall
University of Massachusetts
Amherst, MA 01003
413-545-0895

MD
11.25
Dr. Francis R. Gouin
Dept. of Horticulture
University of Maryland
College Park, MD 20742
301-454-3143

ME
11.26
Dr. Lois Berg Stack
Ornamental Horticulture Specialist
University of Maine
Orono, ME 04469
207-581-2949

MI
11.27
Dr. Jerome Hull, Jr.
Professor
Dept. of Horticulture
Michigan State University
East Lansing, MI 48824
517-355-5194

Cross-reference: 12.503

MN
11.28
Deborah L. Brown
Extension Horticulturist
1970 Folwell Ave.
University of Minnesota
St. Paul, MN 55108
612-624-7419

MO
11.29
Dr. Gary G. Long
Extension Ornamental Plant Specialist
University of Missouri
Columbia, MO 65211
314-882-9625

Cross-references: 12.505, 21.134

MS

11.30
Dr. Richard H. Mullenax
Leader, Extension Horticulturist
Dept. of Horticulture
P.O. Box 5446
Mississippi, MS 39762
601-325-3935

Cross-reference: 12.506

MT

11.31
Dr. James W. Bauder
Extension Tillage Specialist
Plant and Soil Science Dept.
Montana State University
Bozeman, MT 59717-0002
406-994-4605

NC

11.32
Dr. Joseph W. Love
Extension Horticulturist
Dept. of Horticultural Science
North Carolina State University
Raleigh, NC 27695-7609
919-737-3322

Cross-reference: 12.508

ND

11.33
Dr. Ronald C. Smith
Extension Horticulturist
North Dakota State University
Fargo, ND 58105
701-237-8161

Cross-reference: 12.509

NE

11.34
Dr. Donald H. Steinegger
Extension Horticulturist
377 Plant Science Bldg.
University of Nebraska
Lincoln, NE 68503
402-472-2550

Cross-reference: 12.510

NH

11.35
Dr. Charles H. Williams
Extension Ornamentals Specialist
Plant Science Dept.
University of New Hampshire
Durham, NH 03824-3597
603-862-3207

NJ

11.36
Lawrence D. Little, Jr.
Extension Horticulturist
Rutgers—The State University
Blake Hall
P.O. Box 231
New Brunswick, NJ 08903
201-932-9559

Cross-reference: 12.512

NM

11.37
George Dickerson
Extension Horticulturist
New Mexico State University
9301 Indian School Rd., N.E.
Suite 101
Albuquerque, NM 87112
505-292-0097

Cross-reference: 12.513

NV

11.38
William J. Carlos
Extension Agent, Horticulture
1001 E. 9th St.
P.O. Box 1130
Reno, NV 89520
702-328-2650

NY

11.39
Robert E. Kozlowski
Senior Extension Associate
Homes and Grounds
Cornell University
17 Plant Science Bldg.
Ithaca, NY 14853
607-255-1791

Cross-reference: 12.515

OH

11.40
Barbara J. Williams
Extension Horticulturist
Ohio State University
2001 Fyffe Ct.
Columbus, OH 43210
614-292-3852

OK

11.41
Paul J. Mitchell
Extension Ornamental Horticulture
Oklahoma State University
Stillwater, OK 74078
405-624-6593

OR
11.42
Dr. James L. Green
Extension Ornamentals Specialist
Dept. of Horticulture
Oregon State University
Corvallis, OR 97331
503-754-3464

Cross-reference: 21.127

PA
11.43
Dr. J. Robert Nuss
Extension Horticulturist
Dept. of Horticulture
102 Tyson Bldg.
University Park, PA 16802
814-863-2196

Cross-reference: 12.519

RI
11.44
Kathy Mallon Extension Specialist
Dept. of Plant Science
University of Rhode Island
Kingston, RI 02881
401-792-5999

Cross-reference: 12.520

SC
11.45
Dr. Alta R. Kingman
Extension Horticulture
173 P&AS Bldg.
Clemson University
Clemson, SC 29634-0375
803-656-4962

Cross-reference: 12.521

SD
11.46
Dean M. Martin
Extension Horticulturist
South Dakota State University
P.O. Box 2207—C
Brookings, SD 57007
605-688-5136

Cross-reference: 12.522

TN
11.47
Dr. Elmer L. Ashburn
Leader, Plant and Soil Science
University of Tennessee
P.O. Box 1071
Knoxville, TN 37901-1071
615-974-7208

Cross-reference: 12.523

TX
11.48
Dr. William C. Welch
Extension Horticulturist
225 Horticulture/Forestry Bldg.
Texas A & M University
College Station, TX 77843
409-845-7341

Cross-reference: 12.524

UT
11.49
Dr. Gerald R. Olson
Associate Vice-President of Extension and Continued
  Education
UMC 4900
Utah State University
Logan, UT 84322
801-750-2194

VA
11.50
Dr. Paul L. Smeal
Extension Horticulturist
Virginia Polytechnical Institute & State University
Blacksburg, VA 24061
703-961-5609

Cross-references: 12.526, 21.136

VT
11.51
Dr. Leonard L. Perry
Extension Horticulturist
Hills Bldg.
University of Vermont
Burlington, VT 05405
802-656-2630

Cross-reference: 12.527

WA
11.52
Dr. Robert E. Thornton
Extension Horticulturist
Washington State University
Pullman, WA 99164-6414
509-335-2811

WI
11.53
Dr. R. C. Newman
Extension Horticulturist
University of Wisconsin
Madison, WI 53706
608-262-1624

Cross-reference: 12.529

WV
11.54
Dr. Richard K. Zimmerman
Extension Specialist, Horticulture
2088 Agricultural Sciences Bldg.
P.O. Box 6108
West Virginia University
Morgantown, WV 26506-6108
304-293-4801

Cross-reference: 12.530

WY
11.55
Dr. James A. Cook
Extension Horticulturist
University of Wyoming
P.O. Box 3354
University Station
Laramie, WY 82071
307-766-2243

AGRICULTURAL RESEARCH SERVICE
    The A.R.S. is the research arm of the Department of Agriculture. It conducts research on the production of plants, the protection of plants from diseases, and the use and improvement of soil, water, and air resources.

11.56
Information Staff
Agricultural Research Service
B-005, BARC-W
Beltsville, MD 20705
301-344-2264

11.57
U.S. National Arboretum
3501 New York Ave., N.E.
Washington, DC 20002
202-475-4815

Cross-reference: 16.83

The U.S. National Arboretum is the U.S.D.A.'s research
    center for woody ornamental plants. Research is
    directed toward developing better cultivars for home
    use.

AREA DIRECTOR'S OFFICES

11.58
USDA Agricultural Research Service
Beltsville Agricultural Research Center
Room 209, Bldg. 003
BARC-West
Beltsville, MD 20705
301-344-3570

USDA's research for herbaceous ornamentals is carried on
    here.

11.59
USDA Agricultural Research Service
Mid-South Area
P.O. Box 225
Stoneville Rd.
Stoneville, MS 38776
601-686-2311

11.60
USDA Agricultural Research Service
Midwest Area
1815 N. University St.
Peoria, IL 61604
309-685-4011

11.61
USDA Agricultural Research Service
Mountain States Area
2625 Redwing Rd., Suite 350
Fort Collins, CO 80526
303-223-3459

11.62
USDA Agricultural Research Service
North Atlantic Area
600 East Mermaid Ln.
Philadelphia, PA 19118
215-489-6593

11.63
USDA Agricultural Research Service
Northwest Area
809 N.E. Sixth Ave.
Room 204
Portland, OR 97232
503-231-2246

11.64
USDA Agricultural Research Service
Pacific Basin Area
800 Buchanan St.
Albany, CA 94710
415-486-3774

11.65
USDA Agricultural Research Service
South Atlantic Area
P.O. Box 5677
College Station Rd.
Athens, GA 30613
404-250-3311

11.66
USDA Agricultural Research Service
Southern Plains Area
1812 Welsh St., Suite 130
College Station, TX 77840
409-527-1346

AGRICULTURAL EXPERIMENT STATIONS AND
BRANCH UNITS
Agricultural Experiment Stations were created by
the U.S. Department of Agriculture to do scientific
research in the broad field of agriculture, including
the study of plant growth, developing new cultivars,
and disease, insect, and weed control. The results of
this research provide the basis for the fact sheets,
circulars, and bulletins issued by the Cooperative Extension Service of the U.S.D.A.

AK
11.67
Alaska Agricultural and Forestry Experiment Station
School of Agriculture and Land Resources
   Management/University of Alaska
103 Arctic Health Research Bldg.
Fairbanks, AK 99775-0100
907-474-7083; (FAX) 907-474-7439
Cross-reference: 20.63

11.68
Agricultural Research Center
AFES, University of Alaska
533 E. Fireweed
Palmer, AK 99645
907-745-3257

Director: S. H. Restad
Cross-reference: 20.64

AL
11.69
Alabama Agricultural Experiment Station
Auburn University
Auburn University, AL 36849
205-844-2237; (FAX) 205-844-5511

Director: Lowell T. Frobish

11.70
Tennessee Valley Substation
Belle Minna, AL 35615

11.71
Piedmont Substation
Camp Hill, AL 36850

AR
11.72
Arkansas Agricultural Experiment Station
University of Arkansas
Fayetteville, AR 72701
501-575-4446; (FAX) 501-575-7273

Director: Gerald J. Musick

11.73
Fruit Substation
Substation of Arkansas Agricultural Experiment Station
Clarksville, AR 72830
501-754-2406

11.74
Peach Substation
Substation of Arkansas Agricultural Experiment Station
Alma, AR 72921
501-845-1764

11.75
Strawberry Substation
Substation of Arkansas Experiment Station
Baldknob, AR 72010
501-724-3368

11.76
Vegetable Substation
Substation of Arkansas Agricultural Experiment Station
Alma, AR 72921
501-474-0475

AZ
11.77
Arizona Agricultural Experiment Station
University of Arizona
Tucson, AZ 85721
602-621-3859; (FAX) 602-621-7196

Director: C. Colin Kaltenbach

11.78
Maricopa Agricultural Center
Maricopa, AZ 85329
602-437-1366

11.79
Mesa Agricultural Center
State Institution Mesa Branch Section
Mesa, AZ 85201
602-964-1725

11.80
Stafford Agricultural Center
Stafford, AZ 85546
602-428-2432

11.81
Yuma Valley and Yuma Mesa Branch Station
Yuma, AZ 85364
602-782-3836

CA
11.82
California Agricultural and Natural Resources Programs
101 Giannini Hall
University of California
Berkeley, CA 94720
415-642-7171; (FAX) 415-642-6412

Director : Wilford R. Gardener

11.83
California Agricultural and Natural Resources Programs
102 Cooperative Extension Headqrts.
University of California
Davis, CA 95616
916-752-8181; (FAX) 916-752-3757

Director: Kenneth W. Ellis

11.84
California Agricultural and Natural Resources Programs
Kearney Agricultural Center
9240 South Riverbend Ave.
Parlier, CA 93648
209-891-2566; (FAX) 209-891-2593

Director: William R. Hambleton

11.85
California Agricultural and Natural Resources Programs
302 Cooperative Extension Building
University of California
Riverside, CA 92521
714-787-3321; (FAX) 714-787-4675

Director: Allyn D. Smith

11.86
California Citrus Research Center
College of Natural and Agricultural Sciences
University of California
Riverside, CA 92521
714-787-3101; (FAX) 714-787-4190

Director: Seymour D. Van Gundy

11.87
California College of Agricultural and Environmental
  Sciences
University of California
Davis, CA 95615
916-752-1605; (FAX) 916-752-4789

Director: Robert Webster

11.88
California Division of Agriculture and Natural Resources
Agricultural Experiment Station
University of California
300 Lakeside Dr., 6th Floor
Oakland, CA 94612-3560
415-987-0060; (FAX) 415-987-0672

Director: Kenneth R. Farrell

11.89
Colorado Agricultural Experiment Station
Colorado State University
Fort Collins, CO 80523
303-491-5371

Director: Robert D. Heil

11.90
Hotchkiss, Rogers Mesa Research Center
Hotchkiss, CO 81419
303-872-3387

11.91
Orchard Mesa Research Center
Grand Junction, CO 81501
303-434-3264

11.92
San Luis Valley Research Center
Center, CO 81125
303-754-3594

CT
11.93
Connecticut Agricultural Experiment Station
P.O. Box 1106
New Haven, CT 06504
203-789-7214

Director: John F. Anderson
Cross-references: 16.78, 21.120

11.94
Connecticut Agricultural Experiment Station
University of Connecticut
Storrs, CT 06268
203-486-2917; (FAX) 203-486-4128

Director: Kirvin L. Knox

DC
11.95
District of Columbia Agricultural Experiment Station
University of the District of Columbia
4200 Connecticut Ave., N.W.
Washington, DC 20008
202-282-7322

Director: J. R. Allen

DE
11.96
Delaware Agricultural Experiment Station
University of Delaware
Newark, DE 19717-1303
302-451-2501; (FAX) 302-451-1038

Director: Donald F. Crossan

11.97
University of Delaware
College of Agricultural Sciences and Delaware
  Agricultural Experiment Station, Substation
Georgetown, DE 19947
302-856-5254

FL
11.98
Florida Institute of Food and Agricultural Sciences
Agricultural Experiment Station
1022 McCarty Hall
University of Florida
Gainesville, FL 32611
904-392-1971; (FAX) 904-392-6932

Director: Gerald Zachariah

11.99
Agricultural Research Center
Rt. 3, Box 580
Apopka, FL 32703
305-889-4161

11.100
Agricultural Research Center
3205 S.W., 70th Ave.
Fort Lauderdale, FL 33314
305-475-8990

11.101
Agricultural Research Center
P.O. Box 248
Fort Pierce, FL 33454
305-461-4371

11.102
Agricultural Research Center
P.O. Box 728
Hastings, FL 32045
904-692-1792

11.103
Agricultural Research Center
P.O. Box 388
Leesburg, FL 32748
904-787-3423

11.104
Agricultural Research Center
Rt. 4, Box 63
Monticello, FL 32344
904-997-2596

11.105
Central Florida Research and Education Center
P.O. Box 909
Sanford, FL 32911
305-322-4134

11.106
Citrus Research and Education Center
700 Experiment Station Rd.
Lake Alfred, FL 33850
813-956-1151

11.107
Everglades Research and Education Center
P.O. Drawer A
Belle Glade, FL 33430
305-996-3062

11.108
Gulf Coast Research and Education Center
5007 60th St. East
Bradenton, FL 34203
813-755-1568

Cross-reference: 2.70

11.109
North Florida Research and Education Center
Rt. 3, Box 638
Quincy, FL 32351
904-627-9236

11.110
Tropical Research and Education Center
18905 S.W., 280th St.
Homestead, FL 33031
305-247-4624

GA
11.111
Georgia Agricultural Experiment Station
107 Conner Hall
University of Georgia
Athens, GA 30602
404-542-2151

Director: Clive W. Donoho, Jr.

11.112
Georgia Agricultural Experiment Station
Georgia Station
Griffin, GA 30223-1797
404-228-7263; (FAX) 404-228-7270

Director: Gerald F. Arkin

11.113
Georgia Agricultural Experiment Station
Coastal Plain Section
P.O. Box 748
Tifton, GA 31793
912-386-3338

Director: Gale A. Buchanan

GU
11.114
Guam Agricultural Experiment Station
College of Agriculture and Life Sciences
University of Guam/UOG Station
Mangilao, GU 96913
671-734-2579

Director: Jose T. Barcians

HI
11.115
Hawaii Agricultural Experiment Station
College of Tropical Agriculture and Human Resources
University of Hawaii at Manoa
3050 Maile Way, Room 202
Honolulu, HI 96822
808-948-8234; (FAX) 808-948-6442

Director: Noel P. Kefford

11.116
Hilo Substation
461 W. Lanikaula St.
Hilo, HI 96720
808-935-2885

11.117
Kona Substation
P.O. Box 208
Kealakekua, HI 96750
808-322-2718

11.118
Kula Substation
P.O. Box 269
Kula, Maui, HI 96790
808-878-1213

Cross-reference: 20.105

11.119
Poamoho Substation
Poamoho, Oahu, HI 96791
808-948-7138

11.120
Wailua Substation
P.O. Box 278-A
Wailua, Kauai, HI 96746
808-822-4984

11.121
Waimanalo Substation
Waimanalo, Oahu, HI 96795
808-948-7138

IA
11.122
Iowa Agriculture and Home Economics Experiment
    Station
Iowa State University
Ames, IA 50011
515-294-2518

Director: David G. Topel

11.123
Muscatine Station
Muscatine, IA 52761
319-267-8787

11.124
Whiting Station
Whiting, IA 51063
712-458-2660

ID
11.125
Idaho Agricultural Experiment Station
University of Idaho
Moscow, ID 83843
208-885-7173; (FAX) 208-885-6654

Director: Gary A. Lee

11.126
Aberdeen Branch Experiment Station
Aberdeen, ID 83210
208-397-4181

11.127
Research and Extension Center
Kimberly, ID 83341
208-743-3600

11.128
Research and Extension Center
Sandpoint, ID 83864

11.129
Southwest Idaho Research and Extension Center
Parma, ID 83660
208-722-5186

IL
11.130
Illinois Agricultural Experiment Station
211 Mumford Hall
University of Illinois
1301 West Gregory Dr.
Urbana, IL 61801
217-333-0240; (FAX) 217-333-1952

Director: Donald A. Holt

11.131
Dixon Springs Agricultural Center
Simpson, IL 62985

11.132
Kankakee River Valley Sand Field
St. Anne, IL 60964

11.133
St. Charles Horticulture Research Center
St. Charles, IL 60174

IN
11.134
Indiana Agricultural Experiment Station
116 AGAD
Purdue University
West Lafayette, IN 47907
317-494-8362; (FAX) 317-494-0808

Director: Billy R. Baumgardt

11.135
Southwestern Indiana Agricultural Research Center State
    Institution
Vincennes, IN 47907

KS
11.136
Kansas Agricultural Experiment Station
113 Waters Hall
Kansas State University
Manhattan, KS 66506
913-532-6147; (FAX) 913-532-6563

Director: Walter R. Woods

11.137
Colby Experiment Station
Colby, KS 67701
913-462-7575

11.138
Garden City Experiment Station
(KSU Branch Station-General Hort.)
Garden City, KS 67846
316-276-8286

11.139
Horticulture Research Center
(Horticulture Dept.)
Wichita, KS 67233
316-788-0492

11.140
Tribune Experiment Station
(KSU Branch Station-General Hort.)
Tribune, KS 67879
316-376-4761

KY
11.141
Kentucky Agricultural Experiment Station
University of Kentucky
Lexington, KY 40506-0091
606-257-4772; (FAX) 606-258-5842

Director: C. Oran Little

11.142
Eden Shale Experiment Farm Substation
Kentucky Agricultural Experiment Station
Owenton, KY 40359
502-484-5531

11.143
Kentucky Research and Education Center Substation of
    Kentucky Agriculture Experiment Station
P.O. Box 469
Princeton, KY 42445
502-365-7541

11.144
Robinson Substation
Kentucky Agricultural Experiment Station
Quicksand, KY 41363
606-666-2438

LA
11.145
Louisiana Agricultural Experiment Station
Louisiana State University and A & M College
Drawer E, University Station
Baton Rouge, LA 70893-0905
504-388-4181; (FAX) 504-388-4163

Director: Kenneth W. Tipton

11.146
Calhoun Research Station
P.O. Box 10
Calhoun, LA 71225
318-644-2662

11.147
Citrus Research Station
Rt. 1, Box 628
Port Sulphur, LA 71301
504-564-2467

11.148
Hammond Research Station
5925 Old Corington Hwy.
Hammond, LA 70401
504-345-4110

11.149
Idlewild Research Station
Drawer 985
Clinton, LA 70722
504-683-5848

11.150
Pecan Research-Extension Station
P.O. Box 5519
Shreveport, LA 71135
318-797-8034

11.151
Sweet Potato Research Station
P.O. Box 120
Chase, LA 71324
318-435-4584

MA
11.152
Massachusetts Agricultural Experiment Station
University of Massachusetts
Amherst, MA 01003
413-545-2771; (FAX) 413-545-1242

Director: E. Bruce MacDougall

11.153
University of Massachusetts
Agricultural Experiment Station
Cranberry Station
East Wareham, MA 02538
617-295-2212

11.154
University of Massachusetts
Agricultural Experiment Station
Suburban Experiment Station
Waltham, MA 02154
617-891-0650

MD
11.155
Maryland Agricultural Experiment Station
University of Maryland
College Park, MD 20742
301-454-3707

Director: Robert A. Kennedy

11.156
Vegetable Research Farm
Maryland Agricultural Experiment Station
Rt. 5, Quantico Rd.
Salisbury, MD 21801
301-742-8788

11.157
Western Maryland Research and Education Center
Maryland Agricultural Experiment Station
Rt. 1, Box 49B
Keedysville, MD 21756
301-791-2298

ME
11.158
Maine Agricultural Experiment Station
University of Maine
Orono, ME 04469
207-581-3202; (FAX) 207-581-2725

Director: Wallace C. Dunham

MI
11.159
Michigan Agricultural Experiment Station
Michigan State University
East Lansing, MI 48824-1039
517-355-0123; (FAX) 517-353-5406

Director: Robert G. Gast

11.160
Clarksville Horticultural Experiment Station
Clarksville, MI 48815

11.161
Graham Horticultural Experiment Station
Lake Michigan District
N.W. Grand Rapids, MI 49504

11.162
Northwest Michigan Horticultural Research Station
Traverse City, MI 49684

11.163
Sodus Experiment Station
Sodus, MI 49126

11.164
Trevor Nichol Experimental Farm
Fennville, MI 49408

MN
11.165
Minnesota Agricultural Experiment Station
University of Minnesota
220 Coffey Hall
1420 Eckles Ave.
St. Paul, MN 55208
612-625-4211; (FAX) 612-625-0286

Director: C. Eugene Allen

11.166
North Central Experiment Station
Grand Rapids, MN 55744
218-327-1790

11.167
North West Experiment Station
Crookston, MN 56716
218-281-6510

11.168
Sand Plains Experimental Field
Becker, MN 55330
612-261-4063

11.169
South West Experiment Station
Lamberton, MN 56152
507-752-7372

11.170
Southern Experiment Station
Waseca, MN 56093
507-835-3620

11.171
West Central Experiment Station
Morris, MN 56267
612-589-1711

MO
11.172
Missouri Agricultural Experiment Station
University of Missouri
Columbia, MO 65211
314-882-3846; (FAX) 314-882-5127

Director: Roger L. Mitchell

11.173
Missouri Agricultural Experiment Station
Field Station
Portageville, MO 63873
314-379-5431

11.174
New Franklin Horticultural Research Station
New Franklin, MO 65274
816-848-2268

11.175
Powell Horticultural and Natural Resources Center
Rt. 1, Box 90
Kingsville, MO 64061
816-566-2600

MS
11.176
Mississippi Agricultural and Forestry Experiment Station
Mississippi State University
P.O. Drawer ES
Mississippi State, MS 39762
601-325-3005; (FAX) 601-325-1215

Director: Verner G. Hurt

11.177
Alcorn Branch Experiment Station
Alcorn State University
Lorman, MS 39096

11.178
Pontotoc Ridge-Flatwood
Pontotoc, MS 38863

11.179
South Mississippi Branch
Poplarville, MS 39470

11.180
Truck Crops Branch Station
Crystal Springs, MS 39059

MT
11.181
Montana Agricultural Experiment Station
Montana State University
Bozeman, MT 59717-0002
406-994-3681; (FAX) 406-994-6579

Director: James R. Welsh

11.182
Central Montana Branch Station
Moccasin, MT 59462
406-423-5227

11.183
Eastern Montana Branch Station
Sidney, MT 59270
406-482-2208

11.184
Northern Montana Branch Station
Havre, MT 59501
406-265-6115

11.185
Northwestern Montana Branch Station
Kalispell, MT 59901
406-755-4304

11.186
Southern Montana Branch Station
Huntley, MT 59037
406-348-3400

Western Montana Branch Station
Corvallis, MT 59828
406-961-3025

NC
11.187
North Carolina Agricultural Research Service
North Carolina State University
Box 7643
Raleigh, NC 27695-7643
919-737-2718; (FAX) 919-737-3928

Director: Ronald J. Kuhr

11.188
Central Crops Research Station
(NCSU)
Clayton, NC 27520
919-553-6468

11.189
Horticultural Crops Research Station (NCSU)
Castle Hayne, NC 28429
919-675-2314

11.190
Horticultural Crops Research Station (NCDA)
Clinton, NC 28320
919-592-7939

11.191
Mountain Horticultural Crops Research Station (NCSU)
Fletcher, NC 28732
704-684-7197

11.192
Mountain Research Station (NCDA)
Waynesville, NC 28786
704-456-3943

11.193
Oxford Tobacco Research Station
(NCDA)
Oxford, NC 27565
919-693-2483

11.194
Sandhills Research Station (NCSU)
Jackson Springs, NC 27281
919-974-4673

11.195
Upper Piedmont Research Station
(NCSU)
Reidsville, NC 27320
919-349-8347

ND
11.196
North Dakota Agricultural Experiment Station
North Dakota State University
State University Station
Box 5435
Fargo, ND 58105
701-237-7654; (FAX) 701-237-8520

Director: H. Roald Lund

11.197
Carrington Experiment Irrigation Station
Carrington, ND 58421

11.198
Dickinson Experiment Station
Dickinson, ND 58601

11.199
Langdon Experiment Station
Langdon, ND 58249

11.200
Minot Experiment Station
Minot, ND 58701

11.201
Williston Experiment Station
Williston, ND 58801

NE
11.202
Nebraska Agricultural Experiment Station
University of Nebraska
109 Ag. Hall
Lincoln, NE 68583-0704
402-472-2045; (FAX) 402-472-2759

Director: Darrell W. Nelson

11.203
Panhandle Research and Extension Center
Scottsbluff, NE 69361
308-632-1230

11.204
Research and Development Center
Mead, NE 68045
420-624-5935

11.205
West Central Research and Extension Center
North Platte, NE 69101
308-531-3611

NH
11.206
New Hampshire Agricultural Experiment Station
University of New Hampshire
Durham, NH 03824
603-862-1450; (FAX) 603-862-2030

Director: Thomas P. Fairchild

NJ
11.207
New Jersey Agricultural Experiment Station
Rutgers University
P.O. Box 231
New Brunswick, NJ 08903
201-932-9447; (FAX) 201-932-6769

Director: S. Kleinschuster

11.208
Cranberry and Blueberry Research Center
New Lisbon, NJ 08019
609-894-8740

11.209
Rutgers Fruit Research and Development Center
Cream Ridge, NJ 08514
201-932-9711

11.210
Rutgers Research and Development Center
Bridgeton, NJ 08302
609-455-3100

11.211
Soils and Crops Research Center
Adelphia, NJ 07728
201-492-9120

NM
11.212
New Mexico Agricultural Experiment Station
New Mexico State University
P.O. Box 3BF
Las Cruces, NM 88003
505-646-3125; (FAX) 505-646-5975

Director: David W. Smith

11.213
Agricultural Science Center
Alcade, NM 87511
505-852-4241

11.214
Agricultural Science Center
Farmington, NM 87401
505-327-7757

11.215
Agricultural Science Center
Rt. 1, Box 28
Los Lunas, NM 87031

11.216
Agricultural Science Center
Northeastern Branch Station
P.O. Box 689
Tucumcari, NM 88401

11.217
Mora Research Center
Box 357
Mora, NM 87732

NV
11.218
Nevada Agricultural Experiment Station
University of Nevada
Reno, NV 89557
702-784-6611; (FAX) 702-784-4227

Director: Bernard M. Jones

11.219
Fallon Branch Station
111 Scheckler Rd.
Fallon, NV 89406
702-423-2844

11.220
Logandale Branch Station
Box 126
Logandale, NV 89021
702-423-2844

11.221
Pahrump Branch Station
Box 1090
Pahrump, NV 89041
702-727-5532

11.222
Valley Road Branch Station
910 Valley Rd.
Reno, NV 89557
702-784-6600

NY
11.223
New York Agricultural Experiment Station
Cornell University
292 Roberts Hall
Ithaca, NY 14853
607-255-5420; (FAX) 607-255-0788

Director: Brian F. Chabot

11.224
New York Agricultural Experiment Station
State Station
Geneva, NY 14456
315-787-2211; (FAX) 315-787-2397

Director: Robert A. Plane

11.225
Hudson Valley Laboratory
Box 727
Highland, NY 12528
914-691-7231

11.226
Long Island Horticultural Research Laboratory
39 Sound Ave.
Riverhead, NY 11901
516-727-3595

11.227
Vineyard Research Laboratory
Fredonia, NY 14063
716-672-7336

OH
11.228
Ohio Agricultural Research and Development Center
Ohio State University
Columbus, OH 43210
614-292-3897; (FAX) 614-292-3263

Director: Kirklyn M. Kerr

11.229
Ohio Agricultural Research and Development Center
Ohio State University
Wooster, OH 44691
216-263-3703; (FAX) 216-263-3713

Director: Kirklyn M. Kerr

11.230
Mahoning County Branch-OARDC
Canfield, OH 44406

11.231
Muck Crops Branch-OARDC
Willard, OH 44890

11.232
Vegetable Crops Branch-OARDC
Fremont, OH 43420

11.233
Jackson Branch-OARDC
Jackson, OH 45640

OK
11.234
Oklahoma Agricultural Experiment Station
Oklahoma State University
Stillwater, OK 74078-0500
405-744-5398; (FAX) 405-744-5339

Director: C. B. Browning

11.235
Pecan Research Station
Sparks, OK 74869

11.236
Perkins Fruit Research Station
Perkins, OK 74059

11.237
Vegetable Research Station
Bixby, OK 74008

OR
11.238
Oregon Agricultural Experiment Station
Oregon State University
Corvallis, OR 97331
503-754-4251; (FAX) 503-754-3178

Director: Thayne R. Dutson

11.239
Malheur Station
Ontario, OR 97914
503-889-2174

11.240
Mid Columbia Station
Hood River, OR 97031
503-386-2030

11.241
North Willamette Experiment Station
Aurora, OR 97002
503-678-1264

11.242
Southern Oregon Station
Medford, OR 97502
503-772-5165

11.243
Umatilla Station
Branch Station, Columbia Basin Agricultural Center
Pendleton, OR 97801
503-276-5721

PA
11.244
Pennsylvania Agricultural Experiment Station
229 Agriculture Administration Bldg.
Pennsylvania State University
University Park, PA 16802
814-865-2541

Director: Lamartine F. Hood

11.245
Erie County Research Laboratory
462 N. Cemetary Rd.
North East, PA 16428

11.246
Pennsylvania State University
Fruit Research Laboratory
Biglerville, PA 17307
717-677-6116

RI
11.247
Rhode Island Agricultural Experiment Station
University of Rhode Island
Kingston, RI 02881
401-792-2474

Director: William R. Wright

SC
11.248
South Carolina Agricultural Experiment Station
Clemson University
Clemson, SC 29634-0351
803-656-3141; (FAX) 803-656-3779

Director: James R. Fischer

11.249
Coastal Research and Education Center
2865 Savannah Hwy.
Charleston, SC 29407
803-766-3761

11.250
Edisto Research and Education Center
P.O. Box 247
Blackville, SC 29817
803-284-3343

11.251
Pee Dee Research and Education Center
P.O. Box 271
Florence, SC 29503
803-662-3526

11.252
Sandhill Research and Education Center
P.O. Box 280
Elgin, SC 29045
803-788-5700

SD
11.253
South Dakota Agricultural Experiment Station
South Dakota State University
Brookings, SD 57006
605-688-4149; (FAX) 605-688-6065

Director: R. A. Moore

TN
11.254
Tennessee Agricultural Experiment Station
University of Tennessee
P.O. Box 1071
Knoxville, TN 37901-1071
615-974-7121

Director: Don O. Richardson

11.255
Highland Rim Experiment Station
Springfield, TN 37172
615-384-5292

11.256
Martin Branch Station
Martin, TN 38237
901-587-7263

11.257
Middle Tennessee Experiment Station
Spring Hill, TN 37174
615-486-2129

11.258
Plateau Experiment Station
Crossville, TN 38555
615-484-0034

11.259
Tobacco Experiment Station
Greeneville, TN 37743
615-638-6532

11.260
West Tennessee Agricultural Experiment Station
Jackson, TN 38301
901-424-1643

TX
11.261
Texas Agricultural Experiment Station
The Texas A&M University System
College Station, TX 77843
409-845-8484; (FAX) 409-845-0365

Director: Neville P. Clarke

11.262
Texas A&M Agricultural Research and Extension Center
Rt. 3
Lubbock, TX 79401
806-746-7101

11.263
Texas A&M University
Res-Demonstrating Center
Montague, TX 76251
817-894-2906

11.264
Texas A&M University of Agricultural Research and
    Extension Center
Box 292
Stephenville, TX 76401
817-968-4144

11.265
Texas A&M University of Agricultural Research and
    Extension Center
P.O. Drawer 1050
Uvalde, TX 78801
512-278-9151

11.266
Texas A&M University of Agricultural Research and
    Extension Center
2415 East Highway 83
Weslaco, TX 78596
512-968-5585

11.267
Texas A&M University Research and Extension Center
17360 Coit Rd.
Dallas, TX 75252
214-231-5362

11.268
Texas A&M University Research and Extension Center
Drawer E
Overton, TX 75684
214-834-6191

11.269
Texas A&M University Research Center
1380 A&M Circle
El Paso, TX 79927
915-859-9111

11.270
Texas A&M University Vegetable Research Station
Rt. 2, Box 2E
Munday, TX 76371
817-442-4531

UT
11.271
Utah Agricultural Experiment Station
Utah State University
Logan, UT 84322
801-750-2215; (FAX) 801-750-3798

Director: Doyle J. Matthews

11.272
Horticultural Research Station
Farmington, UT 84025
801-451-2763

VA
11.273
Virginia Agricultural Experiment Station
College of Agriculture and Life Sciences
Hutcheson Hall
Virginia Polytechnic Institute and State University
Blacksburg, VA 24061
703-231-6337; (FAX) 703-231-4163

Director: James R. Nichols

11.274
Shenandoah Valley Research Station
Steeles Tavern, VA 24476

11.275
Southern Piedmont Research and Education Center
Blackstone, VA 23824

11.276
Virginia Truck and Ornamentals Research Station
Eastern Shore Branch
Painter, VA 23420

11.277
Virginia Truck and Ornamentals Research Station
Virginia Beach, VA 23458

11.278
Winchester Fruit Research Laboratory
Winchester, VA 22601

VI
11.279
Virgin Islands Agricultural Experiment Station
University of the Virgin Islands
RR # 2-10,000
Kingshill, St. Croix, VI 00850
809-778-0246; (FAX) 809-778-6570

Director: Darshan S. Padda

VT
11.280
Vermont Agricultural Experiment Station
College of Agriculture and Life Sciences
University of Vermont
Burlington, VT 05405
802-656-2980; (FAX) 802-656-8429

Director: Donald L. McLean

WA
11.281
Washington Agricultural Research Center
Washington State University
Pullman, WA 99164
509-335-4563; (FAX) 509-335-2863

Director: James J. Zuiches

11.282
Coastal Washington Research and Extension Unit
Long Beach, WA 98631

11.283
Irrigated Agriculture Research and Extension Center
Prosser, WA 99350

11.284
Northwestern Washington Research and Extension Center
Mount Vernon, WA 98273

11.285
Southwestern Washington Research Unit
Vancouver, WA 98665

11.286
Tree Fruit Research Center
Washington State University
Wenatchee, WA 98801

11.287
Western Washington Research and Extension Center
Puyallup, WA 98371

WI
11.288
Wisconsin Agricultural Experiment Station
University of Wisconsin
1450 Linden Dr.
Madison, WI 53706
608-262-4556; (FAX) 608-262-4930

Director: Leo M. Walsh

11.289
Agricultural Research Station
Hancock, WI 54943
715-249-5961

11.290
Agricultural Research Station
Lancaster, WI 53813
608-723-2580

11.291
Agricultural Research Station
Spooner, WI 54801
715-635-3735

11.292
Agricultural Research Station
Sturgeon Bay, WI 54235
414-743-5406

11.293
Arlington Agricultural Research Station
Arlington, WI 53511
608-846-3761

11.294
Wisconsin Potato Research Station
Rhinelander, WI 54501
715-362-5719

11.295
University Experimental Farm
Ashland, WI 54806
715-682-6844

WV
11.296
West Virginia Agricultural and Forestry Experiment
    Station
West Virginia University
P.O. Box 6108
Morgantown, WV 26506-6108
304-293-2395; (FAX) 304-293-3740

11.297
West Virginia University Experiment Sub Station for Tree
    Fruit
Kearneysville, WV 25430

Director: Robert H. Maxwell

THE NATIONAL PLANT GERMPLASM SYSTEM

11.298
National Seed Storage Laboratory
Steve Eberhart
NSSL
Colorado State University
Fort Collins, CO 80523

Maintains gene bank collections of seed crops and their
    wild relatives.

11.299
Germplasm Services Laboratory
Bldg. 001, Rm. 322
BARC-EAST
Beltsville, MD 20705

The national focal point for the introduction,
    documentation, initial distribution, and foreign
    exchange of germplasm.

11.300
National Small Grains Germplasm Research Facility
Harold Bockelman
P.O. Box 307
Aberdeen ID 83210

Maintains collections of wheat, oats, barley, rye, and rice.

11.301
National Soybean Collection
Randall L. Nelson
1102 South Goodwin Ave.
University of Illinois
Urbana, IL 61801

Maintains collection of soybeans.

11.302
National Cotton Collection
Russell J. Kohel
P.O. Box DN
Texas A&M University
College Station, TX 77841

Maintains collection of cotton.

11.303
National Clonal Germplasm Repository—Brownwood
L. J. Grauke
Pecan Research
Rt. 2, Box 133
Summerville, TX 77879

Maintains collections of pecans and hickories.

11.304
National Clonal Germplasm Repository—Corvallis
Kim Hunter
33447 Peoria Rd.
Corvallis, OR 97333

Maintains collections of pears, filberts, small fruits, hops, and mints.

11.305
National Clonal Germplasm Repository—Davis
Katie Rigert
Dept. of Pomology
University of California
Davis, CA 95616

Maintains collections of grapes, stone fruits, and nuts.

11.306
National Clonal Germplasm Repository—Miami/Mayaguez
Ray Schnell and F. Vasquez
13601 Old Cutler Rd.
Miami, FL 33158

Maintains collections of coffee, cocoa, bananas, pineapples, mangoes, and other tropical and subtropical horticultural crops.

11.307
National Clonal Germplasm Repository—Orlando
Michael L. Cagley
USDA, ARS
NCGR
Rt. 2, Box 375
Groveland, FL 32736

Maintains collection of citrus.

11.308
National Clonal Germplasm Repository—Geneva
Phillip L. Forsline
New York Agricultural Experiment Station
Germplasm Research Unit
Geneva, NY 14456-0462

Maintains collections of apples and American grapes.

11.309
National Clonal Germplasm Repository—Riverside/Brawley
Tim Williams
USDA, ARS
NCGR
1060 Pennsylvania Ave.
Riverside, CA 92507

Maintains collections of citrus and date palms.

11.310
National Clonal Germplasm Repository—Hilo
Francis Zee
461 West Lanikaula St.
Hilo, HI 96720

Maintains collections of macadamias, pineapples, guavas, papayas, and passionfruit.

11.311
National Clonal Germplasm Repository—National Arboretum
Ned Garvey
U.S. National Arboretum
Washington, DC 20002

Maintains collection of woody ornamentals.

11.312
U.S. Plant Introduction Station
Howard E. Waterworth
11601 George Palmer Hwy.
Glenn Dale, MD 20769

Maintains collections of pome and stone fruits and woody ornamentals.

11.313
Northeastern Regional Plant Introduction Station
Jim McFerson
New York Agricultural Experiment Station
Geneva, NY 14456

Maintains collections of perennial clover, onions, peas, broccoli, and timothy.

11.314
Southern Regional Plant Introduction Station
Gilbert R. Lovell
Experiment, GA 30212

Maintains collections of cantaloupes, cowpeas, millet, peanuts, sorghum, and peppers.

11.315
North Central Regional Plant Introduction Station
Raymond L. Clark
Iowa State University
Ames, IA 50011

Maintains collections of alfalfa, corn, sweet clover, beets, tomatoes, and cucumbers.

11.316
Western Regional Plant Introduction Station
S. M. Dietz
59 Johnson Hall
Washington State University
Pullman, WA 99163

Maintains collections of beans, cabbages, fescues, wheat, grasses, lentils, lettuce, safflowers, and chickpeas.

11.317
Interregional Potato Introduction Station
John Banberg
University of Wisconsin
Peninsula Experiment Station
Sturgeon Bay, WI 54235

Maintains collections of Solanum tuberosum and Solanum spp.

PLANT VARIETY PROTECTION OFFICE

Established in 1970, the Plant Variety Protection Office's objective is to protect those who breed, develop, or discover novel varieties of sexually reproduced plants. They issue Certificates of Protection, and record varietal descriptions in a computer retrieval system.

11.318
Plant Variety Protection Office
U.S. Department of Agriculture
Agricultural Marketing Service
NAL Building, Room 500
1031 Baltimore Blvd.
Beltsville, MD 20705-2351
301-344-2518'
Virginia A. Lerch, Examiner

FEDERAL PLANT INTRODUCTION STATIONS

11.319
U.S. Plant Introduction Stations
Box 88
Glenn Dale, MD 20769

This is the principal location in the U.S. for holding foreign plants that have been prohibited entry into the country due to the U.S. plant importation law. They are tested for diseases and insects, and must remain at Glenn Dale for at least five years before they can be released.

REGIONAL PLANT INTRODUCTION STATIONS

11.320
Northeastern Regional Plant Introduction Station
New York State Agricultural Experiment Station
Geneva, NY 14456

11.321
North Central Regional Plant Introduction Station
Iowa State University
Ames, IA 50010

11.322
Southern Regional Plant Introduction Station
University of Georgia
Experiment, GA 30212

11.323
Western Regional Plant Introduction Station
Johnson Hall, Rm. 59
Washington State University
Pullman, WA 99163

11.324
Interregional Potato Introduction Laboratory
University of Wisconsin
Sturgeon Bay, WI 54235

QUARANTINE AND PLANT IMPORTATION PROGRAM

Quarantine regulations protect plants in the U.S. from foreign plant diseases and pests in nursery stock. The rules affect the nursery industry, plant importers, freight forwarders, customs brokers, international freight carriers, hobbyists, and others concerned with importing plant propagative material. Horticulturists who wish to import plants into this country should seek information and permission from the U.S. Department of Agriculture, Animal and Plant Health Inspection Service, Permit Unit, Federal Building, Hyattsville, MD 20782.

The permit unit will send, at no charge, lists of specific regulations governing the species of plants prohibited from importation, size and age limitations of plant materials, packing and shipping instructions, lists of those plant genera subject to post-entry quarantine, those species prohibited under the Endangered Species Convention Regulations, entry status of seeds, and other related information. Importers must furnish complete information about the plants they wish to import (using scientific names) and the point of origin when they request applications for a permit. If permission is granted, importers will receive permits and import labels addressed to one of the Inspection Stations maintained at the following locations.

11.325
Plant Inspection Station
Honolulu International Airport
Terminal Box 57
Honolulu, HI 96819

11.326
Plant Inspection Station
Cordova Border Station
Rm. 172-A
3600 E. Paisano
El Paso, TX 79905

11.327
Plant Inspection Station
102 Terrace Ave., Rm. 116
Nogales, AZ 85621

11.328
Plant Inspection Station
New Border Station, Rm. 505
Lincoln-Juarez Bridge, Bldg. 5
P.O. Box 277
Laredo, TX 78042-0277

11.329
Plant Inspection Station
209 River St.
Hoboken, NJ 07030

11.330
Plant Inspection Station
U.S. Customhouse, Rm. 148
423 Canal St.
New Orleans, LA 70130

11.331
Plant Inspection Station
3500 N.W. 62nd Ave.
P.O. Box 59-2136
Miami, FL 33159

11.332
Plant Inspection Station
J.F. Kennedy International Airport
Cargo Building 80, Rm. 109
Jamaica, NY 11430

11.333
Plant Inspection Station
San Francisco International Airport
San Francisco, CA 94128

11.334
Plant Inspection Station
9610 S. La Cienega Blvd.
Inglewood, CA 90301

11.335
Plant Inspection Station
Luis Munox International Airport
Isla Verde, PR 00913

11.336
Plant Inspection Station
1500 E. Elizabeth St.
Border Services Blvd., Rm. 224
Brownsville, TX 78520

11.337
Plant Inspection Station
P.O. Box 43-L
San Ysidro, CA 92073

11.338
Plant Inspection Station
Federal Office Bldg., Rm. 9014
909 First Ave.
Seattle, WA 98174

SOIL CONSERVATION SERVICE

The S.C.S. helps individuals, groups, organizations, cities and towns, and county and state governments reduce the waste of land and water resources. The S.C.S. mission covers three major areas: soil and water conservation, natural resource surveys, and community resource protection and management.

Of particular interest to those in horticulture, the S.C.S. operates or provides technical assistance to 24 plant materials centers around the U.S. There, scientists seek out native or introduced plants that show promise for reducing erosion and sedimentation and improving water quality. The centers do not sell plants and seed, but instead release their selections to commercial nurseries and seed producers.

11.339
Soil Conservation Service
U.S. Department of Agriculture
P.O. Box 2890
Washington, DC 20013
202-447-4543

NATIONAL AGRICULTURE STATISTICS SERVICE

The N.A.S.S. collects and publishes data on supply, prices, and other items necessary to maintain orderly agricultural operations.

11.340
National Agriculture Statistics Service
U.S. Department of Agriculture
1301 New York Ave., N.W.
Washington, DC 20005-4789

FLORICULTURE SPECIALISTS
Doyle Johnson 202-786-1884
Jim Brewster 202-447-7688

FRUIT AND NUT SPECIALISTS
Jim Brewster 202-447-7688
Jim Smith 202-447-5412
Ben Huang 202-786-1884
Boyd Buxton 202-786-1884

FOREST SERVICE

The Forest Service administers programs for applying sound conservation and utilization practices to the natural resources of the national forests and national grasslands, for promoting these practices on all forest lands, and for carrying out forest and range research.

11.341
Forest Service
U.S. Department of Agriculture
P.O. Box 96090
Washington, DC 20013-6090
202-447-3957

REGIONAL INFORMATION OFFICERS
11.342
USDA—Forest Service (Alaska)
Wayne R. Nicolls
Federal Office Building
P.O. Box 21628
Juneau, AK 99802-1628
907-586-8847

11.343
USDA—Forest Service (Eastern)
Sylvia Brucchi
Henry S. Reuss Federal Plaza
310 West Wisconsin Ave.
Milwaukee, WI 53203
414-291-3640

11.344
USDA—Forest Service (Intermountain)
Patrick J. Sheehan
Federal Building
324 25th St.
Ogden, UT 84401
801-625-5347

11.345
USDA—Forest Service (Northern)
Beth Horn
Federal Building
P.O. Box 7669
Missoula, MT 59807
406-329-3089

11.346
USDA—Forest Service (Pacific Northwest)
John Marker
319 S.W. Pine St.
P.O. Box 3623
Portland, OR 97208
503-221-2971

11.347
USDA—Forest Service (Pacific Southwest)
Regional Information Officer
630 Sansome St.
San Francisco, CA 94111
415-556-1932

11.348
USDA—Forest Service (Rocky Mountain)
Dennis Bschor
11177 West 8th Ave.
P.O. Box 25127
Lakewood, CO 80225
303-236-9660

11.349
USDA—Forest Service (Southern)
Stanford M. Adams
1720 Peachtree Rd., N.W.
Atlanta, GA 30367
404-347-4191

11.350
USDA—Forest Service (Southwestern)
Charles Bazan
517 Gold Ave., S.W.
Albuquerque, NM 87102
505-842-3290

CANADIAN GOVERNMENTAL OFFICES AND PROGRAMS

AGRICULTURE CANADA

Agriculture Canada maintains research facilities, plant protection and seed programs, and other offices that deal with horticulture.

COMMUNICATIONS BRANCH
11.351
Agriculture Canada
Communications Branch
Ottawa, ON K1A 0C7
613-995-5222

Answers public inquiries regarding Agriculture Canada.

CROP DEVELOPMENT DIVISION
This division is the focal point for national initiatives relating to the production and marketing of horticultural grains, oilseeds, and special crops.

11.352
Crop Development Division
Agriculture Development Branch
Sir John Carling Bldg.
Ottawa, ON K1A 0C5
613-995-9554

Horticulture Section
Ken Hunter

CROP PROTECTION DIVISION
The horticulture research program of this division is responsible for maintaining and improving the productivity of the horticultural sector through developing and making available new knowledge and technology.

11.353
Crop Protection—Horticulture
Priorities and Strategies Directorate
Research Branch
Ottawa, ON K1A 0C5
613-995-7084

RESEARCH STATIONS AND EXPERIMENTAL FARMS
The following offices carry out federally sponsored horticultural and agricultural research in Canada.

AB
11.354
Beaverlodge Research Station
Agriculture Canada
Beaverlodge, AB T0H 0C0
403-354-2212

Director: Dr. J. D. McElgunn

11.355
Lethbridge Research Station
Agriculture Canada
Lethbridge, AB T1J 4B1
403-327-4561

Director: Dr. D. G. Dorrell

BC
11.356
Agassiz Research Station
Agriculture Canada
Agassiz, BC V0M 1A0
604-796-2221

Director: Dr. J. M. Molnar

11.357
Saanichton Research and Plant Quarantine Station
Agriculture Canada
8801 East Saanich Rd.
Sidney, BC V8L 1H3
604-656-1173

11.358
Summerland Research Station
Agriculture Canada
Summerland, BC V0H 1Z0
604-494-7711

Director: Dr. D. M. Bowden

11.359
Vancouver Research Station
Agriculture Canada
6660 N.W. Marine Dr.
Vancouver, BC V6T 1X2
Director: Dr. M. Weintraub

MB
11.360
Morden Research Station
Agriculture Canada
Morden, MB R0G 1J0
204-822-4471

Director: Dr. D. K. McBeath
Cross-reference: 16.451

NB
11.361
Fredericton Research Station
Agriculture Canada
P.O. Box 20280
Fredericton, NB E3B 4Z7
506-452-3260

Director: Dr. Y. Martel

NF
11.362
St. John's Research Station
Agriculture Canada
P.O. Box 7098
St. John's, NF A1E 3Y3
709-772-4619

Director: Dr. H. R. Davidson

NS
11.363
Charlottetown Research Station
Agriculture Canada
P.O. Box 1210
Charlottetown, NS C1A 7M8
902-892-5461

Director: Dr. L. B. MacLeod

11.364
Kentville Research Station
Agriculture Canada
Kentville, NS B4N 1J5
902-678-2171

Director: Dr. G. M. Weaver

ON
11.365
Harrow Research Station
Agriculture Canada
Harrow, ON N0R 1G0
519-738-2251

Director: Dr. C. F. Marks

11.366
London Research Centre
Agriculture Canada
London, ON N6A 5B7
519-679-4452

Director: Dr. H. V. Morley

11.367
Ottawa Research Branch Headquarters
Agriculture Canada
Central Experimental Farm
Ottawa, ON K1A 0C5
613-995-7084

Director: Dr. E. J. LeRoux
Cross-references: 15.241, 16.459

11.368
Ottawa Research Station
Agriculture Canada
Central Experimental Farm
Ottawa, ON K1A 0C6
613-995-5287

Director: Dr. A. I. de la Roche

11.369
Smithfield Experimental Farm
Agriculture Canada
P.O. Box 340
Trenton, ON K8V 5R5
613-392-3527

Director: Dr. S. R. Miller

11.370
Vineland Research Station
Agriculture Canada
Vineland Station, ON L0R 2E0
416-562-4113

Director: Dr. D. R. Menzies

PQ
11.371
L'Assomption Experimental Farm
Agriculture Canada
L'Assomption, PQ J0K 1G0
514-589-4775

Director: F. Darisse

11.372
La Pocatière Experimental Farm
Agriculture Canada
P.O. Box 400
La Pocatière, PQ G0R 1Z0
418-856-3141

Director: J. E. Comeau

11.373
Normandin Experimental Farm
Agriculture Canada
1472 St. Cyrille St.
Normandin, PQ G0W 2E0
418-274-3378

Director: J. M. Wauthy

11.374
St. Jean sur Richelieu Research Station
Agriculture Canada
St. Jean sur Richelieu, PQ J3B 6Z8
514-346-4494

Director: Dr. C. B. Aube

PLANT GENE RESOURCES OF CANADA
    P.G.R.C. was established in 1970 to obtain, exchange, preserve, document, and evaluate germplasm of crop plants and their wild relatives. P.G.R.C. also develops, coordinates, and encourages crop germplasm conservation across Canada.

11.375
Plant Gene Resources of Canada
Biosystematics Research Centre
Building # 99
Research Branch
Agriculture Canada
Ottawa, ON K1A 0C6
613-996-1665

11.376
Clonal Repository for Horticultural Crops
Smithfield Experimental Farm
Box 340
Trenton, ON K8V 5R5
613-392-3527

This is the genebank for clonally propagated crop plants.

PRAIRIE FARM REHABILITATION ADMINISTRATION
    The P.F.R.A works in the three prairie provinces to provide technical and financial assistance to farmers, local government organizations, and provincial and other federal agencies for soil conservation and water development projects.

11.377
Prairie Farm Rehabilitation Administration
1901 Victoria Ave.
Regina, SK S4P 0R5
306-780-5081

11.378
Tree Distribution Program
Gordon Howe, Manager
Tree Nursery Division
Indian Head, SK S0G 2K0
306-695-2284

Cross-reference: 16.482

11.379
Soil Conservation Service
Fred Kraft, A/Head
Soil Conservation Planning Section
1901 Victoria Ave.
Regina, SK S4P 0R5
306-780-5159

PLANT PROTECTION DIVISION

The Plant Protection Division's mandate is to prevent the introduction into Canada of foreign pests and diseases of economic significance to agricultural and forestry crops; to detect and control or eradicate plant pests and diseases of economic significance to Canada; and to certify plants and plant products for domestic and export trade.

11.380
Plant Protection Division
Plant Health and Plant Products Directorate
Food and Production Inspection Branch
K.W. Neatby Bldg.
960 Carling Ave.
Ottawa, ON K1A 0C6
613-995-7900

Importation of Plant and Plant Materials
R. Dave Gray, Associate Director
Import Programs

Export Phytosanitary Certification (Except Potatoes) of Plant and Plant Materials
Dianne Hedley, Associate Director
Domestic and Export Programs

Domestic Control of Plant Pests and Diseases (Except Pure, Potato Diseases)
Dianne Hedley, Associate Director
Domestic and Export Programs

SEED DIVISION

The Seed Division is responsible for coordinating the national seed program, the purpose of which is to ensure that imported and domestic seed marketed in Canada is safe, pure, viable, efficacious, accurately represented to maintain identity and avoid fraud, and certified to meet the requirements of importing countries.

11.381
Seed Division
Plant Health Directorate
Food Production and Inspection Branch
K.W. Neatby Bldg.
960 Carling Ave.
Ottawa, ON K1A 0C6
613-995-7900

Variety Registration
Grant R. Watson, Associate Director

Seed Projects
Tom R. Roddy, Associate Director

Plant Breeders' Rights
Valerie Sisson

CANADIAN FORESTRY SERVICE

The objective of the C.F.S is to promote the wise management, conservation, and use of Canada's forests for the economic, social, and environmental benefit of all Canadians.

11.382
Canadian Forestry Service
Communications Branch
Sir John Carling Bldg.
Ottawa, ON K1A 0C5
819-997-1107

REGIONAL CANADIAN FORESTRY SERVICE ESTABLISHMENTS
11.383
Centre Forestier des Laurentides
Service Canadian des Forets
1055 rue du P.E.P.S.
C.P. 3800
Ste.-Foy, PQ G1V 4C7
408-648-4991

11.384
Great Lakes Forestry Centre
Canadian Forestry Service
P.O. Box 490
1210 Queen St.
Saulte Ste. Marie, ON P6A 5M7
705-949-9461

11.385
Maritimes Forestry Centre
Canadian Forestry Service
P.O. Box 4000
MacKay Dr.
Fredericton, NB E3B 5P7
506-452-3500

11.386
Newfoundland Forestry Centre
Canadian Forestry Service
Building 304
Pleasantville
P.O. Box 6028
St. John's, NF A1C 5X8
709-772-6019

11.387
Northern Forestry Centre
Canadian Forestry Service
5320-122nd St.
Edmonton, AB T6H 3S5
403-435-7210

11.388
Pacific Forestry Centre
Canadian Forestry Service
506 West Burnside Rd.
Victoria, BC V8Z 1M5
604-388-0600

11.389
Petawawa National Forestry Institute
Canadian Forestry Service
Chalk River, ON K0J 1J0
613-589-2880

PROVINCIAL GOVERNMENT RESEARCH CENTERS

A number of research stations are sponsored by the provincial departments of agriculture. Many of them are engaged in important horticultural research.

AB
11.390
Alberta Special Crops and Horticultural Research Centre
Plant Industry Division of Alberta Agriculture
Bag Service: 200
Brooks, AB T0J 0J0
403-362-3391

Director: Thomas R. Krahn
Cross-reference: 16.429

NB
11.391
Herve J. Michaud Experimental Farm
Fredericton Research Station
P.O. Box 667
Bouctouche, NB E0A 1G0
506-743-2464

Director: Dr. G. L. Rousselle

11.392
New Brunswick Dept. of Agriculture
Plant Industry Branch
P.O. Box 6000
Fredericton, NB E3B 5H1
506-453-2108

Director: E. T. Pratt

11.393
New Brunswick Dept. of Agriculture
Potato Industry Division
P.O. Box 6000
Fredericton, NB E3B 5H1
506-453-2108

Director: Dr. C. E. Smith

NS
11.394
Horticulture and Biology Services
Nova Scotia Dept. of Agriculture
P.O. Box 550
Truro, NS B2N 5E3
902-895-1571

Director: D. M. Sangster

ON
11.395
College of Agricultural Technology
Ontario Ministry of Agriculture and Food
Ridgetown, ON N0P 2C0
519-674-5456

Director: D. W. Taylor

11.396
Horticultural Experiment Station
Horticultural Research Institute of Ontario, Vineland
  Station
Box 587
Simcoe, ON N3Y 4N5
519-426-7120

Director: A. Loughton

11.397
Horticultural Research Institute of Ontario
Ontario Ministry of Agriculture and Food
Vineland Station, ON L0R 2E0
416-562-4141

Director: Dr. F. C. Eady

PQ
11.398
Crop Protection Research Station
Quebec Ministry of Agriculture
867 L'Ange-Gardien Blvd.
L'Assomption, PQ J0K 1G0
514-589-4780

Director: Dr. G. Edmond

11.399
Deschambault Research Station
Quebec Ministry of Agriculture
Horticulture Section
P.O. Box 123
Deschambault, PQ G0A 1S0
418-286-3351

Director: J. Genest

11.400
Farnham Orchard Protection Section
Quebec Crop Protection Service
1400 Saint Paul St. North
Farnham, PQ J2N 2R5
514-293-6072

11.401
La Pocatière Agricultural Research Station
Quebec Ministry of Agriculture
Horticulture Section
Route 230
La Pocatière, PQ G0R 1Z0
418-856-1110

Director: J. Archambault

11.402
St. Augustin Crop Protection Research Station
Quebec Crop Protection Service
567, RR 138
St. Augustin, PQ G0A 3E0
418-878-2753

Director: M. A. Richard

11.403
St. Hyacinthe Agricultural Research Station
Quebec Ministry of Agriculture
Horticulture Section
3300 Sicotte St.
St. Hyacinthe, PQ J2S 7B8
414-774-0660

Director: P. Lavigne

11.404
St. Lambert de Levis Soil Research Station
Horticulture Section
Quebec Soil Research Service
Rang Saint Patrice
St. Lambert, PQ G0S 2W0
418-889-9950

Director: A. Dube

~ 12 ~

Education

This section lists horticultural, landscape design, and landscape architecture programs at colleges, universities, community and junior colleges, and technical institutions in the United States and Canada. The degrees offered range from certificates to Ph.D. The range of horticultural specialization is wide and varies considerably by school. For detailed information, write the individual departments or the director of admissions for the school.

Certificate programs are usually one year in length, except at Canadian universities where they are normally two-year programs. Generally, associate's degree programs are two years, bachelor's degree programs four years, and master's degree programs two years in length. Ph.D. programs vary by school.

For information on horticultural internships, consult the annual *Internship Directory* published by the American Association of Botanical Gardens and Arboreta (1.5).

HORTICULTURE EDUCATIONAL PROGRAMS

UNITED STATES

AK
12.1
Malanuska-Susitana Community College
Box 899
Palmer, AK 99645

Degree Granted: Associate

12.2
University of Alaska
Dept. of Agriculture
School of Agriculture and Land Resources Management
Fairbanks, AK 99701
907-474-7211

Degree Granted: Bachelor
Cross-reference: 15.1

AL
12.3
Alabama Agricultural and Mechanical University
Normal, AL 35762
205-851-5245

Degrees Granted: Bachelor; Master; Certificate Program
Cross-reference: 12.394

12.4
Auburn University
Auburn University, AL 36849
205-826-4080

Degrees Granted: Bachelor; Master
Cross-references: 12.395, 15.2, 16.1

12.5
Bessemer State Technical College
P.O. Box 308
Bessemer, AL 35020
205-428-6391

Degree Granted: Certificate Program

12.6
Oakwood College
P.O. Box 107
Huntsville, AL 35896
205-837-1630

Degree Granted: Bachelor

12.7
Tuskegee Institute
Tuskegee, AL 36088
308-727-8011

Degree Granted: Bachelor

AR
12.8
Arkansas State University
State University, AR 72467
501-972-3024

Degree Granted: Bachelor
Cross-reference: 15.5

12.9
Mississippi County Community College
Blytheville, AR 72316

Degree Granted: Associate

12.10
Petit Jean Vocational-Technical School
Highway 9 North
Morrilton, AR 72110
501-354-2465

Degree Granted: Certificate Program

12.11
Southern Arkansas University
Magnolia, AR 72110
501-235-4040

Degree Granted: Associate

12.12
University of Arkansas
Fayetteville, AR 72701
501-575-2603

Degrees Granted: Bachelor; Master; Doctorate
Cross-references: 12.396, 15.7

12.13
University of Arkansas
Pine Bluff, AR 71601
501-541-6500

Degree Granted: Bachelor

AZ
12.14
Arizona Western College
Box 929
Yuma, AZ 85364

Degree Granted: Associate

12.15
Central Arizona College
Coolidge, AZ 85228

Degree Granted: Associate

12.16
Glendale Community College
6000 W. Olive Ave.
Glendale, AZ 85301

Degree Granted: Associate

12.17
Mesa Community College
1833 W. Southern Ave.
Mesa, AZ 85201

Degree Granted: Associate

12.18
Pima Junior College
Tucson, AZ 85745

Degree Granted: Associate

12.19
University of Arizona
Tucson, AZ 85721
602-621-1977

Degrees Granted: Bachelor; Master; Doctorate
Cross-references: 4.11, 12.398, 15.12

CA
12.20
American River College
4700 College Oak Dr.
Sacramento, CA 95841

Degree Granted: Certificate Program

12.21
Antelope Valley College
3041 West Ave.
Lancaster, CA 93534

Degree Granted: Associate

12.22
Bakersfield College
1801 Panorama Dr.
Bakersfield, CA 93305
805-395-4100

Degree Granted: Associate

12.23
Butte Junior College
Box 566
Durham, CA 95965

Degree Granted: Associate

12.24
California Polytechnic State University
San Luis Obispo, CA 93407
805-756-2279

Degrees Granted: Bachelor; Master
Cross-references: 12.399, 15.29, 20.68

12.25
California Polytechnic University
Pomona, CA 91768
714-869-2000

Degree Granted: Bachelor
Cross-references: 2.34, 12.400, 20.69

12.26
California State University
Chico, CA 95929
916-895-6023

Degrees Granted: Bachelor; Master
Cross-references: 15.18

12.27
California State University
Fresno, CA 93740
209-294-4240

Degrees Granted: Bachelor; Master
Cross-references: 15.21, 20.70

12.28
Cerritos College
11110 East Alondra Blvd.
Norwalk, CA 90650
213-860-2451

Degree Granted: Associate

12.29
City College of San Francisco
50 Phelan Ave.
San Francisco, CA 94112
415-239-3000

Degree Granted: Associate

12.30
College of San Mateo
1700 W. Hillsdale Blvd.
San Mateo, CA 94402

Degree Granted: Associate

12.31
College of the Desert
43500 Monterey Ave.
Palm Desert, CA 92260

Degree Granted: Certificate Program

12.32
College of the Redwoods
Eureka, CA 95501

Degree Granted: Certificate Program

12.33
College of the Sequoias
915 S. Mooney Blvd.
Visalia, CA 93277
209-733-2050

Degree Granted: Certificate Program

12.34
Cuyamaca College
Ornamental Horticulture Dept.
2950 Jamacha Blvd.
El Cajon, CA 92019
619-670-1980

Degree Granted: Associate

12.35
Feather River College
P.O. Box 1110
Quincy, CA 95971
916-283-0202

Degree Granted: Bachelor

12.36
Fullerton Junior College
321 E. Chapman
Fullerton, CA 92634

Degree Granted: Certificate Program

12.37
Hartnell College
156 Homestead Ave.
Salinas, CA 93901

Degree Granted: Associate

12.38
Kern Joint Junior College
Dept. of Horticulture
1801 Panorama Dr.
Bakersfield, CA 93305

Degree Granted: Associate

12.39
Long Beach City College
1305 E. Pacific Coast Hwy.
Long Beach, CA 90806
213-599-2421

Degree Granted: Associate

12.40
Los Angelos Pierce College
6201 Winnetka Ave.
Woodland Hills, CA 91364
818-347-0551

Degree Granted: Associate

12.41
Mendocino College
P.O. Box 3000
Ukiah, CA 95482
707-468-3137

Degree Granted: Associate

12.42
Merced Community College
3600 M St.
Merced, CA 95340
209-384-6000

Degree Granted: Certificate Program

12.43
Merritt College
12500 Campus Dr.
Oakland, CA 94619
415-531-4911

Degree Granted: Associate

12.44
Mira Costa College
Barnard Dr.
Oceanside, CA 92054

Degree Granted: Associate

12.45
Modesto Junior College
College Ave.
Modesto, CA 95350
209-575-6498

Degree Granted: Associate

12.46
Monterey Peninsula College
980 Fremont St.
Monterey, CA 93940

Degree Granted: Associate

12.47
Moorpark College
7075 Campus Rd.
Moorpark, CA 93021
805-529-2321

Degree Granted: Associate

12.48
Mt. San Antonio College
1100 N. Grand Ave.
Walnut, CA 91789
714-594-5611

Degree Granted: Associate

12.49
Napa Community College
2277 Napa Vallejo Hwy.
Napa, CA 94558
707-253-3249

Degree Granted: Associate

12.50
Ohlone College
43600 Mission Blvd.
Fremont, CA 94539
415-659-6000

Degree Granted: Bachelor

12.51
Orange Coast College
2701 Fairview Rd.
Costa Mesa, CA 92626
714-432-5748

Degree Granted: Certificate Program

12.52
Saddleback Community College
2800 Marguerite Pkwy.
Mission Viejo, CA 92675
714-582-4500

Degree Granted: Associate

12.53
San Bernardino Valley College
701 S. Mount Vernon Ave.
San Bernardino, CA 92403

Degree Granted: Associate

12.54
San Diego Mesa College
7250 Artillery Dr.
San Diego, CA 92111

Degree Granted: Associate

12.55
San Joaquin Delta College
3301 Kensington Way
Stockton, CA 95204

Degree Granted: Associate

12.56
Santa Barbara City College
Landscape Horticulture
721 Cliff Dr.
Santa Barbara, CA 93109-2394
805-965-0581

Degrees Granted: Associate; Certificate Program
Cross-reference: 16.33

12.57
Santa Rosa Junior College
1501 Mendocino Ave.
Santa Rosa, CA 95401
707-527-4408

Degree Granted: Associate

12.59
Shasta Community College
Hwy. 299 E. Old Oregon Trail
Redding, CA 96001

Degree Granted: Associate

12.60
Sierra Community College
5000 Rocklin Rd.
Rocklin, CA 95677

Degree Granted: Associate

12.61
Solano College
Rt. 1, Box 246
Suisun City, CA 94585
707-864-7000

Degree Granted: Bachelor

12.62
University of California at Davis
Davis, CA 95616
916-752-4335

Degrees Granted: Bachelor; Master; Doctorate
Cross-references: 12.402, 15.23, 16.59, 17.33, 19.12

12.63
University of California at Irvine
Irvine, CA 92717
714-856-6701

Degrees Granted: Bachelor; Master; Doctorate
Cross-reference: 16.61

12.64
University of California at Riverside
Riverside, CA 92502
714-787-3411

Degrees Granted: Bachelor; Master; Doctorate
Cross-references: 15.34, 16.19

12.65
Ventura College
4667 Telegraph Rd.
Ventura, CA 93003

Degree Granted: Associate

12.66
Victor Valley College
P.O. Drawer OO
Victorville, CA 92307
619-245-4251

Degree Granted: Certificate Program

12.67
Yuba College
2088 North Beale Rd.
Marysville, CA 95901
916-741-6700

Degree Granted: Associate

CO
12.68
Colorado State University
Fort Collins, CO 80523
303-491-1101

Degrees Granted: Bachelor; Master; Doctorate
Cross-references: 12.403, 12.469, 15.38, 16.69

12.69
Frontrange Community College
3645 W. 112th Ave.
Westminster, CO 80030
303-466-8811

Degree Granted: Associate

12.70
Larimer County Vo-Tech Center
P.O. Box 2397
Fort Collins, CO 80522
303-226-2500

Degree Granted: Associate

12.71
Naropa College
2130 Arapahoe Ave.
Boulder, CO 80302
303-666-0253

Degree Granted: Bachelor

12.72
Northeastern Junior College
Sterling, CO 80751

Degree Granted: Associate

CT
12.73
University of Connecticut
Ratcliffe Hicks School of Agriculture
Storrs, CT 06268
203-486-2924

Degrees Granted: Bachelor; Master; Doctorate;
   Certificate Program
Cross-references: 12.405, 15.42, 20.44

12.74
University of Connecticut at Hartford
West Hartford, CT 06117
203-241-4830

Degrees Granted: Bachelor

DC
12.75
University of the District of Columbia
Environmental Science
Washington, DC 20008
202-282-7379

Degree Granted: Bachelor

DE
12.76
University of Delaware
Newark, DE 19711
302-451-8123

Degrees Granted: Bachelor; Master; Doctorate;
   Certificate Program

FL
12.77
Bradford-Union Vo-Tech School
609 North Orange St.
Starke, FL 32091
904-964-6150

Degree Granted: Certificate Program

12.78
Brevard Community College
Cocoa Campus
1519 Clear Lake Rd.
Cocoa, FL 32992

Degree Granted: Certificate Program

12.79
Broward Community College
Central Campus
3501 Southwest Davie Rd.
Fort Lauderdale, FL 33314

Degree Granted: Associate

12.80
Central Florida Community College
P.O. Box 1388
Ocala, FL 32678
904-237-2111

Degree Granted: Certificate Program

12.81
Daytona Beach Community College
P.O. Box 1111
Daytona Beach, FL 32105
904-255-8131

Degree Granted: Associate
Cross-reference: 20.92

12.82
Edison Community College
College Pkwy.
Fort Myers, FL 33907

Degree Granted: Certificate Program

12.83
Florida A&M University
Tallahassee, FL 32307
904-599-3000

Degree Granted: Bachelor

12.84
Florida Community College at Jacksonville
Fred H. Kent Campus
3939 Roosevelt Blvd.
Jacksonville, FL 32205-8999
904-387-8255

Degree Granted: Certificate Program

12.85
Florida International University
Tamiami Trail
Miami, FL 33130
305-554-3420

Degree Granted: Certificate Program

12.86
Florida Southern College
111 Lake Hollingsworth
Lakeland, FL 33802
813-680-4333

Degree Granted: Bachelor

12.87
Hillsborough Community College
Dale Mabry Campus
P.O. Box 22127
Tampa, FL 33622

Degree Granted: Associate

12.88
Indian River Community College
3209 Virginia Ave.
Fort Pierce, FL 33452

Degree Granted: Certificate Program

12.89
Lake City Community College
Rt. 7, Box 378
Lake City, FL 32055

Degree Granted: Associate

12.90
Manatee Area Vo-Tech Center
5603 34th St., W.
Bradenton, FL 33505
813-755-2641

Degree Granted: Certificate Program

12.91
North Florida Junior College
Turner Davis Dr.
P.O. Box 419
Madison, FL 32340

Degree Granted: Certificate Program

12.92
Okaloosa-Walton Junior College
100 College Blvd.
Niceville, FL 32578

Degree Granted: Certificate Program

12.93
Pasco-Hernando Community College
State Hwy. 41 North
Dade City, FL 33156
904-567-6701

Degree Granted: Associate

12.94
Pasco-Hernando Community College
West Campus
7025 State Rd. 597
New Port Richey, FL 33552

Degree Granted: Associate

12.95
Pensacola Junior College
1000 College Blvd.
Pensacola, FL 32504

Degree Granted: Certificate Program
Cross-reference: 20.98

12.96
Polk Community College
999 Avenue H., N.E.
Winter Haven, FL 33880
813-297-1010

Degree Granted: Certificate Program

12.97
Santa Fe Community College
P.O. Box 1330
Gainesville, FL 32602

Degree Granted: Associate

12.98
South Florida Junior College
600 West College Dr.
Avon Park, FL 33825

Degree Granted: Associate

12.99
University of Florida
Dept. of Ornamental Horticulture
1545 Fifield Hall
Gainesville, FL 32611
904-392-1832

Degrees Granted: Bachelor; Master; Doctorate
Cross-references: 12.407, 12.470, 15.51

12.100
Valencia Community College
P.O. Box 3028
Orlando, FL 32802

Degree Granted: Associate

12.101
Vo-Tech Center
500 N. Appleyard Dr.
Tallahassee, FL 32304

Degree Granted: Certificate Program

12.102
Washington Holmes Vocational Technical Center
209 Hoyt St.
Shipley, FL 32428
904-638-1180

Degree Granted: Certificate Program

GA
12.103
Abraham Baldwin College
Tifton, GA 31794

Degree Granted: Certificate Program

12.104
Berry College
Mount Berry Station
Rome, GA 30149
404-235-4494

Degree Granted: Bachelor

12.105
Fort Valley State College
Horticulture Dept.
Fort Valley, GA 31030
912-825-6447

Degree Granted: Bachelor

12.106
Gwinnett Area Technical School
1250 Atkinson Rd.
Lawrenceville, GA 30245
404-962-7582

Degree Granted: Associate

12.107
North Georgia Technical & Vocational School
Clarksville, GA 30523
404-751-2131

Degree Granted: Certificate Program

12.108
University of Georgia
Dept. of Horticulture
Athens, GA 30602
404-542-2471

Degrees Granted: Bachelor; Master; Doctorate
Cross-references: 12.409, 12.471, 15.53, 15.54, 20.11

HI
12.109
Leeward Community College
Dept. of Horticulture
Pearl City, HI 96782
808-244-9181

Degree Granted: Associate

12.110
Maui Community College
310 Kaahumanu Ave.
Kahului, HI 96732

Degree Granted: Certificate Program

12.111
National Tropical Botanical Garden
Apprentice Gardeners Program
P.O. Box 340
Lawai
Kauai, HI 96765
808-332-7361

Degree Granted: Certificate Program
Cross-reference: 16.121

12.112
University of Hawaii
Hilo, HI 96720
808-933-3315

Degree Granted: Bachelor

12.113
University of Hawaii
Dept. of Horticulture
3190 Maile Way, Rm. 102
Honolulu, HI 96822
808-948-8351

Degrees Granted: Bachelor; Master; Doctorate
Cross-references: 15.57, 16.111

IA

12.114
Des Moines Area Community College
2006 S. Ankeny Blvd.
Ankeny, IA 50021

Degree Granted: Certificate Program

12.115
Hawkeye Institute of Technology
1501 East Orange Rd.
P.O. Box 8015
Waterloo, IA 50704

Degree Granted: Certificate Program

12.116
Indian Hills Community College
9th & College
Ottumwa Industrial Airport
Ottumwa, IA 52501
515-683-5111

Degree Granted: Certificate Program

12.117
Iowa Lakes Community College
South Attendance Center
3200 College Dr.
Emmetsburg, IA 50536
712-852-3554

Degree Granted: Certificate Program

12.118
Iowa State University
Ames, IA 50010
515-294-2751

Degrees Granted: Bachelor; Master; Doctorate
Cross-references: 6.10, 12.410, 15.58, 16.133, 20.109,
  20.276

12.119
Kirkwood Community College
6301 Kirkwood Blvd. S.W.
P.O. Box 2068
Cedar Rapids, IA 52406
319-398-5441

Degree Granted: Certificate Program

12.120
North Iowa Area Community College
500 College Dr.
Mason City, IA 50401

Degree Granted: Certificate Program

12.121
Western Iowa Tech Community College
4647 Stone Ave.
Sioux City, IA 51102

Degree Granted: Certificate Program

ID

12.122
Boise State University
1910 University Dr.
Boise, ID 83701
208-385-3984

Degree Granted: Certificate Program

12.123
Ricks College
Dept. of Horticulture
Rexburg, ID 83490
208-356-2017

Degree Granted: Bachelor
Cross-reference: 20.112

12.124
University of Idaho
Moscow, ID 83843
208-885-6111

Degrees Granted: Bachelor; Master; Doctorate
Cross-references: 12.411, 15.62, 16.136

IL

12.125
Belleville Junior College
2500 Carlyle Rd.
Belleview, IL 62221
618-235-2700

Degree Granted: Associate
Cross-reference: 20.114

12.126
College of Lake County
W. Washington St.
Grayslake, IL 60030

Degree Granted: Associate

12.127
Danville Junior College
2000 E. Main St.
Danville, IL 61832-0821
217-443-1811

Degree Granted: Associate

12.128
DuPage Horticultural School, Inc.
Box 342
W. Chicago, IL 60185

Degree Granted: Certificate Program

12.129
Harper Junior College
Algonquin & Roselle Rds.
Palatine, IL 60067
312-397-3000

Degree Granted: Associate

12.130
Highland Community College
Pearl City Rd.
Freeport, IL 61032
815-235-6121

Degree Granted: Associate

12.131
Illinois Central College
Horticulture Dept.
East Peoria, IL 61611
309-694-5011

Degree Granted: Associate
Cross-reference: 20.117

12.132
Illinois State University at Normal
Normal, IL 61761
309-438-2181

Degrees Granted: Bachelor; Master

12.133
Joliet Junior College
1216 Houbolt Ave.
Joliet, IL 60436-9352
815-729-9020

Degree Granted: Associate

12.134
Kishwaukee College
Malta, IL 60150

Degree Granted: Associate

12.135
McHenry County College
District 528
Crystal Lake, IL 60014
815-455-3700

Degree Granted: Associate

12.136
Sauk Valley College
R.R. 1
Dixon, IL 61021

Degree Granted: Certificate Program

12.137
Southern Illinois University
Carbondale, IL 62901
618-536-4405

Degrees Granted: Bachelor; Master
Cross-references: 15.66, 15.74

12.138
State Community College
417 Missouri Ave.
East St. Louis, IL 62201
618-274-6666

Degree Granted: Associate

12.139
Triton College
2000 Fifth Ave.
River Grove, IL 60171
312-456-0300

Degree Granted: Associate
Cross-reference: 20.119

12.140
University of Illinois
Dept. of Horticulture
125 Mumford Hall
1301 W. Gregory Dr.
Urbana, IL 61801
217-333-0351

Degrees Granted: Bachelor; Master; Doctorate
Cross-references: 12.412, 15.63, 15.68, 20.15

12.141
Wright College
N. Austin Ave.
Chicago, IL 60634

Degree Granted: Certificate Program

IN
12.142
Purdue University
Dept. of Horticulture
West Lafayette, IN 47907
317-494-4600

Degrees Granted: Bachelor; Master; Doctorate
Cross-references: 12.414, 15.79, 16.155

12.143
Vincennes University
1002 N. First St.
Vincennes, IN 47591
812-885-4266

Degree Granted: Associate

KS
12.144
Central Kansas Area Vocational Technical School
Hutchinson Community Junior College
Rt. 2
Hutchinson, KS 67501

Degree Granted: Associate

12.145
Friends University
Wichita, KS 67213
316-261-5842

Degree Granted: Bachelor

12.146
Kansas State University
Dept. of Horticulture
Waters Hall
Manhattan, KS 66502
913-532-6011

Degrees Granted: Bachelor; Master; Doctorate
Cross-references: 12.415, 13.22, 13.23, 15.82, 16.161

12.147
Kaw Area Vocational Technical School
5724 Huntoon
Topeka, KS 66604

Degree Granted: Certificate Program

12.148
Wichita Area Vo-Tech School
301 S. Grove
Wichita, KS 67211
316-833-2400

Degree Granted: Certificate Program

KY
12.149
Berea College
Horticulture Dept.
Berea, KY 40403
606-986-9341

Degree Granted: Bachelor

12.150
Eastern Kentucky University
Agriculture Dept.
Richmond, KY 40475
606-622-2031

Degree Granted: Bachelor

12.151
Jefferson State Vocational Agri-Business Center
2219 Lakeland Rd.
Anchorage, KY 40223

Degree Granted: Certificate Program

12.152
Morehead State University
Horticulture Dept.
Morehead, KY 40351
606-783-2221

Degree Granted: Bachelor

12.153
Murray State University
Murray, KY 42071
502-762-3380

Degree Granted: Bachelor

12.154
University of Kentucky
Dept. of Horticulture & Landscape Architecture
N318 Agriculture Science North
Lexington, KY 40546
606-257-1601

Degrees Granted: Bachelor; Master; Doctorate
Cross-references: 12.416, 15.87, 20.17

12.155
Western Kentucky University
Bowling Green, KY 42101
502-745-2551

Degree Granted: Bachelor

LA
12.156
Delgado College
New Orleans, LA 70119

Degree Granted: Associate

12.157
Louisiana State University and A & M University
Baton Rouge, LA 70803
504-338-1175

Degrees Granted: Bachelor; Master; Doctorate
Cross-references: 12.417, 15.89, 20.128

12.158
Louisiana Tech University
Ruston, LA 71270
318-257-3036

Degree Granted: Bachelor
Cross-reference: 15.90

12.159
McNeese State University
Horticulture Dept.
Lake Charles, LA 70601
318-475-5000

Degree Granted: Bachelor

12.160
Nichols State University
Dept. of Agriculture
University Station
P.O. Box 2016
Thibodaux, LA 70310
504-448-4872

Degree Granted: Associate

12.161
Southern Louisiana University
Hammond, LA 70401

Degree Granted: Bachelor

12.162
Southern University
Dept. of Plant & Soil Sciences
Box 11288
Baton Rouge, LA 70813
504-771-2440

Degree Granted: Bachelor

12.163
University of Southern Louisiana
USL Box 4492
Lafayette, LA 70501
318-231-6640

Degree Granted: Bachelor
Cross-references: 15.88, 16.166

MA
12.164
Becker Junior College
1003 Main St.
Leicester, MA 01524
617-892-3379

Degree Granted: Associate

12.165
Hampshire College
893 West St.
Amherst, MA 01002
413-549-4600

Degree Granted: Bachelor

12.166
Massachusetts Bay Community College
50 Oakland St.
Wellesley, MA 02181
617-237-1100

Degree Granted: Certificate Program

12.167
New England Wildflower Society
Hemenway Rd.
Framingham, MA 01701
508-877-6574

Degree Granted: Certificate Program
Cross-reference: 3.22

12.168
Norfolk County Agricultural School
460 Main St.
Walpole, MA 20281

Degree Granted: Certificate Program

12.169
Endicott College Center for Continuing Education
376 Hale St.
Beverly, MA 01915
508-927-2139

Degree Granted: Certificate Program

12.170
Springfield Technical Community College
Armory Square
Springfield, MA 01105

Degree Granted: Associate

12.171
University of Massachusetts
Dept. of Plant and Soil Sciences
French Hall
Amherst, MA 01003
413-545-2243

Degrees Granted: Associate; Bachelor; Master; Doctorate
Cross-references: 12.424, 12.425, 13.30, 13.99, 15.93, 16.179, 21.56

MD
12.172
Allegany Community College
Cumberland, MD 21502

Degree Granted: Associate

12.173
Charles County Community College
Box 910
Mitchell Rd.
La Plata, MD 20646
301-934-2251

Degrees Granted: Associate; Certificate Program

12.174
Dundalk Community College
7200 Sollers Point Rd.
Baltimore, MD 21222
301-282-6700

Degree Granted: Certificate Program

12.176
Howard Community College
Plant Science Program
Business, Math, Science Division
Little Patuxent Pkwy.
Columbia, MD 21044
301-992-4828

Degree Granted: Certificate Program

12.177
Prince George's Community College
Largo, MD 20772

Degree Granted: Associate

12.178
University of Maryland
Dept. of Horticulture
College Park, MD 20740
301-454-3938

Degrees Granted: Bachelor; Master; Doctorate; Certificate Program
Cross-reference: 15.102

12.179
University of Maryland
Eastern Shore
Princess Anne, MD 21853
301-651-2200

Degree Granted: Bachelor

ME
12.180
Southern Maine Vo-Tech Institute
Portland, ME 04104

Degree Granted: Associate

12.181
Southern Maine Vo-Tech Institute
Fort Rd.
South Portland, ME 04106
207-799-7303

Degree Granted: Certificate Program

12.182
University of Maine
Dept. of Plant & Soil Sciences
Orono, ME 04469
207-581-1110

Degrees Granted: Bachelor; Master; Doctorate
Cross-references: 15.103, 16.193, 20.134

MI
12.183
Kalamazoo Valley Community College
Kalamazoo, MI 49009

Degree Granted: Certificate Program

12.184
Macomb Community College
Center Campus 44575 Garfield
Mt. Clemens, MI 48044
313-286-2058

Degree Granted: Associate

12.185
Michigan State University
Dept. of Horticulture
A164 Plant and Soil Sciences Bldg.
East Lansing, MI 48824-1325
517-355-5180

Degrees Granted: Associate; Bachelor; Master; Doctorate
Cross-references: 12.427, 13.110, 15.110, 16.204

12.186
Oakland Community College
Auburn Hills Campus
2900 Featherstone Rd.
Auburn Heights, MI 48057

Degree Granted: Associate

12.187
Saint Clair Community College
323 Erie St.
Port Huron, MI 48060
313-984-3881

Degree Granted: Associate

12.188
Southeast Oakland Vocational Educational Center
5055 Delemere
Royal Oak, MI 48073
313-280-0600

Degree Granted: Certificate Program

12.189
Southwest Oakland Vocational Educational Center
1000 Beck Rd.
Wixam, MI 48096
313-624-6000

Degree Granted: Certificate Program

12.190
Southwestern Michigan College
Cherry Grove Rd.
Dowagiac, MI 49047

Degree Granted: Associate

12.191
Washtenaw Community College
Washtenaw County, MI 48197

Degree Granted: Associate

MN
12.192
916 Area Vocational Technical Institute
3300 Century Ave., North
White Bear Lake, MN 55110
612-770-2351

Degree Granted: Certificate Program

12.193
Anoka Area Vo-Tech Institute
Landscape Career Center
Box 191
Anoka, MN 55303

Degree Granted: Associate

12.194
Brainerd Area Vo-Tech School
300 Quince St.
Brainerd, MN 56401
218-828-5344

Degree Granted: Certificate Program

12.195
Hennepin Technical Center
District Office, 1820 N. Xenium Ln.
Plymouth, MN 55441
612-425-3800

Degree Granted: Associate

12.196
University of Minnesota
Technical College
Agriculture Division
Crookston, MN 56716
218-281-6510

Degree Granted: Associate

12.197
University of Minnesota
Technical College
St. Paul, MN 55108
612-625-5000

Degrees Granted: Bachelor; Master; Doctorate
Cross-references: 12.430, 15.113, 15.115

12.198
University of Minnesota
Technical College
Waseka, MN 56090
507-835-1000

Degree Granted: Associate

MO
12.199
Central Missouri State University
Warrensburg, MO 64093
816-429-4810

Degree Granted: Bachelor

12.200
East Central Junior College
P.O. Box 529
Union, MO 63084
314-583-5193

Degree Granted: Associate

12.201
Lincoln University
Jefferson City, MO 65101
314-681-5024

Degrees Granted: Bachelor; Master
Cross-reference: 20.145

12.203
Meramec Community College
11333 Big Bend Blvd.
St. Louis, MO 63122
314-966-7710

Degree Granted: Associate

12.204
Northwest Missouri State University
Maryville, MO 64468
816-562-1562

Degree Granted: Bachelor
Cross-reference: 12.431

12.205
Southeast Missouri State University
Cape Girardeau, MO 63701
314-651-2255

Degree Granted: Bachelor
Cross-reference: 20.148

12.206
Southwest Missouri State University
901 S. National
Springfield, MO 65802
417-836-5095

Degree Granted: Bachelor
Cross-reference: 15.117

12.207
University of Missouri
Dept. of Horticulture
1-40 Agriculture Bldg.
Columbia, MO 65211
314-882-7511

Degrees Granted: Bachelor; Master; Doctorate
Cross-references: 12.473, 12.474, 15.118, 16.217, 20.51, 20.149

MS
12.208
Alcorn State University
Lorman, MS 39096
601-877-6147

Degree Granted: Bachelor

12.209
Hinds Junior College
Horticulture Dept.
Raymond, MS 39154
601-857-5261

Degree Granted: Associate

12.210
Jones Junior College
Dept. of Horticultural Technology
Ellisville, MS 39437
601-477-9311

Degree Granted: Associate

12.211
Mississippi Gulf Coast Junior College
Dept. of Horticulture
P.O. Box 47
Perkinston, MS 39573
601-928-5211

Degree Granted: Associate

12.212
Mississippi State University
Dept. of Horticulture
P.O. Drawer T
Mississippi State, MS 39762
601-325-2323

Degrees Granted: Bachelor; Master; Doctorate
Cross-references: 12.432, 15.119, 15.120

MT
12.213
Montana State University
Dept. of Plant and Soil Sciences
Bozeman, MT 59717
406-994-4601

Degrees Granted: Bachelor; Master
Cross-references: 12.433, 15.122, 16.224

NC
12.214
Blue Ridge Technical Institute
Rt. 2
Flat Rock, NC 28731
704-692-3572

Degree Granted: Certificate Program
Cross-reference: 20.152

12.215
Catawba Valley Technical Institute
Highway 64-70
Hickory, NC 28601

Degree Granted: Certificate Program

12.216
Central Piedmont Community College
P.O. Box 4009
Charlotte, NC 28204

Degree Granted: Certificate Program

12.217
Fayette Technical Institute
P.O. Box 35236
Fayetteville, NC 28303
919-323-1961

Degree Granted: Certificate Program

12.218
Forsyth Technical Institute
2100 Silas Creek Pkwy.
Winston-Salem, NC 27103
919-723-1961

Degree Granted: Certificate Program

12.219
Haywood Technical College
P.O. Box 457
Clyde, NC 28721
704-627-2821

Degree Granted: Certificate Program
Cross-reference: 16.226

12.220
Lenoir Community College
Box 188
Kinston, NC 28501

Degree Granted: Certificate Program

12.221
Maryland Technical College
Box 547
Spruce Pine, NC 28777
704-765-7351

Degree Granted: Certificate Program

12.222
North Carolina Agricultural and Technical University
Horticulture Dept.
Greensboro, NC 27411
919-334-7520

Degree Granted: Bachelor
Cross-reference: 12.434

12.223
North Carolina State University
Dept. of Horticulture
Box 7609
Raleigh, NC 27650
919-737-3131

Degrees Granted: Associate; Bachelor; Master; Doctorate
Cross-references: 12.435, 15.125, 16.233

12.224
Piedmont Technical Institute
P.O. Box 1197
Roxboro, NC 27573

Degree Granted: Certificate Program

12.225
Randolph Technical College
Floral Design/Commercial Horticulture
P.O. Box 1009
Asheboro, NC 27204-1009
919-629-1471

Degree Granted: Associate

12.226
Sandhills Community College
Box 182C
Carthage, NC 28327

Degree Granted: Certificate Program
Cross-reference: 16.235

12.227
Surry Community College
P.O. Box 304
Dobson, NC 27017
919-386-8121

Degree Granted: Certificate Program

12.228
Tri-County Community College
P.O. Box 40
Murphy, NC 28906
704-837-6810

Degree Granted: Associate

12.229
Wilkes Community College
P.O. Drawer 120
Wilkesboro, NC 28697
919-667-7136

Degree Granted: Certificate Program

ND
12.230
North Dakota State University
Bottineau Branch and Institute of Forestry
Bottineau, ND 58318

Degree Granted: Associate

12.231
North Dakota State University
Dept. of Horticulture & Forestry
Fargo, ND 58102
701-237-8162

Degrees Granted: Bachelor; Master
Cross-reference: 12.436, 15.127, 20.158

NE
12.232
Central Technical Community College
Box 1024
Hastings, NE 68901
402-463-9811

Degree Granted: Associate

12.233
Metropolitan Technical Community College
P.O. Box 3777
2909 Edward Babe Gomez Ave.
Omaha, NE 68103
402-449-0400

Degree Granted: Associate

12.234
School of Technical Agriculture
University of Nebraska
Curtis, NE 69025
308-367-4124

Degree Granted: Associate

12.235
University of Nebraska
Dept. of Horticulture
377 Plant Science Bldg.
Lincoln, NE 68583-0724
402-472-2854

Degrees Granted: Bachelor; Master; Doctorate
Cross-references: 12.475, 15.129, 16.246, 16.247

NH
12.236
University of New Hampshire
Thompson School of Applied Sciences
Durham, NH 03824
603-862-1025

Degrees Granted: Associate; Bachelor; Master; Doctorate
Cross-references: 15.131, 16.250

NJ
12.237
Bergen County Community College
Horticulture Dept.
Hackensack, NJ 07601
201-447-7225

Degree Granted: Associate

12.238
Cumberland County Community College
P.O. Box 517
Vineland, NJ 08360
609-691-8600

Degree Granted: Associate

12.239
Mercer County Community College
1200 Old Trenton Rd.
P.O. Box B
Trenton, NJ 08690
609-586-4800

Degree Granted: Associate

12.240
Rutgers—The State University of New Jersey
New Brunswick, NJ 08903
201-932-9796

Degrees Granted: Bachelor; Master; Doctorate
Cross-references: 12.437, 15.133, 15.134, 16.261, 20.52,
20.164

12.241
Thomas A. Edison State College
Trenton, NJ 08625
609-984-1150

Degree Granted: Bachelor

NM
12.242
Eastern New Mexico University
Portales, NM 88130
505-562-2178

Degree Granted: Certificate Program

12.243
New Mexico State University
Las Cruces, NM 88003
505-646-3121

Degrees Granted: Bachelor; Master
Cross-references: 15.135, 16.267

12.244
University of New Mexico
Albuquerque, NM 87131
505-277-3430

Degree Granted: Certificate Program
Cross-references: 15.136, 16.265

NV
12.245
Clark Community College
3200 E. Cheyenne Ave.
North Las Vegas, NV 89030
702-564-7484

Degree Granted: Certificate Program

12.246
Luna Vocational Technical Institute
P.O. Box 2055
Las Vegas, NV 87701
505-454-1484

Degree Granted: Certificate Program

12.247
University of Nevada
Dept. of Plant, Soil, & Water Science
Reno, NV 89507
702-784-6600

Degrees Granted: Associate; Bachelor; Master
Cross-reference: 15.137

NY
12.248
Community College of Finger Lakes
Lincoln Hills Rd.
Canandaigua, NY 14424

Degree Granted: Associate

12.249
Cornell University
Dept. of Floriculture and Ornamental Horticulture
20 Plant Science
Ithaca, NY 14853
607-255-1789

Degrees Granted: Bachelor; Master; Doctorate
Cross-references: 12.439, 14.69, 15.142, 15.145, 15.147,
 16.275

12.250
New York Botanic Garden
Bronx Park
New York, NY 10458
212-220-8700

Degree Granted: Certificate Program
Cross-reference: 16.286

12.251
New York City Technical College
City University
300 Jay St.
P 507
Brooklyn, NY 11201
718-643-5110

Degree Granted: Associate

12.252
New York School of Interior Design
155 East 56th St.
New York, NY 10022
212-753-5365

Degree Granted: Certificate Program

12.253
Niagara County Community College
3111 Sanders Settlement Rd.
Sanborn, NY 14132

Degree Granted: Associate

12.254
State University of New York
Agricultural and Technical College at Cobleskill
Cobleskill, NY 12043

Degree Granted: Associate

12.255
State University of New York
Agricultural and Technical College of Morrisville
Morrisville, NY 13408

Degree Granted: Associate

12.256
State University of New York
College of Technology at Alfred
333 N. Main St.
Alfred, NY 14802
607-587-4111

Degree Granted: Associate

12.257
State University of New York
College of Technology at Delhi
Delhi, NY 13753

Degree Granted: Associate

12.258
State University of New York
College of Technology at Farmingdale
Farmingdale, NY 11735

Degree Granted: Associate

12.259
Suffolk County Community College
533 College Rd.
Selden, NY 11784

Degree Granted: Associate

12.260
Ulster County Community College
Stone Ridge Rd.
New Paltz, NY 12484

Degree Granted: Associate

OH
12.261
Cincinnati Technical College
3520 Central Pkwy.
Cincinnati, OH 45223
513-569-1500

Degree Granted: Associate

12.262
Clark Community College
570 E. Leffels Ln.
Springfield, OH 45501

Degree Granted: Associate

12.263
Ohio State University
Dept. of Horticulture
Columbus, OH 43210
614-292-0281

Degrees Granted: Bachelor; Master; Doctorate
Cross-references: 12.442, 15.155, 16.295, 16.309

12.264
Ohio State University
Agricultural Technical Institute
Star Rt. 250
Wooster, OH 44691
216-264-3911

Degree Granted: Associate
Cross-reference: 20.189

OK
12.265
Eastern Oklahoma State College
Wilburton, OK 74578

Degree Granted: Associate

12.266
Murray State Junior College
Tishomingo, OK 73460

Degree Granted: Certificate Program

12.267
Northeastern Oklahoma A&M College
Miami, OK 74354

Degree Granted: Associate

12.268
Northern Oklahoma College
Tonkawa, OK 74653

Degree Granted: Associate

12.269
Oklahoma State University
Dept. of Horticulture
Stillwater, OK 74075
405-744-5414

Degrees Granted: Bachelor; Master
Cross-references: 12.443, 12.476, 15.158

12.270
Oklahoma State University Technical Branch
900 N. Portland
Oklahoma City, OK 73107
405-945-3358

Degree Granted: Associate
Cross-reference: 20.195

OR
12.271
Clackamas Community College
19600 S. MoLalla Ave.
Oregon City, OR 97045
503-657-8400

Degree Granted: Associate
Cross-references: 1.62, 20.196

12.272
Linn-Benton Community College
6500 S.W. Pacific Blvd.
Albany, OR 97405
503-928-2361

Degree Granted: Associate

12.273
Mt. Hood Community College
26000 S.E. Stark St.
Gresham, OR 97030
503-667-7309

Degree Granted: Associate

12.274
Oregon State University
Dept. of Horticulture
Corvallis, OR 97331
503-754-2606

Degrees Granted: Bachelor; Master; Doctorate
Cross-references: 15.161, 16.332

12.275
Treasure Valley Community College
650 College Blvd.
Ontario, OR 97914

Degree Granted: Associate

PA
12.276
Delaware Valley College of Science and Agriculture
Ornamental Horticulture Dept.
Rt. 202
Doylestown, PA 18901
215-345-1500

Degree Granted: Bachelor
Cross-reference: 13.43, 16.344

12.277
Vocational Technical School Eastern Montgomery County
    Area
Willow Grove, PA 19090
215-657-7080

Degree Granted: Associate

12.278
Pennsylvania State University
Dept. of Horticulture
102 Tyson Bldg.
University Park, PA 16802
814-865-5403

Degrees Granted: Bachelor; Master; Doctorate;
    Certificate Program
Cross-references: 12.447, 12.477, 15.169, 20.25, 20.56,
    20.203, 20.288, 21.125

12.279
Temple University
Dept. of Landscape Architecture & Horticulture
Conwell Hall 041-09
Philadelphia, PA 19122-1803
215-787-7200

Degree Granted: Bachelor

12.280
Williamsport Area Community College
1005 W. 3rd St.
Williamsport, PA 17701

Degree Granted: Associate

PR
12.281
Agricultural Residential School
Bario Saltillo
Adjuntas, PR 00601
809-829-5290

Degree Granted: Certificate Program

12.282
University of Puerto Rico
Dept. of Horticulture
Mayaguez, PR 00708
809-832-4040

Degree Granted: Bachelor
Cross-reference: 15.172

12.283
University of Puerto Rico
Río Piedras, PR 00928
809-764-7290

Degree Granted: Associate

12.284
University of Puerto Rico
Utuado, PR 00761

Degree Granted: Associate

RI
12.285
University of Rhode Island
Dept. of Plant & Soil Sciences
Kingston, RI 02881
401-792-5996

Degrees Granted: Bachelor; Master; Doctorate
Cross-references: 12.450, 13.177, 20.208

SC
12.286
Clemson University
Dept. of Horticulture
Clemson, SC 29634
803-656-4964

Degrees Granted: Bachelor; Master; Doctorate
Cross-references: 12.451, 15.176, 16.366, 20.209

12.287
Horry-Georgetown Technical Education Center
P.O. Box 317
Conway, SC 29426

Degree Granted: Associate

12.288
Spartanburg County Technical Education Center
Interstate-85
Spartanburg, SC 29301
803-591-3857

Degree Granted: Associate

12.289
Trident Technical Center
7000 Rivers Ave.
N. Charleston, SC 29405

Degree Granted: Associate

SD
12.290
South Dakota State University
Dept. of Horticulture
Brookings, SD 57006
605-688-5136

Degree Granted: Bachelor
Cross-references: 12.452, 15.179, 16.368

12.291
Southeast Area Vo-Tech School
1001 East 14th St.
Sioux Falls, SD 57104
605-331-7624

Degree Granted: Certificate Program

TN
12.292
Austin Peay State University
Clarksville, TN 37044
615-648-7661

Degree Granted: Bachelor

12.293
Columbia State Community College
Hampshire Blvd.
Columbia, TN 38401
615-388-0120

Degree Granted: Associate

12.294
Jackson State Community College
Jackson, TN 38301

Degree Granted: Associate

12.295
Memphis State University
Memphis, TN 38152
901-454-2101

Degree Granted: Bachelor

12.296
Middle Tennessee State University
Murfreesboro, TN 37132
615-898-2300

Degree Granted: Bachelor

12.297
Tennessee State University
School of Agriculture
Dept. of Plant Science
Nashville, TN 37203
615-320-3335

Degrees Granted: Bachelor; Master

12.298
Tennessee Technological University
School of Agriculture
Box 5034
Cookeville, TN 38501
615-372-3019

Degree Granted: Bachelor
Cross-reference: 13.47, 13.180

12.299
University of Tennessee
Dept. of Ornamental Horticulture & Landscape Design
P.O. Box 1071
Knoxville, TN 37916
615-974-2184

Degrees Granted: Bachelor; Master
Cross-references:12.453, 15.181, 16.375, 20.28, 20.214

12.300
University of Tennessee
Martin, TN 38237
901-587-7020

Degree Granted: Bachelor
Cross-reference: 20.215

12.301
Walker State Community College
Morristown, TN 37813

Degree Granted: Associate

TX
12.302
Alvin Community College
3110 Mustang Rd.
Alvin, TX 77511
713-331-6111

Degree Granted: Associate

12.303
Benz School of Floral Design
Texas A&M University
Dept. of Horticulture
College Station, TX 77840
409-845-3841

Degree Granted: Bachelor

12.304
Blinn College
900 College Ave.
Brenham, TX 77833

Degree Granted: Associate

12.305
Cisco Junior College
Cisco, TX 76437

Degrees Granted: Associate; Bachelor

12.306
Galveston Community College
Galveston, TX 77553

Degree Granted: Associate

12.307
Houston Community College
Houston, TX 77020

Degree Granted: Associate

12.308
Kilgore College
1100 Broadway
Kilgore, TX 75662

Degree Granted: Certificate Program

12.309
Richland College
12800 Abrams
Dallas, TX 75374
214-746-4570

Degrees Granted: Associate; Bachelor

12.310
Sam Houston State University
Dept. of Horticulture
Huntsville, TX 77340
409-294-1215

Degrees Granted: Bachelor; Master

12.311
Southwest Texas State University
Dept. of Horticulture
San Marcos, TX 78667
512-245-2111

Degree Granted: Bachelor

12.312
Stephen F. Austin State University
Nacogdoches, TX 75963
409-568-2504

Degrees Granted: Bachelor; Master
Cross-reference: 15.191

12.313
Tarleton State University
Stephenville, TX 76401
817-968-9125

Degrees Granted: Bachelor; Master

12.314
Tarrant County Junior College—N.W.
4801 Marine Creek Pkwy.
Fort Worth, TX 76179
817-336-7851

Degree Granted: Associate

12.315
Texas A&M University
Dept. of Horticulture
College Station, TX 77843-2133
409-845-1699

Degrees Granted: Bachelor; Master; Doctorate
Cross-references: 12.454, 13.49, 15.188, 16.378, 20.29,
   20.223

12.316
Texas State Technical Institute
Waco, TX 76705

Degrees Granted: Associate; Bachelor

12.317
Texas Tech University
Lubbock, TX 79409

Degrees Granted: Bachelor; Master
Cross-reference: 12.455, 15.184

12.318
Trinity Valley College
Athens, TX 75151

Degree Granted: Associate

12.319
Tyler Junior College
E. Fifth St.
Tyler, TX 75712
214-531-2228

Degree Granted: Associate

UT
12.320
Brigham Young University
Provo, UT 84601
801-378-2507

Degrees Granted: Bachelor; Master
Cross-references: 15.196, 20.224, 20.302

12.321
Utah State University of Agriculture & Applied Sciences
Dept. of Plant Science
U.M.C. 12
Logan, UT 84322
801-750-2236

Degrees Granted: Bachelor; Master; Doctorate;
   Certificate Program
Cross-references: 12.457, 12.478, 15.195

VA
12.322
Christopher Newport College
Newport News, VA 23606
804-594-7015

Degree Granted: Bachelor
Cross-reference: 12.458

12.323
Norfolk School of Horticulture and Landscape Design
Norfolk Botanic Gardens
Airport Rd.
Norfolk, VA 23518
804-853-6972

Degree Granted: Associate

12.324
Northern Virginia Community College
1000 Harry Byrd Hwy.
Sterling, VA 22170

Degree Granted: Associate

12.325
Northern Virginia Community College
8333 Little River Turnpike
Annandale, VA 22003
703-323-3000

Degree Granted: Associate

12.326
Virginia Polytechnic Institute and State University
Dept. of Horticulture
Blacksburg, VA 24061
703-961-6254

Degrees Granted: Bachelor; Master; Doctorate
Cross-references: 12.460, 13.52, 15.201, 16.402, 20.232

12.327
Virginia State University
Dept. of Agriculture
Box F
Petersburg, VA 23803
804-520-5672

Degree Granted: Bachelor

VT
12.328
University of Vermont
Dept. of Plant & Soil Science
Hills Bldg.
Burlington, VT 05405
802-656-2630

Degrees Granted: Bachelor; Master; Doctorate
Cross-references: 15.202, 16.404

WA
12.329
Clark College
P.O. Box 639
Centralia, WA 98531
206-694-6521

Degree Granted: Associate

12.330
Clark College
Horticulture Dept.
1800 E. McLoughlin Blvd.
Vancouver, WA 98663
206-699-0115

Degree Granted: Associate

12.331
Columbia Basin Community College
2600 N. 20th Ave.
Pasco, WA 99301

Degree Granted: Associate

12.332
Edmonds Community College
2000 68th Ave., W.
Lynnwood, WA 98036
206-428-1149

Degree Granted: Associate

12.333
Everett Community College
801 Wetnore Ave.
Everett, WA 98201

Degree Granted: Associate

12.334
Skagit Valley College
Dept. of Agriculture
2405 College Way
Mount Vernon, WA 98273
206-428-1149

Degree Granted: Associate

12.335
South Puget Sound Community College
2011 Mottnan Rd., S.W.
Olympia, WA 98502

Degree Granted: Associate

12.336
South Seattle Community College
550 Mercer St.
Seattle, WA 98109
206-764-5300

Degree Granted: Associate

12.337
Spokane Community College
N. 1810 Green St.
Spokane, WA 99207

Degree Granted: Associate

12.338
University of Washington
Seattle, WA 98195
206-543-9686

Degrees Granted: Bachelor; Master; Doctorate
Cross-references: 1.40, 12.461, 15.205, 16.415

12.339
Washington State University
Dept. of Horticulture & Landscape Architecture
Pullman, WA 99164-6414
509-335-3564

Degrees Granted: Bachelor; Master; Doctorate
Cross-references: 12.462, 12.479, 15.203, 15.204

12.340
Wenatchee Valley College
1300 5th St.
Wenatchee, WA 98801

Degree Granted: Associate

12.341
Yakima Valley College
Yakima, WA 98902

Degree Granted: Associate

WI
12.342
Gateway Technical Institute
3520 30th Ave.
Kenosha, WI 53140
414-656-7551

Degree Granted: Associate

12.343
Milwaukee Area Technical College
Landscape Dept.
5555 Highland Rd.
Mequon, WI 53092
414-242-6500

Degree Granted: Certificate Program

12.344
Milwaukee Area Technical College
1015 N. 6th St.
Milwaukee, WI 55203

Degree Granted: Certificate Program

12.345
University of Wisconsin
Dept. of Horticulture & Landscape Architecture
1575 Linden Dr.
Madison, WI 53706
608-262-1490

Degrees Granted: Bachelor; Master; Doctorate
Cross-references: 12.463, 15.209, 16.420, 16.424

12.346
University of Wisconsin
River Falls, WI 54022
715-425-3100

Degree Granted: Bachelor

12.347
Wisconsin State University
Dept. of Horticulture & Landscape Architecture
Plattville, WI 53818-9989
608-342-1371

Degree Granted: Bachelor

WV

12.348
James Rumsey Vocational Technical Center
Rt. 6, Box 268
Martinsburg, WV 25401
304-754-7925

Degree Granted: Certificate Program

12.349
Potomac State College of West Virginia University
Horticulture Dept.
Keyser, WV 26726
304-788-3011

Degree Granted: Associate

12.350
West Virginia University
College of Agriculture and Forestry
Division of Plant & Soil Sciences
P.O. Box 6108
Morgantown, WV 26506
304-293-6023

Degrees Granted: Bachelor; Master; Doctorate
Cross-references: 12.464, 16.425

WY

12.351
University of Wyoming
Dept. of Plant Science
Laramie, WY 82071
307-766-2243

Degree Granted: Bachelor
Cross-reference: 15.219

CANADA

AB

12.352
Fairview College
P.O. Box 3000
Fairview, AB T0H 1L0

Degree Granted: Certificate Program

12.353
Northern Alberta Institute of Technology
11761-106 St.
Edmonton, AB T5G 2R1

Degree Granted: Certificate Program

12.354
Olds College of Agricultural Technology
Olds, AB T0M 1P0

Degree Granted: Certificate Program
Cross-references: 15.221, 20.252

12.355
University of Alberta
Faculty of Agriculture
Dept. of Plant Science
4-10 Agriculture Centre
Edmonton, AB T6G 2P5
403-492-3283

Degrees Granted: Bachelor; Master; Doctorate
Cross-references: 15.222, 16.432

12.356
University of Calgary
Faculty of Environmental Design
2500 University Dr., N.W.
Calgary, AB T2N 1N4
403-284-7427

Degree Granted: Bachelor
Cross-reference: 15.223

BC

12.357
British Columbia Institute of Technology
300 Willingdon Ave.
Burnaby, BC V5G 3H2
604-434-5734

Degree Granted: Certificate Program

12.358
Capilano College
2055 Purcell Way
North Vancouver, BC V7J 3H5
604-986-1911

Degree Granted: Certificate Program

12.359
Cariboo College
P.O. Box 3010
Kamloops, BC V2C 5N3
604-374-0123

Degree Granted: Certificate Program

12.360
East Kootenay Community College
Cranbrook, BC V1C 5L7

Degree Granted: Certificate Program

12.361
Malaspina College
900 Fifth St.
Nanaimo, BC V9R 5S5
604-753-3245

Degree Granted: Certificate Program

12.362
University of British Columbia
Plant Science
Vancouver, BC V6T 1X2
604-228-4384

Degrees Granted: Bachelor; Master; Doctorate;
   Certificate Program
Cross-references: 12.465, 13.54, 15.226, 16.449, 20.33

MB
12.363
University of Manitoba
Winnipeg, MB R3T 2N2

Degrees Granted: Bachelor; Master; Doctorate
Cross-references: 15.228, 20.34

NS
12.364
Nova Scotia Agricultural College
Plant Science Dept.
Truro, NS B2N 5E3
902-895-1571

Degree Granted: Certificate Program
Cross-reference: 20.256

12.365
Nova Scotia College of Art & Design
5163 Duke St.
Halifax, NS B3J 3J6
902-422-7381

Degree Granted: Bachelor

ON
12.366
Algonquin College of Applied Arts and Technology
Horticultural Program
Colonel By Campus, 140 Main St.
Ottawa, ON K1S 1C2
613-727-7606

Degree Granted: Certificate Program

12.367
Cambrian College of Applied Arts & Technology
1400 Barrydown Rd.
Sudbury, ON P3A 3V8
705-566-8101

Degree Granted: Certificate Program

12.368
College de Technologie Agricole et Alimentaire d'Alfred
C.P. 580
Rue St. Paul
Alfred, ON K0B 1A0
613-679-2218

Degree Granted: Certificate Program

12.369
Durham College of Applied Arts & Technology
P.O. Box 385
Simcoe St., North
Oshawa, ON L1H 7L7
416-576-0210

Degree Granted: Certificate Program

12.370
Fanshawe College of Applied Arts and Technology
1460 Oxford St.
East London, ON N5W 5H1

Degree Granted: Certificate Program

12.371
Humber College of Applied Arts & Technology
205 Humber College Blvd.
Rexdale, ON M9W 5L7
416-675-5085

Degree Granted: Certificate Program

12.372
Kemptville College of Agricultural Technology
Kemptville, ON K0G 1J0
613-258-3411

Degree Granted: Certificate Program

12.373
Kitchener-Waterloo School of Horticulture
39 Durward Pl.
Waterloo, ON N2L 4E5
519-884-1077

Degree Granted: Certificate Program

12.374
Niagara College of Applied Arts & Technology
Horticultural Centre
360 Niagara St.
St. Catharines, ON L2M 4W1
416-688-1380

Degree Granted: Associate

12.375
Niagara College of Applied Arts and Technology
Woodlawn Rd.
P.O. Box 1005
Welland, ON L3B 5S2

Degree Granted: Certificate Program

12.376
Niagara Parks Commission School of Horticulture
Box 150
Niagara Falls, ON L2E 6T2
416-356-8554

Degree Granted: Bachelor
Cross-references: 14.106, 16.469, 20.265

12.377
Ridgetown College of Agricultural Technology
Dept. of Horticulture
Ridgetown, ON N0P 2C0
519-674-5456

Degree Granted: Certificate Program
Cross-reference: 16.466

12.378
Ryerson Polytechnic Institute
350 Victoria St.
Toronto, ON M5B 2K3
416-979-5036

Degree Granted: Associate

12.379
Sault College of Applied Arts & Technology
443 Northern Ave.
P.O. Box 60
Sault Ste. Marie, ON P6A 5L3
705-949-2050

Degree Granted: Bachelor

12.380
Seneca College of Applied Arts and Technology
1750 Finch Ave.
East Willowdale, ON M2J 2X5

Degree Granted: Certificate Program

12.381
Seneca College of Applied Arts and Technology
King Campus, Dufferin St., North
R.R. 3
King City, ON L0G 1K0
416-833-3333

Degree Granted: Certificate Program

12.382
Sheridan College of Applied Arts & Technology
Burlington Campus 2319 Fairview St.
Burlington, ON L7R 2E3
416-632-7081

Degree Granted: Certificate Program

12.383
Sir Sandford Fleming College of Arts and Technology
Frost Campus
Box 8000
Lindsay, ON K9V 5E6
705-324-9144

Degree Granted: Certificate Program

12.384
St. Clair College of Applied Arts and Technology
2000 Talbot Rd., West
Windsor, ON N9A 6S4
519-972-2727

Degree Granted: Certificate Program

12.385
St. Lawrence College
Brockville Campus
2288 Parkdale
Brockville, ON K6V 5X3
613-345-0660

Degree Granted: Certificate Program

12.386
University of Guelph
Dept. of Horticulture
Guelph, ON N1G 2W1
519-824-4120

Degrees Granted: Bachelor; Master; Doctorate;
  Certificate Program
Cross-references: 12.466, 12.480, 15.243, 16.456

PQ
12.387
CEGEP De Victoriaville
475 est, rue Notre-Dame
Victoriaville, PQ G6P 4B3

Degree Granted: Certificate Program

12.388
Institut de Technologie Agricole de St-Hyacinthe
3230, rue Sicotte, C.P. 70
St. Hyacinthe, PQ J2S 2M2
514-773-7401

Degree Granted: Certificate Program

12.389
Laval University
(Université Laval)
Cite Universitaire
Ste.-Foy, PQ G1V 4C7

Degrees Granted: Bachelor; Master; Doctorate
Cross-references: 15.247, 16.478, 20.267

12.390
MacDonald College of Agriculture
Ste. Anne De Bellevue
Quebec City, PQ H9X 1C0
514-398-7802

Degree Granted: Associate

12.391
McGill University
MacDonald College
111 Lakeshore Rd.
Ste. Anne de Bellevue, PQ H9X 1C0
514-457-2000

Degrees Granted: Bachelor; Master; Doctorate
Cross-references: 15.248, 16.480, 20.268

QC
12.392
Université de Montréal
Faculté de l'amenagement
École d'architecture de paysage
5620, ave. Darlington
Montreal, QC H3T 2T1
514-343-7076

Degrees Granted: Bachelor; Master; Doctorate
Cross-references: 15.252, 16.481

SK
12.393
University of Saskatchewan
Saskatoon, SK S7N 0W0
306-343-3756

Degrees Granted: Bachelor; Master; Doctorate
Cross-references: 15.255, 16.483

LANDSCAPE DESIGN AND LANDSCAPE ARCHITECTURE
EDUCATIONAL PROGRAMS

UNITED STATES

AL
12.394
Alabama Agricultural and Mechanical University
P.O. Box 284
Normal, AL 35762
205-851-5245

Degree Granted: Bachelor
Cross-reference: 12.3

12.395
Auburn University
Dept. of Architecture Program
104 Dudley Hall
Auburn, AL 36849-5316
205-844-5424

Degree Granted: Bachelor; Accredited Program
Cross-references: 12.4, 15.2, 16.1

AR
12.396
University of Arkansas
School of Architecture
108 Vol Walker Hall
Fayetteville, AR 72701
501-575-2701

Degree Granted: Bachelor; Accredited Program
Cross-references:12.12, 15.7

AZ
12.397
Arizona State University
Tempe, AZ 85281
602-965-3255

Degree Granted: Bachelor

12.398
University of Arizona
School of Renewable Natural Resources
325 Biological Sciences East Bldg.
Tucson, AZ 85721
602-621-1004

Degree Granted: Bachelor; Accredited Program
Cross-references: 4.11, 12.19, 15.12

CA
12.399
California Polytechnic State University
Dept. of Landscape Architecture
School of Environmental Design
San Luis Obispo, CA 93407
805-756-1319

Degree Granted: Bachelor; Accredited Program
Cross-references: 12.24, 15.29, 20.68

12.400
California Polytechnic University
Dept. of Landscape Architecture
School of Environmental Design
3801 West Temple Ave.
Pomona, CA 91768
714-869-2673

Degrees Granted: Bachelor; Master; Accredited Program
Cross-references: 12.25, 20.69

12.401
University of California at Berkeley
Dept. of Landscape Architecture
College of Environmental Design
202 Wurster Hall
Berkeley, CA 94720
415-642-4022

Degrees Granted: Bachelor; Master; Accredited Program
Cross-references: 12.468, 15.24, 15.36, 16.18

12.402
University of California at Davis
Landscape Architecture Program
School of Architecture and Environmental Design
College of Agriculture and Environmental Sciences
Davis, CA 95616
916-752-6223

Degree Granted: Bachelor; Accredited Program
Cross-references: 12.62, 15.23, 16.59, 17.33, 19.12

CO
12.403
Colorado State University
Dept. of Recreation, Resources and Landscape
   Architecture
College of Forestry
Fort Collins, CO 80523
303-491-6591

Degree Granted: Bachelor; Accredited Program
Cross-references: 12.68, 12.469, 15.38, 16.69

12.404
University of Colorado at Denver
Landscape Architecture Program
School of Architecture and Planning
1200 Larimer St., Box 126
Denver, CO 80204-5300
303-556-4090

Degree Granted: Master; Accredited Program

CT
12.405
University of Connecticut
Plant Science Dept.
Storrs, CT 06268
203-486-3137

Degree Granted: Bachelor
Cross-references: 12.73, 15.42, 20.44

FL
12.406
Florida Agricultural and Mechanical University
Tallahassee, FL 32307
904-599-3796

Degree Granted: Bachelor

12.407
University of Florida
Dept. of Landscape Architecture
College of Architecture
331 Architecture Bldg.
Gainesville, FL 32611
904-392-6098

Degree Granted: Bachelor; Accredited Program
Cross-references: 12.99, 12.470, 15.51

12.408
University of Miami
P.O. Box 248025
Coral Gables, FL 33124
305-284-4323

Degree Granted: Bachelor

GA
12.409
University of Georgia
Dept. of Landscape Architecture
School of Environmental Design
609 Caldwell Hall
Athens, GA 30602
404-542-1816

Degrees Granted: Bachelor; Master; Accredited
  Program
Cross-references: 12.108, 12.471, 15.53, 15.54,
  20.11

IA
12.410
Iowa State University
Dept. of Landscape Architecture
College of Design, Room 146
Ames, IA 50011
515-294-5676

Degree Granted: Bachelor; Accredited Program
Cross-references: 12.118, 15.58, 16.133, 20.109, 20.276

ID
12.411
University of Idaho
Landscape Architecture Dept.
College of Art and Architecture
Moscow, ID 83843
208-885-6272

Degree Granted: Bachelor; Accredited Program
Cross-references: 12.124, 15.62, 16.136

IL
12.412
University of Illinois
Dept. of Landscape Architecture
College of Fine and Applied Arts
214 Mumford Hall
1301 W.Gregory Dr.
Urbana, IL 61801
217-333-0176

Degrees Granted: Bachelor; Master; Accredited Program
Cross-references: 12.140, 15.63, 15.68, 20.15

IN
12.413
Ball State University
Dept. of Landscape Architecture
College of Architecture and Planning
Muncie, IN 47306
317-285-1971

Degrees Granted: Bachelor; Master; Accredited Program

12.414
Purdue University
Landscape Architecture Program
Dept. of Horticulture
Horticulture Building
West Lafayette, IN 47907
317-494-1326

Degree Granted: Bachelor; Accredited Program
Cross-references: 12.142, 15.79, 16.155

KS
12.415
Kansas State University
Dept. of Landscape Architecture
College of Architecture and Design
215 Seaton Hall
Manhattan, KS 66506
913-532-5961

Degrees Granted: Bachelor; Master; Accredited Program
Cross-references: 12.146, 13.22, 15.82, 16.161

KY
12.416
University of Kentucky
Dept. of Horticulture and Landscape Architecture
Agriculture Science Center North
Lexington, KY 40546
606-257-3485

Degree Granted: Bachelor; Accredited Program
Cross-references: 12.154, 15.87, 20.17

LA
12.417
Louisiana State University
School of Landscape Architecture
College of Design Bldg.
Baton Rouge, LA 70803-7020
504-388-1434

Degrees Granted: Bachelor; Master; Accredited Program
Cross-references: 12.157, 15.89, 20.128

MA
12.418
Boston Architectural Center
617 Newbury St.
Boston, MA 02115
617-536-3170

Degree Granted: Bachelor

12.419
Conway School of Landscape Architecture, Inc.
Delabarre Ave.
Conway, MA 01341
413-369-4144

Degree Granted: Master

12.420
Endicott College Center for Continuing Education
376 Hale St.
Beverly, MA 01915
508-927-2139

Degree Granted: Certificate Program

12.421
Essex Agricultural and Technical Institute
562 Maple St.
Hathorne, MA 01937
508-774-0050

Degree Granted: Certificate Program

12.422
Harvard University
Dept. of Landscape Architecture
Harvard Graduate School of Design
409 Gund Hall, 48 Quincy St.
Cambridge, MA 02138
617-495-2573

Degree Granted: Master; Accredited Program
Cross-reference: 16.173

12.423
Radcliffe Seminars
6 Ash St.
Cambridge, MA 02138
617-495-8600

Degree Granted: Certificate Program

12.424
Stockbridge School of Agriculture
225 Whitmore Bldg.
University of Massachusetts
Amherst, MA 01003
413-545-2222

Degree Granted: Associate
Cross-references: 12.171, 12.425, 13.30, 13.99, 15.93, 16.179

12.425
University of Massachusetts
Dept. of Landscape Architecture and Regional
  Planning
Hills North 109
Amherst, MA 01003
413-545-2255

Degrees Granted: Bachelor; Master; Accredited
  Program
Cross-references: 12.171, 12.424, 13.30, 13.99, 15.93,
  16.179

ME
12.426
College of the Atlantic
Bar Harbor, ME 04609
207-288-5015

Degree Granted: Bachelor

MI
12.427
Michigan State University
Landscape Architecture Program
Dept. of Geography
East Lansing, MI 48824-1221
517-353-7880

Degree Granted: Bachelor; Accredited Program
Cross-references:12.185, 13.110, 15.110, 16.204

12.428
University of Michigan
Landscape Architecture
School of Natural Resources
Dana Building, Room 1548
Ann Arbor, MI 48109-1115
313-763-9214

Degree Granted: Master; Accredited Program
Cross-references: 15.111, 16.203

MN
12.429
University of Minnesota
Twin Cities Campus
Minneapolis, MN 55455
612-625-2006

Degree Granted: Bachelor
Cross-reference: 12.472

12.430
University of Minnesota
Dept. of Landscape Architecture
College of Architecture and Landscape Architecture
205 North Hall, 2005 Buford Circle
St. Paul, MN 55108
612-625-8285

Degree Granted: Bachelor; Accredited Program
Cross-references: 12.197, 15.113, 15.115

MO
12.431
Northwest Missouri State University
Maryville, MO 64468
816-562-1562

Degree Granted: Bachelor
Cross-reference: 12.204

MS
12.432
Mississippi State University
Dept. of Landscape Architecture
College of Agriculture and Home Economics
P.O. Box MQ, Montgomery Hall
Room 100
Mississippi State, MS 39762
601-325-3012

Degree Granted: Bachelor; Accredited Program
Cross-references: 12.212, 15.119, 15.120

MT
12.433
Montana State University
Bozeman, MT 59715
406-994-2452

Degree Granted: Bachelor
Cross-references: 12.213, 15.122, 16.224

NC
12.434
North Carolina Agricultural and Technical State
　University
Greensboro, NC 27411
919-334-7946

Degree Granted: Bachelor
Cross-reference: 12.222

12.435
North Carolina State University
Landscape Architecture Dept.
School of Design
P.O. Box 7701
Raleigh, NC 27695-7701
919-737-2206

Degrees Granted: Bachelor; Master; Accredited
　Program
Cross-references: 12.223, 15.125, 16.233

ND
12.436
North Dakota State University
Fargo, ND 58105
701-237-8643

Degree Granted: Bachelor
Cross-references: 12.231, 15.127, 20.158

NJ
12.437
Rutgers—The State University of New Jersey
Dept. of Landscape Architecture
Blake Hall
Cook College, Box 231
New Brunswick, NJ 08903
201-932-9317

Degree Granted: Bachelor; Accredited Program
Cross-references: 12.240, 15.133, 15.134, 16.261, 20.52,
　20.164

NY
12.438
City College of New York
Urban Landscape Architecture Program
School of Architecture and Environmental Studies
138th St. and Convent Ave.
New York, NY 10031
212-690-4118

Degree Granted: Bachelor; Accredited Program

12.439
Cornell University
Landscape Architecture Program
230 East Roberts Hall
Ithaca, NY 14853
607-255-4487

Degrees Granted: Bachelor; Master; Accredited Program
Cross-references: 12.249, 14.69, 15.142, 15.145, 15.147,
　16.275

12.440
Friends World College
Huntington, NY 11743
516-549-1102

Degree Granted: Bachelor

12.441
State University of New York
Faculty of Landscape Architecture
College of Environmental Studies and Forestry
Syracuse, NY 13210
315-470-6541

Degrees Granted: Bachelor; Master; Accredited Program
Cross-reference: 15.146

OH
12.442
Ohio State University
Dept. of Landscape Architecture
School of Architecture
136B Brown Hall, 190 West 17th Ave.
Columbus, OH 43210-1369
614-292-8263

Degrees Granted: Bachelor; Master; Accredited Program
Cross-references: 12.263, 15.155, 16.295, 16.309

OK
12.443
Oklahoma State University
Stillwater, OK 74078
405-744-6876

Degree Granted: Bachelor
Cross-references: 12.269, 12.476, 15.158

12.444
University of Oklahoma
407 West Boyd
Norman, OK 73069
405-325-2151

Degree Granted: Bachelor

OR
12.445
University of Oregon
Dept. of Landscape Architecture
School of Architecture and Allied Arts
216 Lawrence Hall
Eugene, OR 97403
503-686-3634

Degree Granted: Bachelor; Accredited Program
Cross-reference: 15.162

PA
12.446
Pennsylvania State University
University Park Campus
State College, PA 16802
814-865-5471

Degree Granted: Bachelor

12.447
Pennsylvania State University
Dept. of Landscape Architecture
College of Arts and Architecture
210 Engineering Unit D
University Park, PA 16802
814-865-9511

Degree Granted: Bachelor; Accredited Program
    Cross-references: 12.278, 12.477, 15.169, 20.25, 20.56,
    20.203, 20.288

12.448
University of Pennsylvania
Dept. of Landscape Architecture and Regional Planning
    Graduate School of Fine Arts
119 Graduate School
Philadelphia, PA 19104
215-898-6591

Degree Granted: Master; Accredited Program
Cross-references: 15.166, 16.351

RI
12.449
Rhode Island School of Design
Dept. of Landscape Architecture
Division of Architectural Studies
2 College St.
Providence, RI 02903
401-331-3511

Degree Granted: Bachelor; Accredited Program

12.450
University of Rhode Island
Kingston, RI 02881
401-792-9800

Degree Granted: Bachelor
Cross-references: 12.285, 20.208

SC
12.451
Clemson University
Clemson, SC 29631
803-656-2287

Degree Granted: Bachelor
Cross-references: 12.286, 15.176, 16.366, 20.209

SD
12.452
South Dakota State University
Brookings, SD 57007
605-688-4121

Degree Granted: Bachelor
Cross-references: 12.290, 15.179, 16.368

TN
12.453
University of Tennessee
Knoxville, TN 37996
615-974-2184

Degree Granted: Bachelor
Cross-references: 12.299, 15.181, 16.375, 20.28, 20.214

TX
12.454
Texas A&M University
Dept. of Landscape Architecture
College of Architecture and Environmental Design
321 Landford Architecture Center
College Station, TX 77843-3137
409-845-1019

Degrees Granted: Bachelor; Master; Accredited Program
Cross-references: 12.315, 13.49, 15.188, 16.378, 20.29,
    20.223

12.455
Texas Tech University
Dept. of Park Administration and Landscape Architecture
College of Agricultural Sciences
P.O. Box 4169
Lubbock, TX 79409
806-742-2858

Degree Granted: Bachelor; Accredited Program
Cross-references: 12.317, 15.184

12.456
University of Texas at Arlington
Arlington, TX 76019
817-273-2119

Degree Granted: Bachelor

UT
12.457
Utah State University
Dept. of Landscape Architecture and Environmental
  Planning
College of Humanities, Arts, and Social Sciences
Logan, UT 84322-4005
801-750-3471

Degrees Granted: Bachelor; Master; Accredited Program
Cross-references: 12.321, 12.478, 15.195

VA
12.458
Christopher Newport College
Newport News, VA 23606
804-594-7015

Degree Granted: Bachelor
Cross-reference: 12.322

12.459
University of Virginia
Division of Landscape Architecture
School of Architecture
Campbell Hall
Charlottesville, VA 22903
804-924-3957

Degree Granted: Master; Accredited Program

12.460
Virginia Polytechnic Institute and State University
Landscape Architecture Department
College of Architecture and Urban Studies
202 Architecture Annex
Blacksburg, VA 24061
703-231-5506

Degrees Granted: Bachelor; Master; Accredited Program
Cross-references: 12.326, 13.52, 15.201, 16.402, 20.232

WA
12.461
University of Washington
Dept. of Landscape Architecture
College of Architecture and Urban Planning
348 Gould Hall, JO-34
Seattle, WA 98195
206-543-9240

Degrees Granted: Bachelor; Master; Accredited Program
Cross-references: 1.40, 12.338, 15.205, 16.415

12.462
Washington State University
Dept. of Horticulture and Landscape Architecture
College of Agriculture and Home Economics
Johnson Hall 149
Pullman, WA 99164-6414
509-335-9502

Degree Granted: Bachelor; Accredited Program
Cross-references: 12.339, 12.479, 15.203, 15.204

WI
12.463
University of Wisconsin
Dept. of Landscape Architecture
College of Agricultural and Life Sciences
25 Agricultural Hall
1450 Linden Dr.
Madison, WI 53706
608-263-7300

Degree Granted: Bachelor; Accredited Program
Cross-references: 12.345, 15.209, 16.420, 16.424

WV
12.464
West Virginia University
Landscape Architecture Program
Division of Resource Management
College of Agriculture and Forestry
P.O. Box 61
1140 Agricultural Sciences Bldg.
Morgantown, WV 26506-6108
304-293-2141

Degree Granted: Bachelor; Accredited Program
Cross-references: 12.350, 16.425

CANADA

BC
12.465
University of British Columbia
Vancouver, BC V6T 1Z2
604-228-2953

Degree Granted: Bachelor
Cross-references: 12.362, 13.54, 15.226, 16.449, 20.33

ON
12.466
University of Guelph
School of Landscape Architecture
Guelph, ON N1G 2W1
519-824-4120 x3352

Degrees Granted: Bachelor; Master; Accredited Program
Cross-references: 12.386, 12.480, 15.243, 16.456

12.467
University of Toronto
Landscape Architecture Program
230 College St.
Toronto, ON M5S 1A1
416-978-6788

Degree Granted: Bachelor; Accredited Program
Cross-references: 15.237, 15.245

CORRESPONDENCE AND HOME STUDY PROGRAMS

The following is a list of correspondence and home study programs available in the United States and Canada.

UNITED STATES

CA
12.468
University of California Extension
Student Services Supervisor
2223 Fulton St.
Berkeley, CA 94720
415-642-4124

Subject: Horticulture
Level of Instruction: High School, Noncredit
Cross-reference: 12.401

CO
12.469
Colorado State University
Division of Continuing Education
Telecommunications Extended Studies
Spruce Hall
Fort Collins, CO 80523
303-491-5608

Subject: Turf Grass
Level of Instruction: College, Noncredit
Cross-references: 12.68, 12.403

FL
12.470
University of Florida
Division of Continuing Education
Dept. of Independent Study
Gainesville, FL 32611
904-392-1711

Subject: Landscaping
Level of Instruction: College
Cross-references: 12.99, 12.407

GA
12.471
University of Georgia
Associate Director for Academic Credit
Georgia Center for Continuing Education
Athens, GA 30602
404-542-6400

Subject: Horticulture
Level of Instruction: College
Cross-references: 12.108, 12.409

KS
Kansas State University
Manhattan, KS

Subject: Horticultural Therapy
Cross-reference: 13.23

MN
12.472
University of Minnesota
45 Westbrook Hall
77 Pleasant St., S.E.
Minneapolis, MN 55455
612-624-0000

Subject: Landscaping
Level of Instruction: College
Cross-reference: 12.429

MO
12.473
University of Missouri
Columbia, MO 65201
314-882-7651

Subject: Horticulture
Level of Instruction: College
Cross-references: 12.207, 12.474

MO
12.474
University of Missouri-Columbia
Center for Independent Study
136 Clark Hall
Columbia, MO 65211
314-882-6431

Subject: Horticulture
Level of Instruction: High School
Cross-references: 12.207, 12.473

NE
12.475
University of Nebraska
269 Nebraska Center for Continuing Education
33rd and Holdrege Sts.
Lincoln, NE 68583-0900
402-472-1926

Subject: Horticulture
Level of Instruction: High School
Cross-reference: 12.235

OK
12.476
Oklahoma State University
001 Classroom Bldg.
Stillwater, OK 74078
405-744-6390

Subject: Horticulture
Level of Instruction: College
Cross-references: 12.269, 12.443

PA
12.477
Pennsylvania State University
Center for Independent Learning
Mitchell Bldg.
University Park, PA 16802

Subject: Horticulture
Level of Instruction: College
Cross-references: 12.278, 12.447

UT
12.478
Utah State University
Extension Services
Independent Study
Logan, UT 84321
801-750-1105

Subjects: Horticulture; Landscaping; Turf Grass
Level of Instruction: College
Cross-references: 12.321, 12.457

WA
12.479
Washington State University
Independent Study Program
Room 8, Van Doren Hall

Pullman, WA 99164-6242
509-335-3357

Subjects: Horticulture; Turf Grass
Level of Instruction: College
Cross-references: 12.339, 12.462

CANADA

ON
12.480
University of Guelph
Correspondence Study
Office of Continuing Education
Guelph, ON N1G 2W1
519-824-4120

Subjects: Horticulture; Landscaping; Turf Grass
Level of Instruction: College
Cross-references: 12.386, 12.466

MASTER GARDENER PROGRAM

This horticultural education program is sponsored by the U.S. Department of Agricultural Cooperative Extension Service (and in Canada, the Provincial Department of Agriculture). In exchange for a certain number of hours of training, Master Gardeners agree to serve an equal number of hours in some form of community service in horticulture. The first Master Gardener Program was launched in Washington State in 1972. Since then, it has expanded to hundreds of programs in forty-six states and four provinces.

In some states the Master Gardener program is centrally run, in others, a recommended or mandatory curriculum is provided for use by local jurisdictions, and in still others, the training program is developed by the local agent. Specialists from the state land grant universities (or from the Provincial Department of Agriculture) and other local experts usually participate in the instruction. Since programs are tailored to local needs and resources, they vary considerably among jurisdictions. The number of hours of training offered ranges from 30 to 120, and fees charged range from none to several hundred dollars. In some areas, advanced training and refresher courses are offered.

The most common form of volunteer payback is staffing plant clinics and plant information telephone lines. Master Gardeners also take part in community beautification projects, work with children in school or 4-H programs, research and write horticultural fact sheets, test and evaluate new varieties of plants, write gardening columns, speak to citizen groups, and bring horticultural therapy to special groups such as the physically or mentally disabled or non-English speaking communities.

For more information contact your local extension office, the state Master Gardener coordinators listed below, or the Master Gardeners International Corporation (1.79).

UNITED STATES

AK
12.481
Wayne Vandre
2221 E. Northern Lights Blvd., # 118
Anchorage, AK 99508
907-279-6575

Cross-reference: 11.2

AL
12.482
Tony Glover
Montgomery County Extension
4576 South Court St.
Montgomery, AL 36125-0327
205-281-1292

AR
12.483
Gail Lee
P.O. Box 391
Little Rock, AR 72203
501-671-2000

Cross-reference: 11.4

AZ
12.484
Terry H. Mikel
Maricopa County Cooperative Extension
4341 E. Broadway
Phoenix, AZ 85040
602-255-4456

CA
12.485
Dennis Pittenger
Cooperative Extension
4114 Batchelor Hall
UC Riverside
Riverside, CA 92521
714-787-3320

Cross-reference: 11.8

CO
12.486
James R. Feucht
15200 W. 6th Ave.
Golden, CO 80401
303-277-8980

CT
12.487
Edmond Marrotte
Cooperative Extension
1376 Storrs Rd., Box U-67
Storrs, CT 06268
203-486-3435

Cross-reference: 11.10

DC
12.488
Pamela Marshall
CES/UDC
901 Newton St., N.E.
Washington, DC 20017
201-576-6950

Cross-reference 11.11

DE
12.489
Susan S. Barton
Townsend Hall
University of Delaware
Newark, DE 19717-1303
302-451-2532

Cross-reference: 11.12

FL
12.490
Dr. Robert J. Black
2551 Fifield Hall
University of Florida
Gainesville, FL 32611
904-392-1835

Cross-reference: 11.13

GA
12.491
Wayne J. McLaurin
Extension Horticulture
University of Georgia
Athens, GA 30602
404-542-2340

Cross-reference: 11.14

HI
12.492
H. Dale Sato
1420 Lower Campus Rd.
Honolulu, HI 96822
808-956-7138

Cross-reference: 11.16

IA
12.493
Linda Naeve
125 Horticulture
Iowa State University
Ames, IA 50011
515-294-0028

Cross-reference: 11.17

ID
12.494
Susan Bell
5880 Glenwood Ave.
Boise, ID 83714
208-377-2107

IL
12.495
Floyd Giles
College of Agriculture
1201 Plant Sciences Bldg.
Urbana, IL 61801
217-333-2125

Cross-reference: 11.19

IN
12.496
B. Rosie Lerner
Dept. of Horticulture
Purdue University
West Lafayette, IN 47907
317-494-1311

Cross-reference: 11.20

KS
12.497
Charles Marr
Dept. of Horticulture
Waters Hall, Kansas State University
Manhattan, KS 66506
913-532-6173

Cross-reference: 11.21

KY
12.498
Mary Witt
Room N-318 Ag. Sci. Bldg. North
University of Kentucky
Lexington, KY 40546-0091
606-257-3249

Cross-reference: 11.22

LA
12.499
Dr. Tom Koske
LSU Agriculture Center-Knapp Hall
University Station
Baton Rouge, LA 70803
504-388-2222

Cross-reference: 11.23

MA
12.500
Sandra Foss White
State Contact
UMass Suburban Experiment Station
240 Beaver St.
Waltham, MA 02154-8096
617-891-0650

MD
12.501
Denise Sharp
Home & Garden Information Center
12005 Homewood Rd.
Ellicott City, MD 21043
301-531-5556

ME
12.502
Dick Brzozowski
Cumberland County Extension
96 Falmouth St.
Portland, ME 04103
207-780-4205

MI
12.503
Tom Stebbins
A240 Plant & Soil Science Bldg.
Michigan State University
East Lansing, MI 48824
517-353-3774

Cross-reference: 11.27

MN
12.504
Michael E. Zins & Dr. Anne Hanchek
3675 Arboretum Dr.
Chanhassen, MN 55317
612-443-2460

MO
12.505
Barbara Fick
University Extension
1-40 Agriculture Bldg.
Columbia, MO 65211
314-882-7511

Cross-reference: 11.29

MS
12.506
Dr. Milo Burnham
P.O. Box 5446
Mississippi State University
Mississippi, MS 39762
601-325-3935

Cross-reference: 11.30

MT
12.507
Laurence Hoffman
P.O. Box 855
Helena, MT 59601
406-443-1010

NC
12.508
Larry Bass
Box 7609, Kilgore Hall
North Carolina State University
Raleigh, NC 27695-7609
919-737-3113

Cross-reference: 11.32

ND
12.509
Dr. Ronald Smith
Box 5658
North Dakota State Extension
Fargo, ND 58105
701-237-8161

Cross-reference: 11.33

NE
12.510
Susan Schoneweis
377 Plant Science Bldg.
University of Nebraska
Lincoln, NE 68583-0724
402-472-1128

Cross-reference: 11.34

NH
12.511
Margaret Pratt
Hillsborough County Extension
Rt. 13 South
Milford, NH 03055
603-673-2510

NJ
12.512
Eunice Au
Rutgers CES
P.O. Box 231
New Brunswick, NJ 08903-0231
908-932-6633

Cross-reference: 11.36

NM
12.513
Lynn Ellen Doxon
9301 Indian School Rd., N.E.
Albuquerque, NM 87112
505-275-2576

Cross-reference: 11.37

NV
12.514
Richard Post
675 Fairview Dr., Suite 229
Carson City, NV 89702
702-887-2252

NY
12.515
Robert Kozlowski
20 Plant Science Bldg.
Cornell University
Ithaca, NY 14853-5905
607-255-1791

Cross-reference: 11.39

OH
12.516
Jack Kerrigan
Cuyahoga County Cooperative Extension Service
3200 West 65th St.
Cleveland, OH 44102
216-631-1890

OK
12.517
Bill Geer
930 N. Portland
Oklahoma City, OK 73107
405-278-1125

OR
12.518
Ray A. McNeilan
211 SE 80
Portland, OR 97215
503-254-1581

PA
12.519
J. Robert Nuss
102 Tyson Bldg.
University Park, PA 16802
814-863-2196

Cross-reference: 11.43

RI
12.520
Rosanne Sherry
URI CE Education Center
Kingston, RI 02881-0804
401-792-2900

Cross-reference: 11.44

SC
12.521
Robert F. Polomski
Room E-143 B, Poole Agriculture Center
Clemson University
Clemson, SC 29634-0375
803-656-2604

Cross-reference: 11.45

SD
12.522
David Graper
South Dakota State University
Box 2207-C
Brookings, SD 57007-0996
605-688-6253

Cross-reference: 11.46

TN
12.523
A. D. Rutledge
P.O. Box 1071
University of Tennessee
Knoxville, TN 37901-1071
615-974-7208

Cross-reference: 11.47

TX
12.524
Dr. Douglas F. Welch
225 Horticulture/Forestry Science Bldg.
College Station, TX 77843
409-845-7341

Cross-reference: 11.48

UT
12.525
William Varga
1817 N. Main
Farmington, UT 84025
801-451-3204

VA
12.526
David McKissack
Consumer Horticulture
407 Saunders, VPI & SU
Blacksburg, VA 24061-0327
703-231-6254

Cross-reference: 11.50

VT
12.527
Leonard Perry
Hills Bldg.
University of Vermont
Burlington, VT 05405
802-656-2630

Cross-reference: 11.51

WA
12.528
Van Bobbitt
7612 Pioneer Way East
Puyallup, WA 98371-4998
206-840-4500

WI
12.529
Dr. Helen Harrison
Dept. of Horticulture
University of Wisconsin
Madison, WI 53706
608-262-1749

Cross-reference: 11.53

WV
12.530
Richard Zimmerman
P.O. Box 6108, West Virginia University
Morgantown, WV 26506-6108
304-293-4801

Cross-reference: 11.54

WY
12.531
Rodney Davis
2011 Fairgrounds Road
Casper, WY 82604
307-235-9400

CANADA

BC
12.532
Master Gardener Program
c/o VanDusen Botanical Display Garden
5251 Oak Street
Vancouver, BC V6M 4H1
604-266-7194

Cross-reference: 16.450

ON
12.533
Ruth Friendship-Keller
Rural Organizations Consultant
Guelph Agriculture Centre
P.O. Box 1030
Guelph, ON N1H 6N1
519-767-3540

PE
12.534
Margaret Drake
Master Gardener Program
Crops Section
Dept. of Agriculture, P.O. Box 1600
Charlottetown, PE C1A 7N3
902-368-5622

SK
12.535
Bruce Hobin
Master Gardener Program
Division of Extension
University of Saskatchewan
Saskatoon, SK S7N 0W0
306-966-5551

~ 13 ~

Horticultural Therapy

Throughout the United States and Canada, horticulture is increasingly used as a therapeutic tool for the elderly and for people who are physically, mentally, or developmentally disabled, substance abusers, public offenders, or socially disadvantaged. Many horticulture programs have been established to improve the social, educational, psychological, and physical well-being of such persons.

This chapter is divided into three parts. The first section lists contacts for more information on horticultural therapy. The second section lists schools, public gardens, arboreta, and other institutions that offer educational programs and internships in horticultural therapy. The third section lists institutions that offer horticultural therapy to various disadvantaged populations.

For more information on horticultural therapy, contact the American Horticultural Therapy Association (1.17) or the Canadian Horticultural Therapy Association(1.130). The following regional chapter contacts of the American Horticultural Therapy Association also may be able to provide more information about horticultural therapy programs in your area.

REGIONAL CHAPTERS OF THE AMERICAN HORTICULTURAL THERAPY ASSOCIATION

13.1
Carolina Chapter
John Paul Breault, President
A-9 Brookhill
100 Tobacco Rd.
Greenville, NC 27834
919-559-5116

13.2
Central Rocky Mountain Chapter
David Hackenberry, President
2951 E. Hwy. 50
Canon City, CO 81212
303-465-3602

13.3
Delaware Valley Chapter
Linda Ciccantelli, President
211 E. Highland Ave.
Philadelphia, PA 19118
215-587-3422

13.4
Florida Chapter
Terri Martin-Yates, President
950 West Wisconsin Ave.
Deland, FL 32720
904-258-3441

13.5
Horticultural Therapy Association of Greater New York
Glass Garden/Rusk Institute
400 E. 34th St.
New York, NY 10016
212-340-6058

Contact: Nancy K. Chambers

13.6
Kansas State University Chapter
Rita Stevens-LeRoy, President
Kansas State University
Dept. of Horticulture, Waters Hall
Manhattan, KS 66506
913-532-6170

13.7
Michigan Chapter
David H. Houseman, President
Office of Services to the Aging
P.O. Box 30026
Lansing, MI 48909
517-373-9362

13.8
Mid-South Chapter
Deborah Levine, President
460 Downing St.
Jackson, MS 39216
601-939-1221

13.9
New England Chapter
Deborah Krause, President
Perkins School for the Blind
175 N. Beacon St.
Watertown, MA 02172
617-924-3434

13.10
Ohio Chapter
Ellen McCurdy, President
3065 E. Derbyshire
Cleveland Heights, OH 44118
216-721-1600

13.11
Texas Chapter
Terry Quinn, President
2414 Morse # 4
Houston, TX 77019
713-528-6371

HORTICULTURAL THERAPY EDUCATIONAL PROGRAMS AND INTERNSHIPS

UNITED STATES

CA
13.12
Casa Colina Development Center
1324 Arrow Hwy.
LaVerne, CA 91750
714-596-5981

Contact: Micki Spencer
Internship
Cross-reference: 13.58

13.13
Kingsview Hospital
42675 Road 44
Reedley, CA 93654
209-638-2505

Contact: Mary Beth Jantzen
Internship
Cross-reference: 13.63

13.14
Living Desert, The
Palm Desert, CA

Non-degree Program
Cross-reference: 16.34

13.15
Sherman Library and Gardens
Corona Del Mar, CA

Contact: Wade Roberts
Non-degree Program
Cross-reference: 16.55

CO
13.16
Denver Botanic Gardens
Denver, CO

Non-degree Program
Internship
Cross-references: 13.66, 16.65

FL
13.17
Florida State Hospital
Horticultural Therapy Dept.
P.O. Box 103
Chattahoochee, FL 32324
904-663-7393

Contact: Marie Ansley
Internship
Cross-reference: 13.73

GA
13.18
Atlanta Botanical Garden
Atlanta, GA

Contact: Brenda Dreyer
Non-degree Program
Cross-reference: 16.103

IL
13.19
Chicago Botanic Garden
Glencoe, IL

Contact: Eugene Rothert
Non-degree Program
Internship
Cross-references: 13.83, 16.138

KS
13.20
Big Lakes Development Center
1554 Hayes Dr.
Manhattan, KS 66502
913-776-7012

Contact: Debbie Springer
Internship
Cross-reference: 13.89

13.21
Good Samaritan Center
416 West Spruce
Junction City, KS 66441
913-238-1187

Contact: Dorothy Fredericks
Internship
Cross-reference: 13.90

13.22
Kansas State University
Dept. of Horticulture
Manhattan, KS
913-532-6160

Contact: Dr. Richard H. Mattson
Degrees granted: Bachelor; Master
Cross-references: 12.146, 12.415, 13.23

13.23
Kansas State University
Division of Continuing Education
Academic Outreach
311 Umberger Hall
Manhattan, KS 66506
913-532-5686

Contact: Dr. Bill Lockhart
Correspondence Courses
Cross-reference: 13.22

13.24
Meadowlark Hills Skilled Health Care Center
2121 Meadowlark Rd.
Manhattan, KS 66502
913-537-9492

Contact: Linda George
Internship
Cross-reference: 13.91

13.25
Menninger Foundation
Box 829
Topeka, KS 66601

Contact: Timothy Hultquist
Internship
Cross-reference: 13.92

13.26
Merritt Horticultural Center
Kansas City, KS

Contact: Susie Newman
Internship
Cross-references: 7.30, 13.93

13.27
Topeka Veterans Administration Center
220 Gage Blvd.
Topeka, KS 66622
913-272-3111

Contact: Georgia Urish Abbott
Internship
Cross-reference: 13.94

13.28
Youth Center at Beloit
1120 North Hershey
Beloit, KS 67420
913-738-5735

Contact: Ian Draemel
Internship
Cross-reference: 13.96

KY
13.29
Kentucky Botanical Garden
814 ½ Cherokee Rd.
Louisville, KY 40204

Contact: M. Joni Carter
Non-degree Program

MA
13.30
University of Massachusetts
Dept. of Plant & Soil Science
Amherst, MA

Contact: John Tristan
Non-degree Program
Cross-reference: 12.171

MD
13.31
Shepard & Enoch Pratt Hospital
6501 North Charles St.
Baltimore, MD 21285-6815
301-823-8200

Contact: Bruce Weaver
Internship
Cross-reference: 13.107

MI
13.32
Fernwood Garden and Nature Center
Niles, MI

Contact: Stan Beikmann
Non-degree Program
Cross-reference: 16.201

NC
13.33
North Carolina Botanical Garden
Chapel Hill, NC

Contact: Bibby Moore
Non-degree Program
Cross-references: 13.118, 16.232

NE
13.34
Father Flanagan's Boys Home
Horticultural Training Center
Boystown, NE 68010
402-498-1100

Contact: Robert Prucha
Internship
Cross-reference: 13.120

13.35
Richard Young Memorial Hospital
415 South 25th Ave.
Omaha, NE 68103-0434
402-536-6854

Cross-reference: 13.121
Contact: Roger Evans
Internship

NJ
13.36
George Griswold Frelinghuysen Arboretum
Morristown, NJ

Contact: Helen Hesselgrave
Non-degree Program
Cross-references: 13.126, 16.256

NY
13.37
Brooklyn Botanic Garden
Brooklyn, NY

Contact: Lucy Jones
Non-degree Program
Cross-reference: 16.271

13.38
Enid A. Haupt Glass Garden
Rusk Institute of Rehabilitative Medicine
New York, NY

Contact: Nancy Chambers
Internship
Cross-references: 13.134, 16.278

13.39
New York Botanical Garden
Bronx, NY

Contact: Rosemary Kern
Non-degree Program
Cross-reference: 16.286

OH
13.40
Garden Center of Greater Cleveland
Cleveland, OH

Contact: Nancy Stevenson
Non-degree Program
Internship
Cross-references: 13.150, 16.302

13.41
Holden Arboretum
Mentor, OH

Contact: Karen L. Smith
Non-degree Program
Cross-references: 13.151, 16.304

PA
13.42
Ashbury Heights
700 Bower Hill Rd.
Pittsburgh, PA 15243
412-341-1030

Contact: JoAnn McDonald
Internship
Cross-reference: 13.155

13.43
Delaware Valley College of Science and Agriculture
Doylestown, PA

Contact: Dr. John Martin
Non-degree Program
Cross-reference: 12.276

13.44
Friends Hospital
Philadelphia, PA

Contact: Martha C. Straus
Internship
Cross-references: 13.161, 16.342

13.45
Pittsburgh Civic Garden Center
Pittsburgh, PA

Contact: Betty Morgan
Non-degree Program
Internship
Cross-references: 7.52, 13.171

13.46
Temple University
Dept. of Horticulture
Ambler, PA 19002
215-283-1292

Contact: John F. Collins
Non-degree Program

TN
13.47
Tennessee Technological University
Cookeville, TN
615-373-3288

Contact: Dr. Douglas L. Airhart
Non-degree Program
Cross-references: 12.298, 13.180

TX
13.48
Tangram Rehabilitation Network
Rt. 1, Box 155
Maxwell, TX 78656
512-396-0667

Contact: Harlan Shoulders
Internship
Cross-reference: 13.185

13.49
Texas A&M University
Dept. of Horticulture
College Station, TX
409-845-3276

Contact: Dr. Joe Novak
Degrees Granted: Bachelor
Cross-reference: 12.315

UT
13.50
State Arboretum of Utah
Salt Lake City, UT

Contact: Dr. Betty Wullstein
Non-degree Program
Cross-reference: 16.392

VA
13.51
Norfolk Botanical Gardens
Norfolk, VA

Contact: JoAnn Donlan
Non-degree Program
Cross-reference: 16.400

13.52
Virginia Polytechnic Institute and State University
Dept. of Horticulture
Blacksburg, VA

Contact: Dr. P.D. Relf
Non-degree Program
Cross-reference: 12.326

WY
13.53
Cheyenne Botanic Garden
Cheyenne, WY

Contact: Shane Smith
Internship
Cross-references: 13.191, 16.428

CANADA

BC
13.54
University of British Columbia
Botanical Garden
Vancouver, BC

Contact: David Tarrant
Non-degree Program
Cross-reference: 16.449

ON
13.55
Royal Botanical Gardens
Hamilton, ON

Contact: Brian Holley
Non-degree Program
Cross-references: 13.193, 16.473

HORTICULTURAL THERAPY PROGRAMS

UNITED STATES

AK
13.56
Cooperative Extension Service
2221 E. Northern Lights
118
Anchorage, AK 99508
907-279-5582

Contact: Julie Riley
Cross-reference: 11.2

AZ
13.57
Tucson Botanical Gardens
Tucson, AZ

Contact: Marty Eberhardt
Cross-reference: 16.13

CA
13.58
Casa Colina Development Center
1324 Arrow Hwy.
LaVerne, CA 91750
714-596-5981

Contact: Micki Spencer
Cross-reference: 13.12

13.59
Casa Dorinda
300 Hot Springs Rd.
Santa Barbara, CA 93108
805-969-8011

Contact: Joe Franken

13.60
Heather Farms Garden Center
Walnut Creek, CA

Contact: Penny Musanate
Cross-reference: 7.5

13.61
KAINOS Work Activity Center
3631 Jefferson Ave.
Redwood City, CA 94062
415-363-2433

Contact: David Packer

13.62
Kings View Center
42675 Road 44
Reedley, CA 93654
209-638-2505

Contact: Virginia Bollinger

13.63
Kingsview Hospital
42675 Road 44
Reedley, CA 93654
209-638-2505

Contact: Mary Beth Jantzen
Cross-reference: 13.13

13.64
Vets Garden
Brentwood VAMC
Bldg. 208, B 117
11300 Wilshire Blvd.
Los Angeles, CA 90073
213-824-6771

Contact: Bob Vatcher

CO
13.65
Craig Hospital
3425 South Clarkson
Englewood, CO 80110
303-789-8225

Contact: Julia Beems

13.66
Denver Botanic Gardens
Denver, CO

Cross-references: 13.16, 16.65

CT
13.67
Abbey of Regina Laudis
Flanders Rd.
Bethlehem, CT 06751
203-266-7727

Contact: Sister Margaret Patton

13.68
Geer Memorial Health Center
P.O. Box 817
Canaan, CT 06018
203-824-5137

Contact: Tamsin Goggin

13.69
Hockanum Industries Greenhouse
P.O. Box 136
Mansfield Depot, CT 06251
203-429-6697

Contact: Sarah Beardsley

13.70
Institute of Living
Rehabilitation Office
400 Washington St.
Hartford, CT 06106
203-241-6897

Contact: Katherine Heminway

13.71
Waterford Country School
78 Huntsbrook Rd.
Quaker Hill, CT 06375
203-442-9454

Contact: Richard McPherson

FL
13.72
Fairchild Tropical Garden
Miami, FL

Contact: Ann Parsons
Cross-reference: 16.87

13.73
Florida State Hospital
Horticultural Therapy Dept.
P.O. Box 103
Chattahoochee, FL 32324
904-663-7393

Contact: Marie Ansley
Cross-reference: 13.17

13.74
Four Ambassadors
Suite 1170, Tower 4
801 South Bayshore Dr.
Miami, FL 33131
305-324-5315

Contact: M. S. D. Blyth

13.75
G. Pierce Wood Memorial Hospital
Program Services
5100 Hwy. 31, South
Arcadia, FL 33821-9627
813-494-8260

Contact: Mark J. Gallagher

13.76
Grant Center Hospital
20601 S.W. 157th Ave.
Miami, FL 33187
305-251-0710

Contact: Bonnie Zambo

13.77
Sunland Center at Marianna
1008 #3 N. Jefferson St.
Marianna, FL 32446
904-482-9341

Contact: R. Burr Rutland

GA
13.78
Central Board on Care of Jewish Aged, Inc.
3150 Howell Mill Rd., N.W.
Atlanta, GA 30327

13.79
Northwest Georgia Regional Hospital
Bldg. 401
1305 Redmond Rd.
Rome, GA 30161
404-295-6080

Contact: Hugh Lee Edison, Jr.

13.80
Sheperd Spinal Center
2020 Peachtree Rd., N.W.
Atlanta, GA 30309
404-350-7786

Contact: Kurt Baumgardner

HI
13.81
Kalima O Maui Plant Nursery
95 Mahalani St.
Wailuku
Maui, HI 96793
808-244-7093

Contact: Darby Gill

IA
13.82
Handicap Village
Green Therapy Dept.
Box V
Clear Lakes, IA 50428
515-357-5277

Contact: Ron E. Richardson

IL
13.83
Chicago Botanic Garden
Glencoe, IL

Contact: Eugene Rothert
Cross-references: 13.19, 16.138

13.84
Cooperative Extension Service
University of Illinois
5106 South Western Ave.
Chicago, IL 60609
312-737-1179

Contact: Ronald Wolford

13.85
Iron Oaks Environmental Learning Center
2453 Vollmer Rd.
Olympia Fields, IL 60461
312-481-2330

Contact: Kathleen Neer

13.86
Morton Arboretum
Lisle, IL

Contact: Charles Lewis
Cross-reference: 16.145

13.87
Oak Forest Hospital
15900 S. Cicero Ave.
Oak Forest, IL 60452
312-687-7200

Contact: Kathleen Cunningham

IN
13.88
Community Hospital
c/o Rhab OT Dept.
1500 Ritter Ave.
Indianapolis, IN 46219
317-353-5095

Contact: Vicki Scott

KS
13.89
Big Lakes Development Center
1554 Hayes Dr.
Manhattan, KS 66502
913-776-7012

Contact: Debbie Springer
Cross-reference: 13.20

13.90
Good Samaritan Center
416 West Spruce
Junction City, KS 66441
913-238-1187

Contact: Dorothy Fredericks
Cross-reference: 13.21

13.91
Meadowlark Hills Skilled Health Care Center
2121 Meadowlark Rd.
Manhattan, KS 66502
913-537-9492

Contact: Linda George
Cross-reference: 13.24

13.92
Menninger Foundation
Box 829
Topeka, KS 66601

Contact: Timothy Hultquist
Cross-reference: 13.25

13.93
Merritt Horticultural Center
Kansas City, KS

Contact: Susie Newman
Cross-references: 7.30, 13.26

13.94
Topeka Veterans Administration Center
220 Gage Blvd.
Topeka, KS 66622
913-272-3111

Contact: Georgia Urish Abbott
Cross-reference: 13.27

13.95
University of Kansas Medical Center
39th & Rainbow Blvd.
Kansas City, KS 66103
913-588-5181

Contact: David Lewis

13.96
Youth Center at Beloit
1120 North Hershey
Beloit, KS 67420
913-738-5735

Contact: Ian Draemel
Cross-reference: 13.28

KY
13.97
V.A. Medical Center
117-B
Lexington, KY 40511
606-233-4511

Contact: Doug Wachs

MA
13.98
Center for Occupational Awareness & Placement
160 Turnpike Rd.
Chelmsford, MA 01824
617-256-3985

Contact: Wendy Hanson

13.99
Durfee Conservatory
Amherst, MA

Contact: John Tristan
Cross-reference: 16.179

13.100
Farrington House
295 Cambridge Turnpike
Lincoln, MA 01773
617-259-0053

Contact: Daniel Lacey

13.101
Perkins School for the Blind
175 North Beacon St.
Watertown, MA 02172
617-924-3434

Contact: Deborah Krause

13.102
V.A. Medical Center
Horticultural Therapy
Northampton, MA 01060-1288
413-584-1288

Contact: Lorraine Brisson

MD
13.103
Glen Farms
27 E. Parkway
Elkton, MD 21921
301-642-2411

Contact: Howard Woods

13.104
Melwood Horticultural Training Center
5606 Dower House Rd.
Upper Marlboro, MD 20772
301-599-8000

Contact: Earl Copus, Jr.

13.105
National Lutheran Home for the Aged
9701 Viers Dr.
Rockville, MD 20850
301-424-9560

Contact: Shirley Todd

13.106
Providence Center, Inc.
370 Shore Acres Rd.
Arnold, MD 21012
301-757-7215

Contact: Pat Hudson

13.107
Shepard & Enoch Pratt Hospital
6501 North Charles St.
Baltimore, MD 21285-6815
301-823-8200

Contact: Bruce Weaver
Cross-reference: 13.31

13.108
V.A. Medical Center
P.O. Box 279
Perry Point, MD 21902
301-642-2411

Contact: Ellen McLoughlin

MI
13.109
Holy Hospital
4777 East Outer Dr.
Detroit, MI 48234-3281
313-369-9100

Contact: Nancy Rehan

13.110
Michigan State University
East Lansing, MI
517-353-3729

Contact: Patricia Zandstra
Cross-reference: 12.185

MN
13.111
Fairbault Regional Center
802 Circle Dr.
Fairbault, MN 55021
507-332-3216

Contact: Esam Aal

MO
13.112
Men's Garden Club of America
811 West Santa Fe Trail
Kansas City, MO 64145
816-942-8272

Contact: Lloyd Kraft

MS
13.113
Baddour Memorial Center
Highway 51 South
P.O. Box 69
Senatobia, MS 38668

Contact: Ross Leach

13.114
Mississippi Branch of the Choctaw Indians
Rt. 7, Box 21
Philadelphia, MS 39350
601-656-5251

Contact: Kirk R. Morgan

NC
13.115
Appalachian Hall
P.O. Box 5534
Asheville, NC 28813
704-253-3681

Contact: Richard Sackett

13.116
Johnston Community College
Horticulture Dept.
P.O. Box 2350
Smithfield, NC 27577-2350
919-934-3051

Contact: Kitty Thrift

13.117
Johnston County Mental Health Center
P.O. Box 411
Highway 301 N
Smithfield, NC 27577
919-934-5121

Contact: Norma Decker

13.118
North Carolina Botanical Garden
Chapel Hill, NC
Contact: Bibby Moore
Cross-references: 13.33, 16.232

13.119
Southeastern Regional Rehabilitation Center
P.O. Box 2000
Fayetteville, NC 28302
919-323-6303

Contact: Phoebe Hunt

NE
13.120
Father Flanagan's Boys Home
Horticultural Training Center
Boystown, NE 68010
402-498-1100

Contact: Robert Prucha
Cross-reference: 13.34

13.121
Richard Young Memorial Hospital
415 South 25th Ave.
Omaha, NE 68103-0434
402-536-6854

Contact: Roger Evans
Cross-reference: 13.35

NH
13.122
New Medico Highwatch Learning Center
P.O. Box 99
Center Ossippee, NH 03814
603-539-5126

Contact: Keith Badger

NJ
13.123
Allied Clinical Therapists
Carrier Foundation
Box 147
Belle Mead, NJ 08502
201-874-4000

Contact: Frederick Goetz

13.124
Bancroft/Mullica Hill
Rt. 581
Commissioners Pike
Mullica Hill, NJ 08062
609-769-1300

Contact: Rachelle Krause

13.125
Bancroft School
6060 Hopkins Ln.
Haddonfield, NJ 08033
609-429-0010

Contact: Audrey Johnson

13.126
George Griswold Frelinghuysen Arboretum
Morristown, NJ
Contact: Helen Hesselgrave
Cross-references: 13.36, 16.256

13.127
New Jersey Regional Day School
334 Lyons Ave.
Newark, NJ 07112
201-705-3820

Contact: Phyllis D'Amico

13.128
Reeves-Read Arboretum
Summit, NJ

Contact: Lu Rose
Cross-reference: 16.260

13.129
Vineland Development Center
c/o Kimble Cottage
1676 E. Landis Ave.
Vineland, NJ 08360
609-696-6182

Contact: Joe Ann Reamer

13.130
Welkind Rehabilitation Hospital
Pleasant Hill Rd.
Chester, NJ 07930
201-584-8145

Contact: Danna Minch

13.131
Woodbine Development Center
Education Dept.
Woodbine, NJ 08270
609-861-2164

Contact: Patricia Gerew

NY
13.132
Association for Retarded Citizens
Monroe County Chapter
1000 Elmwood Ave.
Rochester, NY 14620
716-271-6813

Contact: Jesse Lucas

13.133
Clinical Campus at Binghamton
P.O. Box 1000
Binghamton, NY 13902
607-770-8617

Contact: Louise K. Stein

13.134
Enid A. Haupt Glass Garden
Rusk Institute of Rehabilitative Medicine
New York, NY

Contact: Nancy Chambers
Cross-references: 13.38, 16.278

13.135
Flowers With Care
23-30 Astoria Blvd.
Astoria, NY 11102
212-726-9790

13.136
Fountainhouse
441 West 47th St.
New York, NY 10036
212-473-2271

Contact: Bodil Drescher Anaya

13.137
Four Winds Hospital
30 Crescent Ave.
Saratoga Springs, NY 12866
518-584-3600

Contact: Janet Altamari

13.138
Hebrew Home for the Aged
5901 Palisade Ave.
Bronx, NY 10471
212-549-8700

Contact: Audrey Weiner

13.139
Kingsbrook Jewish Medical Center
585 Schenectady Ave.
Brooklyn, NY 11203
718-604-5000

Contact: Marion Somers

13.140
Mohawk Valley Psychiatric Center
1213 Court St.
Utica, NY 13502
315-797-6800

Contact: Mildred Kendall

13.141
New Hope Community, Inc.
P.O. Box X, Rt. 52
Loch Sheldrake, NY 12759
914-434-8300

Contact: Lucy Santoro

13.142
New York State Association for Retarded Children, Inc.
3891 County Rd., No. 46
Canandaigua, NY 14424
716-394-7500

Contact: Leonard Kataskas

13.143
New York State Hudson River Psychiatric Center
Branch B
Poughkeepsie, NY 12601
914-452-8000

Contact: Virginia B. Armstrong

13.144
North Towanda City Botanical Gardens
505 Schenck St.
North Towanda, NY 14120
716-695-8530

Contact: Donald A. Gane

13.145
Over the Hill
Box 387, Churchill Rd.
New Lebanon, NY 12125
518-794-9071

Contact: Sherry Boutard

13.146
V.A. Medical Center
OT-ADL Dept., Room 117
Castle Point, NY 12511
914-831-2000

Contact: Peggy Reinfuss

OH
13.147
Cox Arboretum
Dayton, OH

Contact: Robert Butts, Jr.
Cross-reference: 16.297

13.148
Falconskeape Gardens
Medina, OH

Contact: Dr. Karen Murray
Cross-reference: 16.299

13.149
Firelands Community Mental Health Services
2020 Hayes Ave.
Sandusky, OH 44870
419-627-5177

Contact: Bruce Kijowski

13.150
Garden Center of Greater Cleveland
Cleveland, OH

Contact: Nancy Stevenson
Cross-references: 13.40, 16.302

13.151
Holden Arboretum
Mentor, OH

Contact: Karen L. Smith
Cross-references: 13.41, 16.304

13.152
V.A. Administration Medical Center
Room 117
Chillicothe, OH 45601
614-773-1141

Contact: William Goff

OR
13.153
Oregon State University Extension Service
211 S.E. 80th
Portland, OR 97219
503-254-1500

Contact: Jan Powell

13.154
Salem Hospital Rehabilitation Center
2561 Center St., N.E.
Salem, OR 97309
503-370-5986

Contact: Jayanne Teeter

PA
13.155
Ashbury Heights
700 Bower Hill Rd.
Pittsburgh, PA 15243
412-341-1030

Contact: JoAnn McDonald
Cross-reference: 13.42

13.156
Berks County Cooperative Extension
R.D. 1, Box 520
Leesport, PA 19533
215-378-1327

Contact: Judith S. Schwank

13.157
Bryn Mawr Rehabilitation Hospital
414 Paolia Pike
Malvern, PA 19355
215-251-5598

Contact: Patricia Schmidt

13.158
Chateau Nursing & Rehabilitation Center
701A Tose St.
Bridgeport, PA 19405
215-277-8457

Contact: Patricia D. Grzywacz

13.159
Delaware Valley Mental Health Foundation
833 Butler Ave.
Doylestown, PA 18901
215-345-0444

Contact: Suzanne Griffith

13.160
Fair Acres
2430 Larkin Rd.
Boothwyn, PA 19061
215-891-5998

Contact: Judith M. Shuman

13.161
Friends Hospital
Philadelphia, PA

Contact: Martha C. Straus
Cross-references: 13.44, 16.342

13.162
Hamburg Center Greenhouse
Commonwealth of Pennsylvania
Hamburg, PA 19526
215-562-6334

Contact: Marie Hoffmaster

13.163
Harmarville Rehabilitation Center
P.O. Box 11460
Guys Run Rd.
Pittsburgh, PA 15238
412-828-1300

Contact: Pauline Egan

13.164
Human Services Providers, Inc.
St. James Complex
P.O. Box 160
Mansfield, PA 16933
717-662-7871

Contact: Cindy K. Davis

13.165
Jefferson Hospital
3 South West In-Patient
Mental Health
Pittsburgh, PA 15236
412-341-7937

Contact: Gerrie Delaney

13.166
Mayview State Hospital
V.A.S. Greenhouse
1601 Mayview Rd.
Bridgeville, PA 15017-1599
412-257-6440

Contact: Dennis Powell

13.167
Melmark Home, Inc.
Wayland Rd.
Berwyn, PA 19312
215-353-1726

Contact: Paul Krentel

13.168
Norristown State Hospital
Building 2
Norristown, PA 19401
215-270-1594

Contact: Christine Cole

13.169
Pathway School
162 Egypt Rd.
Jeffersonville, PA 19312
215-277-8660

Contact: Benjamin W. Champion

13.170
Philhaven Psychiatric Hospital
283 South Butler Rd.
P.O. Box 550
Mt. Gretna, PA 17064
717-273-8871

Contact: Marilyn Heisey

13.171
Pittsburgh Civic Garden Center
Pittsburgh, PA

Contact: Betty Morgan
Cross-references: 7.52, 13.45

13.172
RITC of York, Inc.
149 South Tremont St.
York, PA 17403
717-848-4312

Contact: David Halliwell

13.173
Reading Rehabilitation Hospital
Rt. 10, Box 250
Reading, PA 19607
215-777-7615

Contact: Dan Fick

13.174
River Crest Horticultural Center
Rt. 29
Mont Claire, PA 19453
215-935-9738

Contact: Susan G. Wieser
Cross-reference: 7.53

13.175
Veteran's Administration
R.D. 4, Shag Bark Ln.
Plains Church
Evans City, PA 16033
412-363-4900

Contact: Christine Zalewski

RI
13.176
Rhode Island Dept. of Corrections
Adult Correctional Institutions
Box 8273, Pontiac Ave.
Cranston, RI 02920
401-464-2688

Contact: Raymond E. Huling

13.177
University of Rhode Island
Kingston, RI
401-792-5996

Contact: Richard Shaw
Cross-reference: 12.285

SC
13.178
Crafts-Farrow State Hospital
7901 Farrow Rd.
Columbia, SC 29205
803-737-7623

Contact: Liz Bethea Fuller

TN
13.179
HCA Parthenon Pavillion Psychiatric Hospital
2401 Murphy Ave.
Nashville, TN 37203
615-321-9232

Contact: Betsy Donegan

13.180
Tennessee Technological University
Cookeville, TN
615-373-3288

Contact: Dr. Douglas L. Airhart
Cross-references: 12.298, 13.47

TX
13.181
Devereaux Foundation
P.O. Box 2666
Victoria, TX 77902
512-575-8271

Contact: J. Bernard Green

13.182
Fort Worth Botanic Garden
Fort Worth, TX

Contact: Jana Johnson
Cross-reference: 16.382

13.183
Moody Gardens
One Hope Blvd.
Galveston, TX 77554
409-744-1745

Contact: Peter Atkins

13.184
Sundown Ranch, Inc.
Rt. 4, Box 182
Canton, TX 75103
214-479-3933

Contact: Jeff Power

13.185
Tangram Rehabilitation Network
Rt. 1, Box 155
Maxwell, TX 78656
512-396-0667

Contact: Harlan Shoulders
Cross-reference: 13.48

VA
13.186
Catawba Hospital
P.O. Box 200
Catawba, VA 24070

Contact: Shelley B. Workman

13.187
Inisfree Village
Rt. 2, Box 506
Crozet, VA 22932
804-823-5400

Contact: Midori Sawada

13.188
St. Albans Psychiatric Hospital
Rt. 11, West
Radford, VA 24143
703-639-2481

Contact: William Litton

VT
13.189
National Gardening Association
Burlington, VT

Contact: Larry Sommers
Cross-reference: 1.89

WA
13.190
Spokane Rehabilitation Center
N. 3128
Spokane, WA 99205
509-325-5451

Contact: Jerry Edwards

WY
13.191
Cheyenne Botanic Garden
Cheyenne, WY

Contact: Shane Smith
Cross-references: 13.53, 16.428

CANADA

ON
13.192
Homewood Sanitarium
150 Delhi St.
Guelph, ON N1E 6K9
519-824-1010

Contact: Virginia Stradiotto

13.193
Royal Botanical Gardens
Hamilton, ON

Contact: Brian Holley
Cross-references: 13.55, 16.473

PQ
13.194
Douglas Hospital Centre
6875 boul. LaSalle
Verdun, PQ H4H 1R3

QC
13.195
Montreal Botanic Garden
Montreal, QC

Contact: Ghyslaine Gagnon
Cross-reference: 16.481

14

Horticultural and Botanical Libraries

This list is comprised of horticultural and botanical libraries in the United States and Canada. Not included are municipal and state public libraries or university and college libraries (unless they have a separate horticultural or botanical collection).

The nature of the collections vary widely by institution. Some libraries have comprehensive national collections, while others have small, specialized collections of local interest. We have highlighted special books or the main emphasis for as many libraries as possible. Many libraries also keep periodical clippings, slide files, photograph files, and video collections. Most of the libraries listed are open to the public for reference work, but few have lending privileges.

A resource for more information on horticultural and botanical libraries is the Council on Botanical and Horticultural Libraries (1.44).

UNITED STATES

AL
14.1
Horace Hammond Memorial Library
Birmingham Botanical Gardens
2612 Land Park Rd.
Birmingham, AL 35223
205-879-1227

Number of books: 3,800; number of periodicals: 125.
Alabama native plants, Japanese gardens, and bonsai.
Lending policy: Circulating.
Cross-reference: 16.2

Mobile Botanical Gardens
Mobile, AL
Cross-reference: 16.4

AR
Arkansas Territorial Restoration
Little Rock, AR
Cross-reference: 17.46

AZ
Arboretum at Flagstaff, The
Flagstaff, AZ
Cross-reference: 16.7

Arizona-Sonora Desert Museum
Tucson, AZ
Cross-reference: 16.8

14.2
Boyce Thompson Southwestern Arboretum Library
P.O. Box AB
Superior, AZ 85273
602-689-2723

Number of books: 3,000; number of periodicals: 60.
Books on lithops, eucalyptus, and cacti and other
    succulents.
Lending policy: Reference only for serious research.
Cross-reference: 16.9

Native Seeds/SEARCH
Tucson, AZ
Cross-reference: 17.29

14.3
Richter Memorial Library
Desert Botanical Garden
1201 N. Galvin Pkwy.
Phoenix, AZ 85008
602-941-1225

Number of books: 35,000.
Rare botanical books and prints including pre-Linnaean
    herbals.
Lending policy: Reference only.
Cross-reference: 16.10

Tohono Chul Park
Tucson, AZ

Cross-reference: 16.12

Tucson Botanical Gardens
Tucson, AZ

Cross-reference: 16.13

CA
American Fuchsia Society
San Francisco, CA

Cross-reference: 2.11

Ardenwood Historic Farm
Fremont, CA

Cross-reference: 17.52

Bio Integral Resource Center
Berkeley, CA

Cross-reference: 1.36

Biological Urban Gardening Services
Citrus Heights, CA

Cross-reference: 1.38

Blake Garden
Berkeley, CA

Cross-reference: 16.18

Cactus and Succulent Society of America
Los Angeles, CA

Cross-reference: 2.33

California Horticultural Society
San Francisco, CA

Cross-reference: 6.3

California Spring Blossom & Wildflower Association
San Francisco, CA

Cross-reference: 3.7

14.4
Fullerton Arboretum Library
Fullerton Arboretum
California State University
Fullerton, CA 92634
714-773-3579

Librarian: Celia Kutcher
Number of books: 1,000; number of periodicals: 20.
3,227 items including 1,227 send catalogs from
    1880-1920.
Lending policy: Reference only.
Cross-reference: 16.24

14.5
Helen Crocker Russell Library
Strybing Arboretum & Botanical Gardens
9th Ave. at Lincoln Way
San Francisco, CA 94122
415-661-1316

Librarian: Barbara M. Pitschel
Number of books: 14,000; number of periodicals: 300.
Rare botanical books, extensive children's collection on
    botany and natural history, ethnobotanical collection,
    and nursery and seed catalogs dating from 1875.
Lending policy: Reference only.
Cross-reference: 16.57

Hortense Miller Garden
Laguna Beach, CA

Cross-reference: 16.30

Indoor Citrus and Rare Fruit Society
Los Altos, CA

Cross-reference: 1.66

14.6
Library & Information Center
San Diego Floral Association
Room 105, Casa del Prado
Balboa Park
San Diego, CA 92103-1619
619-232-5762

Librarian: Elsie M. Topham
Number of books: 3,325; number of periodicals: 35.
Lending policy: Reference only.
Cross-reference: 9.105

14.7
Library of the Huntington
1511 Oxford Rd.
San Marino, CA 91108
818-405-2160

5,000 books and periodicals; herbals, subtropical floras,
    and books on cacti and other succulents.
Lending policy: Reference only for serious research.
Cross-reference: 17.58

Living Desert, The
Palm Desert, CA
Cross-reference: 16.34

14.8
Los Angeles State and County Arboretum Plant Science
    Library
301 N. Baldwin Ave.
Arcadia, CA 91007
818-446-8251; (FAX) 818-445-1217

Number of books: 25,000; number of periodicals: 250.
Collection of horticultural and botanical books featuring
    California plants and gardens as well as Australian and
    South African plants. Small rare book collection
    includes a 1577 herbal.
Lending policy: Reference only.
Cross-reference: 16.35

Mildred E. Mathias Botanical Garden
Los Angeles, CA

Cross-reference: 16.40

Mycological Society of San Francisco
San Francisco, CA

Cross-reference: 2.65

Quail Botanical Gardens
Encinitas, CA

Cross-reference: 16.47

14.9
Rancho Santa Ana Botanic Garden Library
1500 N. College Ave.
Claremont, CA 91711-3101
714-625-8767; (FAX)714-626-7670

Librarian: Bea Beck
76,747 items including books, periodicals, maps, reprints,
catalogs, and microform. Special collections include
evolutionary biology, conservation biology,
ethnobotany, California horticultural history, and
drought-tolerant landscaping and plants. Many rare
books on botany and horticulture. Online computer
library center links collection to 8,000 libraries in U.S.,
Canada, Great Britain, and other European countries.
Lending policy: Open for serious research, reference
only.
Cross-reference: 16.48

San Joaquin County Historical Society and Museum
Lodi, CA

Cross-reference: 17.64

14.10
Santa Barbara Botanic Garden Library
1212 Mission Canyon Rd.
Santa Barbara, CA 93105
805-682-4726

Number of books: 7,500; number of periodicals; 3,000
bound periodicals.
California native plants, California offshore islands,
horticulture, and botany.
Lending policy: Reference only.
Cross-reference: 16.53

Saratoga Horticultural Foundation
San Martin, CA

Cross-reference: 6.4

Seed Saving Project
Davis, CA

Cross-reference: 17.33

14.11
South Coast Plant Science Library
26300 Crenshaw Blvd.
Palos Verdes Peninsula, CA 90274
213-377-0468

Librarian: Virginia Gardner
Number of books: 500; number of periodicals: 10.
Lending policy: Members only, reference only.

14.12
Traub Plant Life Library
The American Plant Life Society
P.O. Box 985
National City, CA 92050
619-477-5333

2,800 volumes and periodicals; Amaryllidaceae and
Liliaceae.
Lending policy: By appointment only for serious research.
Cross-reference: 4.3

University of California Botanical Garden at Berkeley
Berkeley, CA

Cross-reference: 16.60

University of California at Irvine Arboretum
Irvine, CA

Cross-reference: 16.61

14.13
Wallace Sterling Library of Landscape Architecture
Filoli
Canada Rd.
Woodside, CA 94062
415-364-8300

Librarian: Tom Rogers
Number of books: 3,000.
Landscape architecture and garden design.
Lending policy: Open by appointment only to serious
researchers.
Cross-reference: 17.56

CO
American Penstemon Society
Lakewood, CO

Cross-reference: 2.21

14.14
Helen Fowler Library
Denver Botanic Gardens
909 York St.
Denver, CO 80206
303-331-4000

Number of books: 18,000; number of periodicals: 300.
Herbals, extensive bromeliad collection, horticultural
stamp collection, and original watercolors of Colorado
and Oregon wildflowers.
Lending policy: Circulating to members.
Cross-reference: 16.65

Horticultural Art Society Garden
Colorado Springs, CO
Cross-reference: 16.66

International Gladiolus Hall of Fame
Greeley, CO
Cross-reference: 17.24

CT
Connecticut Botanical Society
New Haven, CT
Cross-reference: 3.10

Connecticut Horticultural Society
Wethersfield, CT
Cross-reference: 6.7

14.15
George E. Bye Library
Bartlett Arboretum
University of Connecticut
151 Brookdale Rd.
Stamford, CT 06903-4199
203-322-6971

Number of books: 3,000; number of periodicals: 50.
Books on dwarf conifers and ericaceous plants.
Lending Policy: Circulating.
Cross-reference: 16.72

14.16
Greenwich Garden Center Library
Bible St.
Cos Cobb, CT 06807
203-869-9242

Librarian: Latisha Potter
Number of books: 1,200.
Cross-reference: 7.12

Nurserymen's Gardens
New Haven, CT
Cross-reference: 16.78

Stowe Day Foundation
Hartford, CT
Cross-reference: 17.72

DC
Association for Living Historical Farms and Agricultural
    Museums
Washington, DC
Cross-reference: 17.6

14.17
Botany Library
Smithsonian Institution
10th and Constitution
Washington, DC 20560
202-357-2715

Librarian: Ruth F. Schallart
Number of books: 35,000; number of periodicals: 300.
General collection; grasses, Dawson collection on algae.
Lending policy: By appointment, reference only.
Cross-reference: 15.45

14.18
Dumbarton Oaks Garden Library
1703 32nd St., N.W.
Washington, DC 20007
202-342-3200

Number of books: 14,000.
Rare manuscripts, drawings, prints, and books relating to
    all aspects of the history of gardens, and slides,
    photographs, and microfilms on landscape architecture.
Lending policy: Reference only.
Cross-reference: 17.74

14.19
Landscape Architecture Foundation Library
1733 Connecticut Ave., N.W.
Washington, DC 20015
202-233-6229

Volumes on landscape architecture and design.
Lending policy: Members only, reference only.
Cross-reference: 4.16

14.20
Office of Horticulture Library
Arts & Industries Building Room 2282
Smithsonian Institution
Washington, DC 20560
202-357-1544

Librarian: Susan Gurney
Number of books: 4,000; number of periodicals: 2,000.
Approximately 15,000 seed and nursery catalogs going
    back to 1823, with a concentration in the years
    1890-1930.
Lending policy: Reference only.

Potato Museum
Washington, DC
Cross-reference: 17.77

United States Botanic Garden
Washington, DC
Cross-reference: 16.82

14.21
United States National Arboretum Library
3501 New York Ave., N.E.
Washington, DC 20002
202-475-4815; (FAX) 202-475-5252

Number of books: 6,000; number of periodicals: 250.
Books on plant breeding, bonsai, ikebana, taxonomy, and
    floras.
Lending policy: Reference only.
Cross-reference: 16.83

DE
Hagley Museum and Library
Wilmington, DE
Cross-reference: 17.79

14.22
Wilmington Garden Center Library
Wilmington Garden Center
503 Market St. Mall
Wilmington, DE 19801
302-658-1913

Number of books: 2,000; number of periodicals: 20.
Lending policy: Reference only.
Cross-reference: 7.17

FL
Bonsai Clubs International
Tallahassee, FL

Cross-reference: 1.39

H. P. Leu Botanical Gardens
Orlando, FL

Cross-reference: 16.93

14.23
Marie Selby Botanical Gardens Library
811 South Palm Ave.
Sarasota, FL 34236
813-366-5730

Number of books: 4,800; number of periodicals: 150.
Collection of published folio works on orchids, set of
Curtis's Botanical Magazine, reprint file, slide file,
illustration file, and map file.
Lending policy: Members only, reference only, by
appointment.
Cross-reference: 16.96

14.24
Montgomery Library
Fairchild Tropical Garden
10901 Old Cutler Rd.
Miami, FL 33156
305-667-1651

Number of books: 7,000.
Tropical botany and horticulture.
Lending policy: Reference only.
Cross-reference: 16.87

Simpson Park
Miami, FL

Cross-reference: 16.101

GA
14.25
American Camellia Society Library
American Camellia Society
P.O. Box 1217
Fort Valley, GA 31030
912-967-2358

Number of books: 1,600; number of periodicals: 10.
Horticultural books, prints, and paintings dating from
1669 to the present. Also a large collection of Japanese
camellia books.
Lending policy: Members only, reference only.
Cross-reference: 2.6

14.26
Callaway Gardens Library
Callaway Gardens
Pine Mountain, GA 31822
404-663-5186; (FAX) 404-663-5049

Number of books: 1,000; number of periodicals: 50.
De Renne garden books.
Lending policy: Staff only, reference only.
Cross-reference: 16.104

14.27
Cherokee Garden Library
Atlanta Historical Society
McElreath Hall
3101 Andrews Dr., N.W.
Atlanta, GA 30305
404-238-0654

3,800 books and periodicals; rare books documenting the
history of American horticulture from 1634, including
old seed catalogs.
Lending policy: Reference only.
Cross-reference: 17.93

14.28
Fernbank Science Center Library
Fernbank Science Center
156 Heaton Park Dr., N.E.
Atlanta, GA 30307-1398
404-378-4311

Librarian: Mary Larsen
Numbr of books: 21,000; number of periodicals: 355.
*Gray Herbarium Index, Index to American Botanical
Literature.*
Lending policy: Reference only.
Cross-reference: 16.106

Forest Farmers Association, Inc.
Atlanta, GA

Cross-reference: 1.50

14.29
Sheffield Botanical Library
Atlanta Botanical Garden
Piedmont Park at the Prado
Atlanta, GA 30309
404-876-5859

Librarian: Lu Ann Schwarz
Number of books: 1,300; number of periodicals: 71.
Lending policy: Reference only.
Cross-reference: 16.103

HI
14.30
Bishop Museum Library
Bernice P. Bishop Museum
1525 Bernice St.
P.O. Box 19000-A

Honolulu, HI 96817-0916
808-848-4148

Librarian: Marguerite K. Ashford
Horticultural collection is part of the general collection of
over 90,000 items.
Lending policy: Reference only.
Cross-reference: 15.56

14.31
Lyon Arboretum Reference Collection
Harold L. Lyon Arboretum
3860 Manoa Rd.
Honolulu, HI 96822
808-988-3177

Lending policy: Open by appointment only to serious
researchers.
Cross-reference: 16.111

14.32
National Tropical Botanic Garden Library
Papalina Rd.
Kalaheo
Kauai, HI 96765
808-332-7324

Librarian: Lynwood M. Hume
Number of books: 3,000; number of periodicals: 875.
Tropical botany and horticulture, botanical prints.
Lending policy: Reference only.
Cross-reference: 16.121

14.33
Rock Library
Honolulu Botanic Gardens
50 N. Vineyard Blvd.
Honolulu, HI 96817
808-533-3406

Number of books: 3,000; number of periodicals: 52.
Lending policy: Reference only.
Cross-reference: 16.115

14.34
Waimea Foundation Library
Waimea Falls Park Arboretum & Botanical Gardens
59-864 Kamehameha Hwy.
Haleiwa, HI 96712
808-638-8655

Librarian: Shirley B. Gerum
Number of books: 500; number of periodicals: 30.
Lending policy: Reference only.
Cross-reference: 16.124

IA
14.35
Bickelhaupt Arboretum Library
Bickelhaupt Arboretum
340 S. 14th St.
Clinton, IA 52732
319-242-4771

Number of books: 800; number of periodicals: 30.

Ornamental horticulture, indoor plants, urban forestry,
and ecology.
Lending policy: Circulating.
Cross-reference: 16.125

Des Moines Botanical Center
Des Moines, IA

Cross-reference: 16.126

Dubuque Arboretum & Botanical Gardens
Dubuque, IA

Cross-reference: 16.128

Iowa Arboretum
Madrid, IA

Cross-reference: 16.132

14.36
Men's Garden Clubs of America Lending Library
5560 Merle Hay Rd.
Box 241
Johnston, IA 50131
515-278-0295

Number of books: 2,000.
Lending policy: Circulating to members.
Cross-reference: 5.2

IL
14.37
Chicago Botanic Garden Library
Lake Cook Rd.
P.O. Box 400
Glencoe, IL 60022
708-835-5440; (FAX) 708-835-4484

Number of books: 10,000; number of periodicals: 100.
Small rare book collection, children's collection, and
complete U.S. plant patents publications.
Lending policy: Circulating for members only.
Cross-reference: 16.138

Clayville Rural Life Center and Museum
Pleasant Plains, IL

Cross-reference: 17.103

14.38
Field Museum of Natural History Library
Roosevelt Rd. & Lake Shore Dr.
Chicago, IL 60605
312-922-9410

Number of books: 40,000.
General natural history collection with horticulture and
botany titles.
Lending policy: Reference only.
Cross-reference: 15.65

Illinois Native Plant Society
Westville, IL

Cross-reference: 3.15

Starhill Forest
Petersburg, IL
Cross-reference: 16.147

14.39
Sterling Morton Library
Morton Arboretum
Rt. 53
Lisle, IL 60532
708-968-0074

Number of books: 23,000; number of periodicals: 400.
3,000 rare books dating from the 15th to the 20th
centuries, including a good collection of early herbals.
General collection is devoted to botany and
horticulture, including books on the flora and
vegetation of every state and most European countries,
monographs on plant families and genera, and books on
the practice of gardening in temperate zones.
Lending policy: Circulating to staff, members, and
volunteers only.
Cross-reference: 16.145

Washington Park Botanical Garden
Springfield, IL
Cross-reference: 16.148

IN
Conner Prairie
Noblesville, IN
Cross-reference: 17.112

Eli Lily Botanical Garden of the Indianapolis Museum of
Art
Indianapolis, IN
Cross-reference: 17.113

14.40
Hayes Regional Arboretum Library
801 Elks Rd.
Richmond, IN 47374
317-962-3745

Number of books: 700; number of periodicals: 14.
Slide collection.
Lending policy: Reference only.
Cross-reference: 16.151

KS
14.41
Frank Good Library
Botanica, The Wichita Gardens
701 N. Amidon
Wichita, KS 67203
316-264-0448

Librarian: Amy Kaspar Wolf
Number of books: 2,500; number of periodicals: 30.
Botany, horticulture, Xeriscaping, Kansas native plants,
current garden catalogs, and a video library.
Lending policy: Circulating to members, reference only
for nonmembers.
Cross-reference: 16.157

LA
14.42
American Rose Society Lending Library
P.O. Box 30,000
Shreveport, LA 71130
318-938-5402

Number of books: 500.
All aspects of rose growing and enjoyment.
Lending policy: Circulating for members.
Cross-reference: 2.26

14.43
Garden Library of the New Orleans Town Gardeners
Southeastern Architectural Archive
Tulane University Library
7001 Freret St.
New Orleans, LA 70118
504-865-5699

Number of books: 1,000.
Broad in scope, with an emphasis on Louisiana and the
southeastern U.S., garden history, garden design,
oriental gardens, and flowers.

Longue Vue House and Gardens
New Orleans, LA
Cross-reference: 17.125

14.44
R. S. Barnwell Memorial Gardens and Art Center Library
601 Clyde Fant Pkwy.
Shreveport, LA 71101
318-425-6495

Number of books: 500.
Lending policy: Reference only.
Cross-reference: 17.126

MA
14.45
Arnold Arboretum Library
Arnold Arboretum
125 Arborway
Jamaica Plain, MA 02130
617-524-1718

Number of books: 92,650; number of periodicals: 400.
Early horticultural titles, rare books, extensive collection
of 19th-century botanical and horticultural literature,
and photo archives with 23,000 images.
Lending policy: By appointment, reference only.
Other: Includes the library of the Harvard University
Herbaria, 22 Divinity Ave., Cambridge, MA 02138,
617-495-2365.
Cross-reference: 16.173

14.46
Berkshire Garden Center Library
Berkshire Garden Center
Rts. 102 & 183, P.O. Box 826
Stockbridge, MA 01262
413-298-3926

Number of books: 1,000; number of periodicals: 25.
Lending policy: For members only.
Cross-reference: 16.176

Chesterwood
Stockbridge, MA

Cross-reference: 17.130

Essex Institute
Salem, MA

Cross-reference: 17.132

Fisher Museum of Forestry
Petersham, MA

Cross-reference: 17.133

14.47
Massachusetts Horticultural Society Library
300 Massachusetts Ave.
Boston, MA 02115
617-536-9280

Librarian: Walter T. Punch
Number of books: 35,000; number of periodicals: 300.
80,000 items including rare books, historical periodicals,
   seed and nursery catalogs dating from the early 19th
   century, prints, old flower and plant show programs, an
   early agriculture collection, and a landscape history
   collection.
Lending policy: Circulating to members only.
Cross-reference: 6.16

Naumkeag Museum and Gardens
Stockbridge, MA

Cross-reference: 17.142

New England Botanical Club, Inc.
Cambridge, MA
Cross-reference: 3.21

14.48
New England Wild Flower Society Library
Garden in the Woods
Hemenway Rd.
Framingham, MA 01701
617-237-4924

Number of books: 3,000.
Botany, plant ecology, natural history, and related
   horticultural subjects. Slide library contains 25,000 color
   slides of native plants and their habitats.
Lending policy: Reference only.
Cross-reference: 3.22

Old Sturbridge Village
Sturbridge, MA

Cross-reference: 17.143

14.49
Worcester County Horticultural Society Library
30 Tower Hill Rd.
Boylston, MA 01505
508-869-6111

Librarian: Margot K. Wallin
Number of books: 7,000; number of periodicals: 45.
19th-century fruit books.
Lending policy: Circulating to members, reference only
   to nonmembers.
Cross-reference: 6.17

MD
14.50
Brookside Gardens Library
1500 Glenallan Ave.
Wheaton, MD 20902
301-949-8231

Number of books: 2,000; number of periodicals: 45.
Books on ornamental grasses, roses, viburnums, and azaleas.
Lending policy: Reference only.
Cross-reference: 16.184

14.51
Cylburn Horticultural Library
Cylburn Wild Flower Preserve and Garden Center
4915 Greenspring Ave.
Baltimore, MD 21209-4698
301-367-2217

Librarian: Adelaide Rackemann
Number of books: 2,000.
Horticulture, natural history, rock gardens, and
   wildflowers.
Lending policy: Circulating to members.
Cross-reference: 16.185

Ladew Topiary Gardens
Monkton, MD

Cross-reference: 16.186

14.52
National Agricultural Library
U.S. Dept. of Agriculture
10301 Baltimore Blvd.
Beltsville, MD 20705
301-344-3755

Librarian: Joseph Howard, Director
Number of books: 2,000,000; number of periodicals:
   27,000.
Comprehensive collections on plant science, soils, and
   agriculture; AGRICOLA computerized database of literature citations; and many special collections including
   the M. Truman Fossum collection of personal and business correspondence of the floricultural statistician, the
   organizational records of the Prince Family Nursery
   (1779–1914), the first commercial nursery in the U.S., a
   comprehensive collection of nursery and seed trade catalogs dating back to the 18th century (over 150,000
   items), an extensive collection of pre-Linnaeana, imprints relating to the description of plants as well as
   works by or about Carolus Linnaeus, the Mary Cokely
   Wood Japanese Classical Flower Arrangement rare book
   collection, and the Forest Service historical photo collection (the largest photographic collection on the subject of forestry in the world).
Lending policy: Reference only.
Other: Horticulture Information Center covers technical
   horticultural or botanical questions, economic botany,
   wild plants of possible use, herbs, bonsai, and
   floriculture. It also publishes *Quick Bibliographies*,
   Agri-Topics, and other bibliographies. Coordinator:
   Jayne MacLean (301-344-3704). A major section of the
   Horticulture Information Center is the Floriculture
   Information Connection.

Society for Japanese Irises
Upperco, MD
Cross-reference: 2.84

14.53
Society of American Foresters Library
5400 Grosvenor Ln.
Bethesda, MD 20814
301-897-8720; (FAX) 301-897-3690

Number of books: 1,500.
Forestry.
Lending policy: Reference only.
Cross-reference: 1.114

William Paca Garden
Annapolis, MD
Cross-reference: 17.151

ME
14.54
Thuya Lodge Library
Asticou Terraces Trust
Peabody Dr.
Northeast Harbor, ME 04662
207-276-5130

Number of books: 1,000.
General and historical horticulture works including
    collections of herbals and Linnaeus originals.
Lending policy: Reference only.
Cross-reference: 16.196

MI
14.55
Chippewa Nature Center
400 S. Badour Rd., Rt. 9
Midland, MI 48640
517-631-0803

Librarian: Meg Ulery
Number of books: 2,500; number of periodicals: 86.
General natural sciences collection with horticultural
    titles.
Lending policy: Members only, reference only.

14.56
Detroit Garden Center Library
Detroit Garden Center
1460 E. Jefferson Ave.
Detroit, MI 48207
313-259-6363

Librarian: Margaret E. Grazier
Number of books: 5,800; number of periodicals: 15.
Lending policy: Members only, reference only.
Cross-reference: 17.154

Dow Gardens
Midland, MI
Cross-reference: 16.200

14.57
Fernwood Library
Fernwood Garden and Nature Center
13988 Range Line Rd.
Niles, MI 49120
616-683-8653

Number of books: 4,000; number of periodicals: 50.
Lending policy: Reference only.
Cross-reference: 16.201

14.58
Hidden Lake Gardens Library
M-50
Tipton, MI 49287
517-431-2060

Number of books: 2,025; number of periodicals: 14.
Lending policy: Reference only.
Cross-reference: 16.202

Matthaei Botanical Gardens
Ann Arbor, MI
Cross-reference: 16.203

MN
14.59
Andersen Horticultural Library
Minnesota Landscape Arboretum
3675 Arboretum Dr.
P.O. Box 39
Chanhassen, MN 55317
612-443-2460

Number of books: 9,500; number of periodicals: 350.
Rare book collection featuring botanical illustrations,
    herbals, and other botanical and horticultural books,
    over 100,000 seed and nursery catalogs, Frances
    Williams hosta archives, national archives of the
    American Hemerocallis Society, Minnesota Landscape
    Architects archives, and oral history of Minnesota
    horticulture.
Lending policy: Reference only.
Cross-reference: 16.208

Como Park Conservatory
St. Paul, MN
Cross-reference: 16.205

Minnesota State Horticultural Society
St. Paul, MN
Cross-reference: 6.20

North American Lily Society, Inc.
Owatonna, MN
Cross-reference: 2.72

Stoppel Farm
Rochester, MN
Cross-reference: 17.160

MO

14.60

Missouri Botanical Garden Library
P.O. Box 299
St. Louis, MO 63166
314-577-5100; (FAX) 314-577-9521

Number of books: 110,000; number of periodicals: 1,300.
Pre-Linnaean collection, Linnaean collection, rare book
  collection, bryological library, botanical art and
  illustrations, vegetation maps, pamphlets, maps, and
  manuscript items.
Lending policy: Reference only, by appointment, for
  serious research.
Cross-reference: 16.214

National Council of State Garden Clubs
St. Louis, MO

Cross-reference: 5.3

MS

Crosby Arboretum
Picayune, MS

Cross-reference: 16.218

NC

Old Salem, Inc.
Winston-Salem, NC

Cross-reference: 17.173

14.61

Totten Library
North Carolina Botanical Garden
3375 Totten Center
University of North Carolina—Chapel Hill
Chapel Hill, NC 27599-3375
919-962-0522

Number of books: 1,200; number of periodicals: 12.
Mainly reference works on native plants, general
  horticulture, horticultural therapy, and herbs.
Lending policy: Reference only.
Cross-reference: 16.232

Tryon Palace Restoration
New Bern, NC

Cross-reference: 17.175

University of North Carolina—Charlotte Botanical
  Gardens
Charlotte, NC

Cross-reference: 16.239

NH

Strawberry Banke Museum
Portsmouth, NH

Cross-reference: 17.179

NJ

Acorn Hall
Morristown, NJ

Cross-reference: 17.180

American Dahlia Society
Wayne, NJ

Cross-reference: 2.9

Barclay Farmstead Museum
Cherry Hill, NJ

Cross-reference: 17.181

Clinton Historical Museum
Clinton, NJ

Cross-reference: 17.183

Colonial Park
Somerset, NJ

Cross-reference: 16.252

14.62

Elvin McDonald Horticultural Library
Monmouth County Park System
Deep Cut Park Horticultural Center
352 Red Hill Rd.
Middletown, NJ 07748
201-671-6050

Librarian: Mae H. Fisher
Number of books: 2,400; number of periodicals: 30.
Orchids, cacti and other succulents, and a wide variety of
  seed and nursery catalogs.
Lending policy: Circulating to members only.
Cross-reference: 16.254

14.63

George Griswold Frelinghuysen Arboretum Library
53 East Hanover Ave.
P.O. Box 1295
Morristown, NJ 07962-1295
201-326-7600

Number of books: 2,700; number of periodicals: 30.
Some rare books that date from the 15th century.
Lending policy: Reference only.
Cross-reference: 16.256

Leonard J. Buck Gardens
Far Hills, NJ

Cross-reference: 16.258

Reeves-Reed Arboretum
Summit, NJ

Cross-reference: 16.260

Skylands Botanical Gardens
Ringwood, NJ

Cross-reference: 16.262

United States Golf Association
Green Section
Far Hills, NJ

Cross-reference: 1.120

NM
Talavaya Center
Espanola, NM
Cross-reference: 17.36

NY
14.64
Brooklyn Botanic Garden Library
1000 Washington Ave.
Brooklyn, NY 11225-1099
718-622-4433; (FAX) 718-857-2430

Number of books: 45,000; number of periodicals: 5,000.
Selected botanical flora and monographs.
Lending policy: Reference only.
Cross-reference: 16.271

Buffalo and Erie County Botanical Gardens
Buffalo, NY
Cross-reference: 16.272

Clark Botanic Garden
Albertson, NY
Cross-reference: 16.273

Enid A. Haupt Glass Garden
New York, NY
Cross-reference: 16.278

14.65
Garden Center of Rochester Library
Garden Center of Rochester
5 Castle Park
Rochester, NY 14620
716-473-5130

Librarian: Regina Campbell
Number of books: 4,000; number of periodicals: 24.
Lending policy: Members only, reference only.
Cross-reference: 7.43

14.66
Garden Club of America Library
598 Madison Ave.
New York, NY 10022
212-753-8287; (FAX) 212-753-0134

3,000 volumes and periodicals; conservation, horticulture,
botany, and rare illustrated botanical books of the 17th,
18th, and 19th centuries.
Lending policy: Reference only.
Cross-reference: 5.1

14.67
George Landis Arboretum Library
George Landis Arboretum
Box 186, Lape Rd.
Esperance, NY 12066
518-875-6935

Number of books: 600.
Lending policy: Reference only, by appointment.
Cross-reference: 16.279

Horticultural Alliance of the Hamptons
Bridgehampton, NY
Cross-reference: 6.24

14.68
Horticultural Society of New York Library
128 West 58th St.
New York, NY 10019
212-757-0915

15,000 volumes and periodicals; botany and horticultural
books of the Northeast.
Lending policy: Members only, reference only.
Cross-reference: 6.25

14.69
Liberty Hyde Bailey Hortorium
462 Mann Library
Cornell University
Ithaca, NY 14853
607-255-7781

Number of books: 12,000.
Large collection of reprints, and principal U.S. collection
of seed and nursery catalogs.
Lending policy: Loans made through the A. R. Mann
Library, Cornell University.
Cross-references: 12.249, 12.439, 15.142, 15.145, 15.147,
16.275

14.70
Library of the New York Botanical Garden
200th St. and Southern Blvd.
Bronx, NY 10458-5126
212-220-8700; (FAX) 212-220-6504

Librarian: Bernadette Callery
Over 1,000,000 items including 10,000 seed and
nursery catalogs, 90 films, ethnobotanical artifacts,
historic photographs and glass negatives, rare
book collection including pre-Linnaean and early
herbals, and Darwiniana, botanical art, and numerous
archives.
Lending policy: Reference only.
Cross-reference: 16.286

14.71
Mary Flagler Cary Arboretum Library
Box AB
Millbrook, NY 12545
914-677-5343; (FAX) 914-677-5976

Number of books: 8,000; number of periodicals: 250.
Map collection.
Lending policy: Reference only.
Cross-reference: 16.285

14.72
Monroe County Parks Arboretum Library
Highland Botanical Park
180 Reservoir Ave.
Rochester, NY 14620
716-244-9023

Librarian: James W. Kelly
Number of books: 500; number of periodicals: 23.
Lending policy: Reference only.
Cross-reference: 16.280

14.73
Planting Fields Arboretum Library
Planting Fields Arboretum
Planting Fields Rd.
Oyster Bay, NY 11771
516-922-9206; (FAX) 516-922-0770

Number of books: 5,000; number of periodicals: 25.
General horticultural collection.
Lending policy: Circulating to members only.
Cross-reference: 16.287

Queens Botanical Garden
Flushing, NY
Cross-reference: 16.288

Rose Hybridizers Association
Horseheads, NY
Cross-reference: 2.80

Staten Island Botanical Garden
Staten Island, NY
Cross-reference: 16.292

OH
American Daffodil Society, Inc.
Milford, OH
Cross-reference: 2.8

American Ivy Society
West Carrollton, OH
Cross-reference: 2.19

Carriage Hill Farm
Dayton, OH
Cross-reference: 17.214

14.74
Corning Library
Holden Arboretum
9500 Sperry Rd.
Mentor, OH 44060
216-256-1110; (FAX) 216-256-1655

Number of books: 6,500; number of periodicals: 125.
The Corning Collection of Horticultural Classics.
Lending policy: Reference only.
Cross-reference: 16.304

14.75
Cox Arboretum Library
James M. Cox Arboretum
6733 Springboro Pike
Dayton, OH 45449
513-434-9005

Number of books: 2,300; number of periodicals: 27.
Lending policy: Reference only.
Cross-reference: 16.297

14.76
Dawes Arboretum Library
7770 Jacksontown Rd., S.E.
Newark, OH 43055
614-323-2355

Number of books: 5,000.
General horticultural collection, nature books.
Lending policy: Members only, reference only.
Cross-reference: 16.298

14.77
Eleanor Squire Library
The Garden Center of Greater Cleveland
11030 East Blvd.
Cleveland, OH 44106
216-721-1600

Number of books: 14,746; number of periodicals: 200.
Gardening, horticulture, landscape architecture, rare and
    current herb books, seed and nursery catalogs, and the
    Warren H. Corning Collection of Horticultural Classics.
Lending policy: Circulating to members.
Cross-reference: 16.302

Falconskeape Gardens
Median, OH
Cross-reference: 16.299

Franklin Park Conservatory and Garden Center
Columbus, OH
Cross-reference: 16.301

14.78
Gardenview Horticultural Park Library
16711 Pearl Rd.
Strongsville, OH 44136
216-238-6653

Number of books: 4,000.
Books on gardening with an emphasis on British authors.
Lending policy: Members only, reference only.
Cross-reference: 16.303

14.79
Hoffman Library
Civic Garden Center of Greater Cincinnati
2715 Reading Rd.
Cincinnati, OH 45206
513-221-0981

Librarian: Carol Smith
Number of books: 1,500; number of periodicals: 15.
Gardening, garden design, and individual plant
    categories.
Lending policy: Members have circulating privileges.
Cross-reference: 7.46

14.80
Kingwood Center Library
Kingwood Center
900 Park Ave. West
Mansfield, OH 44906
419-522-0211

Number of books; 800; number of periodicals: 125.
Gardening, nature, and rare books (herbals and illustrated
   flower books).
Lending policy: Circulating.
Cross-reference: 16.307

14.81
Lloyd Library
917 Plum St.
Cincinnati, OH 45202
513-721-3707

Librarian: Rebecca A. Perry
Number of books: 65,000; number of periodicals: 500.
Linnaean literature, herbals (spanning five centuries),
   plant taxonomy, medical botany, and floras.
Lending policy: Reference only.

Ohio Village
Columbus, OH

Cross-reference: 17.217

Slate Run Living Historical Farm
Ashville, OH

Cross-reference: 17.219

14.82
Wegerzyn Horticultural Center
1301 E. Siebenthaler Ave.
Dayton, OH 45414
513-277-6545

Lending policy: Reference only.
Cross-reference: 7.48

OK
14.83
Tulsa Garden Center Library
Tulsa Garden Center
2435 South Peoria
Tulsa, OK 74114
918-749-6401

Librarian: Marina Metevelis
Seed and nursery catalogs, and general gardening
   collection.
Number of books: 4,500; number of periodicals: 27.
Lending policy: Circulating.
Cross-reference: 16.320

OR
Berry Botanic Garden
Portland, OR

Cross-reference: 16.322

Hardy Plant Society of Oregon
Boring, OR

Cross-reference: 2.46

Hoyt Arboretum
Portland, OR

Cross-reference: 16.326

Leach Botanical Garden
Portland, OR

Cross-reference: 16.329

PA
14.84
Academy of Natural Sciences Library
Academy of Natural Sciences of Philadelphia
19th St. & The Parkway
Philadelphia, PA 19103
215-299-1140

Librarian: Linda Rossi
Number of books: 190,000; number of periodicals: 3,200.
Natural sciences including a botany collection, and
   pre-1860 horticultural imprints.
Lending policy: Reference only.
Cross-reference: 15.163

Botanical Society of Western Pennsylvania
Pittsburgh, PA

Cross-reference: 3.39

Bowman's Hill Wildflower Preserve
Washington Crossing, PA

Cross-reference: 16.338

Campus Arboretum of Haverford College
Haverford, PA

Cross-reference: 16.340

14.85
Carnegie Museum of Natural History Library
4400 Forbes Ave.
Pittsburgh, PA 15213
412-622-3264

2,800 botanical titles in a general natural history
   collection.
Lending policy: Reference only.
Cross-reference: 15.165

Colonial Pennsylvania Plantation
Media, PA

Cross-reference: 17.228

14.86
Hunt Institute for Botanical Documentation
Carnegie Mellon University
Pittsburgh, PA 15213
412-268-2434

Number of books: 25,000; number of periodicals: 1,200.
Major taxonomic and floristic studies and literature of the
   history of botany, works published between 1550 and
   1850, especially herbals, Linnaeana, history of botanical
   illustration, and archives of botanists of the 19th and
   20th centuries, botanical biography, and oral history.
Lending policy: Reference only.
Cross-reference: 4.13

International Herb Growers & Marketers Association
Silver Spring, PA

Cross-reference: 1.69

14.87
Library of the Barnes Foundation
P.O. Box 128
Merion Station, PA 19066
215-664-8880

Number of books: 1,500; number of periodicals: 20.
Regional floras.
Lending policy: Reference only.
Cross-reference: 16.335

14.88
Longwood Library
Longwood Gardens
Rt. 1, P.O. Box 501
Kennett Square, PA 19348-0501
215-388-6741

Number of books: 18,500; number of periodicals: 285.
Complete run of *Curtis's Botanical Magazine.*
Lending policy: Reference only.
Cross-reference: 16.348

14.89
Morris Arboretum of the University of Pennsylvania
9414 Meadowbrook Ave.
Philadelphia, PA 19118
215-247-5777

6,000 volumes and periodicals.
Lending policy: Circulating to staff, faculty, students,
    associates, and guides.
Cross-reference: 16.351

Pennsbury Manor
Morrisville, PA

Cross-reference: 17.231

14.90
Pennsylvania Horticultural Society Library
325 Walnut St.
Philadelphia, PA 19106-2777
215-625-8250

Number of books: 14,000; number of periodicals: 200.
Gardening books from the 16th to the 20th centuries,
    herbals, seed and nursery catalogs pertaining to the
    greater Delaware Valley, and a special collection of
    19th-century horticultural books.
Lending policy: Circulating to members only.
Cross-reference: 6.32

Peter Wentz Farmstead
Worcester, PA

Cross-reference: 17.232

14.91
Pittsburgh Civic Garden Center Library
1059 Shady Ave.
Pittsburgh, PA 15232
412-441-4442

Librarian: Jean R. Aiken
Number of books: 2,000; number of periodicals: 13.
Lending policy: Circulating to members.
Cross-reference: 7.52

14.92
Scott Arboretum Horticultural Library
Swarthmore College
500 College Ave.
Swarthmore, PA 19081
215-328-8025

Librarian: Erica Glasener
Number of books: 950; number of periodicals: 124.
Complete collection is focused on horticulture.
Lending policy: Circulating.
Cross-reference: 16.353

RI
Blithewold Gardens and Arboretum
Bristol, RI

Cross-reference: 17.234

Wilcox Park
Westerly, RI

Cross-reference: 16.358

SC
Brookgreen Gardens
Murrells Inlet, SC

Cross-reference: 16.359

TN
Dixon Gallery & Gardens
Memphis, TN

Cross-reference: 17.242

14.93
Minnie Ritchey and Joel Owsley Cheek Library
Tennessee Botanical Gardens and Fine Arts Center at
    Cheekwood
Forrest Park Dr.
Nashville, TN 37205
615-353-2148

Number of books: 3,000; number of periodicals: 61.
Herbs, wildflowers, garden design, landscape architecture,
    natural history, flower arranging, and botanical
    illustration.
Lending policy: Circulating to members, reference only
    to public.
Cross-reference: 17.246

Reflection Riding
Chattanooga, TN

Cross-reference: 16.373

14.94
Sybil G. Malloy Memorial Library
Memphis Botanic Garden
750 Cherry Rd.
Memphis, TN 38117
901-685-1566; (FAX) 901-325-5770

Number of books: 4,700; number of periodicals: 25.
Lending policy: Reference only.
Cross-reference: 16.370

University of Tennessee Arboretum
Oak Ridge, TN
Cross-reference: 16.376

TX
14.95
Botanic Garden Library
Fort Worth Botanic Garden
3220 Botanic Garden Blvd.
Fort Worth, TX 76107
817-870-7682

3,500 books and periodicals; general in scope.
Lending policy: Mainly for staff and volunteers.
Cross-reference: 16.382

Dallas Civic Garden Center
Dallas, TX
Cross-reference: 16.381

Houston Arboretum and Nature Center
Houston, TX
Cross-reference: 16.384

Mercer Arboretum and Botanic Gardens
Humble, TX
Cross-reference: 16.387

National Wildflower Research Center
Austin, TX
Cross-reference: 1.97

Native Prairies Association of Texas
Fort Worth, TX
Cross-reference: 3.48

14.96
W. J. Rogers Memorial Library
Beaumont Council of Garden Clubs
Garden Center in Tyrrell Park
Beaumont, TX 77705
409-842-3135

Librarian: Myra Clay
Number of books: 1,200.
Collection related to Gulf Coast plants.
Lending policy: Members only.
Cross-reference: 7.59

VA
Agecroft Hall
Richmond, VA
Cross-reference: 17.252

American Society for Horticultural Science
Alexandria, VA
Cross-reference: 4.4

Booker T. Washington National Monument
Hardy, VA
Cross-reference: 17.255

Claude Moore Colonial Farm at Turkey Run
McLean, VA
Cross-reference: 17.257

Future Farmers of America
Alexandria, VA
Cross-reference: 1.51

George Washington Birthplace National Monument
Washington's Birthplace, VA
Cross-reference: 17.259

Gunston Hall
Mason Neck, VA
Cross-reference: 17.260

14.97
Harold B. Tukey Memorial Library
American Horticultural Society
7931 East Boulevard Dr.
Alexandria, VA 22308
703-768-5700; (FAX) 703-777-7931

Number of books: 3,000.
Horticulture, botany, and plant exploration.
Lending policy: Members only, reference only, by
  appointment.
Cross-reference: 1.16

Monticello
Charlottesville, VA
Cross-reference: 17.265

Mount Vernon
Mount Vernon, VA
Cross-reference: 17.268

Norfolk Botanical Gardens
Norfolk, VA
Cross-reference: 16.400

14.98
Orland E. White Arboretum Library
Blandy Experimental Farm
P.O. Box 175
Boyce, VA 22620
703-837-1758

Number of books: 1,100.
Lending policy: By appointment, reference only.
Cross-reference: 16.401

Stratford Hall Plantation
Stratford, VA
Cross-reference: 17.271

Vinifera Wine Growers Association
The Plains, VA
Cross-reference: 1.121

Virginia Living Museum
Newport News, VA
Cross-reference: 17.272

Virginia Native Plant Society
Annandale, VA
Cross-reference: 3.50

VT
National Gardening Association
Burlington, VT
Cross-reference: 1.89

WA
14.99
Elisabeth C. Miller Library
Center for Urban Horticulture
University of Washington GF-15
Seattle, WA 98195
206-543-8616

Librarians: Valerie Easton, Laura Lipton
6,000 volumes and periodicals.
Lending policy: Reference only.
Cross-reference: 1.40

Fort Vancouver National Historic Site
Vancouver, WA
Cross-reference: 17.277

Meerkerk Rhododendron Garden
Greenbank, WA
Cross-reference: 16.411

Plant Amnesty
Seattle, WA
Cross-reference: 1.106

14.100
Rhododendron Species Foundation Library
P.O. Box 3798
Federal Way, WA 98063
206-661-9377

Number of books: 1,000; number of periodicals: 7.
Field notes of plant collectors, extensive collection of F.
  Kingdon Ward's publications.
Lending policy: Reference only.
Cross-reference: 2.79

WI
O.J. Noer Research Foundation, Inc.
Milwaukee, WI
Cross-reference: 4.18

Olbrich Botanical Gardens
Madison, WI
Cross-reference: 16.423

Paine Art Center and Arboretum
Oshkosh, WI
Cross-reference: 17.281

14.101
Reference Library of the Boerner Botanical Gardens
5879 South 92nd St.
Hales Corners, WI 53130
414-529-1870

Number of books: 2,000.
Lending policy: Reference only.
Cross-reference: 16.419

Soil Science Society of America
Madison, WI
Cross-reference: 4.23

WV
14.102
Wheeling Garden Center Library
Oglebay Park
Wheeling, WV 26003
304-242-0665

1,100 books and periodicals; general gardening and flower
  arranging collection.
Lending policy: For members only.
Cross-reference: 7.71

WY
Cheyenne Botanic Garden
Cheyenne, WY
Cross-reference: 16.428

CANADA

AB
Calgary Horticultural Society
Calgary, AB
Cross-reference: 6.44

Devonian Botanic Garden
Edmonton, AB
Cross-reference: 16.432

BC
British Columbia Lily Society
Langley, BC
Cross-reference: 2.98

Dr. Sun Yat-Sen Classical Chinese Garden
Vancouver, BC
Cross-reference: 16.439

Grist Mill at Keremeos
Keremeos, BC
Cross-reference: 17.283

Horticulture Center of the Pacific
Victoria, BC
Cross-reference: 16.443

14.103
University of British Columbia
Botanical Garden Library
6804 S.W. Marine Dr.
Vancouver, BC V6T 1W5
604-228-4779; (FAX) 604-228-2016

Number of books: 1,500; number of periodicals: 85.
Lending policy: Reference only.
Cross-reference: 16.449

14.104
VanDusen Botanical Display Garden Library
5251 Oak St.
Vancouver, BC V6M 4H1
604-266-7194

Number of books: 3,100; number of periodicals: 50.
Plant hunting and exploration, Pacific Northwest
   horticulture, and comprehensive botany and
   horticultural collection.
Lending policy: Circulating to members.
Cross-reference: 16.450

NF
Newfoundland Horticultural Society
St. Johns, NF
Cross-reference: 6.49

NS
Cole Harbour Heritage Farm Museum
Cole Harbour, NS
Cross-reference: 17.287

ON
Arboretum of the University of Guelph
Guelph, ON
Cross-reference: 16.456

Balance Life Gardens
Lambeth, ON
Cross-reference: 16.457

Canadian Iris Society
Willowdale, ON
Cross-reference: 2.102

Canadian Rose Society
Scarborough, ON
Cross-reference: 2.106

Canadian Society for Herbal Research
Willowdale, ON
Cross-reference: 1.135

Centre for Canadian Historical Horticultural Studies
Hamilton, ON
Cross-reference: 17.39

14.105
Civic Garden Centre Library
Civic Garden Centre
777 Lawrence Ave., East
North York, ON M3C 1P2
416-444-1552

Librarian: Pamela MacKenzie
Number of books: 6,500; number of periodicals: 65.
Historical collection, orchids.
Lending policy: Circulating to members.
Cross-reference: 7.72

Dominion Arboretum
Ottawa, ON
Cross-reference: 16.459

J.J. Neilson Arboretum
Ridgetown, ON
Cross-reference: 16.466

14.106
Niagara Parks Commission School of Horticulture Library
Niagara Parks Commission School of Horticulture
P.O. Box 150
Niagara Falls, ON L2E 6T2
416-356-8554

Librarian: Shirley Stoner
Number of books: 3,000; number of periodicals: 65.
Lending policy: By appointment only, reference only.
Cross-references: 12.376, 16.469

Ontario Agricultural Museum
Milton, ON
Cross-reference: 17.294

14.107
Royal Botanical Gardens Library
P.O. Box 399
Hamilton, ON L8N 3H8
416-527-1158; (FAX) 416-577-0375

Librarian: Ina Vrugtman
Number of books: 10,000; number of periodicals: 450.
Nursery and seed trade catalog collection, *Gray Herbarium
   Index*, and Canadian historical horticulture.
Lending policy: Reference only.
Cross-reference: 16.473

PQ

14.108
Library of the Montreal Botanical Garden
4101, Sherbrooke St. East
Montreal, PQ H1X 2B2
514-872-1400; (FAX) 514-872-3765

Number of books: 13,000; number of periodicals: 320.
19th-century gardening and botanical magazines, and
contemporary and old books on medicinal plants.

Lending policy: Reference only.
Cross-reference: 16.481

SK

Saskatchewan Perennial Society
Saskatoon, SK

Cross-reference: 2.113

~ 15 ~

Herbaria

A herbarium is an identified collection of dried specimens of plants arranged systematically. The listing in this directory includes all collections in the United States and Canada that number 20,000 specimens or more or those with smaller, specialized collections.

Herbaria are usually open to the public for serious research. Often they are open by appointment only, so researchers should call before they visit the collection. Herbaria also often provide a variety of services to the public such as plant identification, general botanical information, and practical information on plant collection, identification, preservation, and annotation. Herbaria will usually loan their collections to qualified institutions.

The best reference work on herbaria is the *Index Herbariorum* (Antwerp: Bohn, Scheltema, & Holkem, 1981), a comprehensive listing of the herbaria of the world. It is updated and republished approximately every ten years.

UNITED STATES

AK
15.1
University of Alaska Museum Herbarium
Fairbanks, AK 99701
907-479-7108

Number of specimens: 90,000.
Content: Flora of Alaska and other circumpolar areas.
Cross-reference: 12.2

AL
15.2
Auburn University Herbarium
Dept. of Botany and Microbiology
Auburn, AL 36849-5407
205-844-1630; (FAX) 205-844-1645

Contact: John D. Freeman, Curator
Number of specimens: 50,000.
Content: Rare and threatened vascular plants of Alabama.
Cross-references: 12.4, 12.395, 16.1

15.3
University of Alabama Herbarium
Dept. of Biology
Box 870344
Tuscaloosa, AL 35487-0344
205-348-1826; (FAX) 205-348-6544

Contact: Dr. Robert R. Haynes, Director

Number of specimens: 50,000.
Content: Flora of Alabama, aquatic vascular plants of the
 Neotropics.
Cross-reference: 16.5

15.4
University of Alabama Herbarium
Dept. of Biology
University of Alabama
P.O. Box 1927
University, AL 35486
205-348-5966

Number of specimens: 40,000.
Content: Vascular flora of Alabama with an emphasis on
 aquatic vascular plants.

AR
15.5
Arkansas State University Herbarium
Dept. of Biological Science
Arkansas State University
State University, AR 72467
501-972-3082

Number of specimens: 20,000.
Content: Arkansas and southeastern U.S.
Cross-reference: 12.8

15.6
Arkansas Tech University Herbarium
School of Physical and Life Sciences
Arkansas Tech University
Russellville, AR 72801
501-968-0312

Number of specimens: 25,000.
Content: Plants of the southeastern U.S. and Arkansas.

15.7
University of Arkansas Herbarium
Dept. of Botany and Bacteriology
University of Arkansas
Fayetteville, AR 72071
501-575-4901

Number of specimens: 63,000.
Content: Plants of Arkansas and the general area.
Cross-references: 12.12, 12.396

AZ
15.8
Arizona State University Herbarium
Dept. of Botany and Microbotany
Arizona State University
Tempe, AZ 85281
602-965-6162

Number of specimens: 150,000.
Content: Plants of the southwest U.S. and northern
  Mexico.

15.9
Deaver Herbarium
Dept. of Biology
Box 5640
Northern Arizona University
Flagstaff, AZ 86011
602-523-7242

Contact: Dr. Richard Heulg, Acting Curator
Number of specimens: 50,000.
Content: Plants from the Colorado Plateau region.

15.10
Desert Botanical Garden Herbarium
1202 North Galvin Pkwy.
Phoenix, AZ 85008
602-941-1225

Contact: W.C. Hodgson, Curator
Number of specimens: 40,000.
Content: Large collection of southwest U.S. and Mexico
  species.
Cross-reference: 16.10

15.11
Museum of Northern Arizona Herbarium
Rt. 4, Box 720
Flagstaff, AZ 86001
602-774-5211, ext. 68

Number of specimens: 28,000.
Content: Bryophytes and vascular plants of southwest
  U.S., especially northern Arizona.

15.12
University of Arizona Herbarium
Shanz Building # 38, Room 113
Tucson, AZ 85721
602-621-7243; (FAX) 602-621-7196

Contact: Dr. Charles T. Mason, Jr., Curator
Number of specimens: 300,000.
Content: Flora of the Sonoran southwest, grasses,
  Mexican flora.
Cross-references: 12.19, 12.398

CA
15.13
Allan Hancock Foundation Herbarium
University of Southern California
Los Angeles, CA 90007
213-741-7535

Number of specimens: 60,000.
Content: Exclusively marine plants, primarily benthic
  algae, with an emphasis on the eastern Pacific.

15.14
California Academy of Sciences Herbarium
Dept. of Botany
Golden Gate Park
San Francisco, CA 94118
415-221-5100, ext. 265

Number of specimens: 1,400,000.
Content: Worldwide with special emphasis on flora of
  California, western North America, and northern Latin
  America.

15.15
California Dept. of Food and Agriculture Herbarium
1220 N. St.
Sacramento, CA 95814
916-445-4521

Number of specimens: 20,000.
Content: Weeds, especially of California.

15.16
California State University Herbarium
Biology Dept.
California State University
1250 Bellflower Blvd.
Long Beach, CA 90840
213-498-4917

Number of specimens: 19,500.
Content: Flora of southern California, and vascular plants
  of southern California and southeastern U.S.

15.17
California State University Herbarium
Dept. of Biology
Los Angeles, CA 90032
213-224-3518

Number of specimens: 30,000.
Content: Taxa of southern California, Mojave Desert,
  Chihuahuan Desert, Baja California, and Mexico.

15.18
California State University Herbarium
Dept. of Biological Sciences
California State University
Chico, CA 95929
916-895-5381

Number of specimens: 29,000.
Content: Vascular plants and myxomycete flora, primarily
   of northern California.
Cross-reference: 12.26

15.19
Dudley Herbarium of Stanford University
Dept. of Botany
California Academy of Science
Golden Gate Park
San Francisco, CA 94118
415-221-5100

Number of specimens: 850,000.
Content: Vascular plants of western North America
   including Mexico, Mediterranean plants, and plants of
   arctic Alaska.

15.20
Faye A. MacFadden Herbarium
Dept. of Biological Science
California State University
Fullerton, CA 92634
714-773-3614

Contact: Dr. C. Eugene Jones, Chair
Number of specimens: 43,000.
Content: Southern California species.
Cross-reference: 16.24

15.21
Fresno Herbarium
California State University, Fresno
Dept. of Biology
Fresno, CA 93740
209-278-2001

Contact: John C. Stebbins, Curator
Number of specimens: 36,000.
Content: Flora of Central California Valley, Sierra
   Nevada, Coast Range, and Mojave Desert.
Cross-references: 12.27, 20.70

15.22
Humboldt State University Herbarium
Dept. of Biology
Arcata, CA 95521
707-826-4801

Number of specimens: 140,000.
Content: Vascular plants of northwest California and
   southwest Oregon, regional, Mexican, Caribbean, and
   tropical South Pacific bryophytes.

15.23
J. M. Tucker Herbarium and Crampton Herbarium
Botany Dept., Room 262
Robbins Hall
University of California at Davis
Davis, CA 95616
916-752-1091; (FAX) 916-752-4754

Contact: Dr. Grady L. Webster, Director
Number of specimens: 150,000.
Content: Emphasis on California flora, weedy species of
   Mediterranean-type climate regions, *Quercus* of the New
   World, Euphorbiaceae, neotropical material, and alpine
   flora of western North America.
Cross-references: 12.62, 12.402, 16.59, 17.33, 19.12

15.24
Jepson Herbarium
University of California at Berkeley
6701 San Pablo Ave.
Oakland, CA 94608
415-643-7008

Contact: Dr. Lawrence Heckard, Curator
Number of specimens: 85,000.
Content: Only California native species.
Cross-references: 12.401, 15.36, 16.18, 16.60

15.25
Los Angeles State and County Arboretum Herbarium
301 North Baldwin Ave.
Arcadia, CA 91006
213-446-8251

Number of specimens: 50,000.
Content: Exotic woody flora.
Cross-reference: 16.35

15.26
Natural History Museum of Los Angeles Co.
Botany Dept.
900 Exposition Blvd.
Los Angeles, CA 90007
213-744-3378

Contact: Dr. Don Reynolds, Curator
Number of specimens: 250,000.
Content: Marine algae, lichen, and wood-rotting fungi
   collections.

15.27
Pacific Union College Herbarium
Biology Dept.
Pacific Union College
Angwin, CA 94508
707-965-6227

Number of specimens: 20,000.
Content: Plants of Napa county, California, and the
   Klamath Mountains of northern California and southern
   Oregon.

15.28
Rancho Santa Ana Botanic Garden Herbarium
1500 North College Ave.
Claremont, CA 91711
714-625-8767; (FAX) 714-626-7670

Contact: Thomas S. Elias, Director
Number of specimens: 1,000,000.
Content: Nearly complete representation of flowering
plant families and subfamilies and the strongest
collection of southern California flora in the world.
Cross-reference: 16.48

15.29
Robert F. Hoover Herbarium
Biological Science Dept.
California Polytechnic State University
San Luis Obispo, CA 93407
805-756-2043

Contact: Dr. David J. Keil, Director
Number of specimens: 50,000.
Content: Specializes in flora of central coastal California,
Asteraceae, and *Lupinus*.
Cross-references: 12.24, 12.399, 20.68

15.30
San Diego Natural History Museum Herbarium
Balboa Park
P.O. Box 1390
San Diego, CA 92112
619-232-3821

Contact: Dr. Geoffrey A. Levin, Curator
Number of specimens: 128,000.
Content: Native plants of southern California and Baja
California and Mexico, and cultivated plants of San
Diego.
Other: Exhibits on native plants of San Diego county.
Cross-reference: 16.16

15.31
San Francisco State University Herbarium
Dept. of Biological Sciences
San Francisco State University
1600 Holloway Ave.
San Francisco, CA 94132
415-469-2439

Number of specimens: 68,000.
Content: Cryptogamic plants.

15.32
Santa Barbara Museum of Natural History, Herbarium
2559 Puesta Del Sol Rd.
Santa Barbara, CA 93105
805-682-4711

Number of specimens: 29,000.
Content: Plants of southern California and the Channel
Islands of California.

15.33
Sonoma State University Herbarium
Dept. of Biology
Sonoma State University
Rohnert Park, CA 94928
707-664-2189

Number of specimens: 22,000.
Content: Plants of the north coast counties of California.

15.34
UCR Herbarium
Dept. of Botany & Plant Sciences
University of California at Riverside
Riverside, CA 92521
714-787-3601

Contact: Andrew C. Sanders, Curator
Number of specimens: 65,000.
Content: Flora of southwestern U.S. and all of Mexico.
Cross-references: 12.64, 16.19

15.35
University of California Herbarium
Dept. of Biological Science
University of California at Santa Barbara
Santa Barbara, CA 93106
805-961-2506

Number of specimens: 34,500.
Content: Plants of western North America, especially
California and surrounding states.

15.36
University of California Herbarium
Dept. of Botany
University of California at Berkeley
Berkeley, CA 94720
415-642-2465

Number of specimens: 1,452,000.
Content: Worldwide flora, with specialization in the
Americas and the Pacific Basin.
Cross-references: 12.401, 15.24, 16.18, 16.60

15.37
University of California, Los Angeles Herbarium
Rm. 124 Botany Bldg.
618 Circle Dr. South
Los Angeles, Ca 90024-1606
213-825-2714; (FAX) 213-206-3987

Contact: Dr. Arthur C. Gibson
Number of specimens: 180,000.
Content: Flora of the Santa Monica Mountains and
southern California, Apiaceae, *Mentzelia*, *Camissonia*,
Clarkia, and *Dudleya*.
Cross-reference: 16.40

CO
15.38
Colorado State University Herbarium
Dept. of Biology
Colorado State University

Fort Collins, CO 80523
303-491-0496

Contact: Dr. Dieter Wilken, Professor of Biology
Number of specimens: 65,000.
Content: Colorado taxa, including threatened and
  endangered plants.
Other: Bibliographical information on taxonomy of
  Colorado vascular plants, and electronic database of
  specimens in Colorado.
Cross-references: 12.68, 12.403, 16.69

15.39
Forest Pathology Herbarium
U.S. Forest Service
Rocky Mountain Forest and Range Experiment Station
240 West Prospect St.
Fort Collins, CO 80526
303-221-4390

Number of specimens: 12,000.
Content: Forest disease organisms of the Rocky
  Mountains and the southwestern U.S.

15.40
University of Colorado Museum Herbarium
Campus Box 218
Boulder, CO 80302
303-492-6171

Number of specimens: 335,000.
Content: Worldwide flora, especially plants of western
  North America, Europe, Galapagos, Australia, and New
  Guinea.

CT
15.41
Connecticut Botanical Society Herbarium
Osborn Memorial Laboratories
Yale University,
167 Prospect St.
New Haven, CT 06520
203-432-3904

Contact: Donald M. Swan, President
Number of specimens: 35,000.
Content: Flora of Connecticut only.
Cross-references: 3.10

15.42
George Safford Torrey Herbarium
Biological Sciences Group
University of Connecticut
Life Sciences U-43
Storrs, CT 06268
203-486-4150

Number of specimens: 90,000.
Content: Flora of eastern North America and Neotropical
  flora.
Cross-references: 12.73, 12.405, 20.44

15.43
Herbarium of Yale University
Osborn Memorial Laboratories
Room 550, Dept. of Biology
167 Prospect St.
P.O. Box 6666
New Haven, CT 06511
203-432-3904

Contact: Dr. Leo J. Hickey, Curator
Number of specimens: 250,000.
Content: Specializing in flora of Connecticut, New
  England, and North America, rich in ferns and
  bryophytes.
Cross-reference: 16.77

DC
15.44
United States National Arboretum Herbarium
3501 New York Ave., N.E.
Washington, DC 20002
202-475-4843

Contact: Dr. Frederick G. Meyer, Botanist
Number of specimens: 600,000.
Content: Specimens of woody landscape plants of the
  southeastern U.S., Japan, and worldwide.
Cross-reference: 16.83

15.45
United States National Herbarium
Smithsonian Institution
Dept. of Botany, 166 NHB
Washington, DC 20560
(FAX) 202-786-2563

Contact: George F. Russell, Collections Manager
Number of specimens: 4,300,000.
Content: Specializes in plants from the Neotropics.
Cross-reference: 14.17

DE
15.46
Claude E. Phillips Herbarium
Dept. of Agriculture & Natural Resources
Delaware State College
Dover, DE 19901
302-736-5120

Contact: Dr. Arthur O. Tucker
Number of specimens: 106,000.
Content: Flora of the Delmarva Peninsula.

FL
15.47
Fairchild Tropical Garden Herbarium
10901 Old Cutler Rd.
Miami, FL 33156
305-665-2844

Contact: William M. Houghton, Curator
Number of specimens: 60,000.
Content: Specializes in palms and cycads of the world,
  Caribbean flora, and tropical horticultural plants.
Cross-reference: 16.87.

15.48
Florida Atlantic University Herbarium
Dept. of Biological Sciences
Florida Atlantic University
Boca Raton, FL 33431
305-395-5100, ext. 2729

Number of specimens: 23,000.
Content: Flora of southeastern Florida.

15.49
Florida State University Herbarium
Dept. of Biological Sciences
Unit 1
Florida State University
Tallahassee, FL 32306
904-644-6278
Number of specimens: 160,000.
Content: Plants of the southeastern U.S. and tropical
    America.

15.50
Marie Selby Botanical Gardens Herbarium
811 S. Palm Ave.
Sarasota, FL 34236
813-349-0280

Contact: Libby Besse, Curator
Number of specimens: 75,000.
Content: Specializes in tropical plants with an emphasis
    on epiphytes and has an extensive collection of orchids,
    bromeliads, aroids, gesneriads, ferns, heliconias, and
    calatheas. The holdings are primarily from the
    Neotropics, especially from the Andes.
Cross-reference: 16.96

15.51
University of Florida Herbarium
209 Rolfs Hall
University of Florida
322 IFAS
Gainesville, FL 32611-0322
904-392-1767

Contact: Dr. Norris H. Williams, Curator
Number of specimens: 400,000.
Content: Vascular flora of Florida, the southeastern U.S.
    coastal plain, and the Neotropics, especially Haiti,
    southeastern U.S. and tropical bryophytes, Florida
    fungi, and worldwide wood collection.
Cross-references: 12.99, 12.407

15.52
University of South Florida Herbarium
Dept. of Biology
University of South Florida
Tampa, FL 33620
813-974-2359; (FAX) 813-974-5273

Contact: Dr. Richard P. Wunderlin, Director
Number of specimens: 200,000.

GA
15.53
Julian H. Miller Mycological Herbarium
Dept. of Plant Pathology
University of Georgia
Athens, GA 30602
404-542-2571

Number of specimens: 20,000.
Content: Fungi of the southeastern U.S.
Cross-references: 12.108, 12.409, 15.54, 20.11

15.54
University of Georgia Herbarium
Botany Dept.
Plant Sciences Building
University of Georgia
Athens, GA 30602
404-542-1823; (FAX) 404-542-1805

Contact: Michael O. Moore, Curator
Number of specimens: 181,000.
Content: Extensive collections of the vascular flora of
    Georgia and the southeastern U.S.
Cross-references: 12.108, 12.409, 15.53, 20.11

15.55
Valdosta State College Herbarium
Dept. of Biology
Valdosta State College
Valdosta, GA 31698
912-333-5759

Contact: Richard Carter, Curator
Number of specimens: 40,000.
Content: Plants of southeastern U.S. and particularly
    southern Georgia.

HI
15.56
Bernice P. Bishop Museum Herbarium
P.O. Box 1900-A
Honolulu, HI 96817-8968
1525 Bernice St., Honolulu, HI 96817
808-848-4175; (FAX) 808-841-8968

Contact: Dr. S. H. Sohmer, Chairman of Botany Dept.
Number of specimens: 456,400.
Content: Extensive collection of Hawaiian and Pacific
    plants.
Cross-reference: 14.30

15.57
University of Hawaii Herbarium
Dept. of Botany
University of Hawaii
3190 Maile Way
Honolulu, HI 96822
808-948-8369

Number of specimens: 20,000.
Cross-references: 12.113, 16.111

IA
15.58
Ada Hayden Herbarium
Iowa State University
Dept. of Botany
Ames, IA 50011-1020
515-294-9499; (FAX) 515-294-1337

Contact: Deborah Q. Lewis, Curator
Number of specimens: 410,000.
Content: Large holdings of Leguminosae of the United States, Gramineae of the New World, and fungi (especially plant parasitic taxa).
Cross-reference: 12.118, 12.410, 16.133, 20.109, 20.276

15.59
Martin L. Grant Herbarium
Biology Dept.
McCollum Science Hall
University of Northern Iowa
Cedar Falls, IA 50614
319-273-2218

Contact: Kay Klier, Curator
Number of specimens: 33,500.
Content: Specializes in the flora of Iowa. Special collections of the flora of Iran.

15.60
University of Iowa Herbarium
Dept. of Botany
University of Iowa
Iowa City, IA 52242
319-353-5790

Number of specimens: 325,200.
Content: Emphasis on plants of the midwestern U.S., especially Iowa.

ID
15.61
Ray J. Davis Herbarium
Box 8096
Idaho Museum of Natural History
Idaho State University
Pocatello, ID 83209
208-236-3882

Contact: Dr. Karl Holte, Curator
Number of specimens: 68,000.

15.62
University of Idaho Herbarium
Dept. of Biological Sciences
University of Idaho
Moscow, ID 83843
208-885-6798

Number of specimens: 86,000.
Content: Taxa of Idaho and the Pacific Northwest.
Cross-references: 12.124, 12.411, 16.136

IL
15.63
Crop Evaluation Laboratory Herbarium
Dept. of Agronomy

University of Illinois
Urbana, IL 61801
217-333-4373

Number of specimens: 50,000.
Content: Gramineae, especially Andropogoneae and genera of cereals.
Cross-references: 12.140, 12.412, 15.68, 20.15

15.64
E. L. Stover Herbarium of Eastern Illinois University
Botany Dept.
Eastern Illinois University
Charleston, IL 61920
217-581-3525

Contact: Dr. John E. Ebinger, Professor of Botany
Number of specimens: 60,000.
Content: Illinois flora.

15.65
Field Museum of Natural History Herbarium
Dept. of Botany
Roosevelt Rd. at Lake Shore Dr.
Chicago, IL 60605
312-922-9410

Number of specimens: 2,500,000.
Content: South American species.
Other: Hall of Plants in the museum features models of representatives of many plant families.
Cross-reference: 14.38

15.66
Hepatic Herbarium
The American Bryological and Lichenological Society
Dept. of Botany
Southern Illinois University
Carbondale, IL 62901

Number of specimens: 29,000.
Content: *Hepatophyta* and *Anthocerotophyta*, primarily of the U.S.
Cross-references: 4.1, 12.137, 15.74

15.67
Herbarium
Illinois State Museum
Spring & Edward Sts.
Springfield, IL 62706
217-782-2621; (FAX) 217-782-1254

Contact: Dr. Alfred C. Koelling, Curator
Number of specimens: 100,000.
Content: Flora of Illinois

15.68
Herbarium of the University of Illinois
Dept. of Plant Biology
University of Illinois
350 Natural History Building
505 S. Goodwin
Urbana, IL 61801
217-333-2522

Contact: Dr. Almut G. Jones, Curator
Number of specimens: 520,000.
Cross-references: 12.140, 12.412, 15.63, 20.15

15.69
Illinois State Natural History Survey
Center for Biodiversity
607 E. Peabody Dr.
Champaign, IL 61820
217-244-2172; (FAX) 217-333-4949

Contact: Dr. Kenneth R. Robertson, Curator
Number of specimens: 180,000.
Content: Mostly plants of Illinois.
Other: Public lectures.

15.70
Illinois State Natural History Survey Herbarium
172 Natural Resources Building
Urbana, IL 61801
217-333-6886

Number of specimens: 205,000.
Content: Vascular plants and parasitic fungi of Illinois.

15.71
Knox College Herbarium
Dept. of Biology
Knox College
Galesburg, IL 61401
309-343-0112, ext. 305

Number of specimens: 20,000.
Content: Emphasis on plants of western Illinois.

15.72
Morton Arboretum Herbarium
Rt. 53
Lisle, IL 60532
708-719-2425

Contact: William J. Hess, Curator
Number of specimens: 105,000.
Content: Extensive collection of flora of the Chicago region.
Cross-reference: 16.145

15.73
R. M. Myers Herbarium
Dept. of Biological Sciences
Western Illinois University
Macomb, IL 61455

Number of specimens: 26,000.
Content: Taxa of Illinois, especially western Illinois.

15.74
Southern Illinois University Herbarium
Dept. of Plant Biology
Carbondale, IL 62901
618-453-3228

Contact: Donald Ugent, Curator
Number of specimens: 200,000.
Content: Plants of southern Illinois and tropical America, Leguminosae, *Solanum*.
Cross-references: 12.137, 15.66

15.75
University of Illinois Herbarium
Dept. of Biological Sciences
University of Illinois at 1 Chicago Circle
Box 4348
Chicago, IL 60680
312-966-2993

Number of specimens: 25,000.
Content: Flora of the Chicago region, emphasizing Gramineae and Palmae.

IN
15.76
Friesner Herbarium
Dept. of Biological Sciences
Butler University
4600 Sunset
Indianapolis, IN 46208
317-283-9413; (FAX) 317-283-9519

Contact: Dr. Willard F. Yates, Curator
Number of specimens: 100,000.
Content: Specializes in plants of the Ohio River Valley.
Cross-reference: 16.153

15.77
Greene-Nieuwland Herbarium
Dept. of Biological Sciences
University of Notre Dame
Notre Dame, IN 46556
219-239-6684; (FAX) 219-239-7413

Contact: Barbara J. Hellenthal, Curator
Number of specimens: 265,000.
Content: Strong in western North American botany collected around the turn of the century. The Greene Herbarium contains about 5,000 of Edward Lee Greene's type specimens of western American plants.

15.78
Indiana University Herbarium
Dept. of Biology
Jordan Hall
Indiana University
Bloomington, IN 47405
812-855-5007

Contact: Dr. Charles Heiser, Director
Number of specimens: 140,000.

15.79
Purdue University Herbarium
Dept. of Botany and Plant Pathology
Lilly Hall of Life Sciences
Purdue University
West Lafayette, IN 47907
317-494-4623

Contact: Dr. Joe F. Hennen, Curator
Number of specimens: 160,000.
Content: Neotropical species, specialized rust herbarium.
Cross-references: 12.142, 12.414, 16.155

KS

15.80

Elam Bartholomew Herbarium
Fort Hays State Museum
Fort Hays State University
Hays, KS 67601
913-628-5665; (FAX) 913-628-4096

Contact: J.R. Thomasson, Curator
Number of specimens: 545,000.
Content: Largest fossil grass seed collection in the world,
excellent fungi collection.

15.81

Emporia State University Herbarium
Division of Biological Sciences
120 Commercial
Emporia, KS 66801
316-343-5311

Contact: Dr. Thomas Eddy/Dr. Laurie Robbins,
Co-directors
Number of specimens: 80,000.
Content: Good Gramineae collection, also Central and
South American Sapindaceae.

15.82

Herbarium of Kansas State University
Division of Biology
Ackert Hall
Manhattan, KS 66506
Bushnell Hall
913-532-6619; (FAX) 913-532-6653

Contact: T.M. Barkley, Curator
Number of specimens: 219,000.
Content: General herbarium specializing in Great Plains
plants.
Cross-references: 12.146, 12.415, 13.22, 13.23, 16.161

15.83

R. L. McGregor Herbarium
University of Kansas
2045 Constant Ave.
Lawrence, KS 66047
913-864-4493

Contact: Ralph E. Brooks, Curator
Number of specimens: 420,000.
Content: The largest modern collection of specimens
from the central and northern Great Plains region.
Other: Identification workshops.

15.84

Theodore M. Sperry Herbarium
Dept. of Biology
Pittsburg State University
Pittsburg, KS 66762
316-235-4740

Contact: Dr. Steve L. Timme, Director
Number of specimens: 45,000.
Content: Tropical bryophytes.
Other: Community workshops.

KY

15.85

Davies Herbarium
Dept. of Biology
University of Louisville
Louisville, KY 40292
502-588-6771

Contact: Dr. W. S. Davis, Curator
Number of specimens: 30,000.

15.86

Northern Kentucky University Herbarium
Northern Kentucky University
Highland Heights, KY 41076
606-292-5300

Number of specimens: 20,000.
Content: Plants of Kentucky and the central U.S.

15.87

University of Kentucky Herbarium
Morgan 101, Biological Sciences
University of Kentucky
Lexington, KY 40506
606-257-3240; (FAX) 606-257-5889

Contact: William Meijer, Curator
Number of specimens: 50,000.
Content: Flora of Kentucky and surrounding states.
Cross-references: 12.154, 12.416, 20.17

LA

15.88

Herbarium of the University of Southern Louisiana
Dept. of Biology
USL, P.O. Box 42451
Lafayette, LA 70504-2451
318-231-6748; (FAX) 318-231-5834

Contact: Garrie P. Landry, Curator
Number of specimens: 105,000.
Content: Specializes in coastal and wetland species native
to the southeastern U.S.
Cross-references: 12.163, 16.166

15.89

Louisiana State University Herbarium
Dept. of Botany
Baton Rouge, LA 70803
504-388-8555

Contact: Lowell E. Urbatsch, Director
Number of specimens: 150,000.
Content: Rare plants of Louisiana.
Cross-references: 12.157, 12.417, 20.128

15.90

Louisiana Tech Herbarium
Dept. of Botany and Bacteriology
Louisiana Tech University
Ruston, LA 71272
318-257-3204

Number of specimens: 125,000.
Content: Primarily vascular plants and mosses of
Louisiana, Arizona, Mississippi, and Texas.
Cross-reference: 12.158

15.91
Northeast Louisiana University Herbarium
Northeast Louisiana University
Biology Dept., 232 Stubbs Hall
Monroe, LA 71209-0502
318-342-3108
Contact: Dr. R. Dale Thomas, Curator
Number of specimens: 311,570.
Content: Specializes in specimens from Louisiana and surrounding areas. Good collection of adder's-tongue ferns.
Other: Slide shows for local organizations.

15.92
Tulane University Herbarium
Dept. of Biology
311 Dinwiddie Hall
Tulane University
New Orleans, LA 70118
504-865-5191
Contact: Dr. Steven P. Darwin, Director
Number of specimens: 100,000.
Content: Flora of Louisiana, Yucatan specimens.

MA
15.93
Botany Herbarium
Morrill Hall
University of Massachusetts
Amherst, MA 01002
413-545-2775
Contact: Claire M. Johnson, Curator
Number of specimens: 220,000.
Content: New England vascular plants.
Cross-references: 12.171, 12.424, 12.425, 13.30, 13.99, 16.179

15.94
Clark University Herbarium
Dept. of Biology
Clark University
Worcester, MA 01610
617-793-7514
Number of specimens: 70,000.
Content: Vascular plants, lichens, and coralline algae.

15.95
Harvard University Herbarium
22 Divinity Ave.
Cambridge, MA 02138
617-495-2365; (FAX) 617-495-9484
Contact: David E. Boufford, Assistant Director
Number of specimens: 4,650,000.
Content: Worldwide collection of vascular plants with emphasis on New World, especially North America (including Mexico and West Indies), and eastern and southeastern Asia and Malaysia. Collection strong in Orchidaceae of Mexico and Philippines, worldwide nonvascular cryptogams, plants useful or harmful to humans, cultivated plants, and wild plants involved in the origin of cultivated plants.
Other: Includes the Gray Herbarium, the Farlow Herbar-

ium of Cryptogamic Botany, the Ames Orchid Herbarium, portions of the Arnold Arboretum Herbarium, and the New England Botanical Club Herbarium.
Cross-reference: 16.173

15.96
Smith College Herbarium
Dept. of Biological Sciences
Smith College
Clark Science Center
Northampton, MA 01063
413-584-2700

Number of specimens: 53,000.
Content: New England and general eastern North American vascular flora.
Cross-reference: 16.177

15.97
Wellesley College Herbarium
Dept. of Biological Sciences
Wellesley College
Wellesley, MA 02181
617-235-0320

Number of specimens: 92,500.
Content: Lichens, mosses, and angiosperms of the U.S. and especially New England.

MD
15.98
American Type Culture Collection Herbarium
12301 Parklawn Dr.
Rockville, MD 20852

Number of specimens: 27,000.
Cross-reference: 4.8

15.99
Towson State University Herbarium
Smith Hall, Room 200
7900 York Rd.
Towson, MD 21204
301-830-3042; (FAX) 301-830-2604

Contact: Donald R. Windler, Curator
Number of specimens: 50,000.
Content: Good Maryland lichen collection, New World *Crotalaria* collection.

15.100
U.S. National Fungus Collections
Building 011A
BARC-West
Beltsville, MD 20705
301-344-3365

Number of specimens: 825,000.
Content: General collection of fungi.

15.101
U.S. National Seed Herbarium
Plant Taxonomy Laboratory
Room 238, Building 001
BARC-West
Beltsville, MD 20705
301-344-2612

Number of specimens: 100,000.
Content: Seed-fruit collection.

15.102
University of Maryland Herbarium
Dept. of Botany
University of Maryland
College Park, MD 20742
301-454-3812

Number of specimens: 40,000.
Content: Plants of Maryland, the Chesapeake Bay region, the middle Atlantic states, and the intermountain region of the western U.S.
Cross-reference: 12.178

ME
15.103
University of Maine Herbarium
Dept. of Botany and Plant Pathology
University of Maine, Folgar Library
Orono, ME 04473
207-581-7861

Number of specimens: 66,000.
Content: Flora of Maine.
Cross-references: 12.182, 16.193, 20.134

MI
15.104
Alma College Herbarium
Biology Dept.
Alma College
Alma, MI 48801
517-463-2141

Number of Specimens: 20,000.
Content: Vascular plants of central Michigan.

15.105
Billington Herbarium
Cranbrook Institute of Science
500 Lone Pine Rd.
P.O. Box 801
Bloomfield Hills, MI 48013
313-645-3223; (FAX) 313-645-6545

Contact: Dr. James R. Wells, Botanist
Number of specimens: 55,000.
Content: Concentration on Michigan flora.

15.106
Central Michigan University Herbarium
Dept. of Biology
Central Michigan University
Mount Pleasant, MI 48859
517-774-3626

Number of specimens: 20,000.
Content: Local plants.

15.107
Clarence R. Hanes Herbarium
Dept. of Biology
Western Michigan University
Kalamazoo, MI 49008
616-383-1674

Number of specimens: 26,000.
Content: Emphasis on flora of southwest Michigan and on local rare, threatened, and endangered species.

15.108
Eastern Michigan University Herbarium
Dept. of Biology
Eastern Michigan University
West Cross St.
Ypsilanti, MI 48197
313-487-4242

Number of specimens: 20,000.
Content: Regional plants.

15.109
Ford Forestry Center Herbarium
Michigan Technological University
L'Anse, MI 49946
906-524-6181

Number of specimens: 40,000.
Content: Specializes in plants of the Upper Great Lakes region.

15.110
Michigan State University Herbarium
Dept. of Botany and Plant Pathology
Michigan State University
East Lansing, MI 48824
517-355-4696

Number of specimens: 426,304.
Content: Worldwide, with emphasis on vascular plants of North America, Mexico, Guatemala, and the Bahamas, and lichens and bryophytes of North America, West Indies, and south temperate and subantarctic regions.
Cross-references: 12.185, 12.427, 13.110, 16.204

15.111
University of Michigan Herbarium
North University Building
Ann Arbor, MI 48109-1057
313-764-2407; (FAX) 313-763-0369

Contact: William R. Anderson, Director
Number of specimens: 1,700,000.
Cross-references: 12.428, 16.203

MN
15.112
Carleton College Herbarium
Biology Dept.
Carleton College
Olin Hall
Northfield, MN 55057
507-663-4392

Number of specimens: 20,000.
Content: Plants of Minnesota and Costa Rica.

15.113
Dept. of Plant Pathology
Mycological Herbarium
College of Agriculture
University of Minnesota
304 Stakman Hall of Plant Pathology
St. Paul, MN 55108
612-373-1383

Number of specimens: 45,000.
Content: Fungi.
Cross-references: 12.197, 12.430, 15.115

15.114
Olga Lakela Herbarium
University of Minnesota—Duluth
Dept. of Biology
Duluth, MN 55812
218-726-6542

Contact: Larry Hufford, Assistant Professor
Number of specimens: 36,000.

15.115
University of Minnesota Herbarium
Dept. of Plant Biology
220 BioSci Center
University of Minnesota
St. Paul, MN 55108
612-625-1234; (FAX) 612-625-5754

Contact: Dr. John Doebley, Director
Number of specimens: 800,000.
Cross-references: 12.197, 12.430, 15.113

MO
15.116
Missouri Botanical Garden Herbarium
P.O. Box 299
St. Louis, MO 63166
314-577-5100; (FAX) 314-577-9596

Contact: James C. Solomon, Curator
Number of specimens: 3,800,000.
Content: Emphasis on the plants of the Neotropics,
Africa, and North America. The collection is worldwide
in scope.
Cross-reference: 16.214

15.117
Ozarks Regional Herbarium
Dept. of Biology
Southwest Missouri State University
Springfield, MO 65804-0095
417-836-5882; (FAX) 417-836-4538

Contact: Dr. Paul Redfearn, Curator
Number of specimens: 60,000.
Cross-reference:12.206

15.118
University of Missouri Herbarium
Division of Biological Sciences
201 Tucker Hall
Columbia, MO 65211
314-882-6519

Number of specimens: 250,000.
Content: Flora of Missouri, *Lupinus* of the world, and
grasses.
Cross-references: 12.207, 16.217, 20.51, 20149

MS
15.119
Herbarium of Mississippi State University
Dept. of Biological Sciences
Mississippi State University
Drawer GY
Mississippi State, MS 39762
601-325-3120

Number of specimens: 40,000.
Content: Vascular plants of Mississippi.
Cross-references: 12.212, 12.433, 15.120

15.120
Institute for Botanical Exploration Herbarium
Box EN
Mississippi State, MS 39762
601-325-3120

Number of specimens: 72,500.
Content: Worldwide, with an emphasis on plants of the
southeast U.S. and tropical America.
Cross-references: 12.212, 12.432, 15.119

15.121
University of Mississippi Herbarium
Biology Dept.
University of Mississippi
University, MS 38677
601-232-7203

Number of specimens: 53,000.
Content: Vascular flora of Mississippi and the
southeastern U.S.

MT
15.122
Montana State University Herbarium
310 Lewis Hall
Dept. of Biology
Bozeman, MT 59717
406-994-4424

Contact: Matt Lavin, Curator
Number of specimens: 60,000.
Content: Plants of the northern Rocky Mountain
region.
Cross-references: 12.213, 12.433, 16.224

15.123
University of Montana Herbarium
Division of Biological Sciences
University of Montana
Missoula, MT 59812
406-243-4743; (FAX) 406-243-4184

Contact: Kathleen E. Ahlenslager, Curator
Number of specimens: 112,000.
Content: Flora of the northern Rocky Mountains.

NC
15.124
Duke University Herbarium
Dept. of Botany
Duke University
Durham, NC 27706
919-684-3056

Number of specimens: 555,000.
Content: Taxa of the southeastern U.S. and Central
America.
Cross-reference: 16.236

15.125
North Carolina State University Herbarium
Dept. of Botany
North Carolina State University
Raleigh, NC 27695-7612
919-737-2700; (FAX) 919-737-3426

Contact: Dr. James W. Hardin, Curator
Number of specimens: 111,000.
Cross-references: 12.223, 12.435, 16.233

15.126
University of North Carolina Herbarium
Dept. of Biology
CB # 3280, UNC
Chapel Hill, NC 27599-3280
919-962-6931

Contact: Jim Massey, Curator
Number of specimens: 600,000.
Content: Specializes in plants of the Carolinas and plants
of the southeastern U.S.
Cross-references: 13.333, 13.118, 16.227, 16.232

ND
15.127
North Dakota State University Herbarium
Botany Dept.
North Dakota State University
Fargo, ND 58105
701-237-7222

Number of specimens: 165,000.
Content: North Dakota vascular plants with good
representation of the northern Great Plains.
Cross-references: 12.231, 12.436, 20.158

15.128
University of North Dakota Herbarium
Biology Dept., Starcher Hall
University of North Dakota
Grand Forks, ND 58202
701-777-2621

Number of specimens: 25,000.
Content: North Dakota plants.

NE
15.129
University of Nebraska State Museum
W-532 Nebraska Hall
University of Nebraska
Lincoln, NE 68588-0514
402-472-2613

Contact: Margaret R. Bolick, Curator
Number of specimens: 280,000.

Content: Extensive collection of lichens and fungi, many
collections of angiosperms from the 19th century.
Cross-references: 12.235, 16.246, 16.247

15.130
University of Nebraska at Omaha Herbarium
Dept. of Biology
University of Nebraska at Omaha
Omaha, NE 68182
402-554-2641

Number of specimens: 20,500.
Content: Vascular plants of the Great Plains.

NH
15.131
Albion R. Hogdon Herbarium
University of New Hampshire
Dept. of Plant Biology
Nesmith Hall
Durham, NH 03824
603-862-3865

Contact: Dr. Garrett E. Crow, Director
Number of specimens: 147,000.
Content: Primarily New Hampshire and New England
flora, strong marine algae collection, repository for
documentation of rare plants of New Hampshire, and
aquatic plants of the Northeast.
Cross-references: 12.236, 16.250

15.132
Jesup Herbarium
Dept. of Biology
Dartmouth College
Hanover, NH 03755
603-646-2314

Contact: Robert M. Downs
Number of specimens: 70,459.

NJ
15.133
Chrysler Herbarium
Dept. of Biological Sciences
Rutgers—The State University of New Jersey
Nelson Biological Laboratories
P.O. Box 1059
Piscataway, NJ 08855-1059
201-932-2843; (FAX) 201-932-5870

Contact: Janice S. Meyer, Assistant Curator
Number of specimens: 116,000.
Content: Endangered species and vouchers of botanical
research in New Jersey, historical specimens of New
Jersey dating from 1850.
Cross-references: 12.240, 12.437, 15.134, 16.261, 20.52,
20.164

15.134
Chrysler Herbarium
Dept. of Botany
Rutgers University
Nelson Biological Labs
Busch Campus
New Brunswick, NJ 08903
201-932-2843

Number of specimens: 110,100.
Content: Plants of New Jersey.
Cross-references: 12.240, 12.437, 15.133, 16.261, 20.102, 20.224

NM
15.135
New Mexico State University Herbarium
Biology Dept.
New Mexico State University
Las Cruces, NM 88003
505-646-3611

Number of specimens: 50,000.
Content: Plants of New Mexico and northern Mexico.
Cross-references: 12.243, 16.267

15.136
University of New Mexico Herbarium
Dept. of Biology
Room 1
Albuquerque, NM 87131
505-277-3411

Contact: Dr. Timothy Lowry, Curator
Number of specimens: 76,000.
Content: Emphasis on native and naturalized plants of New Mexico. Includes a cactus collection of over 900 specimens.
Other: Atrium in Biology Building houses exotic tropical plants.
Cross-references: 12.244, 16.265

NV
15.137
University of Nevada Reno Herbarium
Dept. of Range, Wildlife, & Forestry
1000 Valley Rd.
Reno, NV 89512
702-784-1105

Contact: Lynda S. Nelson, Herbarium Manager
Number of specimens: 80,000.
Content: Specializes in specimens from Nevada, the Great Basin, and the western U.S., includes over 200 threatened, endangered, and sensitive plants of Nevada.
Other: Some native seed identification.
Cross-reference: 12.247

NY
15.138
Brooklyn Botanic Garden Herbarium
1000 Washington Ave.
Brooklyn, NY 11225-1099
718-622-4433

Contact: Dr. Kerry Barringer, Curator
Number of specimens: 200,000.
Content: Special collections of plants of New York City, the Galapagos Islands, and the western U.S.
Cross-references: 16.271

15.139
Clinton Herbarium
Buffalo Museum of Science
Humboldt Park
Buffalo, NY 14211
716-896-5200

Number of specimens: 60,000.
Content: Flora of the Niagara Frontier region.

15.140
Hoysradt Herbarium
Biology Dept.
Hartwick College
Oneonta, NY 13820
607-432-4200

Contact: Dr. Robert R. Smith, Curator
Number of specimens: 20,000.
Content: Hoysradt Collection (1800–1875), Karl Brook Collection of the flora of the Catskills, and plants of the Bahamas.

15.141
Lamont-Doherty Geological Observatory of Columbia University Herbarium
Palisades, NY 10964
914-359-2900, ext. 229

Number of specimens: 20,000.
Content: Worldwide collection of marine plankton.

15.142
Liberty Hyde Bailey Hortorium
Cornell University
462 Mann Library Building
Ithaca, NY 14853
607-672-2266

Number of specimens: 388,671.
Content: Worldwide with emphasis on cultivated plants and their wild counterparts, economically and ethnobotanically important plants, and the family Palmae.
Cross-references: 12.249, 12.439, 14.69, 15.145, 15.147, 16.275

15.143
New York Botanical Garden Herbarium
Bronx, NY 10458-5126
212-220-8626; (FAX) 212-220-6504

Contact: Dr. Patricia K. Holmgren, Director of
Herbarium
Number of specimens: 5,550,000.
Content: Worldwide collections with greatest strength in
tropical America and North America.
Cross-reference: 16.286

15.144
New York State Museum Herbarium
Biological Survey
3132 CEC
Albany, NY 12230
518-474-5809

Number of specimens: 225,000.
Content: Flora of New York.

15.145
Plant Pathology Herbarium
Cornell University
Ithaca, NY 14853
607-256-3292

Number of specimens: 300,000.
Content: Fungi, particularly of the Americas, the
Neotropics, Macronesia, China, Japan, and southeast
Asia.
Cross-references: 12.249, 12.439, 14.69, 15.142, 15.147.
16.275

15.146
State University of New York Herbarium
College of Environmental Sciences and Forestry
1 Forestry Dr.
Syracuse, NY 13210
315-470-6782; (FAX) 315-470-6779

Contact: Dr. Dudley J. Raynal, Professor
Number of specimens: 65,000.
Content: Wood-decaying fungi.
Cross-reference: 12.441

15.147
Wiegand Herbarium
Cornell University
467 Mann Library Building
Ithaca, NY 14853
607-256-2131

Number of specimens: 323,852.
Content: Plants of North America, especially eastern
regions and the glaciated Allegheny Plateau.
Cross-references: 12.249, 12.439, 14.69, 15.142, 15.145,
16.275

OH
15.148
Bartley Herbarium
Dept. of Botany
Ohio University

Athens, OH 45701
614-593-1128

Contact: Dr. Philip D. Cantino, Curator
Number of specimens: 40,000.

15.149
Bowling Green State University Herbarium
Dept. of Biological Sciences
Bowling Green State University
Bowling Green, OH 43403
419-372-2434

Number of specimens: 20,000.
Content: Local flora of northwestern Ohio.

15.150
Cleveland Museum of Natural History Herbarium
Wade Oval
University Circle
Cleveland, OH 44106
216-231-5919; (FAX) 216-231-5919

Contact: James K. Bissell, Curator
Number of specimens: 54,000.

15.151
Denison University Herbarium
Granville, OH 43023
614-587-6327; (FAX) 614-587-6417

Contact: Juliana C. Mulroy, Director
Number of specimens: 25,000.
Content: Wallace Cathcart fern collection, Licking county
collection.

15.152
Kent State University Herbarium
Dept. of Biological Sciences
Kent State University
Kent, OH 44242
216-672-2266

Number of specimens: 60,000.
Content: Vascular plants of the eastern U.S., especially
Ohio.

15.153
Muskingum College Herbarium
Biology Dept.
Muskingum College
Science Center
New Concord, OH 43762
614-826-8224

Number of specimens: 23,000.
Content: Pteridophyta and Ohio specimens.

15.154
Oberlin College Herbarium
Dept. of Biology
Kettering Hall of Science
Oberlin College
West Lotain St.
Oberlin, OH 44074
216-775-8315

Number of specimens: 34,046.
Content: Plants of Ohio.

15.155
Ohio State University Herbarium
Dept. of Botany
Ohio State University
1735 Neil Ave.
Columbus, OH 43210
614-422-3296

Content: Taxa of Ohio and vascular plants with emphasis
on aquatic groups and the family Compositae.
Cross-references: 12.263, 12.442, 16.295, 16.309

15.156
University of Cincinnati Herbarium
Dept. of Biological Sciences
University of Cincinnati
Cincinnati, OH 45221
513-475-3741

Number of specimens: 125,000.
Content: Vascular plants, especially from the Cincinnati
region.

15.157
Willard Sherman Turrell Herbarium
Dept. of Botany
Miami University
Oxford, OH 45056
513-529-2755

Contact: Michael A. Vincent, Assistant Curator
Number of specimens: 300,000.
Content: *Capsicum* species.

OK
15.158
Herbarium of Oklahoma State University
Dept. of Botany
Oklahoma State University
Stillwater, OK 74078
405-744-9558; (FAX) 405-744-7074

Contact: Ronald J. Tyrl, Curator
Number of specimens: 140,000.
Content: Primarily Oklahoma taxa.
Cross-references: 12.269, 12.443

15.159
Robert Bebb Herbarium
University of Oklahoma
770 Van Vleet Oval, Room 203
University of Oklahoma
Norman, OK 73019
405-325-6443

Contact: James R. Estes, Curator
Number of specimens: 200,000.
Content: Norman Boke Cactus Collection, synoptic
herbarium.

15.160
Southeastern Oklahoma State University Herbarium
Dept. of Biology
Durant, OK 74701
405-924-0121, ext. 209

Number of specimens: 52,000.
Content: Plants of Oklahoma.

OR
15.161
Oregon State University Herbarium
Dept. of Botany and Plant Pathology
Cordley Hall 4082
Corvallis, OR 97331-2910
503-737-4106

Contact: Dr. Kenton L. Chambers, Curator
Number of specimens: 230,000.
Cross-references: 12.274, 16.332

15.162
University of Oregon Herbarium
Dept. of Biology
University of Oregon
Eugene, OR 97403
503-686-3033

Number of specimens: 103,000.
Content: Vascular plants, bryophytes, and lichens of
Oregon and the Pacific Northwest.
Cross-reference: 12.445

PA
15.163
Academy of Natural Sciences Herbarium
Dept. of Botany
19th and Parkway
Philadelphia, PA 19103
215-299-1192

Contact: Dr. Benjamin C. Stone, Chairman
Number of specimens: 1,500,000.
Content: Extremely rich collection of type specimens
(50,000); collection of early material (1790–1850)
including Lewis & Clark Herbarium; good reference
collection for Pennsylvania, New Jersey, and Delaware
flora.
Cross-reference: 14.84

15.164
Bureau of Plant Industry Herbarium
Pennsylvania Dept. of Agriculture
2301 North Cameron St.
Harrisburg, PA 17110
717-787-4843

Number of specimens: 26,000.
Content: Pennsylvania plants.

15.165
Carnegie Museum Herbarium
Section of Botany
Carnegie Museum of Natural History
4400 Forbes Ave.
Pittsburgh, PA 15213
412-622-3253; (FAX) 412-622-8837

Contact: F. H. Utech, Curator
Number of specimens: 550,000.
Content: Focus on Pennsylvania plants.
Cross-reference: 14.85

15.166
Morris Arboretum of the University of Pennsylvania
   Herbarium
9414 Meadowbrook Ave.
Philadelphia, PA 19118
215-247-5727

Contact: Ann F. Rhoads, Chair of Botany
Number of specimens: 22,500.
Content: Flora of Pennsylvania and woody flora of
   temperate regions.
Cross-references: 12.448, 16.351

15.167
Muhlenberg College Herbarium
Dept. of Biology
2400 Chew St.
Allentown, PA 18104
215-821-3251

Contact: Dr. Frank D. Watson, Curator
Number of specimens: 70,000.
Content: Primarily flora of northeastern Pennsylvania.

15.168
North Museum Herbarium
Franklin and Marshall College
P.O. Box 3003
Lancaster, PA 17604
717-291-3941

Contact: Jane Grushow, Associate Curator
Number of specimens: 22,000.

15.169
Pennsylvania State University Herbarium
202 Mueller Lab
University Park, PA 16802
814-865-6201; (FAX) 814-865-3634

Contact: Dr. Carl S. Keener, Curator
Number of specimens: 90,000.
Content: Flora of Pennsylvania.
Cross-references: 12,278, 12.447, 20.25, 20.56, 20.203

15.170
Wayne E. Manning Herbarium
Dept. of Biology
Bucknell University
Lewisburg, PA 17837
717-524-1155

Number of specimens: 20,000.
Content: Vascular plants of Pennsylvania, New England,
   New Jersey, West Virginia, Georgia, South Carolina,
   Florida, and Mexico.

15.171
William Darlington Herbarium
Dept. of Biology
West Chester University
West Chester, PA 19383
215-436-2926

Contact: Dr. Martha A. Potvin, Curator
Number of specimens: 22,000.
Content: The second oldest herbarium in the U.S.

PR
15.172
Herbario
Ficteco Puertorriquena
Dept. de Ciencias Marinas
University of Puerto Rico
Recinto Universitario de Mayaguez
Mayaguez, PR 00708
305-832-4040, ext. 2579

Number of specimens: 50,000.
Content: Marine algae of the Caribbean.
Cross-reference: 12.282

RI
15.173
Brown University Herbarium
Box G—W301
Brown University
Providence, RI 02912
401-863-3435

Contact: Johanna Schmitt, Associate Professor
Number of specimens: 107,100.
Content: Plants of North America, especially Rhode
   Island flora, and collection of fungi.

SC
15.174
Andrew Charles Moore Herbarium
Dept. of Biological Sciences
University of South Carolina
Columbia, SC 29208
803-777-8196

Contact: Dr. John B. Nelson, Curator
Number of specimens: 50,000.
Content: Rare and endangered taxa of South Carolina.

15.175
Charleston Museum Herbarium
360 Meeting St.
Charleston, SC 29403
803-722-2996

Number of specimens: 25,000.
Content: Flora of South Carolina, especially the
   Charleston area.

15.176
Clemson University Herbarium
Dept. of Biological Sciences
Clemson University
Clemson, SC 29634-1903
803-686-3605; (FAX) 803-656-0435

Contact: Dr. Stephen R. Hill, Curator
Number of specimens: 40,000.
Content: Rare plants of South Carolina, and plants of the
   Amazon and Brazil.
Cross-references: 12.286, 16.366, 20.209

SD

15.177

F. L. Bennett Herbarium
Division of Science and Mathematics
Black Hills State College
Spearfish, SD 57783
605-642-6251

Number of specimens: 20,000.
Content: Plants of South Dakota, especially of the Black
    Hills.

15.178

South Dakota Herbarium
Dept. of Biology
University of South Dakota
414 East Clark St.
Vermillion, SD 57069
605-677-6176

Contact: Dr. Frederick J. Peabody, Curator
Number of specimens: 50,000.
Content: Plants of South Dakota and the Black Hills.

15.179

South Dakota State Herbarium
Dept. of Biology & Microbiology
S.D.S.U.
Box 2207B
Brookings, SD 57007-0595
605-688-4552; (FAX) 605-688-6065

Contact: Gary E. Larson, Curator
Number of specimens: 30,000.
Cross-references: 12.290, 12.452, 16.368

TN

15.180

Herbarium of Vanderbilt University
Box 1705, Station B
Vanderbilt University
Nashville, TN 37235
615-322-3548

Contact: Robert Kral, Curator and Director
Number of specimens: 230,000.
Content: Flora of the southeastern U.S., with special
    collections of plants of Alabama and Tennessee.
    Cyperaceae, Poaceae, Eriocaulaceae, and Xyridaceae.

15.181

University of Tennessee Herbarium
Dept. of Botany
University of Tennessee
Knoxville, TN 37996
615-974-6212

Contact: Dr. B. Eugene Wofford, Director
Number of specimens: 450,000.
Content: Plants of the southern Appalachians, eastern
    North America, Mexico, and Central America.
Cross-references: 12.299, 12.453, 16.375, 20.28, 20.214

TX

15.182

Angelo State University Herbarium
Dept. of Biology
Angelo State University
Box 10890, A.S.U. Station
San Angelo, TX 76901
915-942-2189

Number of specimens: 20,000.
Content: Plants of southwest Texas.

15.183

Baylor University Herbarium
Dept. of Biology
Baylor University
Waco, TX 76703
817-755-2911

Number of specimens: 100,000.
Content: Plants of central Texas.

15.184

E. L. Reed Herbarium
Texas Tech University
Lubbock, TX 79409
806-742-2715

Contact: Charles R. Werth, Curator
Number of specimens: 25,000.
Cross-reference: 12.317, 12.455

15.185

Herbarium LL-TEX
Plant Resources Center
Dept. of Botany
127 Main Building
University of Texas
Austin, TX 78713-7640
512-471-5904

Contact: Dr. Guy Nelson, Curator
Number of specimens: 1,000,000.
Content: Extensive collection of Asteraceae.

15.186

Howard Payne University Herbarium
Biology Dept.
Howard Payne University
Brownwood, TX 76801
915-646-2502, ext. 243

Number of specimens: 30,000.
Content: Flora of central Texas.

15.187

Lundell Herbarium
Plant Sciences Dept.
University of Texas at Dallas
Box 688
Richardson, TX 75080
214-690-2279

Number of specimens: 250,000.
Content: Plants of the Neotropics, Texas, and the world.

15.188
S. M. Tracy Herbarium
Dept. of Range Science
Texas A&M University
College Station, TX 77843
409-845-4328; (FAX) 409-845-6430

Contact: Dr. Stephan L. Hatch, Curator
Number of specimens: 200,000.
Content: Many of the grass species of the world.
Cross-references: 12.315, 12.454, 13.49, 16.378, 20.29, 20.223

15.189
South Plains College Herbarium
Science Dept.
South Plains College
Levelland, TX 79336
806-894-9611, ext. 285

Number of specimens: 20,000.
Content: Local flora and flora of the southwest U.S.

15.190
Southern Methodist University Herbarium
6425 Airline Rd.
Dallas, TX 75275-0376
214-692-2257

Contact: Barry Lipscomb, Curator
Number of specimens: 410,000.
Content: Cultivated plants.

15.191
Stephen F. Austin State University Herbarium
P.O. Box 13003, SFA Station
Nacogdoches, TX 75962
409-568-3601

Contact: Elray S. Nixon, Director
Number of specimens: 70,000.
Cross-reference: 12.312

15.192
Sul Ross State University Herbarium
Dept. of Biology
Alpine, TX 79832
915-837-8112

Contact: A. Michael Powell, Director
Number of specimens: 60,000.
Content: Excellent collection of northern California desert plants.

15.193
West Texas State University Herbarium
Biology Dept.
West Texas State University
Canyon, TX 79016
806-656-2262

Number of specimens: 50,000.
Content: Plants of the western U.S. and Alaska.

UT
15.194
Garrett Herbarium
Utah Museum of Natural History
University of Utah
Salt Lake City, UT 84112
801-581-6520

Contact: Dr. Michael Windham, Curator
Number of specimens: 114,000.
Content: Threatened and endangered plants of Utah.
Other: Educational programs.
Cross-reference: 16.392

15.195
Intermountain Herbarium
Dept. of Biology
Utah State University
Logan, UT 84322-5305
801-750-1584; (FAX) 801-750-1575

Contact: Dr. M. E. Barkworth, Director
Number of specimens: 206,000.
Cross-references: 12.321, 12.457

15.196
M. L. Bean Museum
Brigham Young University
Provo, UT 84602
801-378-2289

Contact: Dr. Stanley L. Welsh, Director
Number of specimens: 334,788.
Cross-references: 12.320, 20.224, 20.302

VA
15.197
College of William and Mary Herbarium
Dept. of Biology
The College of William and Mary
Williamsburg, VA 23185
804-253-4240

Number of specimens: 36.042.
Content: Plants of Virginia.

15.198
Harvill-Stevens Herbarium
Longwood College
Farmville, VA 23901
804-395-2571

Contact: Alton M. Harvill, Director/Curator
Number of specimens: 70,000.
Content: Emphasis on flora of Virginia.

15.199
Lynchburg College Herbarium
Dept. of Biology
Lynchburg, VA 24501
804-522-8363

Contact: Dr. Gwynn W. Ramsey, Professor
Number of specimens: 44,000.
Content: Specializes in vascular flora of the Upper Piedmont and the central Virginia Blue Ridge Mountains, Virginia Ranunculaceae, and North American *Cimicifuga*.

15.200
Virginia Military Institute Herbarium
VMI
Lexington, VA 24450
703-463-6247

Number of specimens: 20,000.
Content: Local flora.

15.201
Virginia Polytechnic Institute and State University Herbarium
Biology Dept.
VPI&SU
Blacksburg, VA 24061-0406
703-231-5746

Contact: Dr. Duncan M. Porter, Curator
Number of specimens: 103,000.
Cross-references: 12.326, 12.460, 13.52, 16.402, 20.232

VT
15.202
Pringle Herbarium
Dept. of Botany
Marsh Life Science Building
University of Vermont
Burlington, VT 05405-0086
802-656-3221

Contact: Dr. David Barrington, Curator
Number of specimens: 3000,000.
Content: Mexican and Vermont plants, and early collections from Europe, India, the Philippines, and Australia.
Cross-references: 12.328, 16.404

WA
15.203
Marion Ownbey Herbarium
Washington State University
Pullman, WA 99164-4309
509-335-3250; (FAX) 509-335-3421

Contact: Joy Mastrogiuseppe, Curator
Number of specimens: 304,000.

Content: Specializes in Pacific Northwest plants. Good collections of Umbelliferae and sedges.
Cross-references: 12.339, 12.462, 15.204

15.204
Mycological Herbarium
Dept. of Plant Pathology
Washington State University
Pullman, WA 99164
509-335-1086

Number of specimens: 64,000.
Content: All major fungi groups.
Cross-references: 12.339, 12.462, 15.203

15.205
University of Washington Herbarium
Dept. of Botany
KB-15
26 Hitchcock Hall
University of Washington
Seattle, WA 98195
206-543-1682

Contact: Dr. Melinda F. Denton, Curator
Number of specimens: 500,000.
Content: Flora of the northwestern U.S.
Cross-references: 1.40, 12.338, 12.461, 16.415

15.206
Walla Walla College Natural History Herbarium
Dept. of Biological Sciences
Walla Walla College
Life Science Building 226,
West Whitman Dr.
College Place, WA 99324
509-527-2483

Number of specimens: 20,000.
Content: Primarily Pacific Northwest flora.

15.207
Western Washington State College Herbarium
Biology Dept.
Western Washington University
Bellingham, WA 98225
206-676-3649

Number of specimens: 22,000.
Content: Local vascular plants.

WI

15.208

Forest Products Laboratory Herbarium
Center for Wood Anatomy Research
P.O. Box 5130
Madison, WI 53705
608-264-5742

Number of specimens: 126,000.
Content: Woody plants of the world.

15.209

Herbarium of the University of Wisconsin—Madison
Dept. of Botany, Birge Hall
430 Lincoln Dr.
University of Wisconsin
Madison, WI 53706
608-262-2792

Contact: Dr. Hugh H. Iltis, Director
Number of specimens: 800,000.
Content: Large collection of teosinte and maize
specimens and wild and cultivated *Zea*.
Cross-references: 12.345, 12.463, 16.420, 16.424

15.210

Herbarium of the University of Wisconsin—Milwaukee
Dept. of Biological Sciences
P.O. Box 413
Milwaukee, WI 53201
414-229-6728

Contact: Dr. Donald H. Les, Director of Herbarium
Number of specimens: 75,000.
Content: Wisconsin flowering plants and western North
American flowering plants.

15.211

Milwaukee Public Museum
800 West Wells St.
Milwaukee, WI 53233
414-278-2711; (FAX) 414-278-1396

Contact: Dr. Martyn J. Dibben, Director
Number of specimens: 250,000.
Content: Algae, fungi, lichens, mosses, and liverworts of
Wisconsin, North America, and the Neotropics.

15.212

Museum of Natural History Herbarium
University of Wisconsin—Steven's Point
Steven's Point, WI 54481

Number of specimens: 102,000.
Content: Vascular plants and mosses, grasses, and plants
of Wisconsin.

15.213

University of Wisconsin—La Crosse Herbarium
Dept. of Biology
University of Wisconsin—La Crosse
La Crosse, WI 54601
608-785-8251

Number of specimens: 35,000.
Content: Vascular plants of Wisconsin.

15.214

University of Wisconsin—Oshkosh Herbarium
Biology Dept.
University of Wisconsin—Oshkosh
Oshkosh, WI 54901
414-424-1002

Contact: Neil A. Harriman, Professor
Number of specimens: 86,000.
Content: Many specimens of cultivated plants.

15.215

University of Wisconsin—Superior Herbarium
Biology Dept.
1800 Grand Ave.
University of Wisconsin—Superior
Superior, WI 54880
715-392-8101, ext. 408

Number of specimens: 30,000.
Content: Plants of northern Wisconsin.

WV

15.216

George B. Rossbach Herbarium
West Virginia Wesleyan College
Buckhanon, WV 26201
304-473-8064

Number of specimens: 23,000.
Content: Plants of the northeastern U.S.

15.217

Marshall University Herbarium
Science Hall
Marshall University
Hal Greer Blvd.
Huntington, WV 25701
304-696-6467

Contact: Dr. Dan K. Evans, Curator
Number of specimens: 50,000.
Content: Plants of West Virginia.

WY

15.218

Rocky Mountain Herbarium
College of Agriculture
University of Wyoming
Laramie, WY 82071
307-766-4236

Number of specimens: 60,000.
Content: Grasses of Wyoming, Mexico, and the world.

15.219

Wilhelm G. Solheim Mycological Herbarium
University of Wyoming
Laramie, WY 82071
307-766-2236

Number of specimens: 50,000.
Content: Emphasis on taxa of the Rocky Mountains,
especially plant pathogens.

Cross-reference: 12.351

CANADA

AB

15.220
Mycological Herbarium
Northern Forest Research Center
Canadian Forestry Service
5320 122 St.
Edmonton, AB T6H 3S5
403-435-7210

Number of specimens: 25,000.
Content: Forest fungi of Alberta, Saskatchewan, Manitoba, and the Northwest Territories.

15.221
Olds College Herbarium
Olds, AB T0M 1P0

Contact: Linda Vandermaar, Horticulturist
Number of specimens: 25,000.
Cross-references: 12.354, 20.252

15.222
University of Alberta Herbarium
Dept. of Botany
University of Alberta
Edmonton, AB T6G 2E9
403-432-5518

Number of specimens: 203,000.
Content: Vascular plants and lichens of arctic and cordilleran Canada, bryophytes of the New World and Australia.
Cross-references: 12.355, 16.432

15.223
University of Calgary Herbarium
Dept. of Biology
University of Calgary
Calgary, AB T2N 1N4
403-284-5262

Number of specimens: 100,000.
Content: Vascular and nonvascular plants, especially mosses and lichens, of Alberta, British Columbia, and the Northwest Territories.
Cross-reference: 12.356

BC

15.224
Pacific Forest Research Center Herbarium
Canadian Forestry Service
506 W. Burnside Rd.
Victoria, BC V8Z 1M5
604-388-3811, ext. 214

Number of specimens: 27,000.
Content: Forest pathology, mycology, and ecology.

15.225
Royal British Columbia Museum Herbarium
Botany Unit
675 Belleville St.
Victoria, BC V8V 1X4
604-387-5493; (FAX) 604-387-5360

Contact: Dr. Richard Hebda, Head of Botany Unit
Number of specimens: 175,000.
Content: Comprehensive collection of British Columbia plants including all endemic and rare species, a large collection from northwestern North America, and collections from the Soviet Union.
Other: Museum has exhibits on plants and native uses of plants.
Cross-reference: 16.445

15.226
University of British Columbia Herbarium
3529-6270 University Blvd.
University Campus
Vancouver, BC V6T 2B1
604-228-3344

Number of specimens: 446,000.
Content: Native plants of British Columbia.
Cross-references: 12.362, 12.465, 13.54, 16.449, 24.33

15.227
University of Victoria Herbarium
Dept. of Biology
University of Victoria
P.O. Box 1700
Victoria, BC V8W 2Y2
604-477-6911, ext. 4743

Number of specimens: 30,000.
Content: Plants of Vancouver Island.

MB

15.228
University of Manitoba Herbarium
Dept. of Botany
Winnipeg, MB R3T 2N2
204-474-9368

Number of specimens: 50,000.
Content: Flora of Manitoba.
Cross-references: 12.363, 20.34

NB

15.229
Connell Memorial Herbarium
Biology Dept.
University of New Brunswick
College Hill, P.O. Box 4400
Fredericton, NB E3B 5A3
506-453-4583

Number of specimens: 40,800.
Content: Mainly taxa of New Brunswick; large collection of *Rubus, Amelanchier,* and *Betula* of Maritime Canada.

15.230
New Brunswick Museum Herbarium
277 Douglas Ave.
St. John, NB E2K 1E5
506-693-1196

Number of specimens: 20,000.
Content: Vascular plants of New Brunswick.

NF
15.231
Memorial University of Newfoundland Herbarium
Dept. of Biology
Memorial University of Newfoundland
St. John's, NF A1B 3X9
709-753-1200

Number of specimens: 80,000.
Content: Vascular plants, algae, bryophytes.
Cross-reference: 16.453

NS
15.232
Atlantic Regional Laboratory Herbarium
National Research Council of Canada
1411 Oxford St.
Halifax, NS B3H 3Z1
902-429-6450

Number of specimens: 20,000.
Content: Marine algae of eastern Canada.

15.233
E. C. Smith Herbarium
Biology Dept.
Acadia University
Wolfville, NS B0P 1X0
902-542-2201

Contact: Dr. S. P. Vander Kloet, Curator
Number of specimens: 140,000.
Content: Flora from the Maritime Provinces of Canada.

15.234
Nova Scotia Museum Herbarium
1747 Summer St.
Halifax, NS B3H 3A6
902-429-4610; (FAX) 902-424-0560

Contact: Alex Wilson, Curator of Botany
Number of specimens: 35,000.
Content: Coastal plain species.
Other: Public gallery includes permanent botany exhibits.

ON
15.235
Carleton University Herbarium
Dept. of Biology
Carleton University
Ottawa, ON K1S 5B6
613-231-3871

Number of specimens: 30,000.
Content: Plants of Canada.

15.236
Claude E. Garton Herbarium
Dept. of Biology
Lakehead University
955 Oliver Rd.
Thunder Bay, ON P7B 5E1
807-343-8506

Contact: Claude E. Garton, Curator
Number of specimens: 96,000.
Cross-reference: 16.468

15.237
Cryptogamic Herbarium
Dept. of Botany
University of Toronto
Toronto, ON M5S 1A1
416-978-3543

Number of specimens; 193,000.
Content: Worldwide flora, with an emphasis on plants of
    north-temperate regions.
Cross-references: 12.467, 15.245

15.238
Fowler Herbarium
Biology Dept.
Queen's University
Kingston, ON K7L 3N6
613-547-6675

Number of specimens: 120,000.
Content: Mostly vascular plants, good collection of local
    and arctic bryophytes and lichens.

15.239
Great Lake Forest Research Center Herbarium
Canadian Forestry Service
Box 490
Sault Ste. Marie, ON P6A 5M7
705-949-9461

Number of specimens: 20,500.
Content: Fungal pathogens of forest trees, especially
    Ontario.
Cross-reference: 16.463

15.240
National Herbarium of Canada
Canadian Museum of Natural Sciences
P.O. Box 3443, Station D
Ottawa, ON K1P 6P4
613-990-6439; (FAX) 613-952-9693

Contact: Dr. Michel Poulin, Acting Chief of Botany
    Division
Number of specimens: 911,863.
Content: Excellent collection of arctic vascular plants.

15.241
Ottawa Research Station Herbarium
Agriculture Canada
Central Experimental Farm
Ottawa, ON K1A 0C6

Number of specimens: 700,000.
Content: Native Canadian species.
Cross-reference: 11.367, 16.459

15.242
Royal Botanical Gardens Herbarium
Box 399
Hamilton, ON L8N 3H8
416-527-1158; (FAX) 416-577-0375

Contact: James S. Pringle, Taxonomist
Number of specimens: 65,000.
Content: Vascular plants of southern Ontario.
Cross-reference: 16.473

15.243
University of Guelph Herbarium
Dept. of Botany
University of Guelph
Guelph, ON N1G 2W1
519-824-4120; (FAX) 519-767-1656

Contact: Jack F. Alex, Curator
Number of specimens: 70,000.
Content: Concentrates on flora of Ontario with emphasis
on introduced and weedy plants.
Cross-references: 12.386, 16.456

15.244
University of Waterloo Herbarium
Dept. of Biology
University of Waterloo
Waterloo, ON N2L 3G1
519-885-1211, ext. 3751

Number of specimens: 43,000.
Content: North American vascular plants.

15.245
Vascular Plant Herbarium
Dept. of Botany
University of Toronto
Toronto, ON M5S 1A1
416-978-3542

Number of specimens; 210,000.
Content: Plants of North America, especially Canada and
the adjacent U.S.
Cross-references: 12.467, 15.237

PQ
15.246
Herbarium of L'Ecole d'Agriculture
Rimouski, PQ

Number of specimens: 30,000.
Content: Flora of Quebec, Labrador, and northern
Canada.

15.247
Herbier Louis-Marie
Faculté des Sciences de l'agriculture et de l'alimentation
Université Laval
Ste.-Foy, PQ G1K 7P4
418-656-2613

Number of specimens: 213,200.
Content: North American flora, especially Canada and
Quebec.
Cross-references: 12.389, 16.478

15.248
McGill University Herbarium
MacDonald College
21111 Lakeshore Rd.
Ste. Anne de Bellevue, PQ H9X 1C0
514-398-7864; (FAX) 514-398-7897

Contact: Dr. M. J. Waterway, Curator
Number of specimens; 130,000.
Content: Arctic plants, North American *Carex*.
Cross-references: 12.391, 16.480, 20.268

15.249
Herbier Rolland-Germain
Departement de Biologie
Faculté des Sciences
Universite de Sherbrooke
Sherbrooke, PQ J1K 2R1
819-821-7078; (FAX) 819-821-7921

Contact: Dr. Colette Ansseau, Curator
Number of specimens: 199,421.

15.250
Herbier du Quebec
Ministere de l'Agriculture, des Pecheries et de
l'Alimentation du Quebec
2700 rue Einstein
Ste.-Foy, PQ G1P 3W8
418-644-7268; (FAX) 418-644-3361

Contact: Dr. Dominique Doyon, Curator
Number of specimens: 117,000.
Content: Flora of Quebec and eastern Canada.

15.251
Institut de technologie agricole Herbier
La Pocatiere, PQ

Number of specimens: 33,000.

QC
15.252
Herbier Marie-Victorin
Université de Montreal
4101 est, rue Sherbrooke
Montreal, QC H1X 2B2
514-872-8474

Contact: Luc Brouillet, Curator
Number of specimens; 650,000.
Cross-references: 12.392, 16.481

SK
15.253
Agriculture Canada Research Station
P.O. Box 1030
Swift Current, SK S9H 3X2
306-773-4621; (FAX) 306-773-9123

Contact: John Waddington, Research Scientist
Number of specimens: 20,000.
Content: Western Canada flora.

15.254
University of Regina Herbarium
Biology Dept.
University of Regina
Regina, SK S4S 0A2
306-584-4254

Number of specimens: 25,000.
Content: Vascular plants of Saskatchewan with an
emphasis on grasses and sedges.

15.255
W. P. Fraser Herbarium
University of Saskatchewan
Saskatoon, SK S7N 0W0
306-966-4950; (FAX) 306-373-1025

Contact: Vernon L. Harms, Curator
Number of specimens: 100,000.
Content: Saskatchewan flora including rare plants,
especially strong on North American *Heterotheca* and
Saskatchewan sedges, grasses, and willows.
Cross-references: 12.393, 16.483

16

Botanical Gardens, Arboreta, Conservatories, and Other Public Gardens

The following is a list of gardens and arboreta in the United States and Canada with significant plant collections. It is by no means a comprehensive list of all such gardens, but is an attempt to highlight the more significant ones. Because of space limitations, municipal gardens, collections at colleges and universities, displays at commercial nurseries, gardens at corporate headquarters and theme parks, and state and national parks are not thoroughly represented.

An arboretum is defined as a collection of trees arranged in a naturalized manner, but many arboreta contain shrubs, vines, flowers, and specialty gardens as well. A true botanical garden is a collection of plants arranged systematically according to botanical relationships, native habitat, or economic use. Display gardens or horticultural gardens usually contain various plants arranged for aesthetic or educational effect. Some botanical gardens, display gardens, and arboreta have conservatory greenhouses for the display of more tender plants and for propagation. Many of the facilities below offer a variety of public services such as educational programs, plant identification, libraries, and horticultural information services.

This chapter gives only brief information on the nature of the plant collections and facilities. The following books provide more details and will be very helpful to anyone interested in visiting the gardens listed below.

Everitt L. Miller and Jay S. Cohen, *The American Garden Guidebook*. New York: M. Evans and Co., Inc., 1987.

————, *The American Garden Guidebook West*. New York: M. Evans and Co., Inc., 1989.

These are the best available guidebooks to North American gardens. Gardens are divided into two categories: "excellent" and "don't miss." There are also useful sections on free gardens, family fun gardens, wedding gardens, and winter gardens.

Irene Jacob and Walter Jacob, *Gardens of North America and Hawaii*. Portland, OR: Timber Press, 1985.

This is a more comprehensive listing than Miller and Cohen, but contains fewer details on individual listings.

Mary Helen Ray and Robert P. Nicholls, *The Traveler's Guide to American Gardens*. Chapel Hill, The University of North Carolina Press, 1988.

This is another extensive listing of American gardens, sometimes lacking in botanical detail but very useful for its breadth of coverage.

(Top to bottom) University of California Botanical Garden at Berkeley, California; Dubuque Arboretum & Botanical Gardens, Dubuque, Iowa; International Bonsai Arboretum, West Henrietta, New York; Mercer Arboretum and Botanic Gardens, Humble, Texas.

(Top, left to right) Boyce Thompson Southwestern Arboretum, Superior, Arizona; Missouri Botanical Garden, St. Louis, Missouri; (bottom) Morris Arboretum at the University of Pennsylvania, Philadelphia, Pennsylvania.

The Plant Collections Directory. Swarthmore: American Association of Botanical Gardens and Arboreta, Inc., 1988.

A listing of 118 institutions in the United States and Canada with the most important documented collections of living plants. There is information on the institution's environmental parameters, facilities, familial collections, generic collections, horticultural collections, other collections, research projects, and more.

Directory of Regional Gardening Resources. New York: The Garden Club of America, 1987.

A very useful directory of nurseries, suppliers, gardens, and publications, arranged by region and state.

UNITED STATES

AL

16.1

Auburn University Arboretum
Dept. of Botany, Plant Pathology, and Microbiology
Auburn, AL 36849
205-826-4830

Type of garden: Arboretum.
Pitcher plant bog, sand dune area, alkaline prairie soil garden, and an annual garden.
Cross-references: 12.4, 12.395, 15.2

16.2

Birmingham Botanical Gardens
2612 Lane Park Rd.
Birmingham, AL 35223
205-879-1227

Publication: *Newsletter of the Birmingham Botanical Society*, bimonthly.
Type of garden: Botanical garden.
Rose garden, lily garden, All-America Selections display garden with vegetables and annuals, wildflower garden, fern glen, iris garden, magnolia garden, Southern garden, and a Japanese garden.
Library: 14.1
Seed/specimen exchange.
Other: Classes and lectures.
Cross-reference: 20.66

16.3

Jasmine Hill Gardens
P.O. Box 6001
Montgomery, AL 36106
205-567-9444
Location: Jasmine Hill Rd.

Type of garden: Display garden.
Jasmines, azaleas, and flowering cherry trees.

16.4

Mobile Botanical Gardens
Museum Dr.
P.O. Box 8382
Mobile, AL 36608
205-342-0555

Publication: *Gardenscope*, monthly.
Type of garden: Botanical garden.
Native azaleas, camellias, herbs, ferns, and a fragrance and texture garden for the visually impaired.
Library: Number of books: 500; number of periodicals: 6.
Lending policy: Open to members only.

16.5

University of Alabama Arboretum
Dept. of Biology
Box 870344
Tuscaloosa, AL 35487-0344
205–348–5960
Location: 15th St.

Type of garden: Arboretum.
Native trees, shrubs, and vines, organic demonstration garden, ornamental trees and shrubs, and wildflowers.
Cross-reference: 15.3

AR

16.6

Arkansas State Capitol Rose Garden
Office of the Secretary of State
Information Services Division
State Capitol
Little Rock, AR 72201-1094
501-371-5164

Type of garden: Display garden.
All-America Rose Selections trial garden with over 1,200 rose bushes.

AZ

16.7

Arboretum at Flagstaff, The
P.O. Box 670
S. Woody Mountain Rd.
Flagstaff, AZ 86002
602-774-1441

Publication: *Newsletter*, quarterly.
Type of garden: Arboretum.

520 species of plants native to alpine tundra, coniferous forest, and high desert, including threatened and endangered plants of the Colorado Plateau.
Library: 4,000 items.
Lending policy: Reference only.
Cross-reference: 8.98

16.8

Arizona-Sonora Desert Museum
2021 North Kinney Rd.
Tucson, AZ 85743
602-883-1380; (FAX) 602-883-2500

Publication: *Sonorensis*.
Type of garden: Botanical garden (also zoo and museum).
Regional cactus and succulent collection, Xeriscape collection, hummingbird garden, and many rare species (especially of succulents).
Library: Number of books: 3,600; number of periodicals: 92.
Lending policy: Reference only.
Seed/specimen exchange.
Other: School programs.

16.9

Boyce Thompson Southwestern Arboretum
P.O. Box AB
Superior, AZ 85273
602-689-2723

Publication: *Desert Plants*, annual.
Type of garden: Arboretum.
1,076 acres of desert and woodland with over 1,500 species including agaves, Boojum trees, saguaros, yuccas, Chilean palo verde, and eucalyptus.
Library: 14.2
Lending policy: Staff only, reference only.
Seed/specimen exchange.

16.10

Desert Botanical Garden
1201 North Galvin Pkwy.
Phoenix, AZ 85008
602-941-1225

Publications: *Agave Magazine*, biannual; *Saguaroland Bulletin*, quarterly.
Type of garden: Botanical garden.
Wildflowers, cacti and other succulents, trees and shrubs from the arid regions of Africa, Asia, Australia, and the Americas, and the Xeriscape demonstration garden and vegetable plot.
Library: 14.3
Seed/specimen exchange.
Other: Classes.
Cross-references: 8.99, 15.10

Desert Legume Program
Tucson, AZ

Cross-reference: 4.11

16.11

Mormon Temple Gardens
525 East Main St.
Mesa, AZ 85204
602-964-7164

Type of garden: Display garden.
Trees, shrubs, cacti, palms, citrus, and flowering plants.

16.12

Tohono Chul Park
7366 North Paseo del Norte
Tucson, AZ 85704
602-742-6455

Type of garden: Display garden.
Native and drought-resistant plants.
Library: Number of books: 250; number of periodicals: 10.
Lending policy: Reference only.
Seed/specimen exchange.
Other: Lectures and demonstrations.

16.13

Tucson Botanical Gardens
2150 North Alvernon Way
Tucson, AZ 85712
602-326-9255

Publication: *Tucson Botanical Gardens News*, 11/year.
Type of garden: Botanical garden.
Roses, irises, herbs, and Xeriscape and solar demonstration gardens.
Library: Number of books: 2,800; number of periodicals: 20. Southwestern arid plants.
Lending policy: Open to members only, circulating.
Other: Classes.
Cross-references: 13.57, 17.29

16.14

Valley Garden Center
1809 North 15th Ave.
Phoenix, AZ 85007
602-252-2120

Type of garden: Display garden.
All-America Rose Selections display garden, arboreal garden, Japanese garden, low-maintenance garden, tropical garden, and an edible garden.

CA
16.15

Alice Keck Park Memorial Gardens
City of Santa Barbara Parks Dept.
P.O. Drawer P-P
402 E. Ortega St.
Santa Barbara, CA 93102
805-564-5433
Location: Micheltorena and Santa Barbara Sts.

Type of garden: Display garden.
Flower beds, shrubs, and trees—300 varieties of plants.

All-America Gladiolus Selections
Madera, CA

Cross-reference: 1.3

American Bamboo Society
Springville, CA

Cross-reference: 2.2

American Calochortus Society
Hayward, CA

Cross-reference: 2.5

16.16

Balboa Park
Park & Recreation Dept.
City of San Diego
Balboa Park Management Center
San Diego, CA 92101
619-236-5984
Location: Laurel St. and 6th Ave.

Type of garden: Display garden.
Casa del Rey Moro garden, Alcazar garden, English
garden, rose garden, lily pond, desert garden, and
Botanical Building with tropical and subtropical plants.
Cross-reference: 15.30

16.17

Berkeley Rose Garden
City of Berkeley Dept. of Parks
201 University Ave.
Berkeley, CA 94720
415-644-6530
Location: Euclid Ave. and Bayview Pl.

Type of garden: Display garden.
4,000 rose bushes.

Biological Urban Gardening Services
Citrus Heights, CA

Cross-reference: 1.38

16.18

Blake Garden
Dept. of Landscape Architecture
College of Environmental Design
University of California at Berkeley
Berkeley, CA 94720
415-524-2449
Location: 70 Rancon Rd.

Type of garden: Display and demonstration garden.
Lily pool, formal garden, cut-flower garden, yellow
garden, pink garden, Australian plant garden, and
Mediterranean plant garden with many rare plants and
specimen trees.
Library: Number of books: 300; number of periodicals:
15.
Lending policy: Reference only.
Cross-references: 12.401, 15.24, 15.36, 16.60

16.19

Botanic Gardens
University of California at Riverside
Riverside, CA 92521
714-787-3706; (FAX) 714-787-4437

Type of garden: Botanical garden.
Cacti, succulents, herbs, roses, irises, Australian plants,
South African plants, Mediterranean plants, and
California native plants.
Seed/specimen exchange.
Cross-references: 12.64, 15.34

16.20

C. M. Goethe Arboretum
California State University
6000 J St.
Sacramento, CA 95819
916-278-6494: (FAX) 916-278-5722

Type of garden: Arboretum.
Conifers and temperate woody plants from around the
world and a large display of native California
perennials, trees, and shrubs.

California Rare Fruit Growers
Fullerton, CA

Cross-reference: 2.34

16.21

Descanso Gardens
1418 Descanso Dr.
La Canada Flintridge, CA 91011
818-790-5571

Publication: *Arboreta and Botanic Gardens Newsletter*,
quarterly, free.
Type of garden: Display garden.
The world's largest collection of camellias, plus modern
and old garden roses, and native plants.
Seed/specimen exchange.
Other: Arbor Day tree distribution, classes, lectures, and
seminars.

16.22

Exposition Park Rose Garden
City of Los Angeles Dept. of Parks
City Hall East, Room 1350
200 N. Main St.
Los Angeles, CA 90012
213-748-4772
Location: 701 State Dr.

Type of garden: Display garden.
Over 12,000 bushes representing 200 varieties of roses.

16.23

Firescapes Demonstration Garden
2411 Stanwood Dr.
Santa Barbara, CA 93108
805-564-5702; (FAX) 805-564-5730

Type of garden: Demonstration garden.
Over 200 different plant specimens selected and
landscaped for fire resistance.

16.24

Fullerton Arboretum
California State University
P.O. Box 34080
Fullerton, CA 92634
714-773-3579

Type of garden: Arboretum.
Plants from around the world in temperate, subtropical,
tropical, desert, desert woodland, and Mediterranean
habitats. Also a fruit grove, 16 organic gardening plots,
a carnivorous plant bog, a conifer collection,
Xeriscapes, and heritage kitchen gardens with flowers,
herbs, and vegetables.
Library: 14.4

Cross-references: 15.20, 20.75

16.25

Golden Gate Park
Recreation and Parks Dept.
McLaren Lodge
Fell & Stanyan Sts.
San Francisco, CA 94117
415-661-1316

Type of garden: Display garden and conservatory.
1,000-acre park with a fuchsia garden, camellia garden,
rhododendron garden, Japanese garden, a conservatory
with tropical plants, and the Strybing Arboretum
(16.57). Over 3,000 species of plants.

16.26

Goldsmith Seeds, Inc.
2280 Hecker Pass Hwy.
P.O. Box 1349
Gilroy, CA 95021
408-847-7333

Type of garden: Display garden of a wholesale seed
company. Flower garden, lawn garden, shade garden,
and a grape arbor.

16.27

Greystone Park
905 Lomas Vista Dr.
Beverly Hills, CA 90210
213-285-2537

Type of garden: Display garden.
Formal gardens.

16.28

Grigsby Cactus Gardens
2354 Bella Vista Dr.
Vista, CA 92084
619-727-1323

Type of garden: Display garden of retail nursery.
Three-acre garden and greenhouse with plants from
Madagascar, South Africa, Morocco, India, Mexico, and
the southwestern U.S.

16.29

Hakone Japanese Garden
21000 Big Basin Way
Saratoga, CA 95070
408-264-7024

Type of garden: Japanese garden.
Bamboo garden with 23 species of bamboo, and pines and
azaleas.

16.30

Hortense Miller Garden
3035 Berr Dr.
Laguna Beach, CA 92651
714-497-7692

Publication: *Newsletter*, quarterly.
Type of garden: Display garden.
Over 1,200 species including exotic species from
subtropical regions, native plants including the coastal
sage shrub, and dry climate plants like tree euphorbia,
eucalyptus, and succulents.
Scholarships: Annual scholarship to Orange Coast college
student in horticulture.
Library: Number of books: 500.
Lending policy: Reference only.

Indoor Citrus and Rare Fruit Society
Los Altos, CA

Cross-reference: 1.66

16.31

Kaiser Center Roof Garden
300 Lakeside Dr.
Oakland, CA 94612
415-271-6100

Type of garden: Display garden.
Three and one-half-acre garden on the top of a 28-story
building.

16.32

Lakeside Park Trial and Show Gardens and Garden
Center
666 Bellevue Ave.
Oakland, CA 94612
415-273-3186

Type of garden: Display and demonstration gardens.
Fragrance and herb garden, Xeriscape and composting
demonstration gardens, Japanese garden, tropical
greenhouse, and trial gardens of annuals and
perennials.

16.33

Lifescape Memorial Garden
Santa Barbara City College
721 Cliff Dr.
Santa Barbara, CA 93109-2394
805-965-0581, ext. 244

Type of garden: Display garden.
Ornamental edible plants—trees, shrubs, vines, and
herbaceous perennials—with potential for edible
landscaping.

Cross-reference: 12.56

16.34
Living Desert, The
47900 Portola Ave.
Palm Desert, CA 92260
619-346-5694; (FAX) 619-568-9685

Publication: *Foxpaws.*
Type of garden: Botanical garden and wild animal park.
Nine distinct desert regions—Mojave, Sonora, Baja,
Chihuahua, Colorado, Yuma, a sand dune garden, a
demonstration garden of desert plants for home
landscaping, and an Indian garden with native plants
for food, clothing, medicine, and housing.
Library: Number of books: 1,500; number of periodicals:
10.
Lending policy: Reference only.
Seed/specimen exchange.
Cross-reference: 13.14

16.35
Los Angeles State and County Arboretum
301 North Baldwin Ave.
Arcadia, CA 91007
818-446-8251; (FAX) 818-445-1217

Publication: *Arboreta and Botanic Gardens Newsletter,*
quarterly, free.
Type of garden: Arboretum.
Begonia and tropical greenhouses, herb garden, Australian
garden, water garden, and 4,000 species in the
arboretum.
Library: 14.8
Seed/specimen exchange.
Other: Horticultural information service, Arbor Day tree
distribution, classes, lectures, and seminars.
Cross-references: 15.25, 20.77

16.36
Lummis Gardens
200 East Ave. 43
Los Angeles, CA 90031
213-222-0546

Type of garden: Demonstration garden.
Model residential water-conservation garden with
plantings of California native species and
Mediterranean species.

16.37
Marin Art and Garden Center
P.O. Box 437
Ross, CA 94957-0437
415-454-5597
Location: Sir Francis Drake Blvd. and Laurel Grove Ave.

Type of garden: Display garden.
10-acre garden with wisteria, rhododendrons, orchids,
azaleas, and numerous specimen trees.

16.38
Mendocino Coast Botanical Gardens
18220 North Highway 1
P.O. Box 1143
Fort Bragg, CA 95437
707-964-4352

Publication: *Botanically Speaking,* biannual.
Type of garden: Botanical garden.
3,000 varieties of native and cultivated plants including
heathers, perennials, succulents, ivies, Mediterraneans,
dwarf conifers, heritage roses, camellias, and
rhododendrons.
Cross-reference: 2.19

16.39
Micke Grove Park
San Joaquin County Dept. of Parks and Recreation
4520 W. Eight Mile Rd.
Stockton, CA 95209
209-331-7400

Type of garden: Display garden.
Japanese garden, rose garden, azalea garden, and a
camellia garden.

16.40
Mildred E. Mathias Botanical Garden
University of California at Los Angeles
405 Hilgard Ave.
Los Angeles, CA 90024-1606
213-825-3620; (FAX) 213-206-3987

Type of garden: Botanical garden.
Eucalyptus, aloe, ficus, acacia, California natives,
Hawaiian plants, palms, cycads, desert plants, aquatic
plants, dawn redwoods, and Australian plants.
Scholarships.
Library: Number of books: 3,000.
Lending policy: Reference only.
Seed/specimen exchange.
Other: Plant identification and horticultural information.
Cross-reference: 15.37

16.41
Moorten Botanic Garden
1702 S. Palm Canyon Dr.
Palm Springs, CA 92264
619-327-6555

Type of garden: Botanical garden.
Over 3,000 kinds of desert plants in collections
emphasizing the flora of various desert regions—Sonora,
Mojave, California, Texas, Arizona, Colorado, Baja
California, South America, and Africa.

16.42
Morcom Amphitheater of Roses
Oakland Office of Parks and Recreation
1520 Lakeside Dr.
Oakland, CA 94612
415-273-3090
Location: 700 Jean St.

Type of garden: Display garden.
400 varieties of roses and over 8,000 bushes.

16.43

Municipal Rose Garden
Naglee and Dana Aves.
San Jose, CA 95126

Type of garden: Display garden.
4,000 rose bushes in a five and one-half-acre garden.

16.44

Naiman Tech Center Japanese Garden
9605 Scranton Rd.
San Diego, CA 92121
619-453-9550

Type of garden: Display garden.
Five and one-half acres of gardens, including a Japanese garden.

16.45

Overfelt Botanical Gardens
City of San Jose
Dept. of Parks and Recreation
151 West Mission St.
San Jose, CA 95110
408-259-5477

Location: Park Dr. and McKee Rd.
Type of garden: Botanical garden.
Camellia garden, fragrance garden, All-America Selections display garden, palm grove, roses, irises, and trees.

16.46

Pageant of Roses Garden
3900 S. Workman Mill Rd.
P.O. Box 110
Whittier, CA 90608
213-699-0921

Type of garden: Display garden.
All-America Rose Selections display garden with over 600 varieties and 7,000 bushes of roses. Also a Japanese garden.

16.47

Quail Botanical Gardens
230 Quail Gardens Dr.
Encinitas, CA 92024
619-753-4432

Publications: *Quail Call*, biannual; *Q News*, bimonthly.
Type of garden: Botanical garden.
Demonstration garden of native plants for landscaping, herb garden, subtropical fruits, bamboos, palms, cycads, cacti and other succulents, and hibiscus.
Library: Number of books: 1,500; number of periodicals: 20.
Lending policy: Reference only.
Seed/specimen exchange.
Other: Public information service.
Cross-reference: *2.2*

16.48

Rancho Santa Ana Botanic Garden
1500 North College Ave.
Claremont, CA 91711-3101
714-625-8767; (FAX) 714-626-7670

Publications: *Aliso, Journal of the Rancho Santa Ana Botanic Garden*, 3/year, Newsletter, quarterly.
Type of garden: Botanical garden.
85 acres with 1,500 species of native California plants including trees, shrubs, and succulents, and 95 rare or endangered species.
Library: 14.9
Seed/specimen exchange.
Other: Classes.
Cross-references: 8.100, 15.28

16.49

Regional Parks Botanic Garden
Tilden Park
Berkeley, CA 94708
415-841-8732

Publication: *The Four Seasons*, quarterly.
Type of garden: Botanical garden.
California native plants including rare and endangered plants.
Other: Lectures.

16.50

Rogers Gardens
2301 San Joaquin Hills Rd.
Corona Del Mar, CA 92625
714-640-5800

Type of garden: Display garden at a retail nursery.
Victorian garden, rose garden, perennial garden, and flowering vines.

16.51

San Mateo Arboretum
P.O. Box 1523
San Mateo, CA 94401
415-342-1629
Location: Central Park, Laurel and Fifth Aves.

Type of garden: Arboretum.
60 species of trees, ferns, oriental azaleas, and an All-America Seed Selection garden.

16.52

San Mateo Japanese Garden
San Mateo Park and Recreation Dept.
330 West 20th Ave.
San Mateo, CA 94403-1388
415-377-4700
Location: Central Park, Laurel and Fifth Aves.

Type of garden: Japanese garden.
Over 80 types of trees plus flowering plants.

16.53
Santa Barbara Botanic Garden
1212 Mission Canyon Rd.
Santa Barbara, CA 93105
805-682-4726

Publication:*Newsletter*, biannual, free.
Type of garden: Botanical garden.
1,015 taxa of California native plants including many rare and endangered species.
Library: 14.10
Seed/specimen exchange.
Other: Educational programs.

Saratoga Horticultural Foundation
San Martin, CA

Cross-reference: 6.4

16.54
Saso Herb Gardens
14625 Fruitvale Ave.
Saratoga, CA 95070
408-867-0307

Type of garden: Herb garden.
Biblical garden, medicinal garden, and an astrological garden.

16.55
Sherman Library and Gardens
2647 East Coast Highway
Corona del Mar, CA 92625
714-673-2261

Type of garden: Display garden.
Tropicals, subtropicals, cacti, and succulents.
Cross-reference: 13.15

16.56
South Coast Botanic Garden
26300 Crenshaw Blvd.
Palos Verdes Peninsula, CA 90274
213-377-0468

Publication: *Arboreta and Botanic Gardens Newsletter*, quarterly, free.
Type of garden: Botanical garden.
2,000 species including native California plants, coastal and Mediterranean plants, aquatic plants, and shade and cactus gardens.
Seed/specimen exchange.
Other: Arbor Day tree distribution, and classes.

16.57
Strybing Arboretum and Botanical Gardens
Ninth Ave. and Lincoln Way
San Francisco, CA 94122
415-661-1316

Publication: *Strybing Leaflet*, quarterly.
Type of garden: Arboretum and botanical garden.
Collections of Asian magnolias, Vireya and Medenii rhododendrons, camellias, dwarf conifers, geographic collections from Australia, New Zealand, Chile, and Cape Province, native California plants, collection of endangered species from New World cloud forests of Mexico, and a special collection designed for the visually impaired.
Library: 14.5
Seed/specimen exchange.
Cross-reference: 20.83

16.58
Sunset Magazine Gardens
Willow & Middlefield Rds.
Menlo Park, CA 94025
415-321-3600

Type of garden: Demonstration garden.
Cacti and other succulents, camellias, rhododendrons, and beds of annuals and bulbs, plus native shrubs and trees.
Cross-reference: 21.100

16.59
University Arboretum
University of California at Davis
Davis, CA 95616
916-752-2498

Type of garden: Arboretum.
Collections of oaks, eucalyptus, conifers, California native plants, and acacias, a redwood grove, and demonstration gardens of drought-tolerant flowering perennials.
Cross-references: 12.62, 12.402, 15.23, 17.33, 19.12

16.60
University of California Botanical Garden at Berkeley
Centennial Dr.
University of California at Berkeley
Berkeley, CA 94720
415-642-3343

Publication: *Newsletter*, quarterly.
Type of garden: Botanical garden.
Large collection of rare and endangered California natives, also Asian rhododendrons, cacti and succulents, western herb garden, Chinese medicinal herb garden, ferns, and carnivorous and insectivorous plants.
Library: Number of books: 750; number of periodicals: 20.
Lending policy: Reference only.
Seed/specimen exchange.
Cross-references: 8.101, 12.401, 15.24, 15.36, 16.18

16.61
University of California at Irvine Arboretum
University of California at Irvine
Irvine, CA 92717
714-856-5833

Publication: *Friends of the Arboretum Newsletter*, quarterly.
Type of garden: Arboretum.
Collections of rare African aloes, California native flowering bulbs, Southern African flowering bulbs and irises, and amaryllis.
Library: Number of books: 700; number of periodicals: 5. Plants of Southern Africa, conservation.
Lending policy: Not open to the public.
Seed/specimen exchange.
Cross-reference: 12.63

16.62
University of California, Santa Cruz Arboretum
University of California, Santa Cruz
Santa Cruz, CA 95064
408-427-2998
Location: Bay and High Sts.

Type of garden: Arboretum.
Collections of Australian plants, South African plants, New Zealand plants, and native Californian plants.

16.63
Virginia Robinson Garden
1008 Elden Way
Beverly Hills, CA 91006-2697
213-276-5367

Publication: *Arboreta and Botanic Gardens Newsletter*, quarterly, free.
Type of garden: Display garden.
1,000 species including palms, camellias, roses, gingers, azaleas, annuals, and perennials.
Seed/specimen exchange.
Other: Lectures.

16.64
William Joseph McInnes Botanic Garden
Box 9949
Oakland, CA 94613
415-430-2158

Type of garden: Botanical garden.
California native plants and South African bulbs.

CO
American Penstemon Society
Lakewood, CO
Cross-reference: 2.21

16.65
Denver Botanic Garden
909 York St.
Denver, CO 80206
303-331-4000; (FAX) 303-331-4013

Publications: *Green Thumb Newsletter*, monthly; *Green Thumb Magazine*, biannual.
Type of garden: Botanical garden and conservatory.
Native prairie, tundra, alpine, and wildflower collections, Xeriscape demonstration garden, plains garden, sensory garden, Japanese garden, and tropical and subtropical species in the conservatory.
Library: 14.14
Seed/specimen exchange.
Cross-references: 8.102, 13.16, 13.66, 20.85, 20.272

16.66
Horticultural Art Society Garden
Horticultural Art Society of Colorado Springs
1438 N. Hancock Ave.
Colorado Springs, CO 80903
719-475-0250
Location: Mesa Rd. and Glen Ave.

Type of garden: Demonstration garden.
All-America Selections display garden of vegetables and annuals, fragrance herb garden, perennials, roses, and an aspen grove.
Library.
Cross-reference: 20.86

16.67
Mineral Palace Park
Dept. of Parks
City of Pueblo
800 Goodnight Ave.
Pueblo, CO 81005
719-566-1745
Location: 19th and Santa Fe Sts.

Type of garden: Display garden.
Annuals, English rock garden, roses, and greenhouses with tropical and subtropical plants.

16.68
Vail Alpine Garden
183 Gore Creek Dr.
Vail, CO 81657
303-476-0103; (FAX) 303-476-8702

Publication: *Quarterly Bulletin*, quarterly.
Type of garden: Display garden.
Alpines, mostly natives to the Rocky Mountains. Still under development.
Seed/specimen exchange.

16.69
W. D. Holley Plant Environmental Research Center
Dept. of Horticulture
Colorado State University
Fort Collins, CO 80523
303-491-7019

Type of garden: Demonstration garden.
All-America Selections vegetable and annual flower demonstration gardens, woody ornamentals, perennials, grasses, and fruit.
Cross-references: 12.68, 12.403, 15.38, 20.87

16.70
Washington Park
Dept. of Parks and Recreation
City of Denver
1805 Bryan St.
Denver, CO 80204
303-458-4800
Location: S. Downing and Exposition Sts.

Type of garden: Display garden.
Annuals and a sunken garden.

CT
16.71
Audubon Fairchild Garden of the National Audubon Society
613 Riversville Rd.
Greenwich, CT 06831
203-869-5272

Type of garden: Display garden.
Plants of Connecticut, the Blue Ridge Mountains, Appalachia, and medicinal plants and wildflowers.

16.72
Bartlett Arboretum
University of Connecticut
151 Brookdale Rd.
Stamford, CT 06903-4199
203-322-6971

Publication: *The Bartlett Arboretum Newsletter*, 3/year.
Type of garden: Arboretum.
Dwarf conifers, ericaceous plants, flowering shrubs, small
  ornamental trees, ground covers, herbaceous perennials,
  and native woodlands.
Library: 14.15
Other: Plant diagnostic clinic (May through August).

16.73
Caprilands Herb Farm
534 Silver St.
Coventry, CT 06238
203-742-7244

Type of garden: Display garden at nursery.
Over 300 varieties of herbs.

16.74
Connecticut College Arboretum
5625 Connecticut College
New London, CT 06320
203-447-7700; (FAX) 203-447-7809

Type of garden: Arboretum.
Collections of eastern North American native woody
  plants and special gardens of native azaleas, native
  conifers, and *Kalmia latifolia* cultivars.

16.75
Elizabeth Park
Dept. of Parks and Recreation
25 Stonington St.
Hartford, CT 06106
203-722-6541
Location: Prospect and Asylum Aves.

Type of garden: Display and demonstration gardens.
15,000 rose plants and 1,000 species, also a rock garden,
  annuals, perennials, and an evergreen collection.
Cross-reference: 20.296

16.76
Logee's Greenhouses
55 North St.
Danielson, CT 06239
203-774-8083

Type of garden: Greenhouses at nursery.
Rare and tropical plants, begonias, and geraniums.

16.77
Marsh Botanical Garden
Yale University
227 Mansfield St.
New Haven, CT 06511
203-436-8665

Type of garden: Botanical garden.
Conifers, oaks, hemlocks, and an iris test garden.
Cross-reference: 15.43

16.78
Nurserymen's Gardens
Connecticut Agricultural Experiment Station
123 Huntington, P.O. Box 1106
New Haven, CT 06504
203-789-7272; (FAX) 203-789-7232

Publication: *Frontiers of Plant Science*, biannual, free.
Type of garden: Display garden.
Showcases plants discovered or hybridized and introduced
  to the trade by Connecticut nurserymen, including
  woody and herbaceous perennials.
Library: Number of books: 20,000; number of periodicals:
  1,000.
Lending policy: Reference only.
Other: Soil testing, disease and insect identification, and
  plant problem diagnosis.
Cross-reference: 11.93

16.79
Olive and George Lee Memorial Garden
89 Chichester Rd.
New Canaan, CT 06840

Type of garden: Display garden.
Azaleas representing 175 varieties, rhododendrons, and
  wildflowers.

16.80
Pardee Rose Garden
180 Park Rd.
Hamden, CT 06518
203-787-8142

Type of garden: Display garden.
250 rose bushes and annuals and perennials.

DC
16.81
Kenilworth Aquatic Gardens
National Park Service
Anacostia Ave. and Douglass St., N.E.
Washington, DC 20019
202-426-6905

Type of garden: Aquatic gardens.
14 acres of ponds filled with water lilies, lotus, and a
  variety of bog plants. The only site dedicated to the
  propagation and perpetuation of aquatic plants in the
  National Park System.
Seed/specimen exchange.
Other: Plant exchanges and aquatic gardening workshops.

16.82
United States Botanic Garden
245 First St., S.W.
Washington, DC 20024
202-225-8333

Type of garden: Botanical garden.
Orchids, begonias, aroids, bromeliads, cacti, cycads,
    palms, carnivorous plants, annuals, perennials, and
    herbs.
Library: Number of books: 1,000; number of periodicals:
    50.
Lending policy: Not open to the public.
Other: Plant information service.
Cross-reference: 20.91

16.83
United States National Arboretum
3501 New York Ave., N.E.
Washington, DC 20002
202-475-4815; (FAX) 202-475-5252

Publication: *Friends of National Arboretum Newsletter*,
    bimonthly.
Type of garden: Arboretum.
Bonsai, flowering trees, dwarf conifers, holly, the largest
    designed herb garden in the U.S., and wildflowers.
Library: 14.21
Seed/specimen exchange.
Other: Plant identification, classes.
Cross-references: 1.60, 10.44, 10.66, 10.75, 11.57, 15.44

DE
16.84
Mt. Cuba Center for the Study of Piedmont Flora
P.O. Box 3570
Greenville, DE 19807
302-239-4244

Type of garden: Display garden.
Native plant gardens with wildflowers, and native trees
    and shrubs. Limited access.

FL
16.85
Alfred B. Maclay State Gardens
3540 Thomasville Rd.
Tallahassee, FL 32308
904-893-1301

Type of garden: Display garden.
Camellias, azaleas, oriental magnolias, and native azaleas.
Other: Monthly plant care workshops.

American Ginger Society
Archer, FL

Cross-reference: 2.12

16.86
Bok Tower Gardens
P.O. Box 3810
Lake Wales, FL 33859-3810
813-676-1408

Type of garden: Display garden.
128 acres with over 300 kinds of cultivated plants.
Cross-reference: 8.103

16.87
Fairchild Tropical Garden
10901 Old Cutler Rd.
Miami, FL 33156
305-667-1651

Publication: *Bulletin*, quarterly.
Type of garden: Botanical garden.
The largest tropical garden in the U.S., with collections
    of palms, cycads, and other tropical trees.
Library: 14.24
Seed/specimen exchange.
Cross-references: 8.104, 13.72, 15.47

16.88
Flamingo Gardens
3750 Flamingo Rd.
Ft. Lauderdale, FL 33330
305-473-2955

Publication: *Flamingo Gardens Bulletin*, quarterly.
Type of garden: Botanical garden.
A repository site for *Heliconia* species and related genera,
    also collections of palms, orchids, and fruit trees.
Cross-reference: 2.47

16.89
Florida Citrus Arboretum
Florida Dept. of Agriculture & Consumer Services
3027 Lake Alfred Rd.
Winter Haven, FL 33881
813-294-4267; (FAX) 813-297-6521

Type of garden: Arboretum.
Over 275 specimens of citrus and citrus relatives including
    Australian wild lime, *Atalantia*, African cherry orange,
    Aegle, *Afraegle*, and *Aeglopsis*.
Seed/specimen exchange.

16.90
Florida Cypress Gardens
Box 1
Cypress Gardens, FL 33880
813-324-2111
Location: State Rt. 540.

Type of garden: Display garden.
Over 8,000 kinds of plants, 13 theme gardens including
    All-America Rose Selections garden, oriental gardens,
    gardens of the world, and collections of tropical and
    subtropical plants.

Florida Native Plant Society
Gainesville, FL

Cross-reference: 3.11

16.91
Four Arts Garden
Four Arts Plaza
Royal Palm Way
Palm Beach, FL 33480

Type of garden: Display and demonstration gardens.
    Formal flower garden, rose garden, jungle garden, rock
    garden, herb garden, moonlight garden, and a Chinese
    garden.

16.92
Fruit and Spice Park
24801 S.W. 187 Ave.
Homestead, FL 33031
305-247-5727

Type of garden: Botanical garden.
Over 500 varieties of tropical fruit, vegetables, herbs, and
  spices.
Seed/specimen exchange.
Other: Classes and workshops.

16.93
H. P. Leu Botanical Gardens
1730 N. Forest Ave.
Orlando, FL 32803
407-849-2620

Publication: *The Leu Botanical Quarterly*, quarterly.
Type of garden: Botanical garden.
Over 2,000 varieties of plants including camellias, roses,
  azaleas, and lilies.
Library: Number of books: 1,000; number of periodicals:
  25. Books on genus *Camellia*.
Lending policy: Reference only.
Cross-reference: 20.93

16.94
Heathcote Botanical Gardens
210 Savannah Rd.
Fort Pierce, FL 34982
407-464-4672

Type of garden: Botanical garden.
A new garden, still under development. Native plants and
  a Japanese garden.
Other: Classes.

Heliconia Society International
Fort Lauderdale, FL

Cross-reference: 2.47

16.95
Kanapaha Botanical Garden
North Florida Botanical Society
4625 S.W. 63rd Blvd.
Gainesville, FL 32608
904-372-4981

Type of garden: Botanical garden.
Butterfly garden, bamboo garden, vinery, hummingbird
  garden, rock garden, herb garden, palm hammock,
  spring flower garden, water lily pond, carnivorous plant
  garden, and woodland wildflower garden.
Seed/specimen exchange.

16.96
Marie Selby Botanical Garden
811 South Palm Ave.
Sarasota, FL 34236
813-366-5730

Publications: *Selbyana*, annual; *Bulletin of the Marie Selby
  Botanical Gardens*, bimonthly.
Type of garden: Botanical garden.
Large collection of epiphytes, especially Orchidaceae and
  Bromeliaceae, and Gesneriaceae, ferns, Araceae,

Marantaceae, tropicals, Palmae, Zamiaceae, ficus,
  native shoreline plants, and tropical edible plants.
Library: 14.23
Other: Identification service for bromeliads and orchids.
Cross-references: 15.50, 20.94

16.97
Mounts Horticultural Learning Center
531 North Military Trail
West Palm Beach, FL 33415
305-683-1777

Type of garden: Display garden.
Citrus trees, hibiscus garden, rose garden, lily pond, herb
  garden, fern house, touch garden, native plants, hedge
  materials, and salt-tolerant plants.

16.98
Orchid Jungle
26715 S.W. 157th Ave.
Homestead, FL 33030
305-247-4824

Type of garden: Display garden.
Natural forest with exotic plants on the trees and ground
  including a large collection of orchids.

16.99
Ravine Gardens State Park
P.O. Box 1096
Palatka, FL 32077
904-328-4366

Location: Twigg St.
Type of garden: Display garden.
Azaleas and annuals.

16.100
Sarasota Jungle Gardens
3701 Bayshore Rd.
Sarasota, FL 34234
813-355-5305

Type of garden: Botanical garden.
Banana groves, fern gardens, hibiscus gardens, and
  bougainvilleas, roses, gardenias, and palms.

16.101
Simpson Park
City of Miami Dept. of Parks, Recreation and Public
  Facilities
55 S.W. 17th Rd.
Miami, FL 33129
305-856-6801

Type of garden: Display garden.
Hardwood hammock trees of the Caribbean and southern
  Florida.
Library: Number of books: 500.
Lending policy: Open to members of the Council of
  Garden Club Presidents of Dade County.

16.102
Sunken Gardens
1825 Fourth St. North
St. Petersburg, FL 33704
813-896-3186

Type of garden: Display garden.
Over 5,000 varieties of tropical and subtropical species.

GA
American Camellia Society
Fort Valley, GA

Cross-reference: 2.6

16.103
Atlanta Botanical Garden
P.O. Box 77246
Atlanta, GA 30357
404-876-5859; (FAX) 404-876-7472

Publication: *Clippings*, bimonthly.
Type of garden: Botanical garden.
Rose, Japanese, herb, and vegetable gardens and a
fragrance garden for the visually impaired. 3,500
specimens including Old World succulents, island
palms, carnivorous plants, species orchids, ant plants,
and hardy cacti and succulents.
Library: 14.29
Seed/specimen exchange.
Cross-references: 2.49, 13.18

16.104
Callaway Gardens
Pine Mountain, GA 31822
404-663-5186; (FAX) 404-663-5049

Publications: *Ida Cason Callaway Foundation Newsletter*,
quarterly, free; *Nature Naturally*, 3/year, free.
Type of garden: Display garden, wilderness area, and
conservatory.
The world's largest public display of hollies (more than
450 species), over 700 varieties of azaleas, a wildflower
trail, rhododendron trail, butterfly garden, and
temperate and tropical plants in the conservatory.
Library: 14.26
Cross-references: 20.10, 20.45, 20.101

16.105
Cator Woolford Memorial Garden
1815 Ponce de Leon Ave., N.E.
Atlanta, GA 30307
404-377-3836

Type of garden: Display garden.
Azaleas, dogwoods, bulbs, wildflowers, and a greenhouse.

16.106
Fernbank Science Center
DeKalb County School System
156 Heaton Park Dr.
Atlanta, GA 30307
404-378-4311

Publication: *Fernbank Quarterly*, quarterly, free.
Type of garden: Display garden and greenhouse.
Azaleas, wildflowers, ornamentals, perennials, and
annuals, plus tropical plants, cacti, and succulents in
the greenhouse.
Library: 14.28
Cross-reference: 20.274

16.107
Founders Memorial Garden
The Garden Club of Georgia, Inc.
325 S. Lumpkin St.
Athens, GA 30602
404-542-3631

Publication: *Garden Gateways*, quarterly.
Type of garden: Display garden and arboretum.
Flora of Georgia and the Piedmont in a perennial garden,
a boxwood garden, flower terraces, courtyard gardens,
and an arboretum.

Hoya Society International, Inc.
Atlanta, GA

Cross-reference:2.49

16.108
Lockerly Arboretum
1534 Irwinton Rd.
Milledgeville, GA 31061
912-452-2112

Type of garden: Arboretum.
Tropical foliage plants, cacti and succulents, herbs, a
vineyard and bramble collection, perennial flowers,
rhododendrons, viburnums, hollies, and daylilies.

16.109
Rock City Gardens
1400 Patten Rd.
Lookout Mountain, GA 30750
404-820-2531

Type of garden: Display garden.
Over 400 species of native plants.

16.110
State Botanical Garden of Georgia
2450 South Milledge Ave.
Athens, GA 30605
404-542-1244

Publication: *Garden Leaflet*, quarterly.
Type of garden: Botanical garden.
293-acre garden with collections of roses, wildflowers,
spring bulbs, azaleas, camellias, laurels, dogwoods, and
magnolias and annual, perennial, herb, fragrance,
boxwood, rock, water, and azalea gardens.
Cross-reference: 20.100

HI
16.111
Harold Lyon Arboretum
University of Hawaii, Manoa
3860 Manoa Rd.
Honolulu, HI 96822
808-988-3177; (FAX) 808-988-4231

Publication: *Lyonia*, annual.
Type of garden: Arboretum.
4,200 species, subspecies, taxa, and cultivars including
palm, taro clones, aroids, Zingiberaceae, Costaceae,
heliconias, and economic, native, and endemic
Hawaiian plants.
Library: 14.31

Seed/specimen exchange.
Other: Classes, workshops.
Cross-references: 2.47, 12.113, 15.57

16.112
Hawaii Tropical Botanical Garden
P.O. Box 1415
Hilo, HI 96721
808-964-5233
Location: Onomea Bay.

Type of garden: Botanical garden.
More than 1,800 species including many rare,
   endangered, and unusual plants, and collections of
   heliconias, palms, bromeliads, orchids, and cycads.
Seed/specimen exchange.

16.113
Helani Gardens
Box 215
Hana, HI 96713
Location: North of Wainapana State Park.

Type of garden: Display garden.
Bamboo grove and koi pond.

16.114
Ho'omaluhia
Kane'ohe, HI 96744
808-235-6636
Location: Luluku Rd.

Type of garden: Botanical garden.
Native plants and tropical trees and shrubs from around
   the world. Many endangered species.

16.115
Honolulu Botanic Gardens
50 North Vineyard Blvd.
Honolulu, HI 96817
808-533-3406

Type of garden: Botanical garden.
Hybrid orchid collection, bromeliads, rain forest trees
   from the tropics of the world, and dryland plants from
   Hawaii and elsewhere.
Library: 14.33
Seed/specimen exchange.

16.116
Kahuna Gardens
P.O. Box 45
Hana
Maui, HI 96713
808-248-8912
Location: Hana Hwy.

Type of garden: Display garden.
Plants used for food, medicine, clothing, and shelter by
   Polynesians, and collections of coconuts, breadfruit,
   and Loulo palm.

16.117
Kula Botanical Garden
R.R. 2, Box 288
Kula
Maui, HI 96790
808-878-1715

Type of garden: Botanical garden.
A display of native and tropical plants in a natural setting
   of volcanic rock, with sandalwood, Hawaiian mint, and
   Protea.

16.118
Limahuli Botanical Garden
Box 808
Hanalei, HI 96714
Location: Kuhio Hwy.

Type of garden: Botanical garden.
Wet-area flora including Hawaiian heritage plants.

16.119
Moir's Gardens
R.R. 1, Box 73
Koloa
Kauai, HI 96756
808-742-6411
Location: 2253 Poipu Rd.

Type of garden: Display garden.
4,000-plant collection of aloe, agave, euphorbia, and other
   succulents and cacti; also tropical trees and shrubs.

16.120
Nani Mau Gardens
421 Makalika St.
Hilo, HI 96720
808-959-3541
Type of garden: Display garden.
Over 2,000 varieties of tropical flowering plants, plus
   orchids, tropical fruits and nuts, a ginger garden, fern
   forest, Hawaiian herb garden, and hibiscus and
   anthurium gardens.

16.121
National Tropical Botanical Garden
P.O. Box 340
Lawai, HI 96765
808-332-7324; (FAX) 808-332-9765
Location: Halima Rd.

Publications: *The Bulletin*, quarterly; *Allertonia*, irregular.
Type of garden: Botanical garden.
Rare and endangered species from all tropical regions
   with special collections of palms, legumes, orchids, and
   Hawaiian native plants.
Grants: Research grants.
Library: 14.32
Other: Lectures and workshops.
Cross-references: 2.47, 8.105, 12.111

16.122

Olu Pua Gardens
Box 518
Kalaheo
Kauai, HI 96741
808-332-8182
Location: Rt. 50.

Type of garden: Display garden.
More than 4,000 species of subtropical and tropical plants.

16.123

Wahiawa Botanic Garden
1396 California Ave.
Wahiawa, HI 96786
808-621-7321

Type of garden: Botanical garden.
Hawaiian garden, aroid garden, gingers, heliconias, palms, Hawaiian hibiscus, and other native plants and specimen trees.

16.124

Waimea Falls Park Arboretum & Botanical Gardens
59-864 Kamehameha Hwy.
Haleiwa
Oahu, HI 96712
808-638-8511

Type of garden: Botanical garden.
Over 5,000 types of plants and more than 30 specialty gardens including the Malaysia garden, palm meadow, ginger garden, heliconia garden, hibiscus garden, and lei garden.
Library: 14.34
Cross-reference: 8.106

IA

16.125

Bickelhaupt Arboretum
340 South 14th St.
Clinton, IA 52732
319-242-4771

Type of garden: Arboretum.
Over 800 species of native trees, shrubs, and flowers.
Library: 14.35
Other: Woody plant hotline.

16.126

Des Moines Botanical Center
909 East River Dr.
Des Moines, IA 50316
515-283-4148

Publication: *The Botanical Center Newsletter*, quarterly.
Type of garden: Botanical garden.
Extensive bonsai collection, over 300 species and cultivars of orchids, and over 250 cultivars of begonias.
Library: Number of books: 1,150; number of periodicals: 35. Rare books, ikebana, and herbs.
Lending policy: Reference only.
Other: Classes.
Cross-reference: 20.106

16.127

Des Moines Water Works Crabapple Arboretum
408 Fleur Dr.
Des Moines, IA 50321-1190
515-283-8791

Type of garden: Arboretum.
One of the world's largest collections of flowering crabapple trees with some 300 different varieties and 1,500 trees.

16.128

Dubuque Arboretum & Botanical Gardens
3125 West 32nd St.
Dubuque, IA 52001
319-556-2100

Publication: *Groundcover*, quarterly.
Type of garden: Arboretum and botanical garden.
Dwarf conifer collection, Seed Savers Exchange vegetable display, rose garden, All-America Selections annual and perennial displays, ornamental tree and shrub collections, and water gardens.
Library: Number of books: 1,100; number of periodicals: 12.
Lending policy: Circulating.
Other: Educational programs.
Cross-reference: 20.107

16.129

Early May Seed and Nursery Trial Ground and Show Garden
Shenandoah, IA 51603
712-246-1020
Location: North Elm St.

Type of garden: Demonstration and display gardens of a seed company.
All-America Selections trial gardens of annuals and vegetables, plus other flower and vegetable displays including wildflowers, shade plants, and ornamental grasses.
Cross-references: 20.12, 20.46, 20.108

16.130

Ewing Park Lilac Arboretum and Children's Forest
Dept. of Parks and Recreation
City of Des Moines, 3226 University
Des Moines, IA 50311
515-271-4700
Location: McKinley Rd. and Indianola Rd.

Type of garden: Display garden.
30 acres of 200 varieties of French lilacs and 33 acres of native Iowa trees and introduced species.

16.131

Greenwood Park Rose Garden
Dept. of Parks and Recreation
City of Des Moines
3226 University
Des Moines, IA 50311
515-271-4700
Location: Grant Ave.

Type of garden: Display garden.
All-America Rose Selections display garden with 300 varieties.

16.132
Iowa Arboretum
Box 44A, Rt. 1
Madrid, IA 50156
515-795-3216

Publication: *Iowa Arboretum News*, bimonthly.
Type of garden: Arboretum.
340 acres with several hundred species of trees, shrubs,
grasses, and rare and endangered plants.
Library: Number of books: 800.
Lending policy: Reference only.

Iowa State Horticultural Society
Des Moines, IA

Cross-reference: 6.10

16.133
Iowa State University Horticultural Garden
Iowa State University
Ames, IA 50001
515-294-2751
Location: Elwood Dr. and Haber Rd.

Type of garden: Display garden.
All-America Rose Selections test garden with over 500
varieties, a perennial maze garden, herb garden with
culinary, medical, industrial, dye, beverage, and
fragrant herbs, and peonies, daylilies, and irises.
Cross-references: 12.118, 12.410, 15.58, 20.109, 20.276

Men's Garden Club of America
Johnston, IA

Cross-reference: 5.2

16.134
Noelridge Park
4900 Council St., N.E.
Cedar Rapids, IA 52402-2497
319-398-5101

Type of garden: Display garden.
Rose garden, annual gardens, All-America Selections
display garden of annuals and vegetables, herbs, and
greenhouses with cacti, succulents, ferns, and tropical
plants.
Cross-reference: 20.110

16.135
Stampe Lilac Garden
City of Davenport Dept. of Parks and Recreation
2816 Eastern Ave.
Davenport, IA 52803
319-326-7812
Location: Locust and Marlo Sts.

Type of garden: Display garden.
Over 170 varieties of lilacs plus spring bulbs, peonies, and
ornamental trees.

ID
16.136
Charles Houston Shattuck Arboretum
205 C.E.B.
University of Idaho
Moscow, ID 83843
208-885-6250

Type of garden: Arboretum.
Giant redwoods, incense cedars, Jeffrey pines, and five
special areas—Asian section, western North America
section, European section, eastern North America
section, and display garden with blossoming plants.
Cross-references: 12.124, 12.411, 15.62

16.137
Idaho Botanical Garden
2355 Old Penitentiary Rd.
P.O. Box 2140
Boise, ID 83712
208-343-8649

Type of garden: Botanical garden.
Herb garden, medicinal plant garden, Japanese garden,
iris garden, and seasonal flowers. Still under
development.

IL
All-America Selections
Downers Grove, IL

Cross-reference: 1.2

All-America Rose Selections
Chicago, IL

Cross-reference: 1.4

16.138
Chicago Botanic Garden
P.O. Box 400
Glencoe, IL 60022
708-835-5440; (FAX) 708-835-4484
Location: Lake-Cook Rd.

Publications: *Garden Talk*, monthly; *Green Connection*,
quarterly.
Type of garden: Botanical garden.
Natural Illinois oak-savannah woodland, tropical
economic, houseplant, and arid desert greenhouse
displays, fruit and vegetable display, and a rose
collection.
Library: 14.37
Seed/specimen exchange.
Other: Horticultural therapy service, Green Chicago
intercity horticulture program.
Cross-references: 2.19, 10.2, 13.19, 13.83, 19.26, 20.116,
20.277

16.139
Garfield Park Conservatory
Chicago Park District
Dept. of Public Information
425 E. McFetridge Dr.
Chicago, IL 60605
312-533-1281
Location: 300 N. Central Park Blvd.

Type of garden: Conservatory.
Over 5,000 varieties and species in a palm house, fernery,
aroid house, cactus house, warm and economic plant
house, and show house. Also seasonal displays.

16.140
George L. Luthy Memorial Botanical Garden
Peoria Park District
2218 N. Prospect
Peoria, IL 61603
309-686-3362; (FAX) 309-686-3352

Publication: *Leaves from the Garden*. quarterly.
Type of garden: Botanical garden.
Conservatory of tropicals, an outdoor herb, rose and
    crabapple garden, and a woody plant collection.

International Ornamental Crabapple Society
Lisle, IL

Cross-reference: 2.58

16.141
Ladd Arboretum
2024 McCormick Blvd.
Evanston, IL 60201
312-864-5181

Type of garden: Arboretum.
Cherry trees, other ornamental trees, and grasses.

16.142
Lilacia Park
Lombard Park District
150 S. Park Ave.
Lombard, IL 60148
312-627-1281
Location: Parkside and Park Aves.

Type of garden: Display garden.
275 varieties of lilacs and tulips.

16.143
Lincoln Park Conservatory
2400 N. Stockton Dr.
Chicago, IL 60614
312-294-4770

Type of garden: Conservatory.
Orchids, palms, cycads, ferns, cacti, and other tropical
    plants.

16.144
Merrick Park Rose Garden
Evanston Parks Dept.
2100 Ridge Ave.
Evanston, IL 60204
312-866-2910
Location: Lake Ave. and Oak St.

Type of garden: Display garden.
One-half-acre rose garden.

16.145
Morton Arboretum
Rt. 53
Lisle, IL 60532
708-968-0074

Publication: *The Morton Arboretum Quarterly*, quarterly.
Type of garden: Arboretum.
Flowering crabapples, magnolias, lilacs, and hedge, rose,
    wildflower and native plant, fragrance, and ground
    cover gardens.

Library: 14.39
Seed/specimen exchange.
Other: Classes and plant clinic for woody plant problems.
Cross-references: 13.86, 15.72, 19.27

16.146
Rivendell Botanic Garden
P.O. Box 17
Beardstown, IL 62618

Type of garden: Botanical garden.
1,200 taxa including major collections of prairie, barren,
    sand, boreal, and limestone glade flora. Open by
    appointment.
Seed/specimen exchange.

16.147
Starhill Forest
R.R. 1, Box 272
Petersburg, IL 62675

Type of garden: Arboretum.
600 woody taxa including native species and wild forms of
    exotics.
Library: Number of books: 1,000; number of periodicals:
    10. Antiquarian books, natural history, 10,000
    horticultural slides.
Lending policy: Reference only, open by appointment.
Seed/specimen exchange.

16.148
Washington Park Botanical Garden
Fayette & Chatham Rds.
P.O. Box 5052
Springfield, IL 62705
217-787-2540

Type of garden: Botanical garden and conservatory.
1,227 species in rose, iris, perennial, chrysanthemum, and
    shade gardens and a conservatory.
Library: 1,008 items.
Lending policy: Reference only.
Cross-reference: 20.120

IN
16.149
Christy Woods of Ball State University
200 W. University Ave.
Muncie, IN 47306
317-285-8838

Type of garden: Botanical garden and nature preserve.
Orchid collection with 7,000 orchids and several rare
    species, display gardens of annuals, perennials, and
    shrubs.
Seed/specimen exchange.
Other: Wheeler Orchid Collection and Species Bank
    offers trading of orchid offshoots of parent plants for
    species and/or hybrids not in collection.

Dwarf Iris Society of America
Ossian, IN

Cross-reference: 2.39

16.150
Garfield Park Conservatory
2450 S. Shelby St.
Indianapolis, IN 46203
317-784-3044

Type of garden: Conservatory.
Tropical, arid, and carnivorous plants.

16.151
Hayes Regional Arboretum
801 Elks Rd.
Richmond, IN 47374
317-962-3745

Type of garden: Arboretum.
Woody plants native to the Whitewater River drainage
 basin in western Ohio and eastern Indiana.
Library: 14.40

16.152
Hillsdale Gardens
Hillsdale Landscape Co.
7845 Johnson Rd.
Indianapolis, IN 46250
317-849-2810
Location: 7800 Shadeland Ave.

Type of garden: Display garden.
Formal rose garden, rose display and test garden, and
 lilacs, tulips, bulbs, annuals, perennials,
 chrysanthemums, a rock garden, and a lily pond.

16.153
Holcomb Botanical Garden
Butler University
4600 Sunset Ave.
Indianapolis, IN 46208
317-283-9413; (FAX) 317-283-9519

Type of garden: Botanical garden.
Lilacs, peonies, rhododendrons, gladioluses, hollies, and
 wildflowers.
Cross-reference: 15.76

16.154
Huntington College Arboretum
2303 College Ave.
Huntington, IN 46750
219-356-6000, ext. 2001

Type of garden: Arboretum.
58-acre arboretum with more than 700 species and
 cultivars including a large collection of medicinal plants
 and 19th-century specimens.

16.155
Purdue University Horticultural Gardens
Dept. of Horticulture
Purdue University
West Lafayette, IN 47907
317-494-1296; (FAX) 317-494-0391

Type of garden: Display garden.
Perennials, annuals, and a demonstration vegetable
 garden.
Cross-references: 12.142, 12.414, 15.79

KS
16.156
Bartlett Arboretum
301 N. Line
Belle Plaine, KS 67013
316-488-3451

Type of garden: Arboretum.
20 acres of trees, shrubs, and grasses and formal gardens
 of annuals, perennials, bulbs, roses, and peonies.

16.157
Botanica, The Wichita Gardens
701 Amidon
Wichita, KS 67203
316-264-0448

Publication: *Seedlings*, quarterly.
Type of garden: Botanical garden.
Shakespearean garden, aquatic collection, Xeriscape
 demonstration garden, rose garden, iris and daylily
 collection, peony collection, juniper collection, and a
 pinetum garden.
Library: 14.41
Cross-reference: 20.124

16.158
Dyck Arboretum of the Plains
Hesston College
P.O. Box 3000
Hesston, KS 67062
316-327-8127

Publication: *Newsletter of the Dyck Arboretum*, quarterly,
 free.
Type of garden: Arboretum.
Wildflowers of the Kansas grasslands and other native
 plants.
Other: Educational programs.

16.159
E.F.A. Reinisch Rose and Test Gardens
Gage Park
4320 West 10th St.
Topeka, KS 66604
913-272-6150

Type of garden: Display and test gardens.
Formal All-America Rose Selections garden with almost
 100 beds, 400 varieties and 6,500 rose plants, including
 old-fashioned roses.
Cross-reference: 20.279

16.160
Kansas Landscape Arboretum
Wakefield, KS 67487
Location: Milford Lake.

Type of garden: Arboretum.
Native trees, introduced varieties, and a wildflower
 garden.

16.161
Kansas State University Gardens
Kansas State University
Manhattan, KS 66506
913-532-6250;
Location: U.S. 24.

Type of garden: Display garden and conservatory.
Annuals, perennials, ground covers, and roses plus cacti,
tropical and semitropical plants in the conservatory.
Cross-references: 12.146, 12.415, 13.22, 13.23, 15.82

16.162
Meade Park Botanic Gardens
124 North Fillmore
Topeka, KS 66606
913-232-5493

Type of garden: Botanical garden.
550 varieties of trees and shrubs and 6,000 annual flowers.

KY
16.163
Bernheim Forest Arboretum
Clermont, KY 40110
502-543-2451
Location: Rt. 245

Type of garden: Arboretum.
Collections of rhododendron, azalea, crabapple,
viburnum, redbud, dogwood, nut tree, and holly, and a
sun and shade garden, hedge garden, fragrance garden,
wildflower garden, and a compact plant garden.

LA
American Rose Society
Shreveport, LA

Cross-reference: 2.26

16.164
Briarwood
Foundation for the Preservation of the Caroline Dormon
Nature Preserve
P.O. Box 226
Natchitoches, LA 71457
318-352-2106
Location: State Hwy. 9

Type of garden: Botanical sanctuary.
Displays of Southern wildflowers.

16.165
Hodges Gardens
P.O. Box 900
Many, LA 71449
318-586-3523

Publication: *Hodges Gardens Leaf-let*, bimonthly.
Type of garden: Display garden and conservatory.
Old roses, herb garden, native plants and flowers, and
tropical and desert plants in the conservatory.
Cross-reference: 20.127

16.166
Ira S. Nelson Horticulture Center
University of Southern Louisiana
Lafayette, LA 70504
318-231-5339
Location: Johnson St.

Type of garden: Display garden and conservatory.
Annuals, perennials, flowering trees, roses and Central
and South American plants in the conservatory.
Cross-references: 12.163, 15.88, 20.280

16.167
Jungle Gardens
General Delivery
Avery Island, LA 70513
318-365-8173; (FAX) 318-369-6326

Type of garden: Display garden.
Azaleas, camellias, water lilies, sunken gardens, Louisiana
lilies, wildflowers, Wasi oranges, bamboo, Chinese
wisteria, aralias, and a Chinese garden.

16.168
Laurens Henry Cohn, Sr. Memorial Arboretum
12056 Foster Rd.
Baton Rouge, LA 70811
504-775-1006

Type of garden: Arboretum and greenhouses.
Herbs, native plants, lily pond, wildflowers, flowering
vines and trees, camellias, and greenhouses with
tropical plants, orchids, and bromeliads.

16.169
Louisiana Purchase Gardens and Zoo
P.O. Box 123
Monroe, LA 71202
318-329-2400
Location: Wilson St. and Tichelli Rd.

Type of garden: Display garden.
Formal gardens and seasonal blossoms.

16.170
Zemurray Gardens
Rt. 1, Box 201
Loranger, LA 70446
504-878-6731; (FAX) 504-878-2284

Type of garden: Display garden.
130-acre azalea garden with thousands of azaleas,
camellias, dogwoods, and other native plants.

MA
16.171
Acton Arboretum
472 Main St.
Acton, MA 01720
508-264-9629

Publication: *Acton Arboretum Newsletter*, quarterly, free.
Type of garden: Arboretum.
A new arboretum (established in 1986) with over 200
native plant specimens.

16.172
Alexandra Botanic Gardens and Hunnewell Arboretum at
  Wellesley
108 Central St.
Wellesley, MA 02181
617-235-0320

Type of garden: Botanical garden, arboretum, and
  conservatory.
15 greenhouses with over 1,000 kinds of plants including
  desert, tropical, subtropical, temperate, and aquatic
  plants. The arboretum and botanical gardens contain
  over 500 species of woody plants in 53 families
  including American white and English oaks, lindens,
  tulip trees, bald cypress, and Chinese golden larch.
  The Jennings Biblical Garden displays many of the
  plants mentioned in biblical texts.

16.173
Arnold Arboretum of Harvard University
125 Arborway
Jamaica Plain, MA 02130
617-524-1718; (FAX) 617-524-1418

Publications: *Arnoldia*, quarterly; *The Journal of the Arnold
  Arboretum*, quarterly, free.
Type of garden: Arboretum.
One of North America's largest collections of hardy trees,
  shrub, and vines—approximately 15,000 specimens on
  265 acres. Has special bonsai, lilac, conifer, Chinese
  and Japanese native tree and shrub, and rosaceous plant
  collections.
Grants: Arnold Associate Awards for research using the
  collection and landscapes of the Arnold Arboretum.
Library: 14.45
Seed/specimen exchange.
Other: Adult education programs.
Cross-references: 8.107, 10.14, 10.15, 10.20, 10.26,
  10.27, 10.31, 10.45, 10.49, 10.59, 10.61, 10.74, 10.76,
  15.95, 16.178

16.174
Ashumet Holly Reservation
Massachusetts Audubon Society
286 Ashumet Rd.
East Falmouth, MA 02536
508-563-6390

Type of garden: Botanical preserve.
45-acre reservation of hollies collected by Wilfred
  Wheeler. Good collections of American, European, and
  oriental hollies.

16.175
Barnard's Inn Farm
Rt. 1, Box 538
Vineyard Haven, MA 02568
508-693-0925

Type of garden: Arboretum.
Approximately 1,400 taxa including rhododendrons,
  dogwoods, camellias, crabapples, and magnolias.
Seed/specimen exchange.

16.176
Berkshire Garden Center
Rtes. 102 & 183
P.O. Box 826
Stockbridge, MA 01262
413-298-3926

Publication: *Cuttings*, bimonthly.
Type of garden: Display garden.
Daylilies, primroses, herbs, dwarf conifers, and
  perennials.
Library: 14.46
Cross-reference: 20.130

16.177
Botanic Garden of Smith College
Lyman Plant House
Northampton, MA 01063
413-585-2748; (FAX) 413-585-2075

Type of garden: Botanical garden and conservatory.
Trees, shrubs, rock garden, herbaceous garden, water
  garden, azalea collection, rhododendron collection, herb
  garden, annual garden, rose garden, and woodland
  wildflower area. Approximately 4,500 taxa. The Lyman
  Plant House contains palms, orchids, ferns, bromeliads,
  and begonias.
Seed/specimen exchange.
Cross-reference: 15.96

16.178
Case Estates of the Arnold Arboretum
135 Wellesley St.
Weston, MA 02193
617-524-1718

Type of garden: Display garden.
Perennial garden, herbaceous beds, rhododendron display
  garden, hosta display garden, and ground cover and
  mulch display.
Cross-reference: 16.173

16.179
Durfee Conservatory
French Hall
Dept. of Plant & Soil Sciences
University of Massachusetts
Amherst, MA 01003
413-545-2242

Type of garden: Conservatory and display garden.
Tropical plants, All-America Selections display garden,
  bonsai, and vegetables.
Cross-references: 12.171, 12.424, 12.425, 13.30, 13.99,
  15.93, 20.131

16.180
Mt. Auburn Cemetery
580 Mt. Auburn St.
Cambridge, MA 02138
617-547-7105

Type of garden: Cemetery.
Rare trees, flowering shrubs, perennials, and annuals on
  170 acres.
New England Wild Flower Society
Framingham, MA
Cross-reference: 3.22

16.181
Stanley Park of Westfield, Inc.
400 Western Ave.
P.O. Box 1191
Westfield, MA 01085
413-568-9312

Type of garden: Display garden and arboretum.
Rose garden, Japanese garden, rhododendron display
  garden, and a flowering garden.
Cross-reference: 20.281

16.182
Walter Hunnewell Pinetum
845 Washington St.
Wellesley, MA 02181
617-235-0422

Type of garden: Pinetum.
Eight acres of hardy coniferous trees. Open by
  appointment.

MD
16.183
Baltimore Conservatory
Druid Hill Park
Gwynns Falls Pkwy. and McCulloh St.
Baltimore, MD 21217
301-396-0180

Type of garden: Conservatory.
Tropicals, aroids, orchids.

16.184
Brookside Gardens
1500 Glenallan Ave.
Wheaton, MD 20902
301-949-8231

Type of garden: Display garden.
50 acres with 2,000 taxa and 10,000 woody specimens
  including the Japanese Collection, azalea garden,
  formal gardens, fragrance garden, and a rose garden.
Library: 14.50
Other: Classes, workshops, and computerized vegetable
  planning service.
Cross-references: 2.19, 20.132

16.185
Cylburn Wild Flower Preserve and Garden Center
4915 Greenspring Ave.
Baltimore, MD 21209
301-367-2217

Type of garden: Arboretum and wildflower preserve.
Magnolias, Japanese maples, All-America Selections
  display garden, tree peonies, herb garden, daylilies,
  shade garden, and wildflowers.
Library: 14.51
Other: Lectures, symposia, and workshops.
Cross-reference: 20.133

Holly Society of America, Inc.
Baltimore, MD
Cross-reference: 2.48

16.186
Ladew Topiary Gardens
3535 Jarrettsville Pike
Monkton, MD 21111
301-557-9570

Publication: *Ladew Letter*, biannual.
Type of garden: Topiary garden.
22 acres with topiary, an herb garden, Japanese garden,
  pink garden, rose garden, wild garden, Victorian
  garden, terrace garden, iris garden, and a water lily
  garden.
Library: Topiary.
Lending policy: Reference only.
Other: Lectures.

16.187
Lilypons Water Gardens
6800 Lilypons Rd.
P.O. Box 10
Lilypons, MD 21717
301-874-5133

Type of garden: Display garden of aquatic nursery.
20 acres of pools of water lilies and lotus.

16.188
McCrillis Gardens
6910 Greentree Rd.
Bethesda, MD 20034
301-469-8438

Type of garden: Display garden.
Azaleas, rhododendrons, bulbs, perennials, and
  ornamental trees and shrubs.

16.189
Perkins Garden
Landon School
6101 Wilson Ln.
Bethesda, MD 20817
301-320-3200; (FAX) 301-320-2787

Type of garden: Display garden.
1,000 varieties of azaleas including 300 original Glen
  Duke varieties, rare tree peonies, and *Pieris*.

Society for Japanese Irises
Upperco, MD
Cross-reference: 2.84

16.190
St. John's Herb Garden, Inc.
7711 Hillmeade Rd.
Bowie, MD 20720
301-262-5302

Type of garden: Herb garden.
Various rare herbs. Open by appointment.
Seed/specimen exchange.
Other: Workshops.

ME

16.191

Asticou Azaleas Garden

c/o Ms. D.B. Strauss

Mt. Desert, ME 04660

207-276-3344

Location: Seal Harbor Dr. & Peabody Dr., Northeast
Harbor, ME

Type of garden: Display garden.

Azaleas, rhododendrons, and non-ericaceous trees and
shrubs.

16.192

Deering Oaks Rose Garden

Dept. of Parks and Public Works

55 Portland St.

Portland, ME

207-775-5451

Location: Park Ave.

Type of garden: Display garden.

Roses.

16.193

Fay Hyland Botanical Plantation

Dept. of Botany

University of Maine

Orono, ME 04469

207-581-7461

Type of garden: Arboretum.

Over 200 species of native plants.

Cross-references: 12.182, 15.103, 20.134

16.194

Kennebec Valley Medical Center

8 St. Catherine St.

Augusta, ME 04330

207-626-1322

Type of garden: Herb garden.

Five beds, each planted with herbs used to treat diseases
of a different body system (the hematologic system, the
digestive system, the cardiorespiratory system, the
nervous system, and the genitourinary system).

16.195

Merryspring

Merryspring Foundation

P.O. Box 893

Camden, ME 04843

207-236-9046

Location: U.S. Rt. 1

Type of garden: Display garden.

Maine native plants, wildflowers, trees, and an herb
garden.

16.196

Thuya Garden

Asticou Terraces Trust

Peabody Dr.

Northeast Harbor, ME 04662

207-276-5130

Type of garden: Display garden.

Semiformal herbaceous perennial garden with native and
introduced species.

Library: 14.54

16.197

Wild Gardens of Acadia

Sieur de Monts Spring

Acadia National Park

Bar Harbor, ME 04609

207-228-3338

Type of garden: Display garden.

Plants native to Mount Desert Island, including
wildflowers.

MI

16.198

Anna Scripps Whitcomb Conservatory

Belle Isle

Detroit, MI 48207

313-267-7133

Type of garden: Conservatory.

One of the largest municipally owned orchid collections,
plus ferns, cacti and succulents, and tropical plants.

16.199

Cooley Gardens

P.O. Box 14164

Lansing, MI 48901

517-351-5707

Location: South Capitol Ave. and Main St.

Publication: *Friends of Cooley Gardens & Other Greenspaces
News*, quarterly.

Type of garden: Display garden.

200 species and varieties of woody plants, and 500 species
and varieties of herbaceous plants including peonies,
tree peonies, and grafted topiary ginkgoes.

16.200

Dow Gardens

1018 West Main St.

Midland, MI 48640

517-631-2677; (FAX) 517-631-0675

Type of garden: Display garden.

Native trees, shrubs, and aquatic plants.

Library: Number of books: 1,000; number of periodicals:
34.

Lending policy: Reference only.

Cross-reference: 20.135

16.201

Fernwood Garden and Nature Center

13988 Range Line Rd.

Niles, MI 49120

616-695-6491

Publication: *Fernwood Newsletter*, monthly; *Fernwood Notes*,
bimonthly.

Type of garden: Arboretum.

40-acre arboretum, 5-acre tallgrass prairie, and display
gardens of herbaceous plants. Ferns, native plants,
herbs, roses, alpine plants, dwarf shrubs, vines, and
hostas.

Scholarships: Local school scholarships for programs at
Fernwood.

Library: 14.57

Seed/specimen exchange.

Other: Educational programs.

Cross-references: 13.32, 20.136

16.202
Hidden Lake Gardens
M-50
Tipton, MI 49287
517-431-2060; (FAX) 517-431-9148
Type of garden: Display garden, arboretum, and
   conservatory.
Dwarf and rare conifers, flowering crabapples, azaleas,
   rhododendrons, and tropical, arid, and temperate plants
   in the conservatory.
Library: 14.58
Other: Seminars.
Cross-reference: 20.137

16.203
Matthaei Botanical Gardens
University of Michigan
1800 North Dixboro Rd.
Ann Arbor, MI 48105
313-998-7061
Publication: *Bartlettia*, irregular.
Type of garden: Botanical garden and conservatory.
Rose garden, herb garden, rock garden, and perennial
   garden, plus cacti and tropicals in the conservatory.
Grants: Grant for graduate student research.
Library: Number of books: 2,000; number of periodicals:
   6.
Lending policy: Members, staff only.
Seed/specimen exchange.
Cross-references: 12.428, 15.111

16.204
Michigan State University
Horticulture Gardens
Michigan State University
East Lansing, MI 48823
517-355-0348
Type of garden: Display and test gardens.
All-America Rose Selections test garden, and annuals,
   perennials, bulbs, and water plants.
Cross-references: 12.185, 12.427, 13.110, 15.110, 20.19,
   20.139, 20.297

Professional Plant Growers Association
Lansing, MI
Cross-reference: 1.110

MN
16.205
Como Park Conservatory
Midway Pkwy. and Kaufman Dr.
St. Paul, MN 55103
612-489-5378
Type of garden: Conservatory.
Palm house, fern room, sunken garden, fruit trees,
   Japanese garden, and a bonsai collection.
Library: Number of books: 500.
Lending policy: Not open to the public.

16.206
Eloise Butler Wildflower Garden and Bird Sanctuary
3800 Bryant Ave. South
Minneapolis, MN 55409
612-348-5702; (FAX) 612-348-9354
Location: Theodore Wirth Pkwy. and Glenwood Ave.

Type of garden: Wildflower garden.
13 acres of over 500 herbaceous and woody plant species
   in prairie, bog, and woodland habitats.
Library: Over 200 books.
Lending policy: Reference only.
Scholarships.

16.207
Lyndale Park Gardens
3800 Bryant Ave. South
Minneapolis, MN 55409-1029
612-348-4448; (FAX) 612-348-9354
Location: E. Lake Harriet Pkwy. and Roseway Rd.

Type of garden: Display garden.
All-America Rose Selections test garden with 62 beds of
   roses and annuals and perennials.
Cross-references: 20.141, 20.282

16.208
Minnesota Landscape Arboretum
3675 Arboretum Dr.
P.O. Box 39
Chanhassen, MN 55317
612-443-2460

Publication: *Minnesota Landscape Arboretum News*,
   bimonthly.
Type of garden: Arboretum.
905 acres. 263 varieties of hosta, Northern Lights azalea
   collection, 5 herb gardens, home demonstration garden,
   Japanese garden, rose garden, and collections of
   wildflowers, pines, and trout lilies.
Library: 14.59
Seed/specimen exchange.
Other: Educational programs and horticultural
   information.
Cross-references: 10.38, 20.142

16.209
Sibley Gardens
Parklane and Given St.
Mankato, MN 56001
507-625-3161

Type of garden: Display garden.
Rose garden, rock garden, flowering shrubs and trees,
   annuals, and spring bulbs.

MO
16.210
Cape Girardeau Rose Test Garden
Perry and Rose St., Capaha Park
Cape Girardeau, MO 63701
314-335-0706

Type of garden: Test garden.
Approximately 750 roses and 50 old garden roses.

16.211
Flower and Garden Demonstration Gardens
4521 Pennsylvania Ave.
Kansas City, MO 64111
816-531-5730
Location: 43rd St.

Type of garden: Display garden.
All-America Selections test garden with annual,
perennial, and vegetable beds, and crabapples and
a rose garden.
Cross-references: 20.144, 21.25

16.212
Jewel Box Conservatory
1501 Oakland Ave.
St. Louis, MO 63122
314-535-1503

Type of garden: Conservatory.
A floral conservatory with hydrangeas, cineraias, azaleas,
poinsettias, and other flora.
Other: Terrarium classes.

16.213
Laura Conyers Smith Municipal Rose Garden
Kansas City Dept. of Parks and Recreation
5605 East 63rd St.
Kansas City, MO 64130
816-561-9710
Location: 5200 Pennsylvania Ave.

Type of garden: Display garden.
4,000 rose bushes in this All-America Selections rose
garden.

16.214
Missouri Botanical Garden
P.O. Box 299
St. Louis, MO 63166
314-577-5100
Location: 4344 Shaw Blvd.

Publications: *Annals of the Missouri Botanical Garden*,
quarterly; *Bulletin of Missouri Botanical Garden*,
bimonthly.
Type of garden: Botanical garden, arboretum, and
conservatory.
Rose garden, scented garden, bulb and rock gardens, iris
and daylily gardens, English woodland garden,
Japanese garden, azalea and rhododendron garden, and
a demonstration vegetable garden.
Library: 14.60
Other: Horticulture information service
Cross-references: 8.108, 15.116, 20.146, 20.283

National Council of State Garden Clubs
St. Louis, MO

Cross-reference: 5.3

16.215
Powell Gardens
Rt. 1, Box 90
Kingsville, MO 64061
816-566-2213

Publication: *Garden Horizons*, bimonthly.
Type of garden: Display garden.
All-America Selections display garden.
Cross-reference: 20.147

16.216
Shaw Arboretum
P.O. Box 38
Gray Summit, MO 63039
314-742-3512
Location: State Highway 100

Type of garden: Arboretum.
2,400 acres including 78 acres of restored prairie, a
55-acre pinetum, a collection of conifers, wildflowers,
and undisturbed bottomland forest.
Seed/specimen exchange.
Other: Classes.

16.217
Woodland and Floral Gardens
I-43 Agriculture Bldg.
University of Missouri
Columbia, MO 65211
314-882-7511
Location: I-70.

Type of garden: Display garden.
Rhododendron garden, rock garden, flower beds, and
native plants.
Cross-references: 12.207, 15.118, 20.51, 20.149

MS
16.218
Crosby Arboretum
1801 Goodyear Blvd.
Picayune, MS 39466
501-798-6961

Publications: *Quarterly News Journal*, quarterly; *Arbor
Connection*, monthly, free.
Type of garden: Arboretum.
Plants native to the Pearl River basin of Mississippi and
Louisiana including 400 species of trees, shrubs,
wildflowers, and grasses, rare orchids, and carnivorous
plants.
Grants: Grants to science teachers.
Library: Number of books: 350; number of periodicals:
12. Photo archive.
Lending policy: Circulating.
Seed/specimen exchange.
Other: Native plant identification.

16.219

Forestry Services Laboratory Arboretum
USDA—Forest Service
Southern Forest Experiment Station
201 Lincoln Green, P.O. Box 906
Starkville, MS 39759
601-324-1611

Type of garden: Arboretum.
Woody species native to south central U.S., over 200
trees and shrubs, and wildflowers and native
herbaceous plants. Open by appointment.

Mississippi Native Plant Society
Starkville, MS

Cross-reference: 3.27

16.220

Mynelle Gardens
4738 Clinton Blvd.
Jackson, MS 39209
601-922-4011

Type of garden: Display garden.
Azaleas, irises, daylilies, annuals, perennials, bulbs, and
flowering trees.
Cross-reference: 20.151

16.221

Wister Gardens
P.O. Box 237
Belzoni, MS 39038
601-247-3025
Location: State Rt. 12.

Type of garden: Display garden.
Spring bulbs, azaleas, flowering fruit trees, bedding
plants, chrysanthemums, camellias, and hollies.

MT
16.222

Gatiss Memorial Gardens
4790 Montana 35
Rt. 5
Kalispell, MT 59901
406-755-2950

Type of garden: Display garden.
Succulents, rock garden plants, shrub and miniature
roses, and perennials.

16.223

Memorial Rose Garden
Missoula Dept. of Parks
100 Hickory St.
Missoula, MT 59801
406-721-7275
Location: 700 Brook St.

Type of garden: Display garden.
All-America Rose Selections display garden with over
1,000 bushes.

16.224

Montana State University Gardens
Dept. of Plant and Soil Science
Montana State University
Bozeman, MT 59717

Type of garden: Display garden.
Native alpines.
Cross-references: 12.213, 12.433, 15.122

NC
16.225

Bicentennial Garden
Greensboro Beautiful, Inc.
Drawer W-2
Greensboro, NC 27402
919-373-2558
Location: Holden and Hobbs Rds.

Type of garden: Display garden.
Spring bulbs, flowering trees, azaleas, camellias,
wildflowers, daylilies, roses, and a fragrance garden for
the visually impaired.

16.226

Campus Arboretum of Haywood Technical College
Freelander Dr.
Clyde, NC 28721
704-627-2821, ext. 269

Type of garden: Arboretum.
Rhododendron garden, dahlia garden, terrace garden,
perennial garden, fruit orchard, spring bulbs, and
summer flowers.
Cross-reference: 12.219

16.227

Coker Arboretum
University of North Carolina
Chapel Hill, NC 27514
919-962-8100

Type of garden: Arboretum.
Native and exotic trees and shrubs, and spring bulbs.
Cross-references: 8.109, 13.33, 13.118, 15.126, 16.232

16.228

Daniel Boone Native Garden
Daniel Boone Park
P.O. Box 295
Boone, NC 28607
704-264-2120
Location: Horn in the West Dr.

Type of garden: Display garden.
Collection of native plants (with emphasis on North
Carolina indigenous plants) including azaleas,
dogwoods, and wildflowers.

16.229

Davidson College Arboretum
P.O. Box 121
Davidson College
Davidson, NC 28036
704-892-2119; (FAX) 704-892-2586

Publication: *Elm Row*, quarterly, free.
Type of garden: Arboretum.
Woody plants of the Piedmont of North Carolina.

16.230
Greenfield Gardens
City of Wilmington
Dept. of Parks
P.O. Box 1810
Wilmington, NC 28402
919-763-9871
Location: Central Wilmington.

Type of garden: Display garden.
Fragrance garden, dogwoods, bald cypresses, and azaleas.

16.231
North Carolina Arboretum
P.O. Box 6617
Asheville, NC 28806
704-665-2492
Location: Bent Creek Experimental Forest

Type of garden: Arboretum.
A new arboretum in 424 acres of National Forest.

16.232
North Carolina Botanical Garden
3375 Totten Center
University of North Carolina—Chapel Hill
Chapel Hill, NC 27599-3375
919-962-0522

Publication: *North Carolina Botanical Garden Newsletter*,
   bimonthly.
Type of garden: Botanical garden.
Southeastern native carnivorous species, native ferns,
   Piedmont forest flora, traditional herbs, and shade,
   economic plant, and wildflower gardens.
Library: 14.61
Seed/specimen exchange.
Cross-references: 8.109, 13.33, 13.118, 15.126, 16.227

16.233
North Carolina State University Arboretum
Dept. of Horticultural Science
Box 7609, NCSU
Raleigh, NC 27695-7609
919-737-3132; (FAX) 919-737-7747

Publication: *Newsletter of the Friends of the NCSU
   Arboretum*, 3/year.
Type of garden: Arboretum.
5,000 species of plants including collections of dwarf
   pine, juniper cultivars, an All-America Selections
   display of annuals, a Japanese garden, and a collection
   of shade plants.
Cross-references: 12.223, 12.435, 15.125, 20.153

16.234
Reynolda Gardens of Wake Forest University
100 Reynolda Village
Winston-Salem, NC 27106
919-759-5593

Publication: *Newsletter*, biannual.
Type of garden: Display garden.
Flowering trees, rose garden, and herb, vegetable, and
   flower gardens.
Cross-reference: 20.154

16.235
Sandhills Horticultural Gardens
Sandhills Community College
2200 Airport Rd.
Pinehurst, NC 28374
919-692-6185, ext. 700

Type of garden: Display garden.
Holly garden with over 400 species, also conifer,
   perennial, annual, fruit and vegetable, formal, and
   informal English gardens.
Scholarships: Scholarships for full-time students in the
   Landscape Gardening School.
Cross-reference: 12.226

16.236
Sarah P. Duke Gardens
Duke University
Durham, NC 27706
919-684-3698

Publication: *FLORA*, biannual, free.
Type of garden: Display garden.
Native plants, Asian collection, roses, cherries,
   crabapples, magnolias, and perennials, and an iris
   garden and a fern glen.
Cross-reference: 15.124

16.237
Tanglewood Park
P.O. Box 1040
Highway 158 West
Clemmons, NC 27012
919-766-0591

Type of garden: Display garden.
Flowering trees, flower beds, and an All-America Rose
   Selections display garden.

16.238
University Botanical Gardens at Asheville
151 W. T. Weaver Blvd.
Asheville, NC 28804
704-252-5190

Type of garden: Botanical garden.
Native flora of the southern Appalachian Mountains
   including over 750 species of wildflowers, a heath cove,
   an azalea garden, and a garden for the visually
   impaired.

16.239
University of North Carolina—Charlotte Botanical
   Gardens
Biology Dept.
UNC Charlotte
Charlotte, NC 28223
704-547-4055

Type of garden: Botanical garden and conservatory.
Native plants of the Carolinas, hybrid rhododendrons, and
   orchids, bromeliads, ferns, and rain forest plants in the
   conservatory.
Library: Number of books: 300.
Lending policy: Reference only.
Other: Plant identification.

16.240
Wing Haven Gardens and Bird Sanctuary
248 Ridgewood Ave.
Charlotte, NC 28209
704-331-0664

Type of garden: Display garden.
Native plants and wildflowers.

ND
16.241
Berthold Public School Arboretum
Box 185
Berthold, ND 58718
701-453-3484

Type of garden: Arboretum.
80 varieties of trees and shrubs.

16.242
International Peace Garden, Inc.
P.O. Box 116, Rt. 1
Dunseith, ND 58329
701-263-4390

Type of garden: Display garden.
Perennials, herbs, orchids, All-America Selections display
    garden, and vegetables.
Cross-reference: 20.157

NE
16.243
Alice Abel Arboretum
Nebraska Wesleyan University
5000 St. Paul Ave.
Lincoln, NE 68504-2796
402-466-2371

Type of garden: Arboretum.
Crabapples, native plants, and wildflowers.

16.244
Antelope Park
Dept. of Parks and Recreation
2740 A St.
Lincoln, NE 68502
402-471-7847
Location: 27th St. and Capitol Pkwy.

Type of garden: Display garden.
Sunken garden with lily ponds, annuals, perennials, and
    ornamental shrubs.

16.245
Chet Ager Nature Center
2470 A St.
Lincoln, NE 68502
402-471-7895

Type of garden: Display garden.
Prairie grasses, wildflowers, and an herb garden.

16.246
Earl Maxwell Arboretum
University of Nebraska
East Campus
Lincoln, NE 68583
402-472-2679
Location: 40th and Holdrege Sts.

Type of garden: Arboretum.
Native grasses and forbs, conifers, oaks, prairie plants,
    and wildflowers.
Cross-references: 12.235, 15.129, 16.247

NE
Nebraska Herbal Society
Lincoln, NE

Cross-reference: 2.69

16.247
Nebraska Statewide Arboretum
112 Forestry Sciences Laboratory
University of Nebraska
Lincoln, NE 68583-0823
402-472-2971

Publications: *The Seed*, quarterly; *Leafings*, bimonthly.
Type of garden: Arboreta umbrella organization. An
    association of 40 arboreta in Nebraska.
Scholarships.
Cross-references: 8.110, 12.235, 15.129, 16.246

16.248
Sallows Conservatory and Arboretum
P.O. Drawer D
City of Alliance
Alliance, NE 69301
308-762-7422
Location: 11th and Nebraska Sts.

Type of garden: Arboretum and conservatory.
Locust, elm, hackberry, black walnut, green ash,
    crabapples, fruit trees, and arid and tropical plants in
    the conservatory.

16.249
University of Nebraska West Central
Research & Extension Center
Rt. 4, Box 46A
North Platte, NE 69101
308-532-3611
Location: U.S. 83.

Type of garden: Demonstration garden.
Woody ornamental plants, especially wildflowers.

NH
16.250
Prescott Park Formal Garden
Cooperative Extension Service
University of New Hampshire
Nesmith Hall
Durham, NH 03824
603-862-1520
Location: Prescott Park

Type of garden: Display garden.
All-American Selections demonstration garden with 600
    varieties of plants.
Cross-references: 12.236, 15.131, 20.162

16.251
Rhododendron State Park
Division of Parks and Recreation
State of New Hampshire
P.O. Box 856
Concord, NH 03301
603-271-3254
Location: State Rt. 119

Type of garden: Display garden.
15 acres of native wild rhododendrons.

NJ
16.252
Colonial Park
R.D. 1, Box 49B
Mettler's Rd.
Somerset, NJ 08873
201-873-2459

Publication: *Friends Digest,* quarterly.
Type of garden: Display garden and arboretum.
4,000 rose bushes including All-American Rose
   Selections, heritage roses, and many others, plus dwarf
   conifers, and a fragrance and sensory collection of herbs
   and perennials.
Library: Number of books: 500; number of periodicals:
   100.
Lending policy: Members and staff only, reference only.
Cross-reference: 20.163

16.253
Cora Hartshorn Arboretum and Bird Sanctuary
324 Forest Dr. South
Short Hills, NJ 07078
201-376-3587

Type of garden: Arboretum and nature preserve.
Maple trees, wildflowers, azaleas, laurels, and
   rhododendrons.

16.254
Deep Cut Park Horticultural Center
352 Red Hill Rd.
Middletown, NJ 07748
201-671-6050

Type of garden: Display garden.
Orchard, vineyard, vegetable garden, herb garden,
   rockery, azalea garden, perennial garden, shade garden,
   and a rose garden. Intended as a demonstration area for
   the home gardener.
Library: 14.62
Other: Classes
Cross-reference: 20.165

16.255
Duke Gardens
State Rt. 206 South
Somerville, NJ 08876
201-722-3700

Type of garden: Conservatory.
11 greenhouses each reflecting a particular style, nation,
   or period, including Italian, Colonial, Edwardian,
   French, English, Chinese, Japanese, Indo-Persian,
   tropical, semitropical, and desert gardens.

16.256
George Griswold Frelinghuysen Arboretum
53 East Hanover Ave.
P.O. Box 1295
Morristown, NJ 07962-1295
201-326-7600; (FAX) 201-644-2726

Publication: *Arboretum Leaves,* bimonthly.
Type of garden: Arboretum.
127 acres of flowering trees, azaleas, rhododendrons,
   peonies, ferns, wildflowers, bulbs, roses, and lilacs.
Scholarships: $1,000 award to a New Jersey under-
   graduate or graduate student majoring in horticulture,
   botany, landscape design, or a related field.
Library: 14.63
Cross-references: 13.36, 13.126, 20.166

16.257
Holmdel Park Arboretum
Monmouth County Shade Tree Commission
17 Lafayette Pl., P.O. Box 1255
Freehold, NJ 07728
201-431-7903; (FAX) 201-409-7564

Publication: *County Shade Tree Commission Newsletter,*
   bimonthly, free.
Type of garden: Arboretum.
Trees, shrubs, and other woody plants including weeping
   ginkgo, Kashmir cedar, and a dwarf evergreen garden.

16.258
Leonard J. Buck Gardens
Somerset County Park Commission
R.D. 2, Layton Rd.
Farills, NJ 07931
201-234-2677

Type of garden: Rock garden.
Rhododendrons, ferns, and wildflowers.
Library: Number of books: 3,000; number of periodicals:
   20. Rock gardening.
Lending policy: Not open to public.

16.259
Presby Memorial Iris Gardens
474 Upper Mountain Ave.
Montclair, NJ 07043
201-783-5974

Type of garden: Display garden.
Over 6,000 iris varieties, including historic irises dating
   back to the 16th century.

16.260
Reeves-Reed Arboretum
165 Hobart Ave.
Summit, NJ 07901
201-273-8787

Type of garden: Arboretum.
Perennial, rose, and herb gardens, plus conifers and
   flowering trees.
Library: Horticultural therapy.
Lending policy: Circulating.
Cross-reference: 13.128

16.261
Rutgers Display Gardens
College of Agriculture and Environmental Science
Rutgers—The State University of New Jersey
New Brunswick, NJ 08903
201-932-9639
Location: Ryders Ln.

Type of garden: Display garden.
Azaleas, rhododendrons, evergreens, hedge and vine
display, shrubs, American and Japanese hollies, and
annuals.
Cross-references: 12.240, 12.437, 15.133, 15.134, 20.52,
20.164

16.262
Skylands Botanical Gardens
Box 1304, Sloatsburg Rd.
Ringwood, NJ 07456
201-962-7031

Type of garden: Botanical garden.
Native trees, shrubs, and wildflowers, and perennial,
lilac, peony, summer, and winter gardens.
Library.
Lending policy: Circulating.
Cross-reference: 20.170

16.263
Tourne Park
Morris County Park Commission
P.O. Box 1295
Morristown, NJ 07960-1295
201-326-7600
Location: McCaffery Ln., Boonton.

Type of garden: Display garden.
Wildflowers, shrubs, and ferns.

16.264
Willowood Arboretum
P.O. Box 1295
Morristown, NJ 07960-1295
201-326-7600
Location: Longview Rd.

Type of garden: Arboretum.
3,500 kinds of native and exotic plants including oaks,
maples, willows, redwoods, ferns, lilacs, magnolias,
cherries, conifers, and wildflowers.

NM
16.265
Greenhouse of the Dept. of Biology
University of New Mexico
Albuquerque, NM 87131
505-277-5100
Location: Yale Blvd.

Type of garden: Greenhouse.
Over 340 species including tropical and subtropical plants,
native plants, and desert plants.
Cross-references: 12.244, 15.136

16.266
Living Desert Zoological and Botanical State Park
P.O. Box 100
1504 Skyline Dr.
Carlsbad, NM 88220
505-887-5516

Type of garden: Botanical garden and zoo.
Chihuahuan Desert native plants, and cacti and
succulents from around the world.

16.267
New Mexico State University
Botanical Garden
Dept. of Agronomy and Horticulture
Box 30003, Dept. 3Q
Las Cruces, NM 88003
505-646-3638

Publication: *Newsletter,* quarterly, free.
Type of garden: Botanical garden.
Plants native to southern New Mexico.
Cross-references: 12.343, 15.135

NV
16.268
Wilbur D. May Arboretum
Rancho San Rafael Park
1502 Washington St.
Reno, NV 89502
702-785-4153

Type of garden: Arboretum.
Energy conservation garden, deciduous garden, Xeriscape
garden, songbird garden, rose garden, fragrant garden,
wet clay garden, rock garden, and native plants.

NY
American Willow Growers Network
South New Berlin, NY

Cross-reference: 2.27

16.269
Bailey Arboretum
Bayville Rd. & Leeks Ln.
Lattingtown
Nassau County
Long Island, NY 11560
516-676-4497

Type of garden: Arboretum.
600 species of trees and shrubs, both exotic and native.

16.270
Bayard Cutting Arboretum
P.O. Box 466
Oakdale, NY 11769
516-581-1002
Location: Montauk Highway, Great River, NY

Type of garden: Arboretum.
130 acres of conifers (both tree form and dwarf),
ericaceous plants, hollies, oaks, and hardy deciduous
shrubs and trees.

16.271
Brooklyn Botanic Garden
1000 Washington Ave.
Brooklyn, NY 11225-1099
718-622-4433; (FAX) 718-857-2430

Publications: *Plants & Gardens News,* quarterly; *Plants & Gardens Handbook,* quarterly.
Type of garden: Botanical garden.
Japanese hill-and-pond garden, oriental flowering cherry trees, roses, magnolias, lilacs, and wisteria.
Library: 14.64
Other: Children's garden, workshops, and classes.
Cross-references: 13.37, 15.138, 20.172

16.272
Buffalo and Erie County Botanical Gardens
South Park Ave. and McKinley Pkwy.
Buffalo, NY 14218
716-828-1040

Type of garden: Botanical garden and conservatory.
Palm dome, ferns, cycads, cactus, and various tropical foliage and flowering plants in 12 greenhouses.
Library: Number of books: 75; number of periodicals: 50.
Lending policy: References only.
Cross-reference: 20.173

16.273
Clark Botanic Garden
193 I. U. Willets Rd.
Albertson, NY 11507
516-621-7568

Publications: *Clark (Botanic) Garden Newsletter,* quarterly; *Clark (Botanic) Garden Horticultural Bulletin,* annual.
Type of garden: Botanical garden.
12-acre garden including rose, wildflower, daylily, children's, and seniors' gardens, and flowering trees, azaleas, roses, spring bulbs, perennials, dwarf conifers, bog plants, vegetables, and herbs.
Library: Number of books: 2,000; number of periodicals: 12. Clippings file from the last 20 years of 15 horticultural serials.
Lending policy: Reference only for members.
Seed/specimen exchange.
Other: Plant information service, workshops on botany and ecology.

16.274
Conservatory Garden
Central Park Conservatory
839 Fifth Ave.
New York, NY 10021
212-860-1330
Location: Fifth Ave. at 104th St.

Type of garden: Display garden.
Over 600 plant specimens including woodland plants, perennial gardens, and a spring bulb display.

16.275
Cornell Plantations
One Plantations Rd.
Ithaca, NY 14850
607-255-3020

Publications: *Cornell Plantations,* quarterly; *Notes,* 5/year.
Type of garden: Arboretum and botanical garden.
Trees and shrubs native to New York State, herbs, cut flowers, perennials, heritage and modern vegetables, international crops and weeds, alpine rock garden plants, peonies, flowering ground covers, rhododendrons, and native plants.
Other: Symposia, and lectures.
Cross-references: 12.249, 12.439, 14.69, 15.142, 15.145, 15.147

16.276
Cutler Botanic Garden
840 Front St.
Binghamton, NY 13905
607-772-8953

Type of garden: Botanical garden.
Rock garden, herb garden, wildflower walk, perennials, annuals, vegetables, and rhododendrons.
Cross-references: 20.174

16.277
Dr. E. M. Mills Memorial Rose Garden
Dept. of Parks
City of Syracuse
412 Spencer St.
Syracuse, NY 13204
315-473-4333
Location: Thornden Park

Type of garden: Display garden.
2,350 rose bushes.

16.278
Enid A. Haupt Glass Garden
Rusk Institute of Rehabilitation Medicine
400 East 34th St.
New York, NY 10016
212-340-6058

Type of garden: Display garden.
Bonsai, orchids, begonias, tropical palms, and an aquatic indoor garden.
Library: Number of books: 100. Horticulture and horticultural therapy.
Lending policy: Reference only.
Other: Horticultural therapy.
Cross-references: 13.38, 13.134

16.279
George Landis Arboretum
Box 186, Lape Rd.
Esperance, NY 12066
518-875-6935

Publication: *G.L.A. Newsletter,* quarterly.
Type of garden: Arboretum.
Pines, firs, conifers, bulbs, lilacs, rhododendrons,
    peonies, ferns, roses, perennials, annuals, and southern
    Appalachian plants.
Library: 14.67
Seed/specimen exchange.

16.280
Highland Botanical Park
180 Reservoir Ave.
Rochester, NY 14620
716-244-8079

Type of garden: Arboretum.
Lilacs, rhododendrons, mature pinetum, and a holly test
    garden.
Library: 14.72
Seed/specimen exchange.

16.281
Holtsville Ecology Site
249 Buckley Rd.
Holtsville, NY 11742
516-758-9664

Type of garden: Display garden.
Annuals, perennials, and tropicals.
Other: Free compost, and gardens for senior citizens and
    apartment dwellers.
Cross-reference: 20.175

16.282
International Bonsai Arboretum
1070 Martin Rd.
West Henrietta, NY 14586-9623
716-334-2595

Publications: *International Bonsai,* quarterly; *Student
    Newsletter,* biannual, free.
Type of garden: Bonsai arboretum.
Hundreds of trees including pines, maples, and elms.
    Open by appointment.

16.283
Jackson Garden and Robinson Herb Garden
Union College
Schenectady, NY 12308
518-370-6262

Type of garden: Display garden.
Herbs, tree peonies, orchid test garden, and over 175
    different specimens of trees.

16.284
Maplewood Rose Garden
County of Monroe Dept. of Parks, County Office Bldg.
39 W. Main St.
Rochester, NY 14614
716-428-5301
Location: Lake Ave.

Type of garden: Display garden.
All-America Rose Selections display garden with 4,500
    roses.

16.285
Mary Flagler Cary Arboretum
Institute of Ecosystem Studies
New York Botanical Garden
Box AB
Mill Brook, NY 12545
914-677-5343; (FAX) 914-677-5976

Type of garden: Arboretum.
Herbaceous perennials and woody plants including ferns
    and a rare collection of hardy trees from Russia.
Library: 14.71
Other: Adult education programs.
Cross-reference: 16.286

16.286
New York Botanical Garden
200th St. and Southern Blvd.
Bronx, NY 10458-5126
212-220-8700; (FAX) 212-220-6504

Publications: *The Botanical Review,* quarterly; *Brittonia,*
    quarterly; *Economic Botany,* quarterly; *Mycologia,*
    bimonthly.
Type of garden: Botanical garden and conservatory.
Special collections of begonias, hardy ferns, daylilies,
    roses, and a rock garden. The Enid A. Haupt
    Conservatory includes palms, plants of economic
    importance, ferns, orchids, bromeliads, and insect
    eating plants.
Library: 14.70
Seed/specimen exchange.
Other: Plant information hotline.
Cross-references: 2.44, 8.111, 12.250, 13.39, 15.143,
    19.62

16.287
Planting Fields Arboretum
Planting Fields Rd.
Oyster Bay, NY 11771
516-922-9206; (FAX) 516-922-0770

Publication: *Friends of Planting Fields Newsletter,* quarterly.
Type of garden: Arboretum.
Collections of azaleas and rhododendrons, dwarf conifers,
    flowering magnolias, dogwoods, cherries, and
    crabapples, and a demonstration garden of hardy
    ornamental shrubs.
Library: 14.73

16.288
Queens Botanical Garden
43-50 Main St.
Flushing, NY 11355
718-886-3800

Publication: *Queens Botanical Garden News*, quarterly.
Type of garden: Botanical garden.
Rose, formal display, bird, and bee gardens.
Library: Number of books: 1,000.
Lending policy: Not open to the public.
Cross-reference: 20.179

16.289
Robin Hill Arboretum
11556 Platten Rd.
Lyndonville, NY 14098
716-765-2614

Type of garden: Arboretum.
Over 450 species of trees, plus wildflowers and flower
    gardens.

16.290
Root Glen
107 College Hill Rd.
Clinton, NY 13323
315-859-7193

Type of garden: Display garden.
Trees, shrubs, annuals, and perennials.

Rose Hybridizers Association
Horseheads, NY
Cross-reference: 2.80

16.291
St. Michael's Episcopal Church Botanical Garden
49 Killean Park
Albany, NY 12205
518-456-2958

Type of garden: Botanical garden and greenhouse.
Trees, shrubs, and wildflower, bog, alpine, and herb
    gardens. Cacti, succulents, and carnivorous plants in
    the greenhouse.

16.292
Staten Island Botanical Garden
1000 Richmond Terrace
Staten Island, NY 10301
718-273-8200; (FAX) 718-442-8534

Publications: *Garden Agenda*, quarterly, free; *Garden
    Newsletter*, quarterly.
Type of garden: Botanical garden.
Collections of orchids, herbaceous perennials, and rose
    and herb gardens.
Library: Number of books: 1,500; number of periodicals:
    24.
Lending policy: Reference only.
Seed/specimen exchange.
Other: Classes.

16.293
Vassar College Arboretum
Vassar College
Raymond Ave.
Poughkeepsie, NY 12601
914-452-7000

Type of garden: Arboretum.
450 acres with many specimens of old trees, a
    Shakespeare garden, and a perennial garden.

World Pumpkin Confederation
Collins, NY
Cross-reference: 2.94

OH
American Daffodil Society, Inc.
Milford, OH
Cross-reference: 2.8

American Ivy Society
West Carrollton, OH
Cross-reference: 2.19

16.294
Ault Park
Cincinnati Park Board
950 Eden Park Dr.
Cincinnati, OH 45202
513-352-4080
Location: Heeking Rd.

Type of garden: Display garden.
Rock garden, Japanese garden, herb garden, annuals,
    perennials, roses, ferns, succulents, and exotics.

16.295
Chadwick Arboretum
Ohio State University
2120 Fyffe Rd.
Columbus, OH 43210
614-292-0473; (FAX) 614-292-3263
Location: Fyffe Rd. and Lane Ave.

Publication: *The Chadwick Arboretum Newsletter*, quarterly.
Type of garden: Arboretum.
More than 2,000 types of ornamental plants including
    pines, dogwoods, junipers, maples, oaks, sycamores,
    crabapples, viburnums, annuals, perennials,
    wildflowers, ferns, and ornamental grasses.
Cross-references: 12.263, 12.442, 15.155, 16.309, 20.284

16.296
Cleveland Cultural Gardens
Westgate Legal Bldg.
20088 Center Ridge Rd.
Rock River, OH 44116
216-664-3103
Location: East and Liberty Blvds.

Type of garden: Display garden.
19 displays each representing a different nationality.

16.297

Cox Arboretum
6733 Springboro Pike
Dayton, OH 45449
513-434-9005

Publication: *Cox Arboretum Foundation News,* quarterly.
Type of garden: Arboretum and greenhouse.
700 different specimens of cacti and succulents, a conifer
 collection, herb garden, and an edible landscape
 garden.
Library: 14.75
Seed/specimen exchange.
Other: "PlantLine" diagnostic service.
Cross-references: 2.21, 13.147

16.298

Dawes Arboretum
7770 Jacksontown Rd., S.E.
Newark, OH 43055
614-323-2355

Publication: *The Dawes Arboretum Newsletter,* monthly.
Type of garden: Arboretum.
1,800 species and cultivars on 950 acres, including
 oak, beech, maple, birch, pine, flowering crabapples, and dogwoods. Also azalea, rhododendron,
 and fern gardens, a Japanese garden, and a
 cypress swamp.
Library: 14.76
Other: Educational programs.
Cross-reference: 20.184

16.299

Falconskeape Gardens
7359 Branch Rd.
Medina, OH 44256
216-723-4966

Publication: *Falconskeape Newsletter,* quarterly.
Type of garden: Display garden.
Over 600 lilac hybrids and select cultivars of the U.S.,
 Canada, Europe, and the U.S.S.R., plus apple trees,
 roses, and peonies.
Library: Number of books: 30; number of periodicals: 6.
 General gardening collection.
Lending policy: Members only.
Other: Workshops and seminars.
Cross-reference: 13.148

16.300

Fellows Riverside Gardens of Mill Creek Park
Mill Creek Metropolitan Park
816 Glenwood Ave.
Youngstown, OH 44502
216-743-7275

Type of garden: Display garden.
Roses, azaleas, and rhododendrons.

16.301

Franklin Park Conservatory and Garden Center
1777 East Broad St.
Columbus, OH 43203
614-222-7447

Type of garden: Conservatory.
Collections of orchids, desert plants, bonsai, a tropical
 forest, and seasonal displays. Over 12,000 square
 feet.
Library: Number of books: 1,000; number of periodicals:
 6.
Books on Orchidaceae.
Lending policy: Reference only.
Cross-reference: 20.186

16.302

Garden Center of Greater Cleveland
11030 East Boulevard
Cleveland, OH 44106
216-721-1600

Publication: *The Garden Center Bulletin,* monthly.
Type of garden: Display garden.
600 species, varieties, and cultivars in Japanese, rose,
 herb, perennial, wildflower, and terrace gardens.
Library: 14.77
Other: Gardening information service, and educational
 programs.
Cross-reference: 13.40, 13.150

16.303

Gardenview Horticultural Park
16711 Pearl Rd.
Strongsville, OH 44136
216-238-6653

Type of garden: English cottage gardens.
500 varieties of flowering crabapples, plus hostas, bulbs,
 perennials, annuals, trees, shrubs, vines, and ground
 covers.
Library: 14.78
Cross-reference: 20.187

Herb Society of America
Mentor, OH

Cross-reference: 1.60

16.304

Holden Arboretum
9500 Sperry Rd.
Mentor, OH 44060
216-256-1110; (FAX) 216-256-1655

Publication: *Arboretum Leaves,* quarterly.
Type of garden: Arboretum.
Rhododendrons, conifers, maples, crabapples, viburnums,
 lilacs, native plants, and nut-bearing trees.
Library: 14.74
Seed/specimen exchange.
Other: Educational programs, and horticultural therapy.
Cross-references: 8.112, 13.41, 13.151

16.305
Inniswood Botanical Garden and Nature Preserve
940 Hempstead Rd.
Westerville, OH 43081
614-895-6216

Type of garden: Botanical garden.
Wildflowers, daylilies, hostas, roses, peonies, irises, and
an herb garden.

16.306
Irwin M. Krohn Conservatory
950 Eden Park Dr.
Cincinnati, OH 45202
513-352-4080

Type of garden: Conservatory.
Over 1,500 varieties of rare and exotic plants, including
tropical and desert gardens, a rain forest house, and a
showcase with six changing seasonal displays.
Other: Gardening hotline.

16.307
Kingwood Center
900 Park Ave. West
Mansfield, OH 44906
419-522-0211

Publication: *Kingwood Center News*, monthly.
Type of garden: Display garden.
Herb, shade, perennial, rose, peony, formal, daylily,
daffodil, lily, and chrysanthemum gardens.
Library: 14.80
Other: Lectures, workshops, and gardening information
service.
Cross-reference: 20.188

16.308
Lakeview Rose Garden
Parks and Recreation Dept.
City of Lorain
329 Tenth St.
Lorain, OH 44052
216-244-0705
Location: W. Erie Ave.

Type of garden: Display garden.
48 rose beds.

16.309
Medicinal Plant Greenhouse
College of Pharmacy
Ohio State University
500 W. 12th Ave.
Columbus, OH 43210
614-292-7753

Type of garden: Greenhouse.
Over 200 medicinal, toxic, and narcotic plants.
Other: The College of Pharmacy also has a medicinal
plant garden that is not open to the public.
Cross-references: 12.263, 12.442, 15.155, 16.295

16.310
Mt. Airy Arboretum
5083 Colerain Ave.
Cincinnati, OH 45223
513-541-8176

Type of garden: Arboretum.
Evergreens, wildflowers, rhododendrons, azaleas, shrubs,
ground covers, vines, herbs, and perennials.

16.311
Secrest Arboretum
Ohio Agricultural Research and Development Center
1680 Madison Ave.
Wooster, OH 44691-4096
216-264-3761

Type of garden: Arboretum.
More than 2,000 species, varieties, and cultivars of trees
and shrubs including rhododendrons, a mixed forest
grove, plantation and specimen conifers, crabapples,
azaleas, hollies, junipers and deciduous shrubs, and a
rose garden.

16.312
Stanley M. Rowe Arboretum
c/o Village of Indian Hill
6525 Drake Rd.
Indian Hill, OH 45243
513-561-6500
Location: 4600 Muchmore Rd.

Type of garden: Arboretum.
Conifers, dwarfs, lilacs, crabapples, beeches, oaks, and
viburnums.
Seed/specimen exchange.
Other: Cutting and scion material exchange.

16.313
Stranahan Arboretum
University of Toledo
33 Birckhead Pl.
Toledo, OH 43608
419-882-6806

Type of garden: Arboretum.
Hardy woody ornamentals.

16.314
Toledo Botanical Garden
P.O. Box 7304
5403 Elmer Dr.
Toledo, OH 43615
419-536-8365

Publication: *Toledo Botanical Garden*, bimonthly.
Type of garden: Botanical garden.
Azaleas, rhododendrons, hosta, wildflowers, ferns,
perennials, ground covers, spring bulbs, deciduous and
evergreen trees and shrubs, and a shade garden, herb
garden, back-yard demonstration garden, and pioneer
garden with heirloom varieties of vegetable and farm
crops typical of the mid-19th century.
Other: Classes, workshops, and a horticultural hotline.

16.315
Wartinger Park
1368 Research Park Dr.
Beavercreek, OH 45385
513-426-5100
Location: Kemp Rd.

Type of garden: Display garden.
Herb garden with over 150 species, old roses, and native
   American trees.

16.316
Whetstone Park of Roses
City of Columbus Recreation & Parks Dept.
4015 Olentangy Blvd.
Columbus, OH 43214
614-645-6648

Type of garden: Display garden.
10,000 rose bushes and other woody ornamentals.
Cross-reference: 2.8

OK
16.317
Honor Heights Park
Muskogee Parks and Recreation Dept.
641 Park Dr.
Muskogee, OK 74403
918-682-6602, ext. 302

Type of garden: Display garden.
Extensive azalea collection, daffodils, dogwoods, and
   redbuds.

16.318
Kirkpatrick Center Museum Complex
2100 N.E. 52nd
Oklahoma City, OK 73111
405-427-5461; (FAX) 405-424-1407

Type of garden: Greenhouse and display garden.
Succulents and cacti, tropical plants, hydroponics, and
   bonsai.

16.319
Myriad Gardens
100 Myriad Gardens
Oklahoma City, OK 73102
405-232-1199
Location: Reno St.

Type of garden: Conservatory and display garden.
Japanese koi and desert and tropical plants in the
   conservatory.

16.320
Tulsa Garden Center
2435 S. Peoria Ave.
Tulsa, OK 74114
918-749-6401

Type of garden: Display garden and conservatory.
Azaleas, irises, rock garden, herb garden, chrysanthemum
   test garden and palms, orchids, cacti, and other
   succulents in the conservatory.
Library: 14.83

16.321
Will Rogers Horticultural Gardens
Dept. of Parks and Recreation
Oklahoma City, OK 73102
405-943-0827
Location: 3500 N.W. 36th St.

Type of garden: Botanical garden, arboreta, and
   conservatory.
Over 1,000 varieties in various specialty gardens including
   a rose garden, iris garden, native tree meadow, rock
   garden, peony garden, azalea trail, an arboretum with
   over 900 species and varieties of woody plants, and
   tropical and subtropical plants in the conservatory.

OR
16.322
Berry Botanic Garden
11505 S.W. Summerville Ave.
Portland, OR 97219
503-636-4112

Publication: *Newsletter*, quarterly.
Type of garden: Botanical garden.
Over 4,000 taxa of alpines, lilies, Northwest natives,
   primula, and rhododendron. Many rare and endangered
   plants.
Library: 3,600 items.
Lending policy: Members only.
Seed/specimen exchange.
Other: Classes.
Cross-reference: 8.113

16.323
Cecil and Molly Smith Garden
5065 Ray Bell Rd.
St. Paul, OR 97137
503-246-3710

Publication: *News from the Cecil and Molly Smith Garden*,
   annual, free.
Type of garden: Display garden.
Owned by the Portland chapter of the American
   Rhododendron Society. 600 rhododendron species and
   hybrids.

16.324
Crystal Springs Rhododendron Garden
S.E. 28th and Woodstock
Portland, OR 97202
503-771-8386

Type of garden: Display garden.
5,000 rhododendrons and azaleas.

16.325
Hendrick's Park Rhododendron Garden
Parks and Recreation Dept.
22 W. 7th Ave.
Eugene, OR 97401
503-687-5334
Location: Summit Ave. and Skyline Dr.

Type of garden: Display garden.
Over 5,000 rhododendrons and azaleas, plus magnolias,
   viburnums, witch hazels, and maple trees.

Home Orchard Society
Clackamas, OR
Cross-reference: 1.62

16.326
Hoyt Arboretum
4000 S.W. Fairview Blvd.
Portland, OR 97221
503-228-8732

Publications: *Hoyt Arboretum Newsletter*, monthly, free;
 Hoyt Arboretum Quarterly Bulletin, quarterly.
Type of garden: Arboretum.
Largest species collection of conifers in the U.S.,
 weeping brewers spruce, sequoias and redwoods,
 Alaskan yellow cedar, and a large collection of
 magnolias, oaks, and maples.
Library: Number of books: 500; number of periodicals:
 18.
Lending policy: Reference only.
Seed/specimen exchange.

16.327
International Rose Test Garden
400 S.W. Kingston Ave.
Portland, OR 97201
503-248-4302

Type of garden: Display garden.
60 old garden roses and over 400 varieties of hybrid tea,
 floribunda, climbers, and miniatures. An official testing
 site for All-America Rose Selections.
Cross-reference: 20.286

16.328
Japanese Garden
Japanese Garden Society of Oregon
P.O. Box 3847
Portland, OR 97208-3847
503-223-4070
Location: S.W. Kingston Ave.

Publication: *The Garden Path*, quarterly.
Type of garden: Japanese garden.
Tea garden, strolling pond garden, and a sand and stone
 garden.
Other: Pruning demonstrations, and ikebana and bonsai
 exhibitions.

16.329
Leach Botanical Garden
6704 S.E. 122nd Ave.
Portland, OR 97236
503-761-9503

Publication: *Leach Botanical Garden*, quarterly.
Type of garden: Botanical garden.
Pacific Northwest native plants, viburnums, camellias,
 and ferns. Over 1,500 species and cultivars.
Library: Number of books: 774; number of periodicals:
 30. Books on native plants of Oregon and the Pacific
 Northwest and botany.
Lending policy: Circulating to members.
Seed/specimen exchange.

16.330
Mount Pisgah Arboretum
Box 5621
Eugene, OR 97405
503-747-3817
Location: Mount Pisgah Rd.

Publication: *Tree Time*, bimonthly.
Type of garden: Arboretum.
Douglas fir, incense cedar, Oregon white oak, and other
 native trees, shrubs, and wildflowers.

16.331
Owens Rose Garden
Parks and Recreation Dept.
22 W. 7th Ave.
Eugene, OR 97401
503-687-5333
Location: Jefferson St.

Type of garden: Display garden.
All-America Rose Selections display garden with 100
 old-fashioned varieties and 160 modern varieties, plus
 Japanese irises.

16.332
Peavy Arboretum
School of Forestry
Oregon State University
Corvallis, OR 97331

Type of garden: Arboretum.
Several hundred species of native trees and exotic woody
 plants.
Cross-references: 12.274, 15.161

16.333
Shore Acres State Park and Botanical Garden
Sunset Bay Park District
13030 Cape Arago Hwy.
Coos Bay, OR 97420
503-888-4902

Type of garden: Display garden.
All-America Rose Selections display garden, Japanese
 garden, lily pond, annuals, perennials, dahlias, and
 hydrangeas.

PA
16.334
Ambler Campus of Temple University
Dept. of Landscape Architecture and Horticulture
Ambler, PA 19002-3994
215-283-1292

Type of garden: Arboretum.
300 species of trees and a formal garden with annuals,
 perennials, and bulbs.

16.335

Arboretum of the Barnes Foundation
P.O. Box 128
Merion Station, PA 19066
215-664-8880; (FAX) 215-664-4026
Location: 57 Lapsley Ln.

Type of garden: Arboretum.
12-acre arboretum with over 2,000 plant specimens.
Special collections of dwarf conifers, magnolias,
viburnums, cotoneasters, and peonies.
Library: 14.87

16.336

Awbury Arboretum
Francis Cape House
Philadelphia, PA 19138
215-849-2855
Location: Chew Ave. at Washington Ln.

Type of garden: Arboretum.
57-acre arboretum with an English-style landscape.

16.337

Botanical Society of Western Pennsylvania
Pittsburgh, PA

Cross-reference: 3.39

16.338

Bowman's Hill Wildflower Preserve
Washington Crossing Historic Park
P.O. Box 103
Washington Crossing, PA 18977
215-862-2924

Publication: *Twinleaf Newsletter*, quarterly.
Type of garden: Botanical preserve.
Thousands of specimens representing approximately 750
species of Pennsylvania's native wildflowers, ferns,
vines, trees, and shrubs on 100 acres.
Library: Number of books: 500.
Lending policy: Reference only, by appointment.
Seed/specimen exchange.

16.339

Brandywine Conservancy
P.O. Box 141
Chadds Ford, PA 19317
215-388-7601
Location: Rt. 1, Downtown Chadds Ford.

Type of garden: Three-acre wildflower garden.
Seed/specimen exchange.

16.340

Campus Arboretum of Haverford College
Haverford College
Haverford, PA 19041
215-896-1101; (FAX) 215-896-1224

Publication: *Arboretum Newsletter*, quarterly.
Type of garden: Arboretum.
1,200 specimen trees including flowering dogwood,
Hinoki cypress, loblolly pine, Osage orange, old oaks,
and a pinetum.
Library: Number of books: 200; number of periodicals:
15. Trees, wildflowers, houseplants, ferns.
Lending policy: Circulating.

16.341

Fairmount Park Horticultural Center
Horticulture Dr. near Belmont Ave.
Fairmount Park
Philadelphia, PA 19131
215-686-1776, ext. 81287

Type of garden: Conservatory.
Seasonal displays in the conservatory.

16.342

Friends Hospital
4641 Roosevelt Blvd.
Philadelphia, PA 19124
215-831-4686

Type of garden: Display garden.
Azaleas, roses, dogwoods, and flowers.
Cross-references: 13.44, 13.161

16.343

Henry Foundation for Botanical Research
Box 7, 801 Stoney Ln.
Gladwyne, PA 19035
215-525-2037

Type of garden: Botanical garden.
Native American species including rhododendrons,
magnolias, lilies, and hollies.
Fellowship.
Other: Classes.

16.344

Henry Schmieder Arboretum of Delaware Valley College
Rt. 202 and New Britain Rd.
Doylestown, PA 18901
215-345-1500

Publication: *Henry Schmieder Arboretum Newsletter*, annual,
free.
Type of garden: Arboretum.
Old sassafras, sycamore, herbs, azaleas, rhododendrons,
dwarf conifers, bulbs, wildflowers, a hedge
demonstration garden, and orchids.
Other: Horticultural information service and lectures.
Cross-references: 12.246, 13.43

16.345

Hershey Gardens
P.O. Box B.B., Hotel Rd.
Hershey, PA 17033
717-534-3493

Type of garden: Display garden and arboretum.
Three-acre rose garden with old garden varieties,
miniatures, and others, plus dwarf conifers, Japanese
maples, herbs, tulips, azaleas, dogwoods, cherries, and
magnolias.
Cross-reference: 20.287

16.346

Jenkins Arboretum
631 Berwyn Baptist Rd.
Devon, PA 19333
215-647-8870

Type of garden: Arboretum.
Extensive collection of rhododendron species and hybrids
including native American deciduous azaleas, evergreen
azaleas, hepidote and elepidote rhododendrons, and
native flora.

16.347

John J. Tyler Arboretum
515 Painter Rd.
P.O. Box 216
Lima, PA 19037
215-566-5431

Publication: *Tyler Topics*, 3/year.
Type of garden: Arboretum.
Pines, magnolias, crabapples, cherries, dwarf conifers, a
fragrance garden and a bird habitat garden. 19
specimen trees planted by Jacob and Minshall Painter
during 1840-1860 including a giant sequoia and a cedar
of Lebanon.
Other: Educational programs, workshops, and lectures.

16.348

Longwood Gardens
Rt. 1, P.O. Box 501
Kennett Square, PA 19348-0501
215-388-6741

Type of garden: Botanical garden, arboretum, and
conservatory.
350 acres of formal gardens, 4 acres of conservatories, and
a topiary. Dwarf conifer, rose, rock, heather, herb,
vegetable, wildflower, and annual gardens, and a lily
pond. Arboretum includes numerous trees including
ginkgo, paulownia, Kentucky coffee, larch, and
California incense-cedars.
Library: 14.88
Other: Public education.
Cross-references: 2.19, 10.54, 20.201

16.349

Malcom W. Gross Memorial Rose Garden
Dept. of Parks
2700 Parkway Blvd.
Allentown, PA 18104
215-437-7627
Location: 16th and Linden Sts.

Type of garden: Display garden.
2,500 rose bushes representing 60 varieties.

16.350

Masonic Homes Arboretum
Elizabethtown, PA 17022
717-367-1121

Type of garden: Arboretum.
Hollies, oaks, evergreens, and weeping evergreens.

16.351

Morris Arboretum of the University of Pennsylvania
9414 Meadowbrook Ave.
Philadelphia, PA 19118
215-247-5777

Publication: *Morris Arboretum Newsletter*, bimonthly.
Type of garden: Arboretum.
Special collections of roses, holly, magnolia, native
azaleas, witch hazel, and woody plants of temperate
Asia and eastern North America. Also *Metasequoia* and
Cedrela groves.
Library: 14.89
Seed/specimen exchange.
Other: Adult education, plant clinics, and lectures.
Cross-references: 12.448, 15.166

16.352

Phipps Conservatory
Dept. of Parks
400 City County Bldg.
Pittsburgh, PA 15219
412-255-2370
Location: Schenley Park.

Type of garden: Conservatory.
Tropical jungle display, a water garden, Japanese garden,
orchids, and seasonal displays.

16.353

Scott Arboretum
500 College Ave.
Swarthmore College
Swarthmore, PA 19081
215-328-8025

Publication: *Hybrid*, quarterly.
Type of garden: Arboretum.
Over 5,000 kinds of plants including rhododendrons,
flowering cherries, hollies, and crabapples, plus rock
and woodland gardens.
Library: 14.92
Other: Educational programs.

16.354

Swiss Pines
Charlestown Rd.
R.D. 1, Box 127
Malvern, PA 19355
215-933-6916

Type of garden: Display garden.
Japanese garden, pinetum, rose garden, herb garden, and
crabapples, azaleas, heather, and rhododendrons.

16.355

Taylor Memorial Arboretum
10 Ridley Dr.
Wallingford, PA 19086
215-876-2649; (FAX) 215-353-0517

Type of garden: Arboretum.
Lilacs, magnolias, dogwoods, hollies, azaleas, Southern
oaks, and viburnums.

RI
16.356
Green Animals Topiary Garden
380 Cory's Ln.
Portsmouth, RI 02871
401-683-1267

Type of garden: Display garden.
80 topiary plants, annuals, perennials, boxwood parterre, fruits, and vegetables.
Cross-reference: 20.207

16.357
Roger Williams Park Greenhouse and Gardens
Dept. of Public Parks
Providence, RI 02905
401-785-9450
Location: Broad St.

Type of garden: Greenhouses and display garden.
Rose, Japanese, and flower gardens and seasonal displays in the greenhouses.

16.358
Wilcox Park
71 ½ High St.
Westerly, RI 02891
401-348-8362

Type of garden: Display garden.
Trees (including dwarf conifer collection), shrubs, vines, perennials, bulbs, annuals, and herbs.
Library: Number of books: 200; number of periodicals: 9.
Lending policy: Reference only.
Other: Lectures and workshops.

SC
16.359
Brookgreen Gardens
U.S. 17 South
Murrells Inlet, SC 29576
803-237-4218

Publications: *Brookgreen Journal*, quarterly; *Brookgreen Newsletter*, quarterly.
Type of garden: Display garden.
Over 2,000 kinds of plants including oaks, azaleas, magnolias, hollies, and wildflowers.
Library: Number of books: 5,000; number of periodicals: 200. Flora, fauna, and natural science.
Lending policy: Closed to the public, reference only.
Seed/specimen exchange.
Other: Classes, workshops, and seminars.

16.360
Cypress Gardens
Dept. of Leisure Services
Hampton Park
Charleston, SC 29403
803-553-0515
Location: U.S. 52.

Type of garden: Display garden.
Azalea garden, woodland garden, camellia garden, wisteria, and thousands of bulbs.

16.361
Edisto Memorial Gardens
Dept. of Parks
P.O. Box 1321
Orangeburg, SC 29115
803-534-6376
Location: U.S. 301.

Type of garden: Display garden.
Azaleas, camellias, and a rose garden with 100 varieties.
Cross-reference: 20.289

16.362
Glencairn Garden
Dept. of Parks and Recreation
City of Rock Hill
P.O. Box 11706, 155 Johnson St.
Rock Hill, SC 29731
803-329-5620
Location: U.S. 21.

Type of garden: Display garden.
Azaleas, flowering trees, wisteria, daffodils, and periwinkles.

16.363
Hampton Park
Charleston, SC 29403
803-724-7321
Location: Rutledge Ave. and Moultrie St.

Type of garden: Display garden.
Roses, crape myrtles, magnolias, and azaleas.

16.364
Hopeland Gardens
City of Aiken
P.O. Box 1177
Aiken, SC 29801
803-755-5811
Location: Whiskey Rd. and Mead Ave.

Type of garden: Display garden.
Dogwoods, azaleas, camellias, wisteria, lilies, magnolias, crape myrtles, and roses.

16.365
Kalmia Gardens of Coker College
1624 W. Carolina Ave.
Hartsville, SC 29550
803-332-1381

Type of garden: Botanical garden.
Native pines, oaks, hollies, laurel, and herbaceous species along with plantings of camellias, azaleas, and other ornamentals.
Other: Educational programs.

16.366
South Carolina Botanical Garden at Clemson
Dept. of Horticulture
Clemson University
Clemson, SC 29634-0375
803-656-4949

Type of garden: Botanical garden.
Rare and endangered wildflowers and collections of holly, rhododendron, camellia, and azalea.
Cross-references: 12.286, 12.451, 15.176, 20.209

16.367
Swan Lake Iris Gardens
Dept. of Parks
Sumter, SC 29150
803-775-5811
Location: W. Liberty St.

Type of garden: Display garden.
Extensive collection of Japanese irises, plus azaleas, camellias, lotus, jasmine, gardenias, and wisteria.

SD
16.368
McCrory Gardens
South Dakota State University
Brookings, SD 57007
605-688-5136

Type of garden: Display garden and arboretum.
20 acres consisting of 17 theme gardens and 10 special gardens and a 45-acres arboretum. Hosta, gladiolus, sedum, lily, iris, peony, daylily, roses, native plants, and ornamental grasses.
Cross-references: 12.290, 12.452, 15.179, 20.211

16.369
McKennon Park
Dept. of Parks and Recreation
City of Sioux Falls
600 E. 7th St.
Sioux Falls, SD 57102
605-339-7060
Location: 26th and Phillips Sts.

Type of garden: Display garden.
Formal gardens with annuals and perennials and an Italian Renaissance garden.

TN
16.370
Memphis Botanic Garden
750 Cherry Rd.
Memphis, TN 38117
901-685-1566; (FAX) 901-325-5770

Publication: *The Garden Appeal*, monthly.
Type of garden: Botanical garden and arboretum.
20 different plant collections including an iris garden, wildflower garden, fern glen and gardens of cacti, perennials, conifers, dahlias, and daylilies.
Scholarships: Two scholarships to the Dept. of Ornamental Horticulture and Landscape Design at the University of Tennessee, Knoxville. (12.299).
Library: 14.94
Other: Horticultural information service.
Cross-reference: 20.212

16.371
Opryland Hotel Conservatory
2802 Opryland Dr.
Nashville, TN 37214
615-899-6600

Type of garden: Conservatory.
10,000 plants representing 216 species.
Cross-reference: 20.213

16.372
Racheff Park and Gardens
Tennessee Federation of Garden Clubs, Inc.
6905 Cresthill Dr.
Knoxville, TN 37919
Location: 1943 Tennessee Ave.

Type of garden: Display garden.
Spring bulbs, wildflowers, a butterfly garden, and a cutting garden.

16.373
Reflection Riding
400 Garden Rd.
Chattanooga, TN 37419
615-821-9582

Type of garden: Arboretum and wildflower preserve.
Native American plants of economic importance, butterfly garden, viburnum collection, and wildflowers including a large colony of blue-eyed Mary.
Library: Number of books: 400; number of periodicals: 8.
Lending policy: Reference only.

16.374
Southwestern Arboretum
Southwestern College of Memphis
200 North Pkwy.
Memphis, TN 38112
901-458-0964

Type of garden: Arboretum.
Trees and shrubs of the South.

16.375
Trial Gardens
University of Tennessee
Dept. of Ornamental Horticulture and Landscape Design
P.O. Box 1071
Knoxville, TN 37901-1071
615-974-7324

Type of garden: Trial and demonstration gardens.
Herbs, perennials, annual flowers, and fall garden mums; a shrub collection is being developed.
Cross-references: 12.299, 12.453, 15.181, 20.28, 20.214

16.376
University of Tennessee Arboretum
901 Kerr Hollow Rd.
Oak Ridge, TN 37830
615-483-3571

Publication: *The Leaflet*, bimonthly.
Type of garden: Arboretum.
Korean hollies test garden, rhododendrons, azaleas, magnolias, willows, and pines.
Library: Number of books: 100; number of periodicals: 20.
Lending policy: Reference only.

TX
16.377
Austin Area Garden Center
2200 Barton Springs Rd.
Austin, TX 78746
512-477-8672

Type of garden: Display garden.
Oriental garden, rose garden, Xeriscape garden, herb garden, perennial garden, and a pioneer garden with flowers and vegetables grown with the methods of the pioneers.

16.378
Brazos County Arboretum
Dept. of Horticulture
Texas A&M University
College Station, TX 77843
713-845-2844

Type of garden: Arboretum.
Native plants of central Texas.
Cross-references: 12.315, 12.454, 13.49, 15.188, 20.29, 20.223

16.379
Cactusland
Rt. 3, Box 44
Edinburg, TX 78539
512-383-2996
Location: U.S. 281.

Type of garden: Display garden.
20-acre cactus garden with many varieties of cacti.

Chihuahuan Desert Research Institute
Alpine, TX

Cross-reference: 8.21

16.380
Dallas Arboretum and Botanical Garden
8617 Garland Rd.
Dallas, TX 75218
214-327-3990

Type of garden: Arboretum and botanical garden.
English-style perennial garden with 500 varieties, and 200 varieties of trees and shrubs, 2,000 varieties of azaleas, and seasonal color.
Cross-reference: 20.216

16.381
Dallas Civic Garden Center
P.O. Box 26194
Fair Park
Dallas, TX 75226
214-428-7476
Location: Martin Luther King, Jr. Blvd.

Type of garden: Display garden and conservatory.
Herb and scent garden for the visually impaired, Shakespeare garden, rose garden, water garden, bog garden, azaleas, wildflowers, native trees and shrubs, annuals, and bulbs and tropical plants in the conservatory.
Library.
Cross-reference: 20.217

16.382
Fort Worth Botanical Garden
3220 Botanic Garden Blvd.
Fort Worth, TX 76107
817-870-7686

Publication: *Redbud*, 5/year.
Type of garden: Botanical garden.
Begonias, rose garden, tropicals, Texas native plants, fragrance garden, Japanese garden, and a perennial garden.
Library: 14.95
Other: Workshops and public horticultural information service.
Cross-references: 13.182, 20.218

16.383
Hilltop Herb Farm
Chain-O-Lakes
P.O. Box 218
Romayor, TX 77368
713-592-2150
Location: Daniel Ranch Rd.

Type of garden: Display garden.
Many varieties of edible herbs, medicinal herbs, and ornamental herbs, and roses and flowering ornamentals.

16.384
Houston Arboretum and Nature Center
4501 Woodway Dr.
Houston, TX 77024
713-681-8433

Publication: *Houston Arboretum and Botanical Society Newsletter*, bimonthly.
Type of garden: Arboretum.
Native trees and shrubs.
Library: Number of books: 500. General botanical references with a concentration on literature dealing with the biota of the Southwest.
Lending policy: Reference only.
Other: Classes.

16.385
Houston Garden Center
1500 Hermann Dr.
Houston, TX 77004
713-529-5371

Type of garden: Display garden.
Rose garden, herb garden, wildflower garden, perennials, and bulbs.
Cross-reference: 20.219

16.386
McMurry Gardens
McMurry College
Abilene, TX 79697
915-692-3933
Location: Sales Blvd.

Type of garden: Display garden.
Over 650 varieties of irises.

16.387
Mercer Arboretum and Botanic Gardens
22306 Aldine Westfield Rd.
Humble, TX 77338
713-443-8731

Publication: *Parkscape*, quarterly, free.
Type of garden: Arboretum and botanical garden.
Herb garden, fern garden, iris bog, tropical garden, azalea
walk, and a perennial garden.
Library: Number of books: 2,000; number of periodicals:
10. Texas flora and bamboos.
Lending policy: Circulating.
Cross-reference: 8.114

National Wildflower Research Center
Austin, TX
Cross-reference: 1.97

16.388
Samuel Grand Park
Dept. of Parks and Recreation
City of Dallas
1500 Marilla
Dallas, TX 75201
214-670-4100
Location: 5800 Winslow St.

Type of garden: Display garden.
Rose garden with 300 varieties, azaleas, annuals,
perennials, and irises.

16.389
San Antonio Botanical Gardens
555 Funston Pl.
San Antonio, TX 78209
512-821-5115

Type of garden: Botanical garden and conservatory.
Native Texas plants, plus rose, old-fashioned, sacred, and
herb gardens, a garden for the visually impaired, a
Xeriscape demonstration garden, and a conservatory
with tropical and desert plants.
Cross-references: 8.115, 20.222

16.390
Tyler Municipal Rose Garden
Tyler Parks Dept.
P.O. Box 7039
Tyler, TX 75710
214-531-1212
Location: 420 S. Rose Park Dr.

Type of garden: Display garden and greenhouse.
One of the largest rose gardens in North America with
30,000 bushes and 500 varieties including modern and
heritage roses. Also camellias and a greenhouse with
cycads, bromeliads, ferns, and anthuriums.
Cross-reference: 20.290

UT
16.391
International Peace Gardens
10th South St. at 9th West Ave.
Salt Lake City, UT 84105

Type of garden: Display garden.
Gardens representing 18 nations and Africa with each
garden following a design specific to the country.

16.392
State Arboretum of Utah
Building 436
University of Utah
Salt Lake City, UT 84112
801-581-5322

Publication: *Cultivator*, quarterly.
Type of garden: Arboretum, botanical garden, and
conservatory.
Collections of dwarf conifers, daylilies, irises, beech, plus
rose and cactus gardens. Conservatory contains orchids,
cacti and other succulents, and economic plants.
Cross-references: 8.116, 13.50, 15.194

16.393
Temple Square
50 East North Temple
Salt Lake City, UT 84150
801-240-1000

Type of garden: Display garden.
Roses, irises, bulbs, and annuals.

16.394
Utah Botanical Garden
1817 North Main
Farmington, UT 84025
801-451-3204

Publication: *Green Thumb Connection*, monthly, free.
Type of garden: Botanical garden.
Native plant display garden, collections of irises, roses
(including heritage roses), clematis, and astilbe, a
perennial garden, and vegetables, fruits, and herbs.
Cross-reference: 20.225

VA
American Horticultural Society
Alexandria, VA
Cross-reference: 1.16

16.395
Bryan Park Azalea Gardens
Dept. of Parks
City of Richmond
900 E. Broad St.
Richmond, VA 23219
804-780-8785
Location: Bellevue Ave.

Type of garden: Display garden.
Over 45,000 azaleas plus camellias, crabapples, dogwoods,
crape myrtles, magnolias, and hollies.

16.396
Green Spring Farm Park
4603 Green Spring Rd.
Alexandria, VA 22312
703-642-5173

Type of garden: Display garden and greenhouse.
Rock garden, butterfly garden, native plant trail,
vegetable garden, fruit garden, herb garden, rose
garden, perennials, and irises.
Other: Classes

16.397
James Madison University Arboretum
University Blvd.
James Madison University
Harrisonburg, VA 22807
703-568-6340

Type of garden: Arboretum.
A new arboretum with oak and hickory species, woody
and herbaceous plants, and many native wildflower
species of the mid-Appalachians.

16.398
Lewis Ginter Botanical Garden
7000 Lakeside Ave.
P.O. Box 28246
Richmond, VA 23228-4610
804-262-9887

Type of garden: Botanical garden.
Thousands of species including seasonally changing floral
displays, annual and perennial borders and test gardens,
daylilies, rhododendrons, azaleas, and daffodils.

16.399
Meadowlark Gardens Regional Park
Northern Virginia Regional Park Authority
5400 Ox Rd.
Fairfax, VA 22039
703-352-5900; (FAX) 703-273-0905
Location: 1624 Beulah Rd., Vienna, VA

Publications: *The Meadowlark Gardens Newsletter*, quarterly,
free.
Type of garden: Display garden.
Herb garden, Siberian irises, daylilies, hostas, daffodils,
flowering cherries, lilacs, and azaleas.

National Chrysanthemum Society, Inc.
Fairfax Station, VA

Cross-reference: 2.66

16.400
Norfolk Botanical Gardens
Norfolk, VA 23158
804-853-6972
Location: Airport Rd.

Type of garden: Botanical garden.
Azaleas, camellias, rhododendrons, tulips, annuals, and a
scented garden.

Library: Number of books: 1,950; number of periodicals:
13.
Lending policy: Members only, reference only.
Cross-references: 13.51, 20.231

16.401
Orland E. White Arboretum
Blandy Experimental Farm
P.O. Box 175
Boyce, VA 22620
703-837-1758

Publication: *Arbor Vitae: Newsletter of the Friends of the State
Arboretum*, quarterly.
Type of garden: Arboretum.
43 species of pines, a grove of ginkgo trees, beech,
flowering Chinese tulip trees, and a boxwood memorial
garden.
Grants: Four research grants awarded each summer, two
to postdoctoral or university faculty and two to
undergraduate or graduate students for research in the
area of ecological interaction between plants and
animals, plant ecology, or insect ecology.
Library: 14.98
Seed/specimen exchange.
Other: Lectures and workshops.
Cross-reference: 2.4

Vinifera Wine Growers Association
The Plains, VA

Cross-reference: 1.121

16.402
Virginia Tech Arboretum
Arboretum Committee
Virginia Polytechnic Institute and State University
Blacksburg, VA24601-0324
703-231-5943

Type of garden: Arboretum.
Established in 1989, this new arboretum includes
trees, shrubs, herbaceous plants, and perennial
beds.
Cross-references: 12.326, 12.460, 13.52, 15.201,
20.232

16.403
Winkler Botanical Preserve
4900 Seminary Rd.
9th Floor
Alexandria, VA 22311
703-578-7888
Location: 5300 Roanoke Ave.

Type of garden: Botanical preserve.
Native plants, orchids, and ferns and fern allies.

VT
Terrarium Association
Newfane, VT

Cross-reference: 1.118

16.404
University of Vermont Display Gardens
Dept. of Plant and Soil Science
Hills Bldg., University of Vermont
Burlington, VT 05405
802-656-2630; (FAX) 802-656-0285
Location: Horticultural Research Center, Rt. 7 South, South Burlington.

Type of garden: Display garden.
Hardy woody ornamentals, herbaceous perennials, and an All-America Selections flower display garden.
Cross-references: 12.328, 15.202

WA
16.405
Bloedel Reserve
7571 N.E. Dolphin Dr.
Bainbridge Island, WA 98110-1097
206-842-7631

Type of garden: Botanical preserve and display garden.
Japanese garden, moss garden, reflection garden, glen, and woods on 150 acres.

16.406
Carl S. English, Jr. Gardens
Hiram M. Chittenden Locks
3015 N.W. 54th St.
Seattle, WA 98107
206-783-7059

Type of garden: Display garden.
Over 1,000 trees, shrubs, and herbaceous plants from around the world, including collections of heather, azaleas and rhododendrons, magnolias, Japanese flowering cherries, oaks, maples, and ginkgoes. Also seasonal flower beds.

Center for Urban Horticulture
Seattle, WA

Cross-reference: 1.40

Hardy Fern Foundation
Richmond Beach, WA

Cross-reference: 2.44

16.407
Herbfarm
32804 Issaquah-Fall City Rd.
Fall City, WA 98024
206-784-2222

Type of garden: Display garden of retail nursery.
17 herbal theme gardens including a pioneer garden, shade garden, edible flower garden, thyme lawn, and everlasting flower garden.
Other: Classes.

16.408
John A. Finch Arboretum
W. 3404 Woodland Blvd.
Spokane, WA 99204
509-747-2894

Type of garden: Arboretum.
Over 600 species including rhododendrons, azaleas, lilacs, flowering shrubs, flowering crabapples, hawthorns, conifers, and willows.

16.409
Manito Park and Botanical Gardens
Spokane Parks and Recreation Dept.
West 4-21st Ave.
Spokane, WA 99203
509-456-4331
Location: South Grand Blvd.

Type of garden: Display garden and conservatory.
Old-fashioned rose garden, formal garden, Japanese garden, and bulbs, primroses, perennials, and chrysanthemums plus tropical and foliage plants in the conservatory.

16.410
Master Gardener Demonstration Garden
2543 California Ave. East
Port Orchard, WA 98366
206-871-3344
Location: Fairgrounds Rd. and Tracyton Blvd., Silverdale, WA

Type of garden: Demonstration garden.
Alpine plants, tree fruits, small fruits and berries, vegetables, annuals, and perennials.
Other: Plant diagnostic clinic.

16.411
Meerkerk Rhododendron Garden
P.O. Box 154
3531 Meerkerk Ln.
Greenbank, WA 98253
206-321-6682

Publication: *Goings On at Meerkerk*, monthly.
Type of garden: Display garden.
Large variety of rhododendron species and hybrids.
Library: Number of books: 1,500; number of periodicals: 4.
Lending policy: Open to members of the American Rhododendron Society.

16.412
Ohme Gardens
3327 Ohme Rd.
Wenatchee, WA 98801
509-662-5785

Type of garden: Display garden.
Nine-acre alpine garden with candytufts, dianthus, creeping phlox, ajuga, and baskets-of-gold.

16.413
Pacific Rim Bonsai Collection
P.O. Box 3798
Federal Way, WA 98063-3798
206-661-9377
Location: Weyerhauser Corporate Headquarters.

Type of garden: Bonsai garden.
More than 50 bonsai trees from Canada, Taiwan, Japan, China, and the U.S., some over 500 years old.

16.414

Point Defiance Park
5402 N. Shirley Ave.
Tacoma, WA 98407
206-591-5328

Type of garden: Display garden.
Annual display garden, rose garden, Japanese garden,
    rhododendron garden, dahlia test garden, camellia
    garden, and a native plants garden.

Rhododendron Species Foundation
Federal Way, WA

Cross-reference: 2.79

16.415

University of Washington Medicinal Herb Garden
University of Washington
Dept. of Botany, KB-15
Seattle, WA 98195
206-543-1942

Type of garden: Display garden.
The largest garden of its kind in North America with over
    600 common and rare species of medicinal herbs.
Cross-references: 1.40, 12.338, 12.461, 15.205

16.416

W. W. Seymour Botanical Conservatory
316 G St.
Tacoma, WA 98405
206-591-5330

Type of garden: Conservatory and display garden.
Victorian-style conservatory built in 1908 with orchids,
    cacti, tropical plants, and seasonal displays. Also an
    All-America Selections display garden with annuals on
    the outside.
Cross-reference: 20.238

16.417

Washington Park Arboretum
2300 Arboretum Dr. East
Seattle, WA 98112
206-543-8616

Type of garden: Arboretum.
Hundreds of varieties of maples, camellias, and
    rhododendrons, and major collections of magnolias,
    cherries, conifers, and azaleas, a Japanese garden, a
    winter garden, and a woodland garden.
Cross-reference: 1.40

16.418

Washington State Capitol Grounds
Sunken Garden and Conservatory
Bureau of Facilities Management
OB-2 (PA-11)
Olympia, WA 98504-7611
206-753-1752
Location: 11th and Water Sts.

Type of garden: Display garden and conservatory.
Roses, perennials, and annuals, and tropical and
    subtropical plants, cacti, and succulents in the
    conservatory.

WI

16.419

Boerner Botanical Gardens
Milwaukee County Dept. of Parks, Recreation, and Culture
5879 South 92nd St.
Hales Corner, WI 53130
414-425-1130

Type of garden: Arboretum and botanical garden.
Rose garden, perennial garden, herb garden, crabapple
    collection, tulip collection, trial gardens, wildflower rock
    garden, and native and introduced trees and shrubs.
Library: 14.101
Other: Children's horticultural program and plant doctor.
Cross-references: 2.80, 20.30, 20.240, 20.292

16.420

Botanical Garden of the University of
    Wisconsin—Madison
Dept. of Botany
144 Birge Hall, 430 Lincoln Dr.
Madison, WI 53706
608-262-2235

Type of garden: Botanical garden.
800 species of plants from over 65 families.
Cross-references: 12.345, 12.463, 15.209, 16.424

16.421

Jones Arboretum and Botanical Gardens
U.S. 14
Readstown, WI 54652
608-629-5553

Type of garden: Botanical garden.
Ferns, hostas, lilies, wildflowers, and perennial, herb,
    rose, and Japanese gardens.

16.422

Mitchell Park Conservatory
524 South Layton Blvd.
Milwaukee, WI 53215
414-649-9830

Type of garden: Conservatory.
Rare Madagascar succulents, over 400 species of orchids,
    Hawaiian tree ferns, and tropical economic plants. A
    U.S.D.A. rescue station for orchids and succulents.
Other: Plant culture information service.

North American Gladiolus Council
Milwaukee, WI

Cross-reference: 2.70

16.423

Olbrich Botanical Gardens
3330 Atwood Ave.
Madison, WI 53704
608-246-4551

Publication: *Olbrich Garden News*, quarterly.
Type of garden: Botanical garden.
All-America Selections display garden, and rose, dahlia,
    daylily, hibiscus, chrysanthemum, herb, rock, and
    wildflower gardens.
Library: Number of books: 1,000.
Lending policy: Circulating to members.
Cross-reference: 20.243

16.424

University of Wisconsin Arboretum
1207 Seminole Highway
Madison, WI 53711
608-262-2746

Type of garden: Arboretum.
Prairie, deciduous forest, conifer forest and wetlands
habitats, and ornamental trees and shrubs (especially
lilacs and crabapples), and a viburnum garden.
Cross-references: 12.345, 12.463, 15.209, 16.420

WV

American Chestnut Foundation
Morgantown, WV

Cross-reference: 1.10

16.425

Core Arboretum
Dept. of Biology, P.O. Box 6057
West Virginia University
Morgantown, WV 26506
304-293-5201

Type of garden: Arboretum.
75 acres with 500 types of plants.
Cross-references: 12.350, 12.464

16.426

Davis Park
Municipal Beautification Dept.
City of Charleston
P.O. Box 2749
Charleston, WV 25330
304-341-8000
Location: Lee and Capitol Sts.

Type of garden: Display garden.
Roses, shrubs, and flowers.

16.427

Greenbrier, The
White Sulphur Springs, WV 24986
304-536-1110

Type of garden: Display garden.
Flowering trees, boxwood, rhododendrons, bulbs, azaleas,
lilacs, hydrangeas, and chrysanthemums.

WY
16.428

Cheyenne Botanic Gardens
710 S. Lions Park Dr.
Cheyenne, WY 82001
307-637-6458

Type of garden: Botanical garden.
Tropical plants and food crops tended by senior,
handicapped, and troubled youth volunteers.
Library: Number of books: 500; number of periodicals:
15.
Lending policy: Reference only.
Other: Extensive horticultural therapy program.
Cross-reference: 13.53, 13.191

CANADA

AB
16.429

Alberta Special Crops and Horticultural Research Center
Bag Service 200
Brooks, AB T0J 0J0
403-362-3391; (FAX) 403-362-2554

Publications: *Newsletter*, bimonthly, free; *Research Report*,
annual, free; *Annual Report*, annual, free.
Type of garden: Arboretum.
Prairie hardy plants, demonstration orchard, rock garden,
and junipers.
Cross-references: 11.390, 20.31, 20.248

16.430

Calgary Zoo, Botanical Garden, and Prehistoric Park
P.O. Box 3036, Station "B"
Calgary, AB T2M 4R8
403-232-9342; (FAX) 403-237-7582
Location: Memorial Drive at 12 St. East, St. George's
Island.

Publication: *Dinnies Digest*, quarterly.
Type of garden: Botanical garden and conservatory.
About 2,000 different cultivars and species including a
large outdoor collection of hardy perennials, a rock
garden, a 20,000-sq.-ft. conservatory, a butterfly
garden, and many special collections.
Seed/specimen exchange.
Other: Gardening lectures, and workshops.

16.431

Cascade of Time Gardens
Parks Canada
Box 900
Banff, AB T0L 0C0
403-762-3324
Location: Banff National Park.

Type of garden: Display garden.
Shrubs, flowering plants, and many alpine plants.

16.432

Devonian Botanic Garden
University of Alberta
Edmonton, AB T6G 2E1
403-987-3054

Publication: *Kinnikinick*, quarterly.
Type of garden: Botanical garden.
Alpine garden, native people's garden, collections of iris,
plants from western China and the Himalayan
Mountains, an herb garden, peony display garden,
pinetum, herbaceous perennials, roses, apples, and
native plants.
Library: Number of books: 600.
Lending policy: Members only, reference only.
Seed/specimen exchange.
Other: Public education programs.
Cross-references: 12.355, 15.222

16.433
Devonian Gardens
Toronto-Dominion Square
Calgary, AB T2P 2M5
403-268-5214
Location: Stephen Ave. Mall.

Type of garden: Conservatory.
Enclosed tropical garden with 20,000 plants and 138
    varieties.

16.434
Muttart Conservatory
Box 2359
Edmonton, AB T5J 2R7
403-428-3664; (FAX) 403-428-6757
Location: 19th Ave. and 96A St.

Publication *Muttart Thymes*, quarterly.
Type of garden: Conservatory.
Plant displays in show, arid, temperate, and tropical
    pavilions.
Other: Classes, plant clinics, and telephone plant
    information service.
Cross-reference: 20.251

16.435
Nikka Yuko Japanese Garden
Box 751
Lethbridge, AB T1J 3Z6
403-328-3511
Location: North Parkside Dr.

Type of garden: Japanese garden.

BC
16.436
Bloedel Conservatory
c/o Sunset Nursery
290 E. 51st Ave.
2088 Beach Ave.
Vancouver, BC V5X 1C5
604-873-1133; (FAX) 604-321-7834
Location: Queen Elizabeth Park.

Type of garden: Conservatory.
500 species and varieties of jungle and desert plants.

16.437
Butchart Gardens, Ltd.
Box 4010, Postal Station A
Victoria, BC V8X 3X4
604-652-4422; (FAX) 604-652-3883

Type of garden: Display garden.
Italian garden, Japanese garden, English garden, and a
    sunken garden with spring bulbs, annuals, roses,
    begonias, perennials, trees, and shrubs.

16.438
Crystal Garden
713 Douglas St.
Victoria, BC V8W 2B4
604-381-1213

Type of garden: Conservatory and aviary.
Over 150 species of exotic tropical and subtropical plants
    including palms, orchids, bromeliads, and ficus.
Other: 50 species of birds, and iguanas, monkeys, and
    marmosets.

16.439
Dr. Sun Yat-Sen Classical Chinese Garden
578 Carrall St.
Vancouver, BC V6B 2J8
604-662-3207

Type of garden: Classical Chinese garden.
Library: Chinese garden culture.
Lending policy: Reference only.

16.440
Fable Cottage Estate
5187 Cordova Bay Rd.
Victoria, BC V8Y 2K7
604-658-5741

Type of garden: Display garden.
Orchard gardens and flower beds of perennials and
    annuals.

16.441
Fantasy Garden World
10800 No. 5 Rd.
Richmond, BC V7A 4E5
604-277-7777; (FAX) 604-274-1212

Type of garden: Display garden.
Biblical garden, Chinese garden, and a Japanese garden.

16.442
Friendship Gardens
Parks and Recreation Dept.
600 8th St.
New Westminster, BC V3M 3S2
604-526-4811
Location: Royal Ave. and 4th St.

Type of garden: Japanese garden.
100 Yoshino cherry trees, plus water lilies, and native
    plants.

16.443
Horticulture Center of the Pacific
505 Quayle Rd.
Victoria, BC V8X 3X1
604-479-6162

Type of garden: Display garden.
Junipers, rock and alpine garden, begonias, fuchsia,
    herbs, dahlias, geraniums, perennials, kiwi arbor, and
    vegetable and berry plots.
Library: Number of books: 500.
Lending policy: Reference only.

16.444

Minter Gardens

P.O. Box 40

Chilliwack, BC V2P 6H7

604-794-7191; (FAX) 604-792-8893

Location: 52892 Bunker Rd., Rosedale, BC.

Type of garden: Display garden.

Fragrance garden, rose garden, formal garden, arbor garden, rhododendron garden, fern garden, meadow garden, and an alpine garden.

16.445

Native Plant Garden

Botany Unity

Royal British Columbia Museum

675 Belleville St.

Victoria, BC V8V 1X4

604-387-5493; (FAX) 604-387-5360

Type of garden: Display garden.

Native British Columbia species planted in beds to represent ecological zones of British Columbia.

Other: Lectures.

Cross-reference: 15.225

16.446

Queen Elizabeth Park

Board of Parks and Recreation

2099 Beach Ave.

Vancouver, BC V6G 1Z4

604-872-5513

Location: 33rd Ave. and Cambie St.

Type of garden: Display garden.

Rose gardens, perennial and annual beds, a sunken garden, native and exotic trees and shrubs, and spring bulbs.

16.447

Royal Roads Botanical Garden

F.M.O.

Victoria, BC V0S 1B0

604-380-4526; (FAX) 604-380-5924

Location: Sooke Rd.

Type of garden: Display garden.

500 acres with Italian, rose, and Japanese gardens, and a large collection of old trees.

16.448

Tilford Gardens

North Vancouver Park

1200 Cotton Rd.

North Vancouver, BC V7J 1C1

604-987-9321

Type of garden: Display garden.

A rose garden, oriental garden, and collections of rhododendron, annuals, and native trees and shrubs.

16.449

University of British Columbia Botanical Garden

University of British Columbia

6804 S.W. Marine Dr.

Vancouver, BC V6T 1W5

604-228-4186; (FAX) 604-228-2016

Type of garden: Botanical garden.

Rhododendrons, Asian plants, alpines, native plants, and Japanese, food, winter, and rose gardens, and the Physick Garden (modeled after a 16th-century Italian monastery garden).

Library: 14.103

Seed/specimen exchange.

Cross-references: 2.19, 12.362, 12.465, 13.54, 15.226, 20.33, 20.253.

16.450

VanDusen Botanical Display Garden

5251 Oak St.

Vancouver, BC V6M 4H1

604-266-7194

Publication: *VanDusen Botanical Garden Bulletin*, quarterly.

Type of garden: Botanical garden.

Collections of rhododendron, holly, hydrangea, Sino-Himalayan flora, and Mediterranean flora.

Library: 14.104

Other: Educational programs.

Cross-reference: 12.532

MB

16.451

Agriculture Canada—Morden Research Station

P.O. Box 3001

Morden, MB R0G 1J0

204-822-4471; (FAX) 204-822--6841

Publication: *Focus on Research*, annual, free.

Type of garden: Arboretum.

Ornamentals, hedges, roses, hardy fruit, shade trees, grains, and vegetables. 120 acres.

Seed/specimen exchange.

Cross-reference: 11.360.

16.452

English Garden

Assiniboine Park

2799 Roblin Blvd.

Winnipeg, MB R3R 0B8

204-986-5538

Type of garden: Display garden and conservatory.

Four acres of floral collections with bulbs, annuals, and perennials, and a conservatory with tropical displays and a palm house.

Cross-reference: 20.254

NF

16.453

Memorial University Botanical Garden at Oxen Pond

Memorial University of Newfoundland

St. John's, NF A1C 5S7

709-737-8590

Type of garden: Botanical garden.

Perennials, a cottage garden, heritage plants of Newfoundland, a rock garden, butterfly garden, heather bed, peat beds, and woodland beds.

Cross-reference: 15.231

NS
16.454
Halifax Public Gardens
Friends of the Public Gardens
P.O. Box 3544 South
Halifax, NS B3J 3J2
Location: Spring Garden Rd. and Summer St.

Type of garden: Display garden.
Established in 1867, contains flower beds, lily ponds, and
trees and shrubs.

ON
16.455
Allan Gardens
Dept. of Parks and Recreation
19 Horticultural Ave.
Toronto, ON M5A 2P2
416-392-7288

Type of garden: Display garden and conservatory.
800 species including orchids, jade vine, screw pine,
chestnut dioon, roses, and water lilies.

16.456
Arboretum of the University of Guelph
Guelph, ON N1G 2W1
519-824-4120; (FAX) 519-763-9598

Publication: *Arboretum Program*, quarterly, free.
Type of garden: Arboretum.
2,900 types of trees and shrubs on 250 acres, including
native trees and shrubs, dwarf conifers, roses, and rare
trees and shrubs of Ontario.
Library: Number of books: 800; number of periodicals:
20.
Lending policy: Reference only.
Cross-reference: 10.5, 12.386, 12.466, 15.243

16.457
Balance Life Gardens
R.R. 3
Lambeth, ON N0L 1S0
519-652-9549

Publication: *Balance Life Gardens Calendar*, biannual.
Type of garden: Display garden.
50 acres of herb, fruit, and vegetable gardens.
Library: Number of books: 3,000; Number of periodicals:
50.
Lending policy: Reference only.
Other: Workshops, seminars, and herb dinners.
Cross-reference: 2.109

Canadian Iris Society
Willowdale, ON

Cross-reference: 2.102

Canadian Rose Society
Scarborough, ON

Cross-reference: 2.106

Canadian Society for Herbal Research
Willowdale, ON

Cross-reference: 1.135

16.458
Centennial Botanical Conservatory
Parks & Recreation Dept.
950 Memorial Ave.
Thunder Bay, ON P7B 4A2
807-622-7036; (FAX) 807-623-3420

Type of garden: Conservatory.
Tropical plants, annuals, and perennials with seasonal
displays of lilies, hydrangea, tulips, daffodils, roses,
poinsettias, and azaleas.

16.459
Dominion Arboretum
Agriculture Canada
Building 72
Central Experimental Farm
Ottawa, ON K1A 0C6
613-995-3700; (FAX) 613-992-7909

Type of garden: Arboretum.
2,000 varieties of trees and shrubs, and ornamental
gardens with annuals and perennials, a rock garden,
lilac walk, and roses.
Library.
Cross-references: 11.367, 15.241, 20.259

16.460
Edwards Gardens
777 Lawrence Ave. East
Don Mills, ON M3C 1P2
416-445-1552

Type of garden: Display garden.
27-acre garden with roses, lilies, irises, peonies,
rhododendrons, azaleas, and a rock garden.
Cross-reference: 20.260

16.461
Floral Clock and Lilac Gardens
Niagara Parks Commission
P.O. Box 150
Niagara Falls, ON L2E 6T2
416-356-2241
Location: River Rd.

Type of garden: Display garden.
The world's largest floral clock with 25,000 flowering
plants plus 1,500 shrubs representing 256 varieties of
lilacs.

16.462
Gage Park
Dept. of Parks
Hamilton, ON L8N 3T4
416-526-4627
Location: Main St.

Type of garden: Display garden and greenhouses.
Rose garden and a tropical display in the greenhouses.

16.463
Great Lakes Forestry Research Centre Arboretum
Forestry Canada
Great Lakes Forestry Centre
1219 Queen St. East
Sault Ste. Marie, ON P6A 5M7
705-949-9461; (FAX) 705-759-5700

Type of garden: Arboretum.
A wide variety of native tree and shrub species.
Other: Identification of insects, fungi, and forest insect
and disease problems, and information on tree and
shrub cultural problems.
Cross-reference: 15.239

16.464
Humber Arboretum
205 Humber College Blvd.
Rexdale, ON M9W 5L7
416-675-3111, ext. 445

Type of garden: Arboretum.
Collections of rhododendrons, native trees, and
wildflowers.
Cross-reference: 20.263

16.465
International Friendship Garden
Soroptimist Club of Thunder Bay
Chapples Recreation Park
1800-2000 Victoria Ave. East
Thunder Bay, ON P7C 1E2

Type of garden: Display garden.
Gardens of various ethnic communities including
Canadian, Italian, Slovak, Lithuanian, German, Polish,
and Ukrainian gardens.

16.466
J. J. Neilson Arboretum
Ridgetown College of Agricultural Technology
Ridgetown, ON N0P 2C0
519-674-5456; (FAX) 519-674-3042

Type of garden: Arboretum.
Korean spice, Osage orange, striped maple, Kentucky
coffee, and many other trees.
Scholarships: Horticultural scholarships.
Library: Number of books: 10,000; Number of
periodicals: 100.
General agricultural and horticultural library.
Lending policy: Circulating.
Seed/specimen exchange.
Cross-reference: 12.377

16.467
Jackson Park
Dept. of Parks
2450 McDougall St.
Windsor, ON N8X 3N6
519-255-6276
Location: Tecumseh Rd.

Type of garden: Display garden.
Formal gardens with spring bulbs, summer annuals, and
500 varieties of roses.
Cross-reference: 20.264

16.468
Lakehead University Arboretum
Lakehead University
Thunder Bay, ON P7B 5E1
807-345-2121

Type of garden: Arboretum.
Junipers, arborvitae, hardy conifers, and woody
ornamentals.
Cross-reference: 15.236

16.469
Niagara Parks Commission School of Horticulture
P.O. Box 150
Niagara Falls, ON L2E 6T2
416-356-8554; (FAX) 416-354-6041
Location: 2565 Niagara Pkwy. North.

Type of garden: Display garden and arboretum.
Exotic trees, native wildflowers, a large rose collection,
lilacs, rhododendrons, a woodland garden, rock garden,
vegetable display garden, and an herb garden.
Scholarships.
Library: 14.106
Seed/specimen exchange.
Other: Gardening information service.
Cross-references: 12.376, 20.265

16.470
Northeastern Ontario Arboretum
New Liskeard College of Agricultural Technology
Box "G"
New Liskeard, ON P0J 1P0
705-647-6738; (FAX) 705-647-7008

Type of garden: Arboretum.
Native and cultivated trees and shrubs adapted to
northern Ontario.

Ontario Herbalists Association
Toronto, ON
Cross-reference: 2.109

16.471
Queen Victoria Park
Niagara Parks Commission
P.O. Box 150
Niagara Falls, ON L2E 6T2
416-356-2241
Location: Niagara Pkwy.

Type of garden: Display garden and conservatory.
Spring bulbs, rose garden, perennials, formal gardens, a
rock garden, and tropical plants in the conservatory.

16.472
Rosebud Gardens
770 Brigden Side Rd.
Sarnia, ON N7T 7H3
519-383-8083

Type of garden: Display garden.
Over 2,000 rose bushes and flower, herb, and vegetable
gardens.

16.473
Royal Botanical Gardens
P.O. Box 399
Hamilton, ON L8N 3H8
416-527-1158; (FAX) 416-577-0375
Location: 680 Plains Rd. West.

Publications: *PAPPUS*, quarterly; *Canadian Horticultural History*, annual.
Type of garden: Botanical garden.
Herb garden, native tree and shrub collection, a lilac display, flowering trees, rhododendrons, a pinetum, spring bulbs, summer annuals, irises, perennials, a clematis collection, climbing plants, a rose garden, an annual trial garden, a scented garden, and a medicinal garden.
Library: 14.107
Seed/specimen exchange.
Other: Horticultural therapy and teaching garden
Cross-references: 2.102, 10.71, 13.55, 13.193, 15.242, 17.39

16.474
Sherwood Fox Arboretum
Biological and Geological Sciences Bldg.
University of Western Ontario
London, ON N6A 5B7
519-679-2111; (FAX) 519-661-3292

Type of garden: Arboretum.
3,800 trees and shrubs.

16.475
Walker Botanical Gardens
Rodman Hall Arts Centre
109 St. Paul Crescent
St. Catharines, ON L2S 1M3
416-684-2925

Type of garden: Botanical garden.
Many exotic species, rhododendrons, and azaleas. Still under development.

PE
16.476
Malpeque Gardens
Kensington R.R. 1
Malpeque, PE C0B 1M0
Location: Rt. 20.

Type of garden: Display garden.
Dahlias, begonias, roses, perennials, a sunken garden, and a dwarf orchard.

PQ
16.477
Belle Terre Botanic Garden and Arboretum
Otter Lake, PQ J0X 2P0
819-453-7334

Type of garden: Botanical garden and arboretum.
Herb garden, rose garden, children's garden, rock garden, perennial garden, old-fashioned garden, and vegetable and fruit trial gardens.

16.478
Jardin Roger Van den Hende
Université Laval
Ste.-Foy, PQ G1K 7P4
418-656-3410

Type of garden: Botanical garden.
Magnolias, forsythia, azaleas, rhododendrons, and Japanese maples.
Cross-references: 12.389, 15.247, 20.267.

16.479
Metis Gardens
C.P. 242
Mont-Joli, PQ G5H 3L1
418-775-2221
Location: Rt. 132.

Type of garden: Display garden.
Lilies, rock plants, shrubs, rhododendrons, crabapples, and perennials.
Seed/specimen exchange.

16.480
Morgan Arboretum
P.O. Box 500
MacDonald College
Ste. Anne de Bellevue, PQ H9X 1C0
514-398-7811; (FAX) 514-398-7895

Publication: *Newsletter*, biannual, free.
Type of garden: Arboretum.
Native shrubs and trees and introduced species from the Northern Temperate Zone.
Cross-references: 12.391, 15.248, 20.268

QC
16.481
Montreal Botanical Garden
4101, Sherbrooke St. East
Montreal, QC H1X 2B2
514-872-1400; (FAX) 514-872-3765

Publications: *Delectus Seminum*, annual, free; *Test Garden Report*, annual, free.
Type of garden: Botanical garden and arboretum.
30 gardens including a rose garden, a heath garden, a perennial garden, a flowery brook, a vegetable test garden, an economic plants garden, an annual test garden, a medicinal plant garden, a poisonous plant garden, an alpine rock garden, and a shade garden.
Library: 14.108
Seed/specimen exchange.
Other: Public education programs.
Cross-reference: 12.392, 13.195, 15.252

SK
16.482
PFRA Shelterbelt Centre
Indian Head, SK S0G 2K0
306-695-2284

Type of garden: Arboretum.

Collections of poplar, Japanese elms, hardy perennials, and caragana. Also displays of plants hardy on the plains.
Seed/specimen exchange.
Other: The PFRA Shelterbelt Centre distributes trees and shrubs to farmers, governmental institutions, and nonprofit institutions free of charge.
Cross-reference: 11.378

16.483

Patterson Park Botanical Garden
Dept. of Horticultural Science
University of Saskatchewan
Saskatoon, SK S7N 0W0
306-966-5863

Type of garden: Display garden
Hardy trees, shrubs, and climbers.
Cross-references: 12.393, 15.255

16.484

Saskatoon Civic Conservatory
City of Saskatoon
Building and Grounds Dept.
City Hall
Saskatoon, SK S7K 0J5
306-975-7610
Location: 950 Spadina Crescent East.

Type of garden: Conservatory.
Cacti, succulents, tropicals, orchids, and seasonal displays.

~9 17 ~

Historical Horticulture and Museum Gardens

Historical horticulture, as it is defined in this chapter, includes historic and estate gardens, living historical farms and museums and agricultural museums with historical or heirloom gardens, heritage plant societies, seed saving organizations, historic landscape preservation organizations, garden history societies, and other groups that are involved in the preservation or interpretation of the horticultural heritage of North America.

The chapter is divided into two sections. The first lists societies and associations that are involved in historical horticulture. The second lists gardens, museums, and other facilities that people may visit.

The historical quality of the plant collections in this chapter varies by institution. Some gardens are simply located at a historic site or structure. Others have been re-created with types of plants typical of the period represented. Still others have been planted with heirloom *varieties* of plants typical of the period. The most historically accurate gardens have been planted with heirloom varieties that have been documented to have been grown in the region, or, even better, in the historical garden.

Also included in this chapter are public gardens at museums and a few forest history museums.

The following books provide more information on the topics covered below:

Farm Museum Directory (Lancaster, Pennsylvania: Stemgas Publishing Co., 1988).

Annotated listings of living history farms and agricultural museums of the United States and Canada.

Mary Helen Ray and Robert P. Nicholls, eds., *A Guide to Significant & Historic Gardens of America* (Athens, Georgia: Agee Publishers, 1982).

A reference for travelers. Historic gardens are defined as those in existence for over 75 years that are noteworthy examples of the era or gardens associated with an important person or event.

Carolyn, Jabs, *The Heirloom Gardener* (San Francisco, Sierra Club Books, 1984).

This interesting book on heirloom gardening and gardeners includes appendices on seed exchanges, seed companies offering heirloom varieties, and living historical farms and museums.

Charles Edgar Randall and Henry Clepper, *Famous and Historic Trees* (Washington, D.C.: American Forestry Association, 1976).

A guide to historic and unusual trees.

(Top, left to right) Isabella Stewart Gardner Museum, Boston, Massachusetts; Elizabethan Gardens, Manteo, North Carolina; Hans Herr House, Willow Street, Pennsylvania; (bottom, left to right) George Washington Birthplace National Monument, Washington's Birthplace, Virginia; Forest Resource Center, Lanesboro, Minnesota.

(Top to bottom) Tryon Palace Restoration, New Bern, North Carolina; Dumbarton Oaks, Washington, D.C.; Naumkeag Museum and Gardens, Stockbridge, Massachusetts; Hagley Museum and Library, Wilmington, Delaware; Peter Wentz Farmstead, Worcester, Pennsylvania.

Forest History Museums of the World (Santa Cruz, California: Forest History Society, 1983).

A directory of museums and collections devoted to the history of humanity's use of the forest and its products.

ORGANIZATIONS

UNITED STATES

Abundant Life Seed Foundation
Port Townsend, WA

Cross-reference: 8.9

17.1

Agricultural History Society
Economics Research Service
Room 1232
1301 New York Ave., N.W.
Washington, DC 20005-4788
202-447-8183

Purpose: To stimulate interest in, promote the study of, and facilitate research and publication in the history of agriculture.
Other: Publishes *Agricultural History*, quarterly.

17.2

Alberene Seed Foundation
P.O. Box 271
Keene, VA 22946

Purpose: A living seed bank that collects, preserves, and distributes seed, bulbs, tubers, and starter plants.
Seed/specimen exchange.

17.3

Alliance for Historic Landscape Preservation
82 Wall St.
Suite 1105
New York, NY 10005
617-491-3727

Purpose: To provide a forum of communication for those concerned with landscape preservation, including landscape architects, historians, geographers, horticulturists, architects, and planners.

17.4

American Garden and Landscape History Program
Wave Hill
675 West 252nd St.
Bronx, NY 10471
212-549-3200

Purpose: To gather information about the location and content of archival collections concerned with the American use of the land, to serve as a national clearinghouse for the information, and to promote the preservation and proper management of landscape records.
Other: Maintains the *Catalog of Landscape Records in the United States*, a computer database with extensive information on thousands of landscape documents around the country (such as maps, correspondence, drawings, slides, plans, photographs, film, diaries, postcards, advertisements, plant lists, oral histories, etc.). Also publishes a quarterly newsletter and invites inquiries regarding historical landscapes.
Cross-reference: 17.212

17.5

American Society for Environmental History
c/o John Opie
Center for Technology Studies
New Jersey Institute of Technology
Newark, NJ 07102

Purpose: To promote the study of all aspects of environmental history.
Other: Publishes *Environmental Review*.

17.6

Association for Living Historical Farms and Agricultural Museums
5035 AHB
Smithsonian Institution
Washington, DC 20560
202-357-2813; (FAX) 202-357-1853

Purpose: To provide communication between outdoor living historical farms and indoor museums of agriculture.
Library.
Lending Policy: Reference only.
Other: Publishes the quarterly *Living Historical Farms Bulletin*.

17.7

Association for Preservation Technology
P.O. Box 8178
Fredericksburg, VA 22404
703-373-1621

Purpose: An interdisciplinary membership organization of those involved in the systematic application of the knowledge of methods and materials to the maintenance, conservation, and protection of historic and world heritage buildings, sites, and artifact resources for future use and appreciation.
Other: Publishes the quarterly *APT Bulletin: The Journal of Preservation Technology* (three issues have been devoted to historical horticulture) and *Communique*, a quarterly newsletter.

Botanical Dimensions
Occidental, CA

Cross-reference: 8.17

17.8

Butterbrooke Farm
78 Barry Rd.
Oxford, CT 06483
203-888-2000

Purpose: To promote food and seed self-reliance.
Garden: Heirloom vegetable garden. Numerous heirloom
   vegetable varieties.
Seed/specimen exchange.
Other: Publishes the quarterly *Germinations*; a member of
   the Seed Savers Exchange.

17.9

Central Prairie Seed Exchange
7949 S.W. 21st
Topeka, KS 66604
913-478-4944

Purpose: To find, save, and distribute locally adapted,
   open-pollinated plant varieties of the Central Plains
   area including heirloom and Native American varieties.
Seed/specimen exchange.
Other: Members receive quarterly newsletter *Seeds
   Broadcast*, annual *Broadcast Listings;* fall seed exchange
   and annual meeting; local spring scion and seed
   exchange.

17.10

Corns
Rt. 1, Box 32
Turpin, OK 73950
405-778-3615

Purpose: The preservation of corn varieties by growing
   and distributing seeds.
Garden: 1,000 varieties of corn.
Museum: Several hundred varieties of rare and heirloom
   corn on display plus photographs.
Seed/specimen exchange.
Other: No dues, but all members agree to grow out and
   distribute seeds; members receive annual *Information
   Letter*.

Desert Legume Program
Tucson, AZ

Cross-reference: 4.11

17.11

Flower and Herb Exchange, The
Rt. 3, Box 239
Decorah, IA 52101

Purpose: Dedicated to the preservation and exchange of
   heirloom herbs and flowers.

17.12

Forest History Society
701 Vickers Ave.
Durham, NC 27701
919-682-9319

Purpose: The collection, interpretation, and dissemination
   of the history of forest and all forest-related activities.
Other: Publishes the quarterly *Journal of Forest History*
   and *The Cruiser*, a quarterly newsletter. Also has an

archival collecting program that locates records in
repositories across Canada and the U.S.

17.13

Garden Conservancy, The
Box 219
Cold Spring, NY 10516
914-265-2029

Purpose: To preserve exceptional American gardens and
   to facilitate their transition from private to public
   status.
Other: Publishes a biannual newsletter.

17.14

Grain Exchange, The
2440 East Water Well Rd.
Salinas, KS 67401

Purpose: A network of gardeners and farmers dedicated to
   the preservation of genetic diversity in cereal grains and
   related staple crops.
Seed/specimen exchange.

17.15

Heritage Rose Foundation
1512 Gorman St.
Raleigh, NC 27606
919-834-2591

Purpose: To promote the preservation and study of
   heritage roses.
Other: Members receive quarterly *Heritage Rose
   Foundation News*.

17.16

Heritage Roses Group—North Central
Henry Nagat
6365 Wald Rd.
Monroe, WI 53566

Purpose: To seek out, preserve, and promote heritage
   roses.

17.17

Heritage Roses Group—Northeast
Lily Shohan
R.D. 1, Box 299
Clinton Corners, NY 12514

Purpose: To seek out, preserve, and promote heritage
   roses.

17.18

Heritage Roses Group—Northwest
Judi Dexter
23665 41st South
Kent, WA 98032

Purpose: To seek out, preserve, and promote heritage
   roses.

17.19

Heritage Roses Group—South Central
Mitzi VanSant
810 East 30th St.
Austin, TX 78705

Purpose: To seek out, preserve, and promote heritage
   roses.

17.20

Heritage Roses Group—Southeast
Charles A. Walker, Jr.
1512 Gorman St.
Raleigh, NC 27606

Purpose: To seek out, preserve, and promote heritage
roses.

17.21

Heritage Roses Group—Southwest
Margaret Blodgett
1452 Curtis St.
Berkeley, CA 94702

Purpose: To seek out, preserve, and promote heritage
roses.

17.22

Heritage Roses Group—Southwest
Francis Grate
472 Gibson Ave.
Pacific Grove, CA 93950

Purpose: To seek out, preserve, and promote heritage
roses.

17.23

Historic Iris Preservation Society
4855 Santiago Way
Colorado Springs, CO 80917
719-596-7724

Purpose: To locate, record, preserve, and aid in the
distribution of antique iris cultivars.

17.24

International Gladiolus Hall of Fame
The James A. Michener Library
University of North Colorado
Greeley, CO 80639
303-351-2854

Purpose: To honor and preserve information on the
people and varieties that have contributed to gladiolus
history.
Library: Large archival collection.

17.25

KUSA Society, The
P.O. Box 761
Ojai, CA 93024

Purpose: To provide seed and information on traditional
folk varieties of edible seedcrops to gardeners and
small-scale farmers.
Garden: Test garden of folk varieties and rare strains of
traditional edible seedcrops. Open by appointment in
the late summer.
Seed/specimen exchange.
Other: Publishes *The Cerealist* (bimonthly).

17.26

National Association for Olmstead Parks
5010 Wisconsin Ave., N.W.
Suite 308
Washington, DC 20016
202-363-9511

Purpose: To offer support, advice, research, and
information to those involved in the preservation and
renewal of historic landscapes and the creation of new
ones, and to preserve and enhance our national heritage
of historic parks and landscapes.
Other: Publishes newsletter (3 issues a year).

17.27

National Heirloom Flower Seed Exchange
136 Irving St.
Cambridge, MA 02138
617-576-5065

Purpose: To preserve endangered ornamental annual and
perennial varieties or species and encourage the use of
flowers whose properties have value medicinally, as
insecticides, or esthetically.
Seed/specimen exchange.

National Register of Historic Places
National Park Service
U.S. Department of the Interior
Washington, DC 20013-7127

17.28

National Trust for Historic Preservation
1785 Massachusetts Ave., N.W.
Washington, DC 20036
202-673-4000

Purpose: To encourage preservation of significant
American buildings, sites, and historic districts.
Other: Publishes the bimonthly *Historic Preservation* and
monthly *Preservation News*.

17.29

Native Seeds/SEARCH
2509 N. Campbell Ave., #325
Tucson, AZ 85719
602-327-9123

Purpose: To conserve the seeds of crops grown by Native
Americans in the greater Southwest.
Garden: Demonstration garden at the Tucson Botanical
Gardens (16.13). Various native crops including corn,
squash, beans, chilies, teosinte, devil's claw, indigo,
and melons.
Library: Number of books: 500; number of periodicals:
10.
Lending Policy: Circulating.
Seed/specimen exchange.
Other: Members receive the quarterly *Seedhead News*.

17.30
New England Garden History Society
Massachusetts Horticultural Society
300 Massachusetts Ave.
Boston, MA 02115
617-536-9280

Purpose: To promote the study of the history of New
England gardening, horticulture, and landscape design
and to encourage the preservation of gardens and
landscapes.
Other: This new organization plans to put out a
newsletter and a scholarly journal.

Rare Seed Locators, Ltd.
Berkeley, CA
Cross-reference: 8.62

17.31
Scatterseed Project
c/o Khadighar
P.O. Box 1167
Farmington, ME 04938

Purpose: A program for preserving the diversity of the
American crop heritage. Basically a one-man effort, the
project collects crop plants, preserves and propagates
the varieties, and distributes seeds to interested parties.
Seed/specimen exchange.

17.32
Seed Savers Exchange
R.R. 3, Box 239
Decorah, IA 52101
319-382-5990

Purpose: A genetic preservation project whose members
are dedicated to locating and preserving vegetable and
fruit varieties that are family heirlooms, traditional
Indian crops, garden varieties of the Mennonite and
Amish, varieties dropped from seed catalogs, and
outstanding foreign varieties.
Garden: Heritage Farm contains some 1,200 endangered
vegetable and fruit varieties.
Seed/specimen exchange.
Other: Members receive the annual *Yearbook, Summer
Edition*, and *Harvest Edition*.

17.33
Seed Saving Project
Student Experimental Farm
Dept. of Agronomy, Hunt Hall
University of California at Davis
Davis, CA 95616
916-752-7645

Purpose: To maintain and distribute seeds of rare,
regionally adapted vegetable, flower, and herb varieties
and serve as an educational forum on the importance of
preserving our vegetable heritage.
Garden: Student experimental farm features regionally
adapted varieties of vegetables, flowers, and herbs.
Library: Number of books: 150; number of periodicals:
15.

Lending Policy: Circulating.
Seed/specimen exchange.
Other: Publishes *Seed Saving Project Newsletter* (quarterly),
conducts workshops on seed saving techniques, and has
demonstration garden field days.
Cross-references: 12.62, 12.402, 15.23, 16.59, 19.12

17.34
Society of Architectural Historians
American Garden and Landscape History Chapter
1232 Pine St.
Philadelphia, PA 19107

Purpose: To promote the study of architectural history.

17.35
Southern Garden History Society
Old Salem, Inc.
Drawer F, Salem Station
Winston-Salem, NC 27108
919-724-3125

Purpose: To promote interest in Southern garden and
landscape history, in historical horticulture, and in the
preservation and restoration of historic gardens and
landscapes in the South.
Other: Members receive the quarterly *Magnolia*.

17.36
Talavaya Center
Rt. 2, Box 2
Espanola, NM 87532
505-753-5801

Purpose: To collect and preserve seeds of Native
American plants.
Garden: 15-acre vegetable garden with Native American
species.
Library:
Lending Policy: Reference only.
Other: Members receive the quarterly *Talavaya News* and
a seed catalog.

17.37
Texas Rose Rustlers
9426 Kerrwood
Houston, TX 77080
713-464-8607

Purpose: To search for "lost" roses, identify and preserve
old rose species and varieties, and to establish a rose
identification procedure.
Seed/specimen exchange.
Other: Cutting exchange, publishes the quarterly *Old
Texas Rose*.

17.38
Thomas Jefferson Center for Historic Plants
Monticello
P.O. Box 316
Charlottesville, VA 22902
804-979-5283

Purpose: To collect, preserve, and distribute historic and
native plants, and to educate the public about them.
Other: Offers historic seeds for sale.

CANADA

Canadian Plant Conservation Programme
Guelph, ON
Cross-reference: 8.91

17.39
Centre for Canadian Historical Horticultural Studies
Royal Botanical Gardens
Library, Box 399
Hamilton, ON L8N 3H8
416-527-1158; (FAX) 416-577-0375

Purpose: A national repository and bibliographical center for literature, documents, artifacts, and information relevant to the history of Canadian horticulture.
Library: Library is a part of the Royal Botanical Gardens collections.
Other: Honors requests for information on historical horticulture. Publishes *Canadian Horticultural History*, quarterly.
Cross-reference: 16.473.

17.40
Heritage Seed Program
R.R. 3
Uxbridge, ON L0C 1K0
416-852-7965; (FAX) 416-852-7965

Purpose: To search out, preserve, and promote the enjoyment of heirloom and endangered varieties of vegetables, fruits, grains, herbs, and flowers.
Garden: Heirloom varieties of vegetables and flowers; garden is not open to the public except for a summer garden tour.
Seed/specimen exchange.
Other: Members receive *Heritage Seed Program* 3 times a year; and a seed listing.

17.41
Historic Gardens and Landscapes Committee
ICOMOS Canada
P.O. Box 737, Station B
Ottawa, ON K1P 5R4
613-749-0971

Purpose: Part of the Canadian branch of the International Council on Monuments and Sites, an association of conservation professionals and advocates.

GARDENS

UNITED STATES

AK
17.42
Museum of Alaska Transportation and Industry
P.O. Box 909
Palmer, AK 99645
907-745-449

Museum of transportation and industry.
Garden: Flower beds and vegetable beds.

AL
17.43
Arlington Historic House and Gardens
331 Cotton Ave., S.W.
Birmingham, AL 35211
205-780-5656

19th-century antebellum manor.
Garden: Lawns, flowering trees, azaleas, coleuses, geraniums, a rose garden, and a boxwood garden.

17.44
Bellingrath Gardens and Home
12401 Bellingrath Gardens Rd.
Theodore, AL 36582-9704
205-973-2217

Home of Mr. and Mrs. Walter D. Bellingrath with antiques and art.
Garden: 65-acre garden with azaleas, chrysanthemums, camellias, annuals, perennials, and spring bulbs.
Cross-reference: 20.65

17.45
Oakleigh
350 Oakleigh Pl.
Mobile, AL 36604
205-432-1281

1833 house.
Garden: Azaleas, sunken floral garden, and an herb garden.

AR
17.46
Arkansas Territorial Restoration
3rd & Scott
Little Rock, AR 72201
501-371-2348

Historic site museum.
Garden: Medicinal herb garden and a representative sample of plants from Arkansas's frontier period.
Library.
Lending policy: Reference only

17.47
Heritage Herb Garden
Ozark Folk Center
P.O. Box 500
Mountain View, AR 72560
501-269-3851

Ozark heritage museum.
Garden: "Young Pioneer Garden" for children, dye plant and textile garden, "Yarb" garden, terraced herb garden, garden for the physically handicapped, and bird and butterfly garden. Many heirloom vegetable varieties.
Museum: Exhibits on historical agricultural and historical herb lore and folk remedies.

17.48
Living Farm Museum of the Ozarks
Good Earth Association, Inc.
202 East Church St.
Pocahontas, AR 72455-2899
501-892-9545

Living historical farm.
Garden: Arkansas black apples, Ozark wild grapes.
Seed/specimen exchange.
Other: Ozark Seed Exchange Group gathers and
exchanges seeds from across northern Arkansas and
southern Missouri.

17.49
Stuttgart Agricultural Museum
Grand Prairie Garden Club
921 East Fourth St.
Stuttgart, AR 72160
501-673-7001

Local history museum depicting the story of a prairie farm
family, and rice and soybean history.
Garden: Re-creation of herb garden typical of early prairie
pioneers (circa 1880).
Museum: Photos of prairie wildflowers, exhibit on
medicinal herbs.

AZ
17.50
Century House Museum and Gardens
240 Madison Ave.
Yuma, AZ 85364
602-782-1841

1870 home.
Garden: Garden of native plants and of plants available at
the turn of the century—cacti, mulberry trees, roses,
shrubs, ground covers, and bamboo.

Native Seeds/SEARCH
Tucson, AZ

Cross-reference: 17.29

17.51
Sharlot Hall Museum
415 West Gurley St.
Prescott, AZ 86301
602-445-3122

Frontier buildings including the Governor's Mansion.
Garden: Pioneer herb garden and a memorial rose garden
with 350 bushes honoring Arizona women.

CA
17.52
Ardenwood Historic Farm
34600 Ardenwood Blvd.
Fremont, CA 94555
415-796-0233

200-acre farm depicting life in the late 19th century.
Garden: Four-acre Victorian garden around an 1890
Queen Anne Victorian house, heirloom vegetable
garden, and many old tree specimens on the grounds.

Over 73 tree species including *Taxodium distichum*,
Metasequoia glyptostroboides, and *Quercus macrocarpa*.
Library: Number of books: 100; number of periodicals: 5.
Lending Policy: Reference only, not open to the
public.
Seed/specimen exchange.

17.53
Casa Amesti
Monterey History of Art Association
516 Polk St.
Monterey, CA 93940
408-372-2608

Colonial adobe house.
Garden: Formal Italian-style garden.
Other: A property of the National Trust for Historic
Preservation.

17.54
Cooper Historic Garden
Monterey State Historic Park
525 Polk St.
Monterey, CA 93940
408-649-7118

Garden: Re-creation of pre-1865 vegetable, herb, and
ornamental gardens and orchard, all with heirloom
varieties.

17.55
Dunsmuir House
2960 Peralta Oaks Ct.
Oakland, CA 94605
415-562-7588

Colonial Revival mansion.
Garden: 48 acres of gardens including a cactus garden,
succulent garden, an arbor area, and orchards.

17.56
Filoli
Canada Rd.
Woodside, CA 94062
415-364-8300

Early 19th-century mansion and 654-acre estate.
Garden: 16 acres of formal gardens including a sunken
garden, a walled garden, a woodland garden, and the
Panel Garden with a fruit orchard, daffodils, roses, and
herbs.
Library: 14.13
Other: A property of the National Trust for Historic
Preservation.
Cross-reference: 20.74

Fullerton Arboretum
Fullerton, CA

Cross-reference: 16.24

17.57
Hearst San Simeon State Historical Monument
Dept. of Parks and Recreation
San Simeon Region, Box 8, Hwy. 1
San Simeon, CA 93452-0040
619-452-1950
Location: Micheltorena and Santa Barbara Sts.

Former home of William Randolph Hearst.
Garden: Formal gardens with 300 varieties of annuals, perennials, and tropical plants.

17.58
Huntington, The
1151 Oxford Rd.
San Marino, CA 91108
818-405-2160; (FAX) 818-405-0225

Art museum and library.
Garden: 12 specialized gardens including a Japanese garden, camellia garden, Shakespeare garden, herb garden, rose garden, palm garden, and a desert garden.
Library: 14.7
Seed/specimen exchange.
Cross-reference: 20.271

17.59
J. Paul Getty Museum
Garden of the Villa of the Papyri at Herculaneum
17985 Pacific Coast Highway
Malibu, CA 90265-5799
213-459-7611

Art museum housed in re-creation of ancient Roman country house.
Garden: Re-creation of Roman villa garden and Roman herb garden.
Rare and antique plants.

KUSA Society, The
Ojai, CA
Cross-reference: 17.25

17.60
Leonis Adobe and Plummer House
23537 Calabasas Rd.
Calabasas, CA 91302
818-712-0734

Monterey-style mansion that was the home of Miguel Leonis.
Garden: Orchard, grape arbor, demonstration vegetable garden, and English flower garden.

17.61
Luther Burbank Home and Gardens
Santa Rosa Ave. at Sonoma Ave.
P.O. Box 1678
Santa Rosa, CA 95402
707-576-5445

Modified Greek Revival house where Luther Burbank lived from 1884 to 1906.

Garden: Garden displays many plants that Burbank introduced including *Prunus domestica*, spineless cactus, *Agapanthus*, and *Canna* species.

17.62
Meux Home Museum
1007 R St.
Fresno, CA 93707
209-233-8007

1889 Queen Anne-style home.
Garden: Garden includes plants that were common in the average American garden during 1890–1895: 19th-century roses, peonies, irises, and perennials.

17.63
Rancho Los Alamitos Historic Site and Garden
6400 Bixby Hill Rd.
Long Beach, CA 90815
213-431-3541

Garden: Historic gardens spanning 1880s to 1940s with dominant period of 1920s. Garden areas are: native, desert, rose, herb, "Friendly Garden," geranium walk, front lawn, south garden, old garden, and secret garden. Pair of 100-year-old Moreton Bay fig trees.
Other: Monthly programs.

17.64
San Joaquin County Historical Society and Museum
11793 North Micke Grove Rd.
Lodi, CA 95240
209-368-9154; (FAX) 209-369-2178

Agricultural and local history museum.
Garden: California native plants, and a "Taste, Touch, and Smell Garden" for the visually impaired.
Museum: On plants transferred from gardens of Charles M. Weber circa 1860.
Library: Number of books: 5,000; number of periodicals: 15. Native plants.
Lending Policy: Reference only.

Seed Saving Project
Davis, CA
Cross-reference: 17.33

17.65
Villa Montalvo Arboretum
15400 Montalvo Rd.
P.O. Box 158
Saratoga, CA 95071-0158
408-741-3421

1911 Mediterranean-style villa.
Garden: 175-acre arboretum with oaks, firs, and other mature specimens, and formal gardens with plants from all over the world.

CO

17.66

Cross Orchards Living History Farm
3073 F Rd.
Grand Junction, CO 81504
303-434-9814

A restoration of the 1896–1923 farm—one of the largest
apple orchards and packaging facilities in Colorado.
Garden: 150 apple trees. Black twig, Gano, Jonathan,
Rome, beauty, winesap, and Ben Davis; are in the
process of replacing semi-dwarf trees planted in 1950s
with standard trees of original varieties.
Museum: Interprets the Cross Ranch apple operation.

17.67

Four Mile Historic Park
715 S. Forest St.
Denver, CO 80202
303-399-1859

Living history museum dealing with many aspects of early
Colorado and agricultural life.
Garden: Heirloom field crops and ornamental and kitchen
gardens with historic varieties of herbs and Victorian
roses.

CT

17.68

American Indian Archaeological Institute
P.O. Box 260
Curtis Rd.
Washington, CT 06793
203-868-0518

Research institute and museum of Indian New England.
Garden: Native trees used for food, beverage, medicine,
technology, dyes, and for other reasons by Native
Americans.

Butterbrooke Farm
Oxford, CT

Cross-reference: 17.8

17.69

Harkness Memorial State Park
275 Great Neck Rd.
Waterford, CT 06385
203-443-5725

234-acre estate and mansion.
Garden: Italian garden, oriental garden, rock garden, and
flower gardens.

17.70

Heckscher Farm
Stamford Museum and Nature Center
39 Scofieldtown Rd.
Stamford, CT 06903
203-322-1646

Restored 1750s New England farm.
Garden: Vegetable and herb gardens. Salad garden,
edible flowers, annual herbs, cold crops, heirloom
vegetables, hybrid vegetables, and perennial herbs.

17.71

Ogden House
1520 Bronson Rd.
Fairfield, CT 06430
203-259-6356

Mid-18th-century farmhouse.
Garden: Demonstration kitchen and herb garden.

17.72

Stowe Day Foundation
77 Forest St.
Hartford, CT 06105
203-522-9258

Restored Harriet Beecher Stowe house.
Garden: Victorian-style round and oval gardens,
cottage-style "Gertrude Jekyll" perennial gardens. No
cultivar plantings date after 1870; period roses;
*Mertensia virginica, Chionanthus virginicus, Cladrastis
lutea, Macleaya cordata.*
Library: 19th-century seed and plant catalogs.

DC

17.73

Bishop's Garden
Washington National Cathedral
Mount St. Albans
Washington, DC 20016
202-537-6200
Location: Wisconsin and Massachusetts Aves., N.W.

Close garden adjacent to the National Cathedral.
Garden: Enclosed garden of trees, shrubs, herbs, and
perennials; includes herbs grown during the 9th
century, and a boxwood grown from a cutting of Dolly
Madison's bridal bouquet.

17.74

Dumbarton Oaks
1703 32nd St., N.W.
Washington, DC 20007
202-342-3200

Research center and museum for Byzantine civilization,
pre-Columbian cultures, and the history of landscape
architecture.
Garden: 14 acres of formal and naturalized gardens with
numerous species and mature specimens of black oak,
katsura, and beech.
Grants: Research grants.
Scholarships: Fellowships in the history of landscape
architecture and garden history.
Library: 14.18
Other: Symposia, lectures, and seminars.

17.75

Franciscan Monastery
1400 Quincy St., N.E.
Washington, DC 20017
202-526-6800

Monastery established in 1900.
Garden: Daffodils, lilies, roses, azaleas, and
rhododendrons.

17.76

Hillwood
4155 Linnean Ave., N.W.
Washington, DC 20008
202-686-5807

Museum of decorative art in the former mansion of
Marjorie Merriweather Post.
Garden: Rose garden, Japanese garden, parterre gardens,
rhododendrons, and a greenhouse with an orchid
collection.

17.77

Potato Museum
704 North Carolina Ave., S.E.
Washington, DC 20003
202-544-1558

Museum: Exhibits on the history and use of the potato
with over 200 items.
Library: Number of books: 500; number of periodicals:
200.
Lending Policy: Reference only.
Other: Publishes the bimonthly *Peelings*.

DE
17.78

Delaware Agricultural Museum
866 N. Du Pont Highway
Dover, DE 19901
302-734-1618

Agricultural museum.
Garden: Vegetable and flower garden with varieties
typical of those grown in the late 19th century.

17.79

Hagley Museum and Library
P.O. Box 3630
Wilmington, DE 19807
302-658-2400
Location: State Rt. 141

A museum of American industrial history and a historic
site on the property where the du Pont gunpowder
yards were established in 1802.
Garden: Historic gardens of the 1830s, 1880s, and 1930s
including a French-style ornamental garden and a
worker's vegetable garden. Heirloom varieties of
flowers, bulbs, and fruit trees, native wildflowers, and
trees.
Library: Number of books: 180,000; number of
periodicals: 225. Limited collections of garden history
and horticulture.
Lending Policy: Reference only.

17.80

Homestead, The
12 Dodds Ln.
Henlopen Acres
Rehoboth Beach, DE 19971
302-227-8408

1743 home.
Garden: Boxwood garden, herb garden, sundial garden,
and foxglove, begonia, roses, sedums, and bulbs.

17.81

Nemours Mansion and Gardens
1600 Rockland Rd.
Wilmington, DE 19803
302-651-6905; (FAX) 302-651-4019

Early 20th-century mansion built by Alfred I. du Pont.
Garden: 300 acres of formal gardens with annuals,
perennials, conifers, and other woody specimens.

17.82

Rockwood Museum
610 Shipley Rd.
Wilmington, DE 19809
302-571-7776

Gothic-style estate of the mid-19th century.
Garden: 201 varieties of trees and shrubs, a walled
garden, and a pleasure garden.

17.83

Winterthur Museum and Gardens
Winterthur, DE 19735
302-888-4600; (FAX) 302-888-4700

Mansion of Henry Francis du Pont housing decorative
and fine arts.
Garden: 60-acre garden with a naturalistic style of
landscape design. Azaleas, magnolias, winter hazels,
sycamores, wildflowers, ferns, shrubs, perennials,
Japanese dogwoods, peonies, and a pinetum.

FL
17.84

Cummer Gallery of Art
829 Riverside Dr.
Jacksonville, FL 32204
904-356-6857

Art museum.
Garden: English and Italian formal gardens.

17.85

Eden State Gardens
Point Washington, FL 32454
904-231-4214

1895 Greek Revival mansion.
Garden: 11 acres with oaks, camellias, and azaleas.

17.86

Forest Capital State Museum
204 Forest Park Dr.
Perry, FL 32347
904-584-3227

Forest history museum.

17.87

Hemingway House
907 Whitehead St.
Key West, FL 33040
305-294-1575

Spanish Colonial-style home of Ernest Hemingway, built
in 1851.
Garden: Tropical garden.

17.88
Opal Lily Gourd Museum
Coral Farms Rd.
P.O. Box 265
Florahome, FL 32635
904-659-2121
Gourd museum.

17.89
Ringling Museums
P.O. Box 1838
Sarasota, FL 33578
813-355-5101
Location: U.S. 41.
Art and circus museum and a restored theater.
Garden: Rose garden, parterre garden, and a secret garden.

17.90
Thomas Edison Winter Home and Botanical Gardens
2350 McGregor Blvd.
Fort Meyers, FL 33901
813-334-3614
Winter home of Thomas Edison and his rubber research
  laboratory.
Garden: Over 400 species of tropical and subtropical flora
  including palms, tropical fruit, cycads, ficus, flowering
  trees, and vines. There are many rare trees including a
  sausage tree, a dynamite tree, a banyan tree, and a
  sloth tree of the West Indies.

17.91
Vizcaya
3251 S. Miami Ave.
Miami, FL 33129
305-854-6559
Location: Biscayne Bay.
Early 20th-century classical Italian mansion and 10-acre
  estate.
Garden: Classical Italian garden.

17.92
Walt Disney World
P.O. Box 40
Dept. GL
Lake Buena Vista, FL 32830
305-824-8000
Location: U.S. 192.
Theme park.
Garden: Flower beds, palms, bamboos, flowering trees,
  and a Garden of the World Showcase with gardens of
  nine countries.
Cross-references: 20.9, 20.99, 20.273.

GA
17.93
Atlanta Historical Society
3101 Andrews Dr., N.W.
Atlanta, GA 30305
404-238-0654
A 30 acre museum complex.
Garden: Seven distinct gardens within a woodland
  setting. Each garden represents a part of the history of
  land use and gardening in the Atlanta area by its style

and choice of plant material. Colonies of endangered
native plants, southeastern wildflowers and native
azaleas, and plants commonly grown in the 1840s.
Library: 14.27

17.94
Southern Forest World
Rt. 5, Box 406B
Waycross, GA 31501
912-285-4260
Forest museum.

HI
17.95
Contemporary Museum and Garden
2411 Makiki Heights Dr.
Honolulu, HI 96822
808-526-1322
House of Alice Cooke Spalding, now a private art gallery.
Ground covers, tropical shade plants, and macadamia,
  kopsia, and banyan trees.

17.96
Kipahulu District, Living Farm Area
P.O. Box 97
Hana, HI 96713
Living historical farm.
Garden: Plants used by Hawaiians before the arrival of
  Captain Cook in 1778.

Limahuli Botanical Garden
Hanalei, HI
Cross-reference: 16.118

17.97
Queen Emma Summer Palace and Garden
2913 Pali Hwy.
Honolulu, HI 96817
808-595-6291
Museum of Hawaiian history in a 19th-century mansion.
Garden: Spider lilies, native plants, and specimen trees.

17.98
Senator Fong's Plantation and Garden
47-285 Pulama Rd.
Kaneohe, HI 96744
808-239-6755
Estate of lawyer, businessman, and former U.S. Senator
  Hiram Fong.
Exotic ornamentals plus fruit and nut trees.

IA
17.99
Brucemoore
2160 Linden Dr., S.E.
Cedar Rapids, IA 52403
319-362-7375
Queen Anne-style mansion with lawns, formal gardens,
  and a duck pond.
Garden: Rose and perennial borders in continuous cultiva-
  tion since 1915. Iris, roses, peonies, daylilies and lilies.
Other: A property of the National Trust for Historic
  Preservation.

17.100

Living History Farms
2600 N.W. 111th
Des Moines, IA 50322
515-278-5286

600-acre open-air agricultural museum.
Garden: Historic gardening and farming practices
demonstrated with a 1700 Ioway Indian farm, an 1850
pioneer farm, and a 1900 farm with heirloom varieties.
Orchards, herb gardens, and vegetable beds.

17.101

Old Fort Madison
City of Fort Madison
811 Ave. E
Fort Madison, IA 52627
319-372-7700

Reconstructed 1808 U.S. Army fort and trading post.
Garden: Heirloom European herb, vegetable, Native
American, and utility garden areas. Over 110 plant
specimens, including 15 different tea plants, ethnic
vegetables, famous North American plants like
Echinacea, *Angelica purpurea*, and *Lobelia inflata*, dye
plants, insect-repelling strewing herbs, and several
Georgian-era everlasting flowers.
Museum: Period-style drying rack exhibit with seasonally
harvested herbs and plants.
Seed/specimen exchange.
Other: Rose bush, wild plant, and mushroom
identification.

Seed Savers Exchange
Decorah, IA
Cross-reference: 17.32

IL
17.102

Cantigny Gardens
1S151 Winfield Rd.
Wheaton, IL 60187
708-668-5161; (FAX) 708-668-5332

Former estate of Robert R. McCormick with 500 acres.
Garden: Ten acres of gardens designed in 1966 by Franz
Lipp at the former estate of Robert R. McCormick.
Seventeen area groupings including a columnar tree
collection.
Cross-reference: 20.115

17.103

Clayville Rural Life Center and Museum
R.R. 1
Pleasant Plains, IL 62677
217-626-1132

Rural life museum with inn/farmhouse built in late 1820s.
Garden: 1850s heirloom vegetable garden. Vegetable
varieties from the 1850s.
Library.
Lending Policy: Reference only.
Other: Classes for children on organic and historic
gardening.

17.104

Garfield Farm Museum
3N016 Garfield Rd.
Box 403
LaFox, IL 60147
708-584-8485

1840s northern Illinois living history farm and museum.
Garden: 19th-century heirloom vegetable garden.
Other: Member of the Seed Savers Exchange.

17.105

Lake of the Woods Botanic Garden
Early American Museum
P.O. Box 336
Mahomet, IL 61853
217-586-3360

Location: State Rt. 47
Pioneer life museum.
Garden: Antique rose garden, herb garden, dye plant
garden, a conservatory with tropicals, and daylilies,
peonies, irises, and wildflowers.
Cross-reference: 20.118

17.106

Lincoln Log Cabin Historical Site
R.R. 1, Box 175
Lerna, IL 62440
217-345-6489

The last home of Abraham Lincoln's father and
stepmother, Thomas and Sarah Bush Lincoln, with
1840s living historical farm.
Garden: Vegetable garden, orchard, and crops utilizing
19th-century varieties.

17.107

Lincoln Memorial Garden and Nature Center
2301 E. Lake Dr.
Springfield, IL 62707
217-529-1111

Garden: Trees, shrubs, grasses, and flowering plants
present in central Illinois in the 1850s.
Scholarships: Scholarship for college students who do
internship at the garden.

17.108

Naper Settlement
201 West Porter Ave.
Naperville, IL 60540

Living historical site re-creating the 1830s.
Garden: Flower and herb gardens with plants used in the
1830s.

17.109

Shakespeare Garden
Garden Club of Evanston
2703 Euclid Pl.
Evanston, IL 60201
312-864-0655
Location: Sheridan Rd.

Shakespeare garden.
Garden: Flowers and herbs mentioned in Shakespeare's
plays and poems.

IN
17.110
Billie Creek Village
R.R. 2, Box 27
Rockville, IN 47872
317-569-3430

Re-created turn-of-the-century village and farmstead.
Garden: Old-style herb garden and grape arbor.

17.111
Buckley Homestead
3606 Belshaw Rd.
Lowell, IN 46356
219-769-PARK

160-acre living history farms of 1840s and 1910s.
Garden: Historic vegetable garden and herb garden.

17.112
Conner Prairie
13400 Allisonville Rd.
Noblesville, IN 46060
317-776-6000; (FAX) 317-776-6014

1836 living history museum, village, and estate.
Garden: Re-creation of kitchen and other gardens
   common to central Indiana in the 1830s.
Library: Number of books: 3,500; number of periodicals:
   30.
Microfilm newspapers of 1820s–1840s.
Lending Policy: Reference only.

17.113
Eli Lilly Botanical Garden of the Indianapolis Museum of
   Art
1200 West 38th St.
Indianapolis, IN 46208
317-923-1331; (FAX) 317-926-8931

Art museum.
Garden: 120 acres of landscaped grounds, and gardens of
   perennials and annuals and a historic garden designed
   by the Olmsted Brothers in 1920.
Library: Number of books: 2,000.
Lending Policy: Circulating.

Huntington College Arboretum
Huntington, IN
Cross-reference: 16.154

17.114
Lincoln Boyhood National Memorial
Lincoln City, IN 47552
812-937-4541

Boyhood home where Abraham Lincoln lived from
   1816–1830.
Garden: Re-creation of 1820s garden and field crop
   plantings. Yellow gourd corn, old red chaff bearded
   wheat, Tennessee green seed cotton, old Connecticut
   pumpkins, and heirloom apples, peaches, plums, and
   pears.

17.115
Robert Lee Blaffer Trust
Box 581
New Harmony, IN 47631
812-682-4431

Utopian community of the early 19th century.
Garden: Period herb gardens, plus rose gardens and a
   labyrinth.

KY
17.116
Ashland
Sycamore Rd.
Lexington, KY 40502
606-266-8581

19th-century home of Henry Clay.
Garden: Parterre garden and herb garden. Boxwood, holly,
   hornbeam trees, herbaceous borders, roses, herbs.

17.117
Farmington Historic Home Museum
3033 Bardstown Rd.
Louisville, KY 40205
502-452-9920

Home designed by Thomas Jefferson and built in 1810.
Garden: Kitchen garden, herb garden, perennial garden,
   and orchard. Gardens restored with plants known to be
   used before 1820.

17.118
Homeplace 1850
TVA/ Land Between the Lakes
100 Van Morgan Dr.
Golden Pond, KY 42211-9001
502-924-5602

Living history museum.
Garden: Garden crops and field crops representative of
   those grown in the region in the mid-19th century.
   Heirloom field crops, including Swedish oats, red May
   wheat, white and yellow flint corn, white Swedish oats,
   white gourd seed corn, and an heirloom orchard with
   plums, peaches, and apples.

17.119
Locust Grove Historic Home
561 Blankenbaker Ln.
Louisville, KY 40207
502-897-9845

1790 Georgian mansion.
Garden: 18th-century border garden with flowers and
   herbs.

LA
17.120
Acadian House Museum
P.O. Box 497
St. Martinville, LA 70582
Location: 1200 North Main.

Plantation home.
Garden: Extensive gardens with many rare and unusual
   plants of the 18th century including heirloom varieties
   of peppers, squash, okra, cotton, and more.

17.121
Audubon State Commemorative Area
P.O. Box 546
St. Francisville, LA 70775
504-635-3739
Location: State Highway 965.

Former home of John James Audubon.
Garden: 40 acres of historical gardens containing plants
from the early 1800s including annuals, camellias, and
azaleas, a vegetable garden with heirloom crops, and an
1858 hothouse with period tools.

17.122
Biedenharn House and Elsong Gardens
2006 Riverside Dr.
Monroe, LA 71201
318-387-5281

1914 home of Joseph August Biedenharn and Bible
museum.
Garden: White garden, oriental garden, woodland garden,
and a "Four Seasons Garden."

17.123
Hermann-Grima Historic House
820 St. Louis St.
New Orleans, LA 70112
504-525-5661

Restored 1831 house of a German merchant.
Garden: Restored New Orleans French Quarter parterre
garden; exotic perennials, climbing roses, and jasmines,
most available prior to 1860.

17.124
Live Oak Gardens and Joseph Jefferson Home
284 Rip Van Winkle Rd.
New Iberia, LA 70561
318-367-3485

19th-century Georgian-style home.
Garden: English garden, Alhambra garden, oriental
garden, old and new camellia gardens, rose garden,
magnolia garden, tropical garden, woodland garden, and
annuals and perennials.

17.125
Longue Vue House and Gardens
7 Bamboo Rd.
New Orleans, LA 70124
504-488-5488; (FAX) 504-486-7015

1936 Greek Revival-style mansion with eight acres of
gardens and fountains.
Garden: Series of interconnecting gardens. The South
Lawn was designed to reflect Louisiana's Spanish
heritage. *Camellia, Magnolia, Rosa*, native plants.
Museum: "Art in Flowers," an exhibit of botanically
correct flora of Louisiana's marshes and woodlands.
Library: Number of books: 200; number of periodicals:
100. 12-volume collection (1849–1860) of *Nouvelle
Iconographie des Camellias.*
Other: Horticultural programs.

17.126
R. S. Barnwell Memorial Garden and Art Center
601 Clyde Fant Parkway
Shreveport, LA 71101
318-425-6495

Garden and art museum.
Garden: Plant conservatory, outdoor gardens, small
fragrance garden for handicapped. Orchidaceae,
Bromeliaceae, tropicals, and ornamental trees and
shrubs.
Library: 14.44
Other: Horticultural seminars.

17.127
Rosedown Plantation
P.O. Box 1816
St. Francisville, LA 70775
504-635-3332
Location: State Highway 10.

19th-century mansion and 2,000-acre estate.
Garden: Formal French 17th-century garden with azaleas,
camellias, hydrangeas, cryptomerias, gardenias, crape
myrtles, and deutzias. Also a kitchen garden, herb
garden, flower gardens, and old and new roses.

17.128
Shadows-on-the-Teche
317 East Main St.
New Iberia, LA 70560
318-369-6446

19th-century Classic-Revival manor house.
Garden: *Quercus virginiana*, Spanish moss, *Camellia.*
Other: A property of the National Trust for Historic
Preservation.

MA
17.129
Adams National Historic Site
135 Adams St.
P.O. Box 351
Quincy, MA 02269
617-773-1177

Home of John Quincy Adams and subsequent generations
of Adamses.
Garden: 19th-century flower garden and a 1731 boxwood
hedge.

17.130
Chesterwood
P.O. Box 827
Box 248
Stockbridge, MA 01262
413-298-3579

1898 studio of sculptor Daniel Chester French.
Garden: 1900–1931 New England gardens designed by
French at his Georgian revival house.
Library: Number of books: 684.
Pre-1931 books on gardens and landscaping.
Other: A property of the National Trust for Historic
Preservation.

17.131

Codman House
Codman Rd.
P.O. Box 429
Lincoln, MA 01773
617-259-8843

1730 Georgian mansion.
Garden: Restored Italian garden.

17.132

Essex Institute
132 Essex St.
Salem, MA 01970
508-744-3390

Local history museum with several gardens.
Garden: 1800 garden based on a plan of E. H. Derby's
   garden and a Colonial Revival summer flower garden at
   Ropes Mansion.
Museum: Exhibit on the restoration and planning of the
   1800 Derby garden.
Grants: Fellowship in history and culture, not specific to
   any topic.
Library: Many books from 17th–19th centuries on
   landscape history and horticulture.
Lending Policy: Reference only.

17.133

Fisher Museum of Forestry
Harvard University
Harvard Forest
Petersham, MA 01366
508-724-3302; (FAX) 508-724-3595

Forest, museum, and research facility.
Garden: 200-hectare research forest; one of the oldest
   continually managed forests in the United States.
Museum: 23 dioramas portraying the history of central
   New England forests, from primeval forest, through
   agricultural clearing, reforestation, and forest
   management.
Grants: Bullard Fellowships for advanced study and
   research in forest-related fields.
Library: Number of books: 24,500; number of periodicals:
   150.
Forest biology, forest management, forest soils, and
   international forestry.
Lending Policy: Reference only.
Other: Nature and hiking trails.

17.134

Glen Magna
Danvers Historical Society
P.O. Box 381
Danvers, MA 01923
617-774-9165
Location: 57 Forest St.

Home of the Peabody-Endicott families.
Garden: Formal and informal gardens.

17.135

Hancock Shaker Village
P.O. Box 898
Pittsfield, MA 01202
413-443-0188

Restored Shaker village.
Garden: Demonstration garden of over 100 varieties of
   herbs the Shakers grew for their herb and seed
   business. Aconite, maidenhair fern, Jack-in-the-pulpit,
   backache brake, belladonna, black cohosh, jimson
   weed, hops, poppy, vervain, coltsfoot, and pleurisy
   root.

17.136

Heritage Plantation of Sandwich
Box 566
Grove St.
Sandwich, MA 02563
508-888-3300

Diversified museum of Americana.
Garden: 75 acres of over 10,000 plants. Dexter
   rhododendrons; 550 species and cultivars of
   Hemerocallis, *Ilex*, and herbs.

17.137

Isabella Steward Gardner Museum
280 The Fenway
Boston, MA 02115
617-566-1401

Fine arts museum.
Garden: Inner court garden and outside gardens.
   Jasminum, Orchidaceae, azaleas.

17.138

Jeremiah Lee Mansion
16 Beacon St.
Marblehead, MA 01495
617-631-1069

1768 mansion.
Garden: Period herb garden, sunken garden, and flower
   beds with plants of the Colonial period.

17.139

John Whipple House
Ipswich Historical Society
53 South Main St.
Ipswich, MA 01938
617-356-2811

Restoration of 1640 house.
Garden: Rose garden with heritage roses, formal flower
   beds, and herb garden with plants used in the 17th
   century.

17.140

Longfellow National Historic Site
105 Brattle St.
Cambridge, MA 02138
617-876-4491

Georgian-style house of Henry Wadsworth Longfellow.
Garden: Shrubs, perennials, annuals, and an herb garden.

17.141
Lyman Estate, The
154 Main St.
Waltham, MA 02154
617-891-7095

Museum, greenhouse, and grounds.
Garden: Greenhouses (the oldest known to be standing in New England) and an English-style landscape. 19th- and early 20th-century exotic plants, black Hamburg grapes, green muscat, *Camellia japonica*.

17.142
Naumkeag Museum and Gardens
The Trustees of Reservations
Prospect Hill Rd.
P.O. Box 792
Stockbridge, MA 01262
413-298-3239

Mansion designed by Stanford White in 1885.
Garden: Garden designed by Fletcher Steele and Mabel Choate; terraced gardens, formal flower beds.
Library: Correspondence between Mabel Choate and Fletcher Steele.

17.143
Old Sturbridge Village
1 Old Sturbridge Village Rd.
Sturbridge, MA 01566
508-347-3367; (FAX) 508-347-5375

Living history museum of 1830s rural New England town.
Garden: Re-created flower, vegetable, and herb plantings depict New England gardens of the early 19th century. Also a large formal exhibit garden with over 300 herb varieties. Majority of plantings are documented 19th-century varieties.
Library: Number of books: 30,800; Number of periodicals: 100. Collection of 19th-century garden advice books, periodicals, seed and nursery catalogs.
Lending Policy: Reference only.
Seed/specimen exchange.

17.144
Plimoth Plantation
P.O. Box 1620
Warren Ave.
Plymouth, MA 02360
508-746-1622; (FAX) 508-746-4978

Living history museum of 17th-century Plymouth.
Garden: 13 kitchen gardens. Vegetables, flowers, herbs, shrubs, and trees of the 17th century.

17.145
St. Mary's Church
3055 Main St.
Barnstable, MA 02630
508-362-3977

Garden: Daffodils, herb garden, wall gardens, annual garden.

17.146
Sedgwick Gardens
The Trustees of Reservations
572 Essex St.
Beverly, MA 01915
617-922-1536

114-acre estate with gardens originally designed by Mabel Cabot Sedgwick.
Garden: Hosta garden, lotus pool, gray garden, tree peony garden, cut-flower garden, flowering trees, azaleas, and rhododendrons.

MD
17.147
Carroll County Farm Museum
500 S. Center St.
Westminster, MD 21157
301-848-7775; (FAX) 301-848-0003

Farm museum in former almshouse.
Garden: Old-fashioned rose garden with 26 varieties of pre-1900 rose species. Austrian copper, Father Hugo's rose, apothecary rose, rose mundi, white rose of York, Konigin von Danemark, York and Lancaster, celsiana, Jacques Cartier, Comte de Chambord, old blush, blush noisette, souvenir de la Malmaison.

17.148
Hampton National Historic Site
535 Hampton Ln.
Towson, MD 21204
301-962-0688

18th-century Georgian-style mansion and estate.
Garden: Formal gardens, exotic trees, herb garden, and 19th-century English-style landscape.

17.149
London Town Publik House and Gardens
839 Londontown Rd.
Edgewater, MD 21037
301-956-4900

18th-century inn and tavern.
Garden: 12 acres of gardens with native plants and exotics including magnolias, rhododendrons, cherries, Japanese irises, wildflowers, a winter garden, herbs, and 18th-century field crops.

17.150
National Colonial Farm
3400 Bryant Point Rd.
Accokeek, MD 20607
301-283-4201

Re-creation of a middle class, southern Maryland tobacco plantation of the mid-18th century.
Garden: Field crops, herb garden, and kitchen garden. Heirloom vegetables including Virginia gourd seed corn, red May wheat, and several bean varieties.

17.151
William Paca Garden
1 Martin St.
Annapolis, MD 21401
301-267-6656

Garden: Restored 18th-century pleasure garden and 18th-century species rose parterre. European garden plants of the 18th century and native species popular in that era.
Museum: Small display introduces the horticulture of Paca's day.
Library: Number of books: 300; number of periodicals: 400. Books on native plants and 18th-century horticulture, plant catalogs, and color slides of antique plants and native plants.

ME
17.152
Hamilton House
Vaughan's Ln.
South Berwick, ME 03908
207-384-5269

Restored 1785 mansion.
Garden: Terraced formal gardens.

MI
17.153
Cranbrook House and Gardens
380 Lone Pine Rd.
Box 801
Bloomfield Hills, MI 48013
313-645-3149

Turn-of-the-century English manor owned by George Booth.
Garden: Formal gardens including a sunken garden, herb garden, rose garden, and an English garden.

17.154
Detroit Garden Center
1460 E. Jefferson
Detroit, MI 48207
313-259-6363

Garden center.
Garden: Re-creation of 1850 garden for modest home. *Hamamelis virginiana, Asarum canadense, Hepatica, Polygonum cuspidatum, Cimicifuga racemosa, Trillium grandilforum, Arisaema triphyllum, Cerastium tomentosum, Arabis caucasica, Mertensia virginica, Malva alcea*, and *Geranium sanguineum.*
Library: 14.56

17.155
Grand Hotel
Mackinac Island, MI 49757
906-847-3331

Victorian hotel.
Garden: Lilac garden, gazebo garden, English gardens and begonias, daylilies, and bedding plants.

17.156
Greenfield Village
P.O. Box 1970
Dearborn, MI 48121

313-271-1620
Location: Village Rd. and Oakwood Blvd.
Historical homes, shops, and buildings in a re-created village museum of American industrial and rural history.
Garden: Vegetable and flower gardens in styles of the early 19th century.

MN
17.157
Forest History Center
Minnesota Historical Society
2609 County Rd. 76
Grand Rapids, MN 55744
218-327-4482

Forest history museum.
Other: Woodland trail.

17.158
Oliver H. Kelley Farm
15788 Kelley Farm Rd.
Elk River, MN 55330
612-441-6896

Home and living history farm of Oliver Kelley of the mid-19th century.
Garden: Fields, orchards, and vegetable gardens with varieties from 1850–1876.

17.159
Schell Mansion and Gardens
Jefferson St. South
New Ulm, MN 56073
507-354-5528

19th-century mansion of August Schell.
Garden: Formal Victorian gardens.

17.160
Stoppel Farm
Olmsted County Historical Society
P.O. Box 6411
Rochester, MN 55903
507-282-9447

Living history farm.
Garden: Heritage vegetable garden.
Library: Number of books: 51,000; number of periodicals: 50.
Lending Policy: Reference only.

MO
17.161
Chance House and Gardens
The Chance Foundation
123 N. Rollins St.
P.O. Box 128
Centralia, MO 65240
314-682-5711

Turn-of-the-century Victorian house, now a historical museum.
Garden: Formal garden with roses, perennials, and annuals.

335

17.162
Missouri Town 1855
Jackson County Parks and Recreation
Independence Square Courthouse
Room 205
Independence, MO 64050
816-881-4431

1850s pioneer village.
Garden: Historical herb garden, flower garden, and dye garden.

17.163
Watkins Woolen Mill State Historic Site
R.R. 2, Box 270M
Lawson, MO 64062
816-296-3357

Living history farm interpreting the 1870s.
Garden: Heirloom vegetable garden.

MS
17.164
Beauvoir
200 W. Beach Blvd.
Biloxi, MS 39531
601-388-1313

Restored plantation house of Jefferson Davis.
Garden: Azaleas and camellias.

17.165
Florewood River Plantation
P.O. Box 680
Greenwood, MS 38930
601-455-3821

1850s cotton plantation.
Garden: Orchard, vegetable garden, and field crops.

17.166
Palestinian Gardens
Rt. 9, Box 792
Lucedale, MS 39452
601-947-8422

Garden: Garden is a scale model of the Holy Land with shrubs and plants mentioned in the Bible.

NC
17.167
Biltmore Estate
One Biltmore Plaza
Asheville, NC 28803
704-274-1776

250-room French chateau.
Garden: 25 acres of formal gardens including a 4-acre English garden, and collections of native laurel, azaleas, rhododendrons, and hollies.

17.168
Duke Homestead State Historic Site
2828 Duke Homestead Rd.
Durham, NC 27706
919-477-5498

Original farm home of Washington Duke.
Garden: 1870s herb garden with herbs for flavoring food, fragrance, and natural dyeing.

17.169
Elizabethan Gardens
P.O. Box 1150
Manteo, NC 27954
919-473-3234

A garden memorial to the first English colonists in the New World.
Garden: Shakespearean herb garden with 45 herbs, wildflower garden with plants indigenous to the Outer Banks, and 500 varieties of camellias. English garden design of the 16th century.

17.170
Horne Creek Living History Farm
Box 118-A
Rt. 2
Pinnacle, NC 27043
919-722-9346

1900s living history farm.
Garden: Fields, orchard, and vegetable garden with 19th-century varieties of field crops (including tobacco and field corn), fruit trees, and vegetables.

17.171
Marvin Johnson's Gourd Museum
P.O. Box 666
Fuquay-Varina, NC 27526
919-639-2894

Gourd museum.

17.172
North Carolina State Museum of Natural Sciences
P.O. Box 26747
Raleigh, NC 27611
919-733-7450

Science museum.
Garden: Butterfly garden.

17.173
Old Salem, Inc.
Drawer F, Salem Station
Winston-Salem, NC 27108
919-721-7300

Re-created landscape surrounding restored buildings in 1766–1858 Moravian congregation town.
Garden: Kitchen gardens, orchards and fields with historically correct plants; and native plant arboretum. Many 18th- and early 19th-century vegetables, flowers, herbs, and fruit including fiber flax, African storage gourds, Virginia gourdseed corn, and hops.
Library: Number of books: 125; number of periodicals: 6.

17.174
Orton Plantation Gardens
R.F.D. 1
Winnabow, NC 28479
919-371-6851

Garden: 20 acres of gardens including *Camellia*, azaleas, fruit trees, *Quercus virginiana*, and *Wisteria*.

17.175
Tryon Palace Restoration
P.O. Box 1007
610 Pollock St.
New Bern, NC 28563
919-638-1560

Restored 18th-century palace.
Garden: 18th-century-style ornamental gardens with
    pre-1770 native and imported plants. *Gordonia
    lasianthus, Camellia sinensis, Laurus nobilis, Quercus cerris,
    Halesia carolina, Cedrus libani.*
Library: Number of books: 200; number of periodicals: 12.
Lending Policy: Reference only.
Other: Monthly garden workshops.

NE
17.176
Arbor Lodge Arboretum
Rt. 2
Nebraska City, NE 68410
402-873-7222

Neocolonial mansion that was the home of J. Sterling
    Morton, the originator of Arbor Day.
Garden: Italian terraced garden, pine grove, tree trail, and
    prairie plants garden. Mature stand of American
    chestnut trees, and old specimens of Osage orange.

NH
17.177
Fuller Gardens
10 Willow Ave.
North Hampton, NH 03862
603-964-5414

Early 20th-century estate.
Garden: All-America Rose Selections display garden with
    1,500 rose bushes, All-America Selections hosta display
    garden, and annual and perennial plantings.
Cross-reference: 20.161

17.178
Mofatt-Ladd House
154 Market St.
Portsmouth, NH 03801
603-436-8221

18th-century house.
Garden: Formal flower beds, terraces, peonies, herbs,
    roses, and perennials all typical of the period.

17.179
Strawberry Banke Museum
454 Court St.
Portsmouth, NH 03801
603-433-1100

35 historic houses ranging from 1695–1945.
Garden: Four period gardens: circa 1720s raised bed herb
    and vegetable garden, 1830s landscape, 1860s Victorian
    garden, and 1900 Colonial Revival garden of Thomas
    Bailey Aldrich. Heirloom vegetables, a variety of herbs,
    Victorian annuals, and old roses.

Museum: Exhibit on the excavation of the c.1720s
    Sherborne Garden showing artifacts, pollen and seed
    analysis, and post hole evidence.
Library: Number of books: 6,000.
Lending Policy: Reference only.
Seed/specimen exchange.

NJ
17.180
Acorn Hall
Morris County Historical Society
68 Morris Ave.
Box 170M
Morristown, NJ 07963
201-267-3465

Victorian-style house built in 1853.
Garden: Restored Victorian gardens retaining original
    plants and flowers, and shrubs appropriate for the
    period: old-fashioned roses, original hydrangeas, mock
    oranges, Korean azaleas, and a Katsura tree.
Library: Number of books: 2,500; number of periodicals:
    100. Victorian Research Library has some horticultural
    books from the Victorian era.
Lending Policy: Reference only.

17.181
Barclay Farmstead Museum
Barclay Ln.
Cherry Hill, NJ 08034
609-795-6225

Garden: Early American herb garden of plants that played
    an important role in the life of a 19th-century Quaker
    farm family. Medicinal, culinary, dye, and fragrance
    plants.
Library.
Lending Policy: Reference only.

17.182
Botto House National Landmark
American Labor Museum
83 Norwood St.
Haledon, NJ 07508
201-595-7953

1908 home of Italian immigrant silk workers.
Garden: Vegetable garden, fruit trees, ornamental beds,
    and grape arbor.

17.183
Clinton Historical Museum
56 Main St.
Clinton, NJ 08809
201-735-4101

Rural history museum.
Garden: 25-sq. ft. formal colonial herb garden.
Library.
Lending Policy: Reference only.

17.184
Fosterfields Living Historical Farm
P.O. Box 1295
Morristown, NJ 07962-1295
201-326-7645

Living historical farm representing 1880–1910.
Garden: Historic orchard, vegetable garden, and field
  crops.

17.185
Georgian Court College
Lakewood Ave.
Lakewood, NJ 08701
201-364-2200

Location: U.S. 9.
Former estate of George Jay Gould.
Garden: Italian garden, sunken garden, formal garden,
  and a Japanese garden.

17.186
Historic Allaire Village
Rt. 524
Allaire, NJ 07727
201-938-2253

Re-created 1830s village.
Garden: Vegetable garden with period varieties.

17.187
Leaming's Run Gardens and Colonial Farm
1845 Rt. 9 North
Cape May Court House, NJ 08210
609-465-5871

Replica of a 1695 New Jersey farm.
Garden: Vegetable and herb garden and an annual
  garden.

17.188
Oakside Bloomfield Cultural Center
240 Belleville Ave.
Bloomfield, NJ 07003
201-429-0960

31-room Victorian mansion.
Garden: Water garden, kitchen garden, rose garden with
  historical roses, and a solarium garden.
Other: Community vegetable garden.

17.189
Wicks House
Morristown National Historical Park
Morristown, NJ 07961
201-543-4030

Restored 18th-century farmhouse.
Garden: 18th-century-style herb and vegetable garden
  using plants typical of the period.

NM
17.190
El Rancho de las Golondrinas—A Living History Museum
Rt. 14, Box 214
Sante Fe, NM 87505
505-471-2261

Museum: Depicts Spanish colonial life in New Mexico
  from 1700 to mid-1800s.
Garden: Indigenous crops of the Southwest mountains
  and crops that were introduced by Spanish colonization;
  also a Mountain Village Herb Garden of native and
  introduced herbs. Gourds, Chimayo melon, sorghum,
  San Juan melon squashes, Navajo blue corn, New
  Mexico bolitas, "Rainbow" corn, and maiz concho.
Other: Limited seed exchange for local organizations, and
  herb workshops.

Talavaya Center
Espanola, NM
Cross-reference: 17.36

NY
17.191
Abigail Adams Smith Museum
421 East 61st St.
New York, NY 10021
212-838-6878

18th-century coach house.
Garden: 18th-century-style flower and herb garden.

17.192
Astor Chinese Garden Court
Metropolitan Museum of Art
Fifth Ave. at 82nd St.
New York, NY 10028
212-879-5500; (FAX) 212-570-3879

Astor Garden Court was modeled after a small courtyard
  originally constructed during the Ming dynasty
  (1368–1644) in the Garden of the Master of the Fishing
  Nets in Suchou, China.
Garden: Ophiopogon (mondo) grass, *Cymbidium* and
  Phalaenopsis orchids, banana tree, and golden bamboo.

17.193
Bartow-Pell Mansion Museum
Pelham Bay Park
New York, NY 10464
212-885-1461

Neoclassic stone mansion with 19th-century furnishings
  and decorative arts.
Garden: Formal gardens, herb garden, kitchen garden,
  and orchard.

17.194
Boscobel Restoration
Rt. 9-D
Garrison, NY 10524
914-265-3638

19th-century Federal-style mansion.
Garden: English gardens, rose garden, herb garden,
  boxwood garden, apple orchard, and orangery.

17.195

Cloisters, The
Fort Tryon Park
New York, NY 10040
212-923-3700; (FAX) 212-795-3640

Museum of medieval art.
Garden: Three cloister gardens with 300 species of European flora known to have grown in Europe during the Middle Ages including Madonna lilies, old roses, quince trees, espalier pear tree, topiary myrtle, and rosemary.
Other: A branch of the Metropolitan Museum of Art.

Cornell Plantations
Ithaca, NY
Cross-reference: 16.275

17.196

Ellwanger Garden
Landmark Society of Western New York
130 Spring St.
Rochester, NY 14608
716-546-7028

Location: 625 Mt. Hope Ave.
Garden: Renovated Victorian garden with perennials, specimen trees, and new and old-fashioned roses.

17.197

Farmers' Museum, The
P.O. Box 800
Lake Rd.
Cooperstown, NY 13326
607-547-2593

Living historical farm village.
Garden: Heirloom vegetable garden and medicinal herb garden.

17.198

Genesee Country Museum
P.O. Box 1819
Rochester, NY 14603

Several heirloom gardens; sponsors a 19th-century agricultural fair.
Garden: Heirloom varieties.

17.199

George Eastman House
900 East Ave.
Rochester, NY 14607
716-271-3361

House and estate of George Eastman, now a photography museum.
Garden: Historic landscape with a terrace garden, library garden, rock garden, grape arbor, and a sunken garden.

17.200

Hammond Museum Stroll Gardens
P.O. Box H
North Salem, NY 10560
914-669-5135

Arts and humanities museum.
Garden: Oriental stroll garden, waterfall garden, azalea garden, garden of the Rakan, fruit garden, and red maple terrace.

17.201

Lyndhurst
635 South Broadway
Tarrytown, NY 10591
914-631-0046; (FAX) 914-631-6825

Gothic Revival-style mansion.
Garden: Historic landscape: trees, shrubs, lawns, meadows; one of the few extant examples of mid-19th century landscape design in the Hudson River Valley.
Other: Internships in landscape and tree maintenance; a property of the National Trust for Historic Preservation.

17.202

Mohonk Mountain House
Lake Mohonk
New Paltz, NY 12561
914-255-1000

19th-century hotel.
Garden: Gardens begun in 1869, Victorian style predominates. Plantings of annuals, perennials, an herb garden, a rock garden, a fern and wildflower garden, and a cutting garden.
Cross-reference: 20.177

17.203

Muscoot Farm Park
Rt. 100
Katonah, NY 10536
914-232-7118

Turn-of-the-century living history farm.
Garden: Grape arbor, vegetable garden, and herb garden.

17.204

Old Bethpage Village Restoration
Round Swamp Rd.
Old Bethpage, NY 11804
516-420-5281

Reconstructed pre-Civil War village.
Garden: Six demonstration gardens with historic varieties of vegetables, each garden representing a different time period from the 1760s to 1866. Early red tomato, champion of England pea, Prince Albert pea, and early Bassano beet.

17.205

Old Westbury Gardens
Box 430, 71 Old Westbury Rd.
Old Westbury, NY 11568
516-333-0048; (FAX) 516-333-6807

Turn-of-the-century mansion and an English-style country estate.
Garden: Cottage garden, rose garden, vegetable garden, evergreen garden, Japanese garden, demonstration perennial garden, All-America Rose Selections test garden, lilac walk, perennials, and 300 European beeches.
Cross-reference: 20.300

17.206
Roosevelt Rose Garden and Gravesite
Home of Franklin D. Roosevelt
National Historic Site
249 Albany Post Rd.
Hyde Park, NY 12538
914-229-8114
Location: U.S. 9.

Home and gravesite of Franklin D. Roosevelt.
Garden: Rose garden designed in 1912.

17.207
Schuyler Mansion State Historic Site
32 Catherine St.
Albany, NY 12202
518-474-0834

18th-century mansion of Phillip Schuyler.
Garden: Fruit trees, herb garden, and 18th-century
parterre garden.

17.208
Shakespeare Garden
Garden Club of New Rochelle
1530 North Ave.
New Rochelle, NY 10804
914-636-0715

Shakespeare garden.
Garden: Contains only plants mentioned in Shakespeare's
works. Ivy started from cuttings taken from Ann
Hathaway's cottage and Shakespeare's garden in
Stratford, England in 1937.

17.209
Sonnenberg Gardens
151 Charlotte St.
Canandaigua, NY 14424
716-394-4922

Turn-of-the-century garden estate with ten gardens, a
conservatory, and a mansion.
Garden: Includes pansy garden, Italian garden, Japanese
garden, rose garden, Colonial garden, and a rock
garden. *Pinus densiflora* 'Umbraculifera', *Sciadopitys
verticillata*, *Ulmus glabra* 'Camperdownii'.
Cross-reference: 20.180

17.210
Stone-Tolan House
Landmark Society of Western New York
130 Spring St.
Rochester, NY 14608
716-546-7029
Location: 2370 East Ave.

Restoration of 1790s farmstead.
Garden: Kitchen garden, orchard, annuals, and perennials
in period style.

17.211
Vanderbilt Historic Site
Frederick W. Vanderbilt Garden Association
P.O. Box 239
Hyde Park, NY 12538
914-229-7770
Location: U.S. 9.

Mansion built in 1898.
Garden: Rose garden, terraced flower gardens, cherry
walk, and greenhouses.

17.212
Wave Hill
675 W. 252 St.
Bronx, NY 10471
212-549-2055

1848 house and 28-acre estate.
Garden: English-style wild garden, rose garden, herb
garden, aquatic garden, annuals, perennials, and
greenhouses with cacti, succulents, and tropicals.
Other: Maintains the *Catalog of Landscape Records in the
United States* as a part of its American Gardens and
Landscape History Program.
Cross-reference: 17.4

OH
17.213
Adena State Memorial
Adena St. Rd.
P.O. Box 831-A
Chillicothe, OH 45601
614-772-1500

19th-century mansion.
Garden: Period garden of roses, annuals, and perennials.

17.214
Carriage Hill Farm
Montgomery County Park District
7860 Shull Rd.
Dayton, OH 45424
513-879-0461

An 1880s living historical farm.
Garden: Heirloom kitchen garden and truck patch.
Family heirloom varieties of yellow eye beans, goose
craw beans, little greasy beans, green mountain
potatoes, and grease back beans.
Library: Number of books: 1,500. Collection primarily for
staff for research.
Lending Policy: Reference only.
Other: Member of the Seed Savers Exchange.

17.215
Dayton Museum of Natural History
2629 Ridge Ave.
Dayton, OH 45414

Restoration of a Native American fort.
Garden: Gardens with heirloom varieties including many
believed to have been grown in the 12th century.
Varieties of corn, including Iroquois, Papago,
Longfellow yellow flint, Tama flint, and Tamaroa flint.

17.216
Hale Farm and Village
2686 Oak Hill Rd.
P.O. Box 296
Bath, OH 44210-0296
216-666-3711

Outdoor living history museum depicting life in the
Western Reserve of Ohio between 1825–1850.
Garden: The Goldsmith Garden is of the "ancient style"
quadrant design characteristic of an upper class home in
the Western Reserve of the 1830s. There is also an
heirloom apple orchard, and there are heirloom varieties
of vegetables including scarlet runner beans, pod corn,
and Boston marrow squash.

17.217
Ohio Village
Ohio Historical Society
1982 Velma Ave.
Columbus, OH 43211
614-297-2680; (FAX) 614-297-2411

Re-created 19th-century village.
Garden: 19th-century kitchen garden and medicinal herb
gardens. Some heirloom varieties.
Library: Number of books: 150,000. Extensive collection
of horticultural and agricultural books.
Other: Heirloom seeds sales.

17.218
Oldest Stone House
Lakewood Historical Society
14710 Lake Ave.
Lakewood, OH 44107
216-221-7343

1830 pioneer house.
Garden: Herb garden for culinary, scented, and dye
plants, and an old-fashioned flower garden.

17.219
Slate Run Living Historical Farm
9130 Marcy Rd., Rt. 1
Ashville, OH 43103
614-833-1880

Operating 1800s central Ohio Farm.
Garden: 1880-1890s orchard and gardens with heirloom
varieties: apples, herbs, and vegetables.
Library: Number of books: 1,000; number of periodicals: 5.
Lending policy: Not open to the public.
Seed/specimen exchange.
Other: Member of the Seed Savers Exchange.

17.220
Stan Hywet Hall Foundation, Inc.
714 N. Portage Path
Akron, OH 44303
216-836-0576

Tudor-style mansion.
Garden: Japanese, rose, grape, cutting, and elliptical gardens designed by Warren Manning and Ellen Shipman.

Toledo Botanical Garden
Toledo, OH
Cross-reference: 16.314

OK
17.221
Anthony Oklahoma Heritage Gardens
201 Northwest 14th St.
Oklahoma City, OK 73103
405-235-4458

1917 mansion of Judge R. A. Hefner.
Garden: Conservatory with tropical and subtropical plants
and trees and flowering shrubs.

Corns
Turpin, OK
Cross-reference: 17.10

17.222
Philbrook Museum of Art
P.O. Box 52510
Tulsa, OK 74152
918-749-7941
Location: 2727 South Rockford Rd.

Art museum in 1926 Italian Renaissance-style mansion.
Garden: 23 acres of formal and informal gardens including
topiary, annual, perennial, and test gardens.

OR
17.223
Bishop's Close
11800 S.W. Military Ln.
Portland, OR 97219
503-636-5613

Manor house of Peter Kerr.
Garden: Magnolias, native trees, rhododendrons, azaleas,
wood hyacinths, irises, and red-flowering currants.

PA
17.224
1870 Miller's House Garden
Historic Bethlehem, Inc.
459 Old York Rd.
Bethlehem, PA 18018
215-691-5300

Restored 18th-century industrial area and houses.
Garden: Circa 1870 vegetable, herb, and flower garden.
Includes varieties dating back to 1782.

17.225
Appleford
The Appleford Committee
P.O. Box 182
Villanova, PA 19085
215-525-9430

Location: 770 Mt. Moro Rd.
18th-century house.
Garden: Roses, perennials, annuals, wildflowers,
rhododendrons, fruit trees, and wisteria.

17.226
Bartram's Garden
John Bartram Association
54th St. and Lindbergh Blvd.
Philadelphia, PA 19143
215-729-5281

The oldest surviving botanical garden in America and the
home of American botanist John Bartram.
Garden: 11-acre botanical garden. *Franklinia alatamaha*,
Ginkgo biloba, Cladrastis lutea.

17.227
Cliveden
6401 Germantown Ave.
Philadelphia, PA 19144
215-848-1777

18th-century mid-Georgian-style manor house.
Garden: Historic landscape. Azaleas, rhododendrons,
Franklinia, and old specimen trees.
Other: A property of the National Trust for Historic
Preservation.

17.228
Colonial Pennsylvania Plantation
Ridley Creek State Park
Media, PA 19063
215-566-1725

Colonial plantation.
Garden: A working colonial kitchen garden that supplies
food, dyes, and medicinal preparations; plantation is in
the process of replanting orchard using heirloom
varieties of fruit and nut trees used in Pennsylvania in
the late 18th century.
Library.
Lending Policy: Reference only.

17.229
Hans Herr House
1849 Hans Herr Dr.
Willow Street, PA 17584
717-464-4438

Medieval German-style farmhouse.
Garden: Historic garden and orchard typical of a
Pennsylvania German farmer of 1719–1750. Apples:
Baldwin, Roxbury russet, doctor of Germantown,
Tollman sweet, Spitzenberg, and many others.

17.230
Heckler Plains Farmstead
474 Main St.
Harleysville, PA 19438
215-822-7422

Living historical farm.
Garden: Working garden designed from 18th-century
layouts of a Pennsylvania German four-square garden;
18th-century historic herb garden. Herbs and
vegetables grown between 1800–1850.

17.231
Pennsbury Manor
400 Pennsbury Memorial Rd.
Morrisville, PA 19067
215-946-0400

Re-creation of manor built by William Penn.
Garden: Ornamental courts, kitchen garden, and orchards.
Numerous herbaceous perennials, herbs, and
vegetables. 4,000 specimens.
Library: Number of books: 1,000; number of periodicals:
15. John Loudon's *Garden Vegetables, Trees and Shrubs,
Home Garden* magazine, works by Luther Burbank.
Lending Policy: Reference only.

17.232
Peter Wentz Farmstead
P.O. Box 240
Worcester, PA 19490
215-584-5104

Mid-18th-century restored house, barn, and farm.
Garden: 18th-century German kitchen garden with herbs
and vegetables.
Library: Number of books: 450.
Lending Policy: Reference only.

17.233
Wyck
6026 Germantown Ave.
Philadelphia, PA 19144
215-848-1690

Oldest house in Philadelphia, built in the 17th century.
Garden: Rose collection with varieties dating back to the
18th century, and wildflowers, wisteria, rare shrubs,
fruit trees, and vegetables.

RI
17.234
Blithewold Gardens and Arboretum
101 Ferry Rd.
Bristol, RI 02809
401-253-2707

Turn-of-the-century mansion built in the style of a
17th-century English manor house.
Garden: Historic landscape following the original design.
About 2,000 taxa of native and exotic trees, flowers,
and shrubs.
Library: Number of books: 500; number of periodicals:
12. Turn-of-the-century American gardening.
Lending Policy: Reference only.
Cross-reference: 20.206

17.235
Preservation Society of Newport County
118 Mill St.
Newport, RI 02840
401-847-1000; (FAX) 401-847-1361

Owns nine properties: Breakers, Rosecliff, The Elms,
Chateau-Sur-Mer, Marble House, Kingscote, Hunter
House, Rovensky Park, and Green Animals.
Garden: Trees, shrubs, and a rose garden.

17.236
Shakespeare's Head
Providence Preservation Society
24 Meeting St.
Providence, RI 02903
401-831-7440
Location: 21 Meeting St.

Garden: Restored Colonial garden with boxwood, quince trees, flowers, and an herb garden.

SC
17.237
Boylston Gardens
Governor's Mansion
Columbia, SC 29201
803-737-1710

Governor's mansion.
Garden: Restored formal boxwood garden.

17.238
Drayton Hall
3380 Ashley River Rd.
Charleston, SC 29414
803-766-0188

Georgian Palladian mansion.
Garden: Wildflowers, ferns, shrubs, trees, and vines.
Other: A property of the National Trust for Historic Preservation.

17.239
Magnolia Plantation and Gardens
Rt. 4, Highway 61
Charleston, SC 29414
803-571-1266

17th-century plantation house and estate.
Garden: 50 acres of gardens with azaleas, camellias, summer blossoms, a biblical garden, a 17th-century herb garden, and a topiary garden.

17.240
Middleton Place
Rt. 4
Charleston, SC 29407
803-556-6020

18th-century plantation.
Garden: The oldest landscaped garden in the U.S. Formal gardens include a moon garden, octagonal garden, terraced gardens, secret garden, bamboo grove, a rare camellia garden, and a rose garden. Also azaleas, magnolias, a thousand-year-old live oak, and kalmias.

TN
17.241
Blount Mansion
200 West Hill Ave.
P.O. Box 1703
Knoxville, TN 37901
615-525-2375

18th-century home of Governor William Blount.

Garden: Colonial revival garden with 18 plant specimens. Alabama snow wreath, Ben Franklin tree, chaste tree, Jackson vine, and more.

17.242
Dixon Gallery & Gardens
4339 Park Ave.
Memphis, TN 38117
901-761-5250; (FAX) 901-682-0943

Art museum.
Garden: Perennial wildflowers, hardy ferns, native and evergreen azaleas, viburnum, and camellias.
Library: Number of books: 1,100; number of periodicals: 40.
Lending Policy: Open to members only.

17.243
Hermitage
4580 Rachel's Ln.
Hermitage, TN 37076-1331
615-889-2941

Home of Andrew Jackson.
Garden: One-acre period garden with irises, peonies, roses, hyacinths, pinks, jonquils, crapemyrtles, herbs, and other plants available during Jackson's lifetime.

17.244
Historic Travellers Rest
636 Farrell Pkwy.
Nashville, TN 37220
615-832-8169

House museum that interprets life in Middle Tennessee during 1789–1833.
Garden: Flower, herb, and vegetable garden, meadow, and orchard.
Other: Public lectures on historical horticulture and historic plant sales.

17.245
Magevney House
198 Adams Ave.
Memphis, TN 38103
901-526-4464

The oldest private residence still standing in Memphis.
Garden: Medicinal and cooking herbs used in the 1850s.

17.246
Tennessee Botanical Gardens & Fine Arts Center at Cheekwood
Forrest Park Dr.
Nashville, TN 37205
615-353-2148

Art museum in a Georgian-style mansion built around 1930.
Garden: Roses, herbs, daffodils, perennials, wildflowers, irises, peonies, azaleas, boxwood, magnolias, dogwoods, a garden of scent and taste, a Japanese garden, a tropical atrium, and greenhouses with orchids and camellias.
Library: 14.93

TX
Austin Area Garden Center
Austin, TX
Cross-reference: 16.377

17.247
Bayou Bend Garden
1 Westcott St.
P.O. Box 13157
Houston, TX 77219
713-529-8773

Art museum in 1927 mansion.
Garden: Eight specialized gardens including a
  17th-century English parterre garden, an English
  garden, a topiary garden, and a butterfly garden.

17.248
Heritage Farmstead
1900 West 15th St.
Plano, TX 75075
214-424-7874

Re-created 19th-century farmstead.
Garden: Field crops, orchards, and a kitchen garden with
  herbs and vegetables.

17.249
Jourdan Bachman Pioneer Farm
11418 Sprinkle Cut Off Rd.
Austin, TX 78754
512-837-1215

Living history farm depicting farming in central Texas
  circa 1880.
Garden: Heritage plants grown in gardens and crop fields.
  Longhorn okra, Texas gourdseed corn, black valentine
  beans, Melrose brown cotton, Martinka orange
  sorghum, and yellow oxhart tomatoes.
Seed/specimen exchange.

17.250
Judge Roy Bean Visitor Center
P.O. Box 160
Langtry, TX 78871
915-291-3340; (FAX) 915-291-3366
Location: Loop 25 off U.S. 90.

19th-century courtroom-saloon and a museum on Roy
  Bean and his era.
Garden: Two-acre cactus garden with native cacti and
  other desert plants labeled with their medicinal and
  practical uses.

17.251
Varner-Hogg State Historical Park
P.O. Box 696
West Columbia, TX 77486
409-345-4656
Location: State Rt. 35.

Greek revival mansion built in 1836.
Garden: Orchards and antebellum formal and informal
  gardens.

VA
17.252
Agecroft Hall
The Agecroft Association
4305 Sulgrave Rd.
Richmond, VA 23221
804-353-4241

Tudor manor house with 23 acres of woodland, lawn, and
  ornamental garden.
Garden: Thematic gardens reflecting the style of early
  17th-century England.
Library: Number of books: 350.

17.253
Ash Lawn—Highland
Rt. 6, Box 37
Charlottesville, VA 22901
804-293-9539

Home of James Monroe, situated on a 535-acre estate.
Garden: Vegetable, herb, and flower gardens, and English
  and American boxwoods.

17.254
Blue Ridge Farm Museum
The Blue Ridge Institute
Ferrum College
Ferrum, VA 24088
703-365-4416

1800 German-American farm museum.
Garden: 1800 period vegetable garden and a 1900 Blue
  Ridge farm garden and an herb garden.
Unicorn plant, gherkin cucumber, scarlet runner bean,
  citron melon, bushel gourd, fava beans, vegetable
  marrow squash, and hops.
Other: No official seed exchange program but will
  exchange seeds.

17.255
Booker T. Washington National Monument
Rt. 3, Box 310
Hardy, VA 24101
703-721-2094; (FAX) 703-721-8311

Living historical farm.
Garden: Heirloom gardens demonstrating the field and
  cash crops grown in Piedmont, Virginia, in the 1850s,
  and the herbs, flowers, and vegetables common to
  kitchen gardens of the same period. Gourdseed corn,
  hickory king corn, Oranoka lizard tail tobacco.
Library: Number of books: 250.
Lending Policy: Reference only.

17.256
Carter's Grove Plantation
Colonial Williamsburg Foundation
P.O. Box C
Williamsburg, VA 23185
804-229-1000

1750 mansion and plantation.
Garden: Over 700 acres with an annual garden, and
  period vegetable and herb gardens.

17.257
Claude Moore Colonial Farm at Turkey Run
6310 Georgetown Pike
McLean, VA 22101
703-442-7557

Living history farm portraying a small, low-income family farm of the late Colonial period.
Garden: Kitchen gardens and corn and tobacco fields. White dent corn and tobacco from Venezuela.
Library: Number of books: 2,100. Volumes on domestic trade and agriculture of late Colonial America.
Lending Policy: Open to members only, by appointment, reference only.
Seed/specimen exchange.

17.258
Colonial Williamsburg
P.O. Drawer C
Williamsburg, VA 23187
804-220-7095

Extensive 18th-century restoration with 500 buildings and 100 gardens on 178 acres.
Garden: Kitchen, herb, formal, fruit, and parterre gardens of the Colonial period. European and Asian species imported by the colonists, and native flora cultivated by plantsmen.
Seed/specimen exchange.
Other: Tours explain the beginnings of ornamental horticulture in America.

17.259
George Washington Birthplace National Monument
R.R. 1, Box 717
Washington's Birthplace, VA 22443
804-224-1732

Garden: Heirloom gardens planted with 18th-century varieties and laid out in a style typical of early Colonial Virginia tobacco plantations. Washington family heirloom figs.
Museum: Archeological artifacts associated with horticulture.
Library: Number of books: 1,500; number of periodicals: 3.
Lending Policy: Reference only.
Seed/specimen exchange.
Other: Extensive living history programs demonstrate early gardening methods.

17.260
Gunston Hall
10709 Gunston Rd.
Mason Neck, VA 22079
703-550-9220

Plantation home of George Mason on 556 acres.
Garden: Re-creation of an 18th-century garden. English boxwood allee originally planted by Mason and extensive formal boxwood garden.
Museum: One small exhibit with gardening tools, an 18th-century botanical book, and other agricultural items.
Library: The Elizabeth L. Frelinghuysen Collection of rare books on gardening and culinary arts.
Lending Policy: Reference only.

17.261
Kenmore
Kenmore Association, Inc.
1201 Washington Ave.
Fredericksburg, VA 22401
703-373-3381

Colonial mansion that was the home of George Washington's only sister, Betty.
Garden: Boxwoods and trees, herbs and shrubs of the 18th century.

17.262
Mary Washington House Garden
Mary Washington Branch, Association
for the Preservation of Virginia Antiquities
1200 Charles St.
Fredericksburg, VA 22401
703-373-1569

House of Mary Ball Washington, the mother of George Washington.
Garden: English-style flower garden, kitchen garden, herbs, and shrubs.

17.263
Maymont Foundation
1700 Hampton St.
Richmond, VA 23220
804-358-7166

100-acre Victorian estate and mansion.
Garden: Italian garden, daylily garden with over 150 cultivars, herb garden, English courtyard garden, Japanese stroll garden, and a Victorian vegetable garden with 200 plants.

17.264
Meadow Farm Museum
P.O. Box 27032
Richmond, VA 23140
804-672-5106

Farm museum.
Garden: 19th-century kitchen garden. China rose radish, black Spanish radish, hollow grown parsnips, Tailor's dwarf shell beans, Reid's yellow dent, potato onions, and turkey red bearded wheat.

17.265
Monticello
Thomas Jefferson Memorial Foundation
P.O. Box 316
Charlottesville, VA 22902
804-296-4800

Thomas Jefferson's 1769–1826 home that he designed himself.
Garden: Restored flower gardens, vegetable gardens, orchard, and oval lawn. Historic varieties of fruits, flowers, and vegetables.
Library: Number of books: 150; number of periodicals: 5. Garden history.
Lending Policy: Reference only.
Seed/specimen exchange.
Other: Workshops.

17.266
Montpelier
P.O. Box 67
Montpelier Station, VA 22957
703-672-2728; (FAX) 703-672-0411

Circa 1755 estate of James Madison.
Garden: *Juglans nigra, Liriodendron tulipifera, Quercus alba, Cunninghamia lanceolata, Abies pinsapo, Abies nordmanniana, Cedrus libani,* , and *Buxus sempervirens.*
Other: Horticultural educational program; a property of the National Trust for Historic Preservation.

17.267
Morven Park
Rt. 3, Box 5
Leesburg, VA 22075
703-777-2414
Location: State Rt. 7.

18th-century estate.
Garden: Boxwood garden.

17.268
Mount Vernon
Mount Vernon Ladies Association
Mount Vernon, VA 22121
703-780-7262

500-acre estate of George Washington from 1735–1799.
Garden: Three gardens that were established by George Washington: a kitchen garden, a botanical garden, and a pleasure garden. All plants are appropriate to the 18th century.
Library: Number of books: 10,000; number of periodicals: 65. Pre-1800 rare books including Washington's library.
Lending Policy: Reference only.
Seed/specimen exchange.

17.269
Oatlands Plantation
Rt. 2, Box 352
Leesburg, VA 22075
703-777-3174

Circa 1800 Classic Revival mansion with 1827 Greek Revival portico.
Garden: Terraced gardens, herb garden, and a rose garden. Early boxwood, Victorian roses, and early 1800s English oak. 2,500 specimens.
Other: Garden lectures; a property of the National Trust for Historic Preservation.

17.270
Pope Leighey House
9000 Richmond Highway
Alexandria, VA 22309
703-780-3264

House designed by Frank Lloyd Wright in 1940.
Garden: Pink lady's-slippers, native shrubs, wildflowers, and indigenous ferns.
Other: A property of the National Trust for Historic Preservation.

17.271
Stratford Hall Plantation
Robert E. Lee Memorial Association
Stratford, VA 22558
804-493-8038

18th-century mansion, birthplace of Robert E. Lee.
Garden: 18th-century flower beds, vegetable gardens, and orchard.
Library: 250 horticulture books.
Lending policy: Not open to the public.

17.272
Virginia Living Museum
524 J. Clyde Morris Blvd.
Newport News, VA 23601
804-595-1900; (FAX) 804-595-1900

Nature museum.
Garden: Plants native to the southeastern coastal plain of North America.
Library: Number of books: 800; number of periodicals: 25.
Lending Policy: Reference only.

17.273
Woodlawn Plantation
P.O. Box 37
Mount Vernon, VA 22121
703-780-4000

Federal-style estate that was originally part of George Washington's Mount Vernon estate.
Garden: 19th-century-style gardens, two parterres, and a heritage rose garden.
Other: A property of the National Trust for Historic Preservation.

VT
17.274
Hildene
Friends of Hildene, Inc.
P.O. Box 377
Manchester, VT 05254
802-362-1788

Georgian Revival mansion that was the home of Robert Todd Lincoln.
Garden: Formal gardens based on 1907 design. Fringe peonies, tree peonies, and over 20 different varieties of single and double peonies.

17.275
Park-McCulough House
North Bennington, VT 05257
802-442-2747

Location: West St.
Victorian mansion.
Garden: Colonial garden of annuals and perennials, an herb garden, a vegetable garden, a lily pond, and a greenhouse.

17.276
Shelburne Museum
Rt. 7
Shelburne, VT 05482
802-985-3344

Museum of American crafts.
Garden: Medicinal herb garden, garden of culinary and
dye plants, and 900 varieties of lilacs.

WA
17.277
Fort Vancouver National Historic Site
National Park Service
612 East Reserve St.
Vancouver, WA 98661
206-696-7655

Fort Vancouver was the headquarters of the Hudson Bay
Company's operations west of the Rocky Mountains
from 1825 to 1849.
Garden: 1840 period heirloom garden with over 100
specimens and historic orchard. Cardoon, wormwood,
lemon lily, scented geraniums, heliotrope, *Humulus
lupulus, Lavatera trimestris, Iris X germanica* var.
florentina, common moss rose, Damask rose, and
climbing musk rose.
Library: Number of books: 50; number of periodicals: 5.
Complete collections of *The Beaver, Oregon Historical
Quarterly*, and *Washington Historical Quarterly*. Complete
collection of the Hudson Bay Company's Records.
Seed/specimen exchange.

17.278
Hulda Klager Lilac Gardens
115 South Pekin Rd.
Woodland, WA 98674
206-225-8996

Former estate of hybridizer Hulda Klager.
Garden: Lilacs, camellias, azaleas, rhododendrons, bulbs,
and exotic trees.

17.279
Pomeroy Living History Farm
20902 N.E. Lucia Falls Rd.
Yacolt, WA 98675
206-686-3537

1920s-era living history farm.
Garden: Native plants.

WI
17.280
Old World Wisconsin Outdoor Ethnic Museum
S103 W37890 Highway 67
Eagle, WI 53119
414-594-2116

Living history village.
Garden: 19th-century heirloom gardens reflecting the
tastes and economic status of Yankee and European
immigrants in Wisconsin. Wide variety of heirloom
vegetables, flowers, and herbs.
Seed/specimen exchange.
Other: Heirloom Garden Fair each fall.

17.281
Paine Art Center and Arboretum
1410 Algoma Blvd.
Oshkosh, WI 54901
414-235-4530

1920s Tudor Revival mansion of lumber baron Nathan
Paine.
Garden: 1,500 specimens representing 350 taxa on 15
acres. Circa 1600 bur oak, native and exotic trees,
18th-century English garden, herb garden, and
prairie-woodland restoration area.
Museum: Occasional exhibits with botanical prints and
landscape paintings.
Library: Number of books: 5,000; number of periodicals:
50.
Lending Policy: Reference only.

CANADA

BC
17.282
British Columbia Forest Museum
R.R. 4
Trans-Canada Highway
Duncan, BC V9L 3W8
604-746-1251

Forestry museum.

17.283
Grist Mill at Keremeos
R.R. 1, Upper Bench Rd.
Keremeos, BC V0X 1N0
604-499-2888

Historic 1877 mill.
Garden: Heritage gardens and fields. Heirloom wheats,
flowers, and vegetables.
Library: Books on historic agriculture in British Columbia.
Lending Policy: Reference only.
Other: Gene bank of historic varieties adapted to local
climate, slide presentations on heritage gardens, and
seed saving.

17.284
University of British Columbia Botanical Garden
Vancouver, BC

Cross-reference: 16.449

MB
17.285
Sandilands Forest Centre
Manitoba Forestry Association, Inc.
900 Cordyon Ave.
Winnipeg, MB R3M O4Y
204-453-3182

Forestry museum.
Other: Nature trails, nursery.

NF
Memorial University Botanical Garden at Oxen Pond
St. John's, NF

Cross-reference: 16.453

NS
17.286
Annapolis Royal Historic Gardens
P.O. Box 278
Annapolis Royal, NS B0S 1A0
902-532-7018
Location: 441 St. George St.

Historical theme gardens and reclaimed marshlands.
Garden: "Innovative Garden" demonstrates innovations
in plant materials and gardening techniques, rose
garden with heritage roses, winter garden, Acadian
vegetable garden, spring color garden, rock garden,
perennial garden, Colonial Governor's garden, Victorian
garden, and a knot garden.

17.287
Cole Harbour Heritage Farm Museum
Cole Harbour Rural Heritage Society
471 Poplar Dr.
Cole Harbour, NS B2W 4L2
902-434-0222

Farm museum on two and one-half acres.
Garden: Vegetable and flower gardens and small fruits.
An old rose they call the Bissett rose that is supposed
to have come with an early family.
Museum: Occasional horticultural exhibits, many on the
history of local agriculture.
Library: Number of books: 200; number of periodicals: 5.
Lending Policy: Reference only.

17.288
Haliburton House
Windsor, NS
902-798-2915
Location: Rt. 101.

Home of Thomas Haliburton, built in 1836.
Garden: 25-acre estate with annuals, perennials, and a
rose garden.

17.289
Prescott House
Starr's Point, NS
902-542-3984

House built in 1799 by horticulturist Charles Ramage
Prescott, who introduced apples to Nova Scotia.
Garden: Sunken rock garden, bulbs, annuals, perennials,
and rare shrubs and trees.

17.290
Ross Farm Museum
Rt. 12
New Ross, NS B0J 2M0
902-689-2210

19th-century living history farm.
Garden: Field crops and an herb garden. Some heirloom
varieties.

ON
17.291
Bellevue House National Historic Site
35 Centre St.
Kingston, ON K7L 4E5
613-545-8666

Garden: Representative of a country villa ornamental and
vegetable garden and orchard of 1849.

17.292
Cullen Gardens and Miniature Village
300 Taunton Rd. West
Whitby, ON L1N 5R5
416-294-7965

Historic village with over 100 miniature buildings.
Garden: 22 acres of gardens with daffodils, tulips,
annuals, roses, and chrysanthemums.
Cross-reference: 20.258

17.293
Eldon House
325 Queens Ave.
London, ON N6B 3L7
519-432-7166

19th-century home.
Garden: Victorian period garden.

Heritage Seed Program
Uxbridge, ON

Cross-reference: 17.40

17.294
Ontario Agriculture Museum
P.O. Box 38
Milton, ON L9T 2Y3
416-878-8151; (FAX) 416-876-4530

Agricultural museum with 30 buildings on 80 acres.
Garden: 1830s and 1860s heirloom plants and garden
arrangements.
Library: Number of books: 15,000; number of periodicals:
20,000.
Lending Policy: Reference only.

17.295
Spruce Lane Farm
Bronte Creek Provincial Park
1219 Burloak Dr.
Burlington, ON L7R 3X5
416-335-0023

100-acre operating turn-of-the-century farm.
Garden: Period farm crops, vegetable garden, herb
garden, and flower beds.
Seed/specimen exchange.

17.296
Upper Canada Village
R.R. 1
Morrisburg, ON K0C 1X0
613-543-3704; (FAX) 613-543-2847

19th-century village.
Garden: 1860s-style vegetable gardens with varieties of
the 19th century.

17.297
Victorian Garden
St. James Park
Dept. of Parks & Recreation
City Hall
Toronto, ON M5H 2N2
416-947-7251

Garden: Victorian garden with shrubs, perennials, and
annuals.

18

Other Horticultural Displays

ZOOLOGICAL PARKS

As animals's native habitats are increasingly destroyed, zoos have become more interested in attempting to ensure the preservation of certain animal species. Vital to this is the use of more plants, and plants native to an animal's habitat in order to re-create natural environments. Zoos are also offering new plant-related exhibits addressing topics such as biological diversity and the interdependence between plants and animals.

Although plants have been included in zoos for some time, zoological horticulture is relatively new as a distinct field of study and practice. The Association for Zoological Horticulture (1.28) was founded only in 1980. There are presently fifty institutional members.

The following list by no means includes all of the zoological parks in North America with plant collections. Listed instead are the zoos that have been the most active in zoological horticulture with the most interesting plant collections.

UNITED STATES

AK
18.1
Alaska Zoo
4731 O'Malley Rd.
Anchorage, AK 99516
907-346-2133

Mostly native plants including a noteworthy collection of native mosses.
Other: Summer lecture walking tours and winter horticulture classes.

AZ
Arizona-Sonora Desert Museum
Tucson, AZ

Cross-reference: 16.8

18.2
Phoenix Zoo
3810 E. Van Buren
Phoenix, AZ 85008
602-273-1341

Native Sonoran desert flora including saguaros, Boojum trees, and organ pipe cactus. Also Australian shrubs and trees.

CA
18.3
Butterfly World
Marine World Africa, U.S.A.
Marine World Parkway
Vallejo, CA 94589

Butterfly garden.

18.4
Fresno Zoo
894 W. Belmont Ave.
Fresno, CA 93728
209-498-1549

Plant collection includes a number of unusual palm tree species.
Over 140 species of plants in the zoo. The rainforest exhibit contains hundreds of tropical South American plants.

Living Desert, The
Palm Desert, CA

Cross-reference: 16.34

Santa Barbara Zoological Gardens, Santa Barbara, California.

San Diego Zoo, San Diego, California.

18.5
Los Angeles Zoo
5333 Zoo Dr.
Los Angeles, CA 90027
213-666-4650

Large collection of succulents including Baja California
   succulents, unusual specimens of aloe, and an African
   succulent collection. Also dragon trees and huge
   specimens of elephant-foot trees.
Other: Occasional lectures and tours.

18.6
Sacramento Zoo
3930 West Land Park Dr.
Sacramento, CA 95822
916-449-5166

Several species of oak estimated to be 200 to 300 years
   old, eight species of palm, and two species of cycads.

18.7
San Diego Zoo
P.O. Box 551
San Diego, CA 92112
Location: Balboa Park.

Extensive collections of tropicals and subtropicals.
   Primary collection groups include ficus, erythrina,
   orchids, bromeliads, aloes, euphorbias, cycads, and
   palms.
Cross-reference: 2.34

18.8
Santa Ana Zoo
1801 E. Chestnut Ave.
Santa Ana, CA 92701
714-647-6575

15 clusters of *Strelitzia nicolai* (giant bird-of-paradise).

18.9
Santa Barbara Zoological Gardens
500 Ninos Dr.
Santa Barbara, CA 93103
805-962-5339

A large collection of mature palms and other tropical and
   subtropical plants, including specimens of Canary
   Islands date palms, Chilean wine palm, dragon tree,
   Mediterranean fan palm, and Morton bay fig.

18.10
Sea World of California
1720 South Shores Rd.
Mission Bay
San Diego, CA 92109-7995
619-282-7716

Approximately 5,000 kinds of plants including palms,
   cycads, succulents, trough and rock garden plants, plus
   rose gardens, and annual color beds.

CO
18.11
Cheyenne Mountain Zoological Park
4250 Cheyenne Mountain Zoo Rd.
Colorado Springs, CO 80906
719-633-0917

Colorado native wildflowers, shrubs, and grasses.

18.12
Denver Zoological Gardens
City Park
Denver, CO 80205
303-331-4100

Over 125 species of exotic plants located in five bird exhibits.

DC
18.13
National Zoological Park
Smithsonian Institution
Washington, DC 20008
202-673-4670

Bamboo, oaks, butterfly bushes, nut-bearing tree species,
   and butterfly attractants. Also an educational vegetable
   garden.
Other: Volunteer guide program and plant/animal
   relationship interpretive lessons.

FL
18.14
Butterfly World
3600 West Sample Rd.
Coconut Creek, FL 33073

Butterfly garden.

18.15
Discovery Island Zoological Park
P.O. Box 10,000
Lake Buena Vista, FL 32830
407-824-2875

Many unusual palms, flowering trees, bamboos, and
   tropicals.

GA
18.16
Zoo Atlanta
800 Cherokee Ave., S.E.
Atlanta, GA 30315
404-624-5653

Over 167,000 flowering bulbs, tropical vegetation
   including over 250 bananas of 10 cultivars, areca palms,
   scheffleras, fiddleleaf figs, and Benjamin figs.

HI
18.17
Waikiki Aquarium
University of Hawaii
2777 Kalakaua Ave.
Honolulu, HI 96815
808-923-9741

Collection of native Hawaiian coastal plants including the
   endangered 'ohai plant (*Sesbania tomentosa*).

IN
18.18
Indianapolis Zoological Society
1200 West Washington St.
Indianapolis, IN 46222
317-630-2001

Plants hardy to Zone 4, endangered and threatened
plants, tropical plants, and arid plants with emphasis on
Caudiciform and Pachycaul succulents.

KS
18.19
Emporia Zoo
P.O. Box 928
Emporia, KS 66801
316-342-7306

Plants include century plant, white buds, ornamental
kale, mimosa, Harry Landis walkingstick, Chinese
fringe tree, Chinese pistachio, Japanese maple
'Burgundy Lace', ginkgo, and several varieties of
hostas.

18.20
Lee Richardson Zoo
P.O. Box 499-301 North 8th St.
Garden City, KS 67846
316-276-1250

Cactus garden and native grass display. Aspen, pussy-
willow, bamboo, and the state champion lacebark elm.

18.21
Sedgwick County Zoo & Botanical Garden
5555 Zoo Blvd.
Wichita, KS 67212
316-942-2213

120 acres of mixed prairie grass, nearly 100 herbs, 250
specimens of mature tropicals, and 600 species of
annuals, perennials, and woody landscape plants.
Other: Classes, workshops, and volunteer program.

18.22
Topeka Zoological Park
635 Gage Blvd.
Topeka, KS 66606
913-272-5821

Includes an enclosed rainforest with bromeliads, orchids,
hibiscus, an assortment of palms, and other plants
cohabitating with rainforest birds and animals.

LA
18.23
Audubon Park & Zoological Gardens
P.O. Box 4327
6500 Magazine St.
New Orleans, LA 70178
504-861-2537

Large collection of orchids and bromeliads, and many
endangered plants.

Louisiana Purchase Gardens and Zoo
Monroe, LA

Cross-reference: 16.169

MA
18.24
Franklin Park Zoo
Boston, MA 02121
617-442-0991

The African Tropical Forest contains the largest tropical
plant collection in New England with 3,000 plants
representing 150 species.

MD
18.25
Baltimore Zoo
Druid Hill Park
Baltimore, MD 21217
301-396-7102

A wide variety of plant species dating back to the 1870s,
including 16 *Acer palmatum* 'Dissectum', 100 Indian
currants, and more than 20 varieties of ornamental
grasses. Also an eight-acre Maryland Wilderness section
in the Children's Zoo which contains more than 100
species of Maryland native plants.

18.26
National Aquarium in Baltimore
501 E. Pratt St.
Baltimore, MD 21202
301-576-3800

Several unusual aroids, including *Anthurium warocqueanum*,
A. croatii, *A. crystallinum*, *A. gladifolium*, *Monstera dubia*,
and *Philodendron verrucosum*.
Other: Extensive volunteer program where volunteers
gain experience in growing tropical plants, plant
identification, pruning, and propagation.

MN
18.27
Minnesota Zoological Gardens
12101 Johnny Cake Rd.
Apple Valley, MN 55124
612-431-9200

A one-half-acre Tropics Trail Building heavily planted
with Southeast Asian tropical plants.

NC
18.28
North Carolina Zoological Park
Rt. 4, Box 83
Asheboro, NC 27203
919-879-7000; (FAX) 919-879-2891

604 species, 14,715 plants, African plant collection,
tropical plants, and grass collection.
Other: Provides park tours to garden clubs and
horticultural organizations.

ND
18.29
Roosevelt Zoological Park
Minot Park District
P.O. Box 538
Minot, ND 58702
701-839-1300

Lily ponds, flower beds, roses, perennials, and climbing
vines.

NE
18.30
Folsom Children's Zoo and Botanical Gardens
2800 "A" St.
Lincoln, NE 68502
402-475-6741

Papaw, *Phellodendron amurense*, devil's-walking-stick, American beech, dawn redwood, bald cypress, and maidenhair tree.
Other: Classes, programs, and demonstrations on botany, plants, and habitats.
Cross-reference: 2.69

NM
Living Desert Zoological and Botanical State Park
Carlsbad, NM

Cross-reference: 16.266

18.31
Rio Grande Zoological Park
903 10th St., S.W.
Albuquerque, NM 87102
505-843-7413.

Hardy cacti and succulents, hardy palms and eucalyptus, and a large tropical aviary.

NY
18.32
Buffalo Zoological Gardens
Delaware Park
Buffalo, NY 14214
716-837-3900; (FAX) 716-837-0738

110 species of subtropical, 9 species of carnivorous, 138 species of native woody, 11 species of bedding, and 6 species of aquatic plants; 12 species of mesembs, and 6 species of flowering bulbs.

18.33
New York Aquarium
West 8th St. and Boardwalk
Brooklyn, NY 11224
718-265-3432

Saltwater macroalgae including Atlantic kelp and salt marsh grasses such as cordgrass (500 specimens) and salt marsh hay (50 specimens). Aquatic plants have only recently been used in displays of marine environment.

18.34
New York Zoological Park
(Bronx Zoo)
185th St. and Southern Blvd.
Bronx, NY 10460
212-220-5100

Hardy trees, tropical plants, hardy bamboo, native grasses, wildflowers, and aquatic plants.

18.35
Ross Park Zoo
Southern Tier Zoological Society
185 Park Ave.
Binghamton, NY 13903
607-724-5461

The Butterfly Garden includes buddleia, *Asclepias tuberosa*, monarda, salvias, 'Autumn Joy' sedums, Russian-sage, erigeron, aquilegia, *Achillea*, *Myosotis*, and *Centaurea cyanus*. The Rare Plant Garden has plants rare and endangered in New York including northern monkshood, hart's-tongue fern, bloodroot, Jack-in-the-pulpit, globeflower, white lady's-slipper, American ginseng, and several species of rare ferns. There is also an ornamental grass garden and a garden of heaths, heathers, and old roses.

18.36
Seneca Park Zoo
2222 St. Paul St.
Rochester, NY 14621
716-266-6846

Rare monkey puzzle tree and native endangered plants, exotic specimens in animal exhibits, rich collection of native/cold climate species.

OH
18.37
Cincinnati Zoo and Botanical Garden
3400 Vine St.
Cincinnati, OH 45220
513-281-4701

One of the first public gardens in the Midwest—includes exotic trees over 100 years old, and over 2,200 species and cultivars of plants, 500,000 spring bulbs, and a large collection of ornamental grasses, bamboo, and perennials. Botanical Center has interpretive displays.
Other: Annual Spring Floral Festival in April, classes and lectures, and garden volunteer program.

18.38
Cleveland Metroparks Zoo
3900 Brookside Dr.
Cleveland, OH 44109
261-661-6500.

Around 2,000 plant species and varieties with a large collection of ornamental grasses and other herbaceous perennials. Also a large collection of plants from Australasia, particularly China.
Other: Tours of grounds and gardens.

18.39
Columbus Zoological Gardens
9990 Riverside Dr.
P.O. Box 400
Powell, OH 43065
614-645-3400

150 tropical plant species, bamboo, bromeliads, wildflowers, and a prairie plant collection.
Other: Plant adoption program, and tree and shrub walk guides.

18.40
Toledo Zoo
2700 Broadway
Toledo, OH 43609
419-385-5721

Over 700 species of tropical foliage plants and succulents,
and a rose garden, a tropical water lily pool, ornamental
trees, shrubs, grasses, bamboo, perennials, and annuals.

OK
18.41
Oklahoma City Zoological Park
2101 N.E. 50th St.
Oklahoma City, OK 73111
405-424-3344

A large collection of unusual plants native to Oklahoma
including devil's-walking-stick, *Penstemon oklahomensis*,
Mamillaria vivipara, coral cactus, Indian pink, and
dwarf palmetto.
Other: Horticultural tours of the grounds.

18.42
Tulsa Zoological Park
5701 E. 36th St. N.
Tulsa, OK 74115
918-596-2400

Carnivorous plants, South American succulents, and
Southwest desert plants. The zoo also has a living
museum with exhibits representing habitats of the
Arctic tundra, Southwest desert, Eastern forest, and
Southern lowlands.

OR
18.43
Wildlife Safari
P.O. Box 1600
Winston, OR 97496
503-679-6761

Over 100 species of trees and shrubs found in the Pacific
Northwest landscape.
Other: The Safari has botanical displays in the
educational center, a one-half-mile nature trail, and
interpretive gardens. They also do research on
zoological horticulture and wildlife management, and
on forestry demonstration sites.

PA
18.44
Philadelphia Zoological Garden
34th St. and Girard Ave.
Philadelphia, PA 19104
215-243-1100

1,800 tree specimens including an English elm over 200
years old, an American elm 100 years old, and a
centennial ginkgo planted in 1876; also native flora and
tropical species.

18.45
Pittsburgh Aviary
Allegheny Commons West
Pittsburgh, PA 15212
412-323-7235

Over 5,000 specimens. Exhibits feature birds and plants
in geographical and ecological displays.

18.46
Pittsburgh Zoo
Hill Rd., Highland Park
Pittsburgh, PA 15206
412-665-3641

4,500 different types of trees, shrubs, and grasses. The
African savanna exhibit features plants that mimic
species from the African habitat.
Other: Tours for local garden clubs and educational
workshops.

18.47
Zooamerica North American Wildlife Park
100 West Hersheypark Dr.
Hershey, PA 17033
717-534-3860

The plant collection consists of all native North American
species. Plants are exhibited from five habitats:
Florida's everglades, Canada's evergreen forest, Great
Plains, Sonoran desert, and Eastern broadleaf
woodlands.

RI
18.48
Roger Williams Park Zoo
Roger Williams Park
Providence, RI 02905
401-785-9450, ext. 80

The zoo features a tropical American exhibit and a large
wetlands area with native flora.

TN
18.49
Knoxville Zoological Gardens, Inc.
P.O. Box 6040
Knoxville, TN 37914
615-637-5331

Native trees, ornamental shrubs, wildflowers, and
thousands of daffodils and other bulbs.
Other: Plant sales and educational programs.

18.50
Memphis Zoo & Aquarium
2000 Galloway
Memphis, TN 38112
901-726-4787

30 species of bamboo, 15 species of other ornamental
grasses, and herbs and native plants.
Other: Affiliated with several garden clubs and plant
rescues.

18.51

Zoological Society of Middle Tennessee
P.O. Box 25187
Nashville, TN 37202
615-327-2724.

This group is working to establish the Nashville Zoological Park. It plans to include "lookalike" hardy plants to simulate various exotic environments, restoration and habitat enhancement for native wildlife, and agricultural exhibits from various parts of the world. The projected opening for the first phase is 1993.

TX

18.52

Abilene Zoological Gardens
P.O. Box 60
Abilene, TX 79604
915-672-9771.

Southwestern native plant exhibit.

18.53

Central Texas Zoo
Rt. 10, Box 173-E
Waco, TX 76708
817-752-0363

Many types of woody plants, including specimens rare in the area such as the Ohio buckeye, French mulberry, Texas persimmon, Durand oak, and the blackhaw arrowhear.

Other: The Central Texas Zoo will relocate to a new facility at Cameron Park in the spring of 1992. Cameron Park is 500 acres of natural forest.

18.54

Dallas Zoo
621 East Clarendon Dr.
Dallas, TX 75203-2996
214-670-6825

Landscaped animal habitats which simulate the major African environmental regions; an African plant species collection is being developed.

18.55

El Paso Zoo
Evergreen & Paisano
El Paso, TX 79905
915-544-2402

Cacti, palm trees, live oaks, California cedars, ginkgo trees, desert willows, and evergreens.

18.56

Fossil Rim Wildlife Center
Rt. 1, Box 210
Glen Rose, TX 76043
817-897-2960.

Texas madrone, big bluestem, bigtooth maple, and 60–80 species of Texas native plants.

Other: Educational programs on water conservation with native plants.

18.57

Gladys Porter Zoo
500 Ringgold St.
Brownsville, TX 78520
512-546-7187

A varied collection of native, subtropical, and tropical plants, including a cactus garden

18.58

San Antonio Zoological Park and Aquarium
3903 N. St. Mary's
San Antonio, TX 78212
512-734-7184

Over 24 species of bamboo, 20 species of bromeliads, 25 species of orchids, 12 species of tropical fruit, 30 species of hibiscus and cultivars, and many specimens of *Crinum*, and zephyranths bulbs. The 50-acre park has a diverse landscape including native Texas plants, tropicals, a butterfly garden, and biogeographical representations of many areas of the world.

Other: Docent training, garden club tours, and honors high school study program. Has put on programs such as "Landscaping for Wildlife," "Plant-animal Relationships," and "Biological Pest Control and the Zoo."

WA

18.59

Woodland Park Zoological Gardens and Rose Garden
5500 Phinney Ave. North
Seattle, WA 98103
206-789-7919

Bamboos and All-America Rose Selections garden.
Cross-reference: 20.291

CANADA

AB

Calgary Zoo, Botanical Garden, and Prehistoric Park
Calgary, AB
Cross-reference: 16.430

BC

Crystal Garden
Victoria, BC
Cross-reference: 16.438

ON

18.60

Metropolitan Toronto Zoo
P.O. Box 280
West Hill, ON M1E 4R5
416-392-5900

Location: Highway 401 and Meadowvale Rd., Scarborough.

Carolinian woodland species on northern limit of hardiness and extensive indoor plant collection.

PQ
18.61
Jardin zoologique du Quebec
8191 Avenue du Zoo
Charlesbourg, PQ G1G 4G4
418-622-0313

Trees, perennial, and annual flowers.

SK
18.62
Forestry Farm Park and Zoo
Sutherland, Sub. P.O.
Saskatoon, SK S7N 2H0
306-975-3382

Prairie marsh exhibit, native prairie plants, and extensive collection of hardy ornamentals.
Other: Horticultural workshops.

CEMETERIES WITH NOTABLE GARDENS

The following list of cemeteries with notable gardens includes:

1. Cemeteries with major areas that were designed (usually by a notable landscape architect) to be gardens, as well as cemeteries. Among these are (a) the Victorian garden (or rural) cemeteries that were modeled after Mount Auburn Cemetery of Cambridge, Massachusetts, and (b) the memorial park type of cemetery, wherever the grounds are worthy of visits by persons interested in plants, and wherever notable examples of landscape architecture are present.

2. Cemeteries that could qualify as arboreta, because they have notable collections of trees and shrubs with some kind of identification.

3. Cemeteries that currently employ gardeners trained in ornamental horticulture (as distinct from expert grounds maintenance) and emphasize good display of various plants and garden design.

This list is based upon one compiled by Robert Grierson for the 1982 edition of *North American Horticulture.*

AL
18.63
Elmwood Cemetery
600 Montevallo Rd.
Birmingham, AL 35210

18.64
Magnolia Cemetery
Virginia St. and Ann
Mobile, AL 36604

18.65
Oakwood Cemetery
829 Columbus St.
Montgomery, AL 36104

AZ
18.66
Greenwood Memorial Park
2300 West Van Buren St.
Phoenix, AZ 85009

CA
18.67
Cypress Lawn Cemetery
593 Market St.
San Francisco, CA 94105

18.68
Forest Lawn Memorial Parks
1712 South Glendale Ave.
Glendale, CA 91205

18.69
Hollywood Memorial Park Cemetery
6000 Santa Monica Blvd.
Los Angeles, CA 90038

18.70
Inglewood Park Cemetery
720 East Florence Ave.
Inglewood, CA 90302

18.71
Mountain View Cemetery
5000 Piedmont Ave.
Oakland, CA 94611

18.72
Rose Hills Memorial Park
3900 South Workman Mill Rd.
Whittier, CA 90608-0110

CO
18.73
Fairmount Cemetery
East Alameda and Quebec Sts.
Denver, CO 80231

CT
18.74
Cedar Hill Cemetery
453 Fairfield Ave.
Hartford, CT 06114

18.75
Grove Street Cemetery
227 Grove St.
New Haven, CT 06502

18.76
Mountain Grove Cemetery
2675 North Ave.
Bridgeport, CT 06604

DC
18.77
Oak Hill Cemetery
3001 R St., N.W.
Washington, DC 20007

18.78
Rock Creek Cemetery
Rock Creek Church Rd. and Webster St., N.W.
Washington, DC 20011

DE
18.79
Wilmington and Brandywine Cemetery
701 Delaware Ave.
Wilmington, DE 19801

GA
18.80
Bonaventure Cemetery
330 Bonaventure Rd.
Savannah, GA 31404

18.81
Colonial Park Cemetery
East Oglethorpe Ave.
Savannah, GA 31401

18.82
Laurel Grove Cemetery
802 West Anderson
Savannah, GA 31401

18.83
Oakland Cemetery
248 Oakland Ave., S.E.
Atlanta, GA 30312

18.84
Oconee Hill Cemetery
297 Cemetery St.
Athens, GA 30601

18.85
Rose Hill Cemetery
Macon, GA 31204

18.86
Westview Cemetery
1680 Gordon St., S.W.
Atlanta, GA 30310

IA
18.87
Aspen Grove Cemetery
2043 Sunnyside Ave.
Burlington, IA 52601

18.88
Glendale Cemetery
4909 University Ave.
Des Moines, IA 50311

18.89
Woodland Cemetery
2019 Woodland Ave.
Des Moines, IA 50309

IL
18.90
Graceland Cemetery & Crematory
4001 North Clark St.
Chicago, IL 60613

18.91
Oak Ridge Cemetery
1441 Monument Ave.
Springfield, IL 62702

18.92
Oakwoods Cemetery
1035 East 67th St.
Chicago, IL 60637

18.93
Rosehill Cemetery & Mausoleum
5800 Ravenswood Ave.
Chicago, IL 60660

IN
18.94
Crown Hill Cemetery
700 West 38th St.
Indianapolis, IN 46208

KS
18.95
Oak Hill Cemetery
1605 Oak Hill Ave.
Lawrence, KS 66044

KY
18.96
Cave Hill Cemetery
701 Baxter Ave.
Louisville, KY 40204

18.97
Lexington Cemetery
833 West Main St.
Lexington, KY 40508

LA
18.98
Metairie Cemetery
5100 Pontchartrain Blvd.
New Orleans, LA 70124

MA
18.99
Forest Hills Cemetery
95 Forest Hills Ave.
Jamaica Plain, MA 02130

18.100
Harmony Grove Cemetery
Salem, MA 01970

18.101
Mount Auburn Cemetery
580 Mount Auburn St.
Cambridge, MA 02138
Cross-reference: 16.180

18.102
Newton Cemetery
791 Walnut St.
Newton Centre, MA 02159

18.103
Rural Cemetery
180 Grove St.
Worcester, MA 01605

18.104
Woodlawn Cemetery
302 Elm St.
Everett, MA 02149

MD
18.105
Fort Lincoln Cemetery
3401 Bladensburg Rd.
Brentwood, MD 20722

18.106
Greenmount Cemetery
Greenmount Ave. at Oliver
Baltimore, MD 21202

18.107
Loudon Park Cemetery
3801 Frederick Ave.
Baltimore, MD 21229

ME
18.108
Laurel Hill Cemetery
293 Beach St.
Saco, ME 04072

18.109
Mount Hope Cemetery
Outer State St.
Bangor, ME 04401

MI
18.110
Holy Sepulchre Cemetery
25800 West Ten Mile Rd.
Southfield, MI 48037

18.111
Woodlawn Cemetery
19975 Woodlawn Ave.
Detroit, MI 48203

MN
18.112
Lakewood Cemetery
3600 Hennepin Ave.
Minneapolis, MN 55408

18.113
Oakland Cemetery
75 East Sycamore Ave.
St. Paul, MN 55117

MO
18.114
Bellafontaine Cemetery
4947 West Florissant Ave.
St. Louis, MO 63115

18.115
Oak Grove Cemetery
7800 St. Charles Rd.
St. Louis, MO 63114

NE
18.116
Forest Lawn Memorial Park
40th and Forest Lawn Ave.
Omaha, NE 68112

18.117
Wyuka Cemetery
3600 O St.
Lincoln, NE 68510

NH
18.118
Blossom Hill Cemetery
207 North State St.
Concord, NH 03301

NY
18.119
Albany Rural Cemetery
Cemetery Ave.
Albany, NY 12204

18.120
Ferncliff Cemetery
Secor Rd.
Hartsdale, NY 10530

18.121
Forest Lawn Cemetery
1411 Delaware Ave.
Buffalo, NY 14209

18.122
Greenwood Cemetery
Fifth Ave. at 25th St.
Brooklyn, NY 11232

18.123
Kensico Cemetery
Valhalla, NY 10595

18.124
Mount Hope Cemetery
791 Mount Hope Ave.
Rochester, NY 14620

18.125
Sleepy Hollow
Tarrytown, NY 10591

18.126
Woodlawn Cemetery
Webster Ave. at 233rd St.
New York, NY 10470

OH
18.127
Cemetery of Spring Grove
4521 Spring Grove Ave.
Cincinnati, OH 45232

18.128
Ferncliff Cemetery
501 West McCreight Ave.
Springfield, OH 45504

18.129
Green Lawn Cemetery
1000 Greenlawn Ave.
Columbus, OH 43223

18.130
Lake View Cemetery
12316 Euclid Ave.
Cleveland, OH 44106

18.131
Spring Grove Cemetery & Arboretum
4521 Spring Grove Ave.
Cincinnati, OH 45232
Cross-reference: 20.191

18.132
Woodlawn Cemetery
1502 West Central Ave.
Toledo, OH 43606

OK
18.133
Rose Hill Burial Park
6001 N.W. Grand Blvd.
Oklahoma City, OK 73118

OR
18.134
Riverview Abbey Mausoleum
0319 S.W. Taylors Ferry Rd.
Portland, OR 97219

PA
18.135
Allegheny County Memorial Park
1600 Duncan Ave.
Wilson Park, PA 15101

18.136
Allegheny Cemetery
4734 Butler St.
Pittsburgh, PA 15201

18.137
Calvary Cemetery
718 Hazlewood Ave.
Pittsburgh, PA 15217

18.138
Jefferson Memorial Park
Curry Hollow Rd.
Pleasant Hills Borough
Pittsburgh, PA 15236

18.139
Laurel Hill Cemetery
3822 Ridge Ave.
Philadelphia, PA 19132

RI
18.140
Swan Point Cemetery
585 Blackstone Blvd.
Providence, RI 02906

SC
18.141
Magnolia Cemetery
No. 52 and No. 78
Charleston, SC 29405

TN
18.142
Forest Hill Cemetery
1661 South Elvis Presley Blvd.
Memphis, TN 38134

18.143
Memorial Park
5668 Poplar Ave.
Memphis, TN 38117

18.144
Mount Olivet Cemetery
1101 Lebanon Rd.
Nashville, TN 37210

18.145
Spring Hill Cemetery
5110 Gallatin Rd.
Nashville, TN 37216

TX
18.146
Forest Park Cemetery
6900 Lawndale Ave.
Houston, TX 77023

18.147
Hillcrest Memorial Park
7403 West Northwest Hwy.
Dallas, TX 75225

18.148
Oakwood Cemetery
1601 Navasota St.
Austin, TX 78702

VA
18.149
Arlington National Cemetery
Fort Meyer, VA 22211

18.150
Holywood Cemetery
4125 Cherry St.
Richmond, VA 23220

VT
18.151
Lake View Cemetery
353 North Ave.
Burlington, VT 05401

WA
18.152
Lake View Cemetery
1554 15th Ave. East
Seattle, WA 98112

WI
18.153
Forest Hill Cemetery
One Speedway Rd.
Madison, WI 53705

18.154
Forest Home Cemetery
2405 West Forest Home Ave.
Milwaukee, WI 53215

CANADA

ON
18.155
Assumption Cemetery
Mississauga, ON

18.156
Holy Cross Cemetery
Thornhill, ON

~9 19 ~9

Community Gardens

Since the early 1970s, skyrocketing food prices, the growing awareness of the nutritional merits of fresh food, concerns about commercial pesticide residues, the appreciation of gardening as a source of relaxing exercise, and neighborhood pride have greatly increased the popularity of community gardening nationwide.

Community gardens are defined as gardens used by more than one household. Currently more than two million families in North America who would otherwise have no land available are able to "grow their own" on these plots, reaping the benefits of fresh food, exercise, and neighborhood beauty and pride. The gardens can be used as a place for educating youth, for horticultural therapy rehabilitation for the mentally and physically handicapped, as an answer to the nutritional and economic needs of the elderly and those on fixed incomes, and as a catalyst for other community development.

Community gardening programs have been established in some 200 towns and cities in the United States and Canada. Some are municipally managed, while others are sponsored by private nonprofits. There are also twenty-six programs funded by the federal government through the 26 Cities Program. If your city is not listed below, contact other community garden programs in the state. They may be able to direct you to a community garden close by. Also, the local Cooperative Extension office (11.2–11.55) should have information on community gardens in your area.

Another source of information is the American Community Gardening Association (1.11)—founded in 1979 to support public gardening and greening in urban, suburban, and rural America, and to create a network to help those involved in gardening organizing. Toward these ends, the Association established an annual National Community Gardening and Open Space Conference and a quarterly publication, the *Journal of Community Gardening*.

UNITED STATES

AK
19.1
Cooperative Extension Service
2221 E. Northern Lights #240
Anchorage, AK 99508
907-279-5582
Contact: Julie Riley
Cross-reference: 11.2

19.2
Fairbanks Community Garden
P.O. Box 73488
Fairbanks, AK 99707
Contact: John Kent

AZ
19.3
Cooperative Extension Service
Urban Gardening Program
4341 E. Broadway
Phoenix, AZ 85040
602-255-4456
Contact: Sue Whitsitt

CA
19.4
Alhambra Community Garden
111 S. First St.
Alhambra, CA 91801
818-570-3244
Contact: Kathy Voltz

19.5
California Community Garden Council
Box 1715
Los Gatos, CA 95030

19.6
Common Ground Urban Gardening Program
2615 S. Grand Ave. #400
Los Angeles, CA 90007
213-744-4346; (FAX) 213-745-7513

Contact: Sharl Hopkins

19.7
Fruition Project, The
P.O. Box 872
Santa Cruz, CA 95061
408-458-3365

19.8
Oakland Land Project
5848 Foothill Blvd.
Oakland, CA 94605
415-568-8595

Contact: Mitch Hardin

19.9
San Francisco League of Urban Gardeners
2540 Newhall St.
San Francisco, CA 94124
415-468-0110

Contact: Cynthia Hall

19.10
Saratoga Community Gardens
14500 Fruitvale Ave.
Box 756
Saratoga, CA 95070

19.11
Southland Farmer's Market Association
1010 S. Flower St., Rm. 402
Los Angeles, CA 90015
213-749-9551

Contact: Merion Kow

19.12
University of California at Davis
Center for Design Research
Davis, CA 95616
916-752-6031

Contact: Mark Francis
Cross-references: 12.62, 12.402, 15.23, 16.59, 17.33

CO
19.13
Denver Urban Community Garden Program
227 S. Grant St.
Denver, CO 80209
303-692-5600

Contact: Chris Cordts

CT
19.14
Bridgeport Urban Gardening Program
264 North State
Ansonia, CT 06401
203-775-3800

Contact: Bill Urban

19.15
Horticulture Ranger Program
c/o Pardee Rose Garden
East Rock Park
180 Park Rd.
Hamden, CT 06517
203-787-8142

Contact: Cindy Long

19.16
Knox Parks Foundation
150 Walbridge Rd.
West Hartford, CT 06119
203-523-4276

Contact: Jack Hale and Suzanne Gerety

DC
19.17
Garden Resources of Washington
1419 V St., N.W.
Washington, DC 20009
202-234-0591

19.18
Washington Youth Gardens
1250 I St., N.W., Suite 500
Washington, DC 20005
202-789-2900

19.19
Youth and Urban Gardens Program
D.C. Dept. of Recreation and Parks
3149 16th St., N.W.
Washington, DC 20010
202-576-6257

Contact: Patricia Pyle

DE
19.20
Wilmington Garden Center
503 Market Street Mall
Wilmington, DE 19801
302-658-1913

Contact: Michael Shurilla
Cross-reference: 7.17

FL
19.21
End World Hunger Inc.
11431 S.W. 112th Ave.
Miami, FL 33176
305-251-3371

Contact: John Waterhouse

19.22
Jacksonville Urban Gardening Program
1010 North McDuff Ave.
Jacksonville, FL 32205
904-384-2001

Contact: Marcie L. Kelt

GA
19.23
E.O.A./Community Gardens Program
Box 1353
Savannah, GA 31402
912-232-2165

19.24
Georgia Extension Service
2390 Wildcat Rd.
Decatur, GA 30034
404-241-7444

Contact: John Arnold/Steve Rogers

HI
19.25
Oasis Group, The
P.O. Box 330094
Kahului, HI 96733
808-877-8016

IL
19.26
Chicago Botanic Garden
Community Gardening Program
Box 400
Glencoe, IL 60022
312-835-5440

Cross-reference: 16.138

19.27
Morton Arboretum
23 W. 177 Indian Hill Dr.
Lisle, IL 60532
312-219-2433

Contact: Charles Lewis
Cross-reference: 16.145

19.28
Open Lands Project
220 S. State, Suite 1880
Chicago, IL 60604
312-427-4256

Contact: Kathy Dickhut

19.29
PCCEO, Inc.
1000 S.W. Adams
Peoria, IL 61602
309-671-3906

Contact: Ron Cooley

19.30
University of Illinois Cooperative Extension Service
36 S. Wabash, Suite 1012
Chicago, IL 60603
312-996-3535; (FAX) 312-996-3957

Contact: Vernon Bryant

19.32
Urban Gardening Program
Cooperative Extension Service
4141 West Belmont
Chicago, IL 60641
312-286-6767

IN
19.33
Capital City Gardens Project
Marion County C.E.S.
9245 N. Meridian St., Suite 118
Indianapolis, IN 46260
317-848-7351

Contact: Tom Tyler

KY
19.34
Jefferson County Cooperative Extension Service
8012 Vinecrest Ave.
Louisville, KY 40222
502-425-4482

Contact: Brett R. Mills

LA
19.35
Urban Garden Program
Delgado College, Bldg. 15
615 City Park Ave.
New Orleans, LA 70119
504-486-5929

MA
19.36
Boston Urban Gardeners
33 Harrison Ave., 5th Floor
Boston, MA 02111
617-423-7497

Contact: Gayle Knight

19.37
Human Services
Endicott Gardens
P.O. Box 1305
Dedham, MA 02026
617-326-1004

Contact: Charles Clinton

19.38
Safer, Inc.
189 Well Ave.
Newton, MA 02159

19.39
University of Massachusetts
Suffolk County Cooperative Extension Service (CFNR)
150 Causeway St., Room 803
Boston, MA 02114
617-727-4107

Contact: Linda Bowman

MD
19.40
Baltimore City Extension Urban Gardening Program
17 S. Gay St.
Baltimore, MD 21202
301-396-1888

Contact: Jon Traunfield

19.41
Enterprise Foundation, The
Suite 505
American City Building
Columbia, MD 21044
301-964-1230

Contact: Patricia Rouse

MI
19.42
4-H Gardening Program
640 Temple
Detroit, MI 48215
313-721-6550

Contact: Jocelyn Fitzpatrick

19.43
Office of Services to the Aging
P.O. Box 30026
Lansing, MI 48909
517-373-9362

Contact: David Houseman

19.44
Project Grow Community Gardens
P.O. Box 8645
Ann Arbor, MI 48107

MN
19.45
Duluth Plant-A-Lot
Community Gardening Program
206 W. 4th St.
Duluth, MN 55806
218-722-4583

Contact: Mary Beth Nevers

19.46
Minnesota Green
Minnesota Horticultural Society
1970 Fowell Ave., 161 Alderman Hall
St. Paul, MN 55108
612-624-7752

Contact: Rick Bonlender
Cross-reference: 6.20

19.47
Self-Reliance Center
1916 2nd Ave. South
Minneapolis, MN 55403
612-870-4255

Contact: Margaret Vaillancourt

19.48
St. Anthony Park Community Council
890 Comwell
St. Paul, MN 55114
612-292-7884

Contact: Roberta Megard

MO
19.49
Community Garden Coalition
P.O. Box 7051
Columbia, MO 65205
314-443-7190

Contact: Crystal Frey, Executive Director

19.50
Community Garden Project
3200 Wayne
Kansas City, MO 64109
816-924-2055

Contact: Ira Harritt

19.51
Kansas City Community Gardens
3200 Wayne
Kansas City, MO 64109
816-931-3877

Contact: Ben Sharda

19.52
Missouri Cooperative Extension Service
724 North Union Blvd.
Saint Louis, MO 63108
314-367-2585

Contact: Lois M. Laster

19.53
Urban Gardening Program
University Extension Center
727 North Union
St. Louis, MO 63108
314-367-2585

Contact: Lois M. Laster, Director

NC
19.54
M.A.G.I.C. Community Garden Program
P.O. Box 168
Asheville, NC 28802
704-251-5666

Contact: Tom Youngblood-Peterson

NH
19.55
Town & Country Gardens
P.O. Box 113
Marlon, NH 03456
603-446-7196

NJ
19.56
Camden County Park Dept.
Camden City Garden Club
P.O. Box 4210
Cherry Hill, NJ 08002
609-795-7275

19.57
Greater Newark Conservancy
303-9 Washington St.
5th Floor, Rm. 3
Newark, NJ 07102
201-642-4646

Contact: Marleny Franco

19.58
Isles Inc.
126 N. Montgomery St.
Trenton, NJ 08608
609-393-7153

Contact: Elizabeth Johnson

19.59
New Jersey Regional Day School Garden
334 Lyons Ave.
Newark, NJ 07112
201-705-3829

Contact: Phyllis D'Amico

19.60
Rutgers Urban Gardening
249 University Ave., Rm. 201
Newark, NJ 07102
201-648-5948

NY
19.61
Battery Park City Parks Corp.
40 West St., Gate 3
New York, NY 10006
212-248-4990

Contact: Tessa Huxley

19.62
Bronx Greenup
New York Botanical Garden
Watson Bldg., Rm. 214
Southern Blvd.
Bronx, NY 10458-5126
212-220-8995

Contact: Terry Keller
Cross-reference: 16.286

19.63
Capital District Community Garden
83 4th St.
Troy, NY 12180
518-274-8685

Contact: Maria Trabka

19.64
City of White Plains
Dept. Recreation & Parks
85 Gedney Way
White Plains, NY 10605
914-682-4336

Contact: Margaret Carter

19.65
Cornell University Cooperative Extension Service
15 E. 26th St., 5th Floor
New York, NY 10010
212-340-2930

Contact: Urban Horticulture

19.66
Council on the Environment of New York City
51 Chambers St., Rm. 228
New York, NY 10007
212-566-0990

Contact: Lenny Librizzi
Cross-reference: 8.24

19.67
Green Guerillas
625 Broadway, 2nd Floor
New York, NY 10012
212-674-8124

Contact: Barbara Earnest

19.68
L.I.V.E. Program
c/o Central Park Conservancy
The Arsenal, Central Park
New York, NY 11215
212-860-1336

Contact: Mary Scent

19.69
New York City of Parks and Recreation
The Arsenal, Rm. 203
830 Fifth Ave.
New York, NY 10021
212-360-8203

Contact: Debra Lerer, Director of Horticulture

19.70
New York State Office of Community Dept. of
  Agriculture & Markets
1 Winner's Circle
Albany, NY 12235
512-457-6468

Contact: Paul S. Winkeller

19.71
Neighborhood Open Space Coalition
72 Reade St.
New York, NY 10007
212-513-7555

19.72
Operation Green Thumb
49 Chambers St., Rm. 1020
New York, NY 10007
212-233-2926

Contact: Jane Weissman

19.73
Parks Council
457 Madison Ave.
New York, NY 10022
718-768-0733

Contact: Joseph Pupello

19.74
Public Environment Center, Inc.
One Milligan Pl.
New York, NY 10011
212-691-4877

19.75
Town of Babylon
Environmental Commission
278 Waldo St.
Copiague, NY 11726
516-422-7640

Contact: Carole Wilder

19.76
Trust for Public Land
666 Broadway
New York, NY 10024
212-677-7171

Contact: Andrew Stone

19.77
Union Settlement Association
237 E. 104th St.
New York, NY 10029
212-360-8818

Contact: Sally Yarmolinsky

OH
19.78
Civic Garden Center of Greater Cincinnati
2715 Reading Rd.
Cincinnati, OH 45206
513-221-0991

Contact: Benjamin Long
Cross-reference: 7.46

19.79
Cuyahoga Cooperative Extension Urban Gardening
   Program
3200 West 65, Rm. 216
Cleveland, OH 44102
216-631-1890

Contact: Dennis Rinehart

19.80
Grow With Your Neighbors
Wegerzyn Horticulture Center
1301 E. Siebenthaler Ave.
Dayton, OH 45414
513-277-6545

Contact: Lorka Munoz
Cross-reference: 7.48

OR
19.81
Portland Community Gardens
6437 S.E. Division St.
Portland, OR 97206
503-248-4777

Contact: Leslie Pohl-Kosbail

PA
19.82
Citiparks Urban Gardening
2005 Beechwood Blvd.
Pittsburgh, PA 15217
412-422-6532

Contact: Roberta Greenspan

19.83
Neighborhood Garden Association
325 Chestnut St., Suite 411
Philadelphia, PA 19106
215-625-8264

19.84
Penn State Urban Gardening Program of Delaware
   County
P.O. Box 693
Chester, PA 19013
215-872-8210

Contact: Susan Goldsworthy

19.85
Penn State Urban Gardening
4601 Market St., 3rd Fl.
Philadelphia, PA 19139
215-560-4166

Contact: Terry Mushovic

19.86
Pennsylvania Horticultural Society
325 Walnut St.
Philadelphia, PA 19106-2777
215-625-8256

Contact: Peggy Grady
Cross-reference: 6.32

19.87
Western Pennsylvania Conservancy
316 4th Ave. RN.
Pittsburgh, PA 15222
412-288-2771

Contact: Jeff Gerson

RI
19.88
Southside Community Land Trust
288 Dudley St.
Providence, RI 02907
401-273-9419

Contact: Deborah Schimberg

TN
19.89
Chattanooga Food Bank
P.O. Box 608
Chattanooga, TN 37401-0608
615-265-6300

Contact: Nona Minton-Harp

19.90
Greenthumb Community Gardens
Knoxville City Community Action Commission
2247 Western Ave.
Knoxville, TN 37921
615-546-3500

Contact: Pat Bing

19.91
Memphis Urban Garden Program
5565 Shelby Oaks Dr.
Memphis, TN 38134
901-521-2946

Contact: William Vasser

TX
19.92
Interfaith Hunger Coalition
Community Gardening Program
Houston Metropolitan Ministries
3217 Montrose Blvd.
Houston, TX 77006
713-520-4620

Contact: Robert Randall

UT
19.93
Wasatch Fish and Garden Project
347 South 400 East
Salt Lake City, UT 84103
801-364-7765

Contact: Nick Hershenow

VT
19.94
Burlington Area Community Gardens
Burlington Dept. of Parks
216 Leddy Park Rd.
Burlington, VT 05401
802-864-0123

Contact: Maggie Leugers

19.95
National Gardening Association
180 Flynn Ave.
Burlington, VT 05401
802-863-1308

Contact: Tim Parsons
Cross-reference: 1.89

WA
19.96
Seattle P-Patch Program
Alaska Building, 5th Floor
618 Second Ave.
Seattle, WA 98104-2222
206-684-0264

Contact: Barbara Donette and Nancy Allen

19.97
Trust for Public Lands
Smith Tower/Suite 2123
506 2nd Ave.
Seattle, WA 98104

Contact: Jodi Olson

WI
19.98
C.A.C. Garden Program
1810 S. Park St.
Madison, WI 53713-1214
608-266-9730

Contact: Judy Siegfried/Joe Mathers

19.99
Shoots 'n Roots Program
1333 N. 12th St.
Milwaukee, WI 53205
414-342-0300

Contact: Barbara Pacey

WY
19.100
Cheyenne Community Botanic Gardens
1213 Richardson Court
Cheyenne, WY 82001
307-635-9340

CANADA

BC
19.101
B.C. Housing Management Commission
Suite 1701, 4330 Kingway
Burnaby, BC V5H 4G7
604-433-1711

Contact: Joyce Fitz-Gibbon

ON
19.102
Harrowsmith Magazine
7 Queen Victoria Rd.
Camden East, ON K0K 1J0
613-378-6661

Contact: Jennifer Bennett
Cross-reference: 21.49

20

Test and Demonstration Gardens

ALL-AMERICA SELECTIONS

Each year a group of new flower and vegetable cultivars are announced as the All-America Selections (1.2) winners for the coming gardening season. These new cultivars have been evaluated by a council of expert judges at trial grounds across the United States and Canada and rated as the best plant selections for North American gardens. Experienced gardeners like to try these new cultivars in their home gardens, but beginners should also consider them because they have not only been bred for award-winning flowers or fruit, but also for their ruggedness, disease resistance, and adaptability.

The annual awards constitute the culmination of a long process of development and evaluation. Each year seed breeders, governmental institutions, and private plant breeders from around the world send seed samples, descriptions, and photographs of their best new flower and vegetable cultivars to the All-America Selections office for evaluation. About eighty new cultivars are entered in the trials each year, the majority submitted by seed companies. Each entry is grown and evaluated in side-by-side trials with an appropriate comparison cultivar suggested by the breeder and selected by a panel of experts. The plants are evaluated and scored on the basis of home garden performance. On the average, six to ten awards are made each year.

TRIAL GROUNDS AND JUDGES

A trial grounds is a large spread of trial rows averaging twenty to thirty feet in length. The All-America Selections section of any trial grounds is generally only a small part of the total, because most seedsmen, universities, and public gardens try many other non-AAS varieties as well. Each flower or vegetable trial grounds has an All-America Selections judge.

If you are interested in the fate of a particular AAS entry, check with the trial grounds after February 15th of the year following your visit. The judge will have a list of designated winners. Varieties that do not win an award are usually not introduced into commercial production.

Seed company trial grounds are open to the public seasonally on weekdays, but guided tours are not offered to individuals. Guided tours for groups are available by prior arrangement.

ALL-AMERICA SELECTIONS FLOWER TRIAL GROUNDS

UNITED STATES

CA
20.1
Bodger Seeds, Ltd.
1851 W. Olive
Lompoc, CA 93436

20.2
Denholm Seeds
222 North A St.
Lompoc, CA 93436

Cross-reference: 20.72

20.3
Ferry-Morse Seed Co.
2191 San Juan Hollister Rd.
San Juan Bautista, CA 95045

Cross-references: 20.37, 20.73

20.4
Goldsmith Seeds, Inc.
2280 Hecker Pass Rd.
Gilroy, CA 95020

20.5
Sakata Seed America, Inc.
105 Boronda Rd.
Salinas, CA 93912

Cross-reference: 20.40

20.6
Sunseeds Genetics
9800 Fairview Rd.
Hollister, CA 95024

Cross-reference: 20.41

20.7
Waller Flowerseed Co.
400 Obispo St.
Gaudalupe, CA 93434

CT
20.8
Comstock, Ferre & Co.
263 Main St.
Wethersfield, CT 06109

Cross-references: 20.43, 20.89

FL
20.9
Walt Disney World
Tree Farm Nursery
Lake Buena Vista, FL 32830

Cross-references: 17.92, 20.99

GA
20.10
Callaway Gardens
Hwy. 27
Pine Mountain, GA 31822

Cross-references: 16.104, 20.45, 20.101

20.11
University of Georgia
Dept. of Horticulture
Athens, GA 30602

Cross-references: 12.108, 12.409, 15.53, 15.54

IA
20.12
Earl May Seed & Nursery
208 North Elm St.
Shenandoah, IA 51603

Cross-references: 16.129, 20.46, 20.108

IL
20.13
Ball Seed Co.
250 Town Rd.
West Chicago, IL 60185

Cross-reference: 20.48

20.14
Pan American Seed Co.
728 Town Rd.
West Chicago, IL 60185

20.15
University of Illinois
201a Ornamental Horticulture Bldg.
Urbana, IL 61801

Cross-references: 12.140, 12.412, 15.63, 15.68

20.16
Vaughan's Seed Co.
5300 Katrine Ave.
Downers Grove, IL 61801

Cross-reference: 20.49

KY
20.17
University of Kentucky
Corner of Cooper Dr. and Limestone St.
Lexington, KY 40546

Cross-references: 12.154, 12.416, 15.87

MA
20.18
University of Massachusetts
Suburban Experiment Station
240 Beaver St.
Waltham, MA 02254

372

MI
20.19
Michigan State University
Horticulture Bldg.
East Lansing, MI 48824

Cross-reference: 16.204

MN
20.20
Northrup King Co.
Rt. 1
Stanton, MN 55081

NE
20.21
Bluebird Nursery, Inc.
515 N. Cherry St.
Clarkson, NE 68629

Cross-reference: 20.159

NY
20.22
Harris Moran Seed Co.
3670 Buffalo Rd.
Rochester, NY 14624

Cross-reference: 20.53

OR
20.23
Daehnfeldt, Inc.
1100 S.E. Jackson St.
Albany, OR 97321

PA
20.24
Abbot & Cobb, Inc.
4151 Street Rd.
Trevose, PA 19047

Cross-reference: 20.55

20.25
Pennsylvania State University
Dept. of Horticulture
101 Tyson Bldg.
University Park, PA 16802

Cross-references: 12.278, 12.447, 15.169, 20.56, 20.203,
20.288

20.26
W. Atlee Burpee Co.
300 Park Ave.
Warminster, Pa 18974

Cross-references: 20.57, 20.205

SC
20.27
George W. Park Seed Co., Inc.
Hgwy. 254
Greenwood, SC 29647

Cross-reference: 20.58

TN
20.28
University of Tennessee
Dept. of Ornamental Horticulture and Landscape
    Design
Neyland Dr.
Knoxville, TN 37901

Cross-references: 12.299, 12.453, 15.181, 16.375,
20.214

TX
20.29
Texas A&M University
Houston St. on campus
College Station, TX 77843

Cross-references: 12.315, 12.454, 13.49, 15.188, 16.378,
20.223

WI
20.30
Boerner Botanical Gardens
5879 S. 92nd St.
Hales Corners, WI 53130

Cross-references: 16.419, 20.240

CANADA

AB
20.31
Alberta Special Crops and Horticultural Research Centre
3 miles east of Brooks on Hwy. #1
Brooks, AB T0J 0J0

Cross-references: 16.429, 20.248

20.32
Alberta Tree Nursery & Horticultural Centre
RR 6
Edmonton, AB T5B 4K3

BC
20.33
University of British Columbia
next to Dept. of Plant Science
Vancouver, BC V6T 2A2

Cross-references: 12.362, 12.465, 13.54, 15.226,
16.449

MB
20.34
University of Manitoba
Dept. of Plant Science
Winnipeg, MB R3T 2N2

Cross-references: 12.363, 15.228

ON
20.35
Stokes Seeds Ltd.
39 James St.
St. Catherines, ON L2R 6R6

Cross-reference: 20.60

PQ
20.36
W.H. Perron & Co.
2000 rue DuBois
Bois Briand, PQ H7V 2T3

Cross-references: 20.62, 20.269

ALL-AMERICA SELECTIONS VEGETABLE TRIAL GROUNDS

UNITED STATES

CA
20.37
Ferry-Morse Seed Company
2191 San Juan Hollister Rd.
San Juan Bautista, CA 95045

Cross-references: 20.3, 20.73

20.38
Northrup King Co.
5355 Monterey Rd.
Gilroy, CA 95020

20.39
Petoseed Co., Inc.
Rt. 4
Woodland, CA 95695

Cross-reference: 20.79

20.40
Sakata Seed America, Inc.
105 Boronda Rd.
Salinas, CA 93912

Cross-reference: 20.5

20.41
Sunseeds Genetics
9800 Fairview Rd.
Hollister, CA 95024

Cross-reference: 20.6

CO
20.42
Rocky Mountain Seed Co.
1325-15th St.
Denver, CO 80217

CT
20.43
Comstock, Ferre & Company
263 Main Street
Wethersfield, CT 06109

Cross-references: 20.8, 20.89

20.44
University of Connecticut
Dept. of Plant Science U-67
Rt. 195
Storrs, CT 06268

Cross-references: 12.73, 12.405, 15.42

GA
20.45
Callaway Gardens
Highway 27
Pine Mountain, GA 31822

Cross-references: 16.104, 20.10, 20.101

IA
20.46
Earl May Seed & Nursery
208 North Elm Street
Shenandoah, IA 51603

Cross-references: 16.129, 20.12, 20.108

ID
20.47
Musser Seed Co., Inc.
Rt. 4
Twin Falls, ID 83301

IL
20.48
Ball Seed Co.
250 Town Rd.
West Chicago, IL 60185

Cross-reference: 20.13

20.49
Vaughan's Seed Co.
5300 Katrine Ave.
Downers Grove, IL 60515

Cross-reference: 20.16

ME
20.50
Johnny's Selected Seeds
310 Foss Hill Rd.
Albion, ME 04910

MO
20.51
University of Missouri
1-40 Agriculture Bldg.
Columbia, MO 65201

Cross-references: 12.207, 15.118, 16.217, 20.149

NJ
20.52
Rutgers University Research & Development Center
Northville Rd.
Bridgeton, NJ 08302

Cross-references: 12.240, 12.437, 15.133, 15.134, 16.261,
20.164

NY
20.53
Harris Moran Seed Co.
3670 Buffalo Rd.
Rochester, NY 14624

Cross-reference: 20.22

OH
20.54
Liberty Seed Co.
near Schoenbrunn Village on East High St.
New Philadelphia, OH 44663

PA
20.55
Abbott & Cobb, Inc.
4151 Street Rd.
Trevose, PA 19047

Cross-reference: 20.24

20.56
Pennsylvania State University
Dept. of Horticulture
103 Tyson Bldg.
University Park, PA 16802

Cross-references: 12.278, 12.447, 15.169, 20.25, 20.203,
20.288

20.57
W. Atlee Burpee Co.
300 Park Ave.
Warminster, PA 18974

Cross-references: 20.26, 20.205

SC
20.58
George W. Park Seed Co., Inc.
Highway 254
Greenwood, SC 29647

Cross-reference: 20.27

WA
20.59
Alf. Christianson Seed Co.
Railroad & Milwaukee
Mt. Vernon, WA 98273

CANADA

ON
20.60
Stokes Seeds Limited
39 James St.
St. Catherines, ON L2R 6R6

Cross-reference: 20.35

PE
20.61
Vesey's Seeds
York, PE C0A 1P0

PQ
20.62
W.H. Perron & Co.
2000 rue DuBois
Bois Briand
Quebec, PQ H7V 2T3

Cross-references: 20.36, 20.269

ALL-AMERICA SELECTIONS DISPLAY GARDENS

Display gardens arrange All-America Selections winners in landscape
situations. Some beds appear as part of the landscaping around public parks
and buildings. Where display gardens are operated in conjunction with trial
grounds, they are usually set off in a special place and identified with a red,
white, and blue AAS sign.

Display gardens are not judged, because the flowers and vegetables
featured in them have already been designated as award winners at trial
grounds. When planning a visit to a display garden, you should check with
the manager of the garden first. Flowers are usually in peak bloom for about
one month, but the best time to see them can range from May or June in the
Deep South through late August in the far North. Cool-weather flowers are
ready thirty to forty-five days before summer flowers. Vegetables mature on
about the same schedule as flowers but are at their best when the plants are
just starting to bear.

UNITED STATES

AK
20.63
Experiment Station Farm
Noatak Dr.
University of Alaska
Fairbanks, AK 99775

Cross-reference: 11.67

20.64
University of Alaska
Agricultural Experiment Station
533 E. Firewood
Palmer, AK 99645

Cross-reference: 11.68

AL
20.65
Bellingrath Gardens
12401 Bellingrath Gardens Rd.
Theodore, AL 36582

Cross-reference: 17.44

20.66
Birmingham Botanical Gardens
2612 Lane Park Rd.
Birmingham, AL 35223

Cross-reference: 16.2

AZ
20.67
Elena's Flowers
Hwy. 75 in York, 20 miles North of Duncan
Duncan, AZ 85534

CA
20.68
California Polytechnic State University
Ornamental Horticulture Dept.
San Luis Obispo, CA 93407

Cross-references: 12.24, 12.399, 15.29

20.69
California Polytechnic University
3801 W. Temple Ave.
Pomona, CA 91768

Cross-references: 12.25, 12.400

20.70
California State University
Plant Science Dept.
Fresno, CA 93740

Cross-references: 12.27, 15.21

20.71
City of Lompoc
100 Civic Center Plaza
Lompoc, CA 93438

20.72
Denholm Seeds
222 North A St.
Lompoc, CA 93436

Cross-reference: 20.2

20.73
Ferry-Morse Seed Company
2191 San Juan Hollister Rd.
San Juan Bautista, CA 95045

Cross-references: 20.3, 20.37

20.74
Filoli
Canada Rd.
Woodside, CA 94062

Cross-reference: 17.56

20.75
Fullerton Arboretum
California State University
Yorba Linda Blvd. & Associate Rd.
Fullerton, CA 92634

Cross-reference: 16.24

20.76
Howe Homestead Park
2950 Walnut Blvd.
Walnut Creek, CA 94596

20.77
Los Angeles State and County Arboretum
301 N. Baldwin Ave.
Arcadia, CA 91006

Cross-reference: 16.35

20.78
Lane Publishing
80 Willow Rd.
Menlo Park, CA 94025

20.79
Petoseed Co., Inc.
Rt. 4
Woodland, CA 95695

Cross-reference: 20.39

20.80
Pier 39 Inc.
Beach and Embarcadero
San Francisco, CA 94133

20.81
Prusch Farm Park
647 S. King Rd.
San Jose, CA 95116

20.82
Soquel Future Farmers of America
401 Old San Jose Rd.
Soquel, CA 95073

20.83
Strybing Arboretum and Botanical Gardens
Golden Gate Park
9th & Lincoln Way
San Francisco, CA 94122
Cross-reference: 16.57

CO
20.84
City of Aurora Parks Dept.
151 Potomac St.
Aurora, CO 80012

20.85
Denver Botanic Gardens
909 York St.
Denver, CO 80206
Cross-reference: 16.65

20.86
Horticultural Arts Society
Monument Valley Park
222 W. Mesa Rd.
Colorado Springs, CO 80909
Cross-reference: 16.66

20.87
W. D. Holly Plant Environmental Research Center
630 W. Lake
Fort Collins, CO 80523
Cross-reference: 16.69

CT
20.88
4-H Farm Resource Center
Simsbury Rd. (Rt. 185)
Bloomfield, CT 06002

20.89
Comstock, Ferre & Company
263 Main St.
Wethersfield, CT 06109
Cross-references: 20.8, 20.43

DC
20.90
Potomac Center #1
D.C. Village Ln. at Jr. College
Washington, DC 20032

20.91
United States Botanic Garden
245 First St., S.W.
Washington, DC 20024
Cross-reference: 16.82

FL
20.92
Daytona Beach Community College
1200 Volusia Ave.
Daytona Beach, FL 32014
Cross-reference: 12.81

20.93
H. P. Leu Botanical Gardens
1730 N. Forest Ave.
Orlando, FL 32803
Cross-reference: 16.93

20.94
Marie Selby Botanical Garden
811 S. Palm Ave.
Sarasota, FL 34236
Cross-reference: 16.96

20.95
Medical Center Clinic Gardens
8333 N. Davis Hwy.
Pensacola, FL 32514

20.96
Miami-Dade Community College
South Campus
11011 S.W. 104 St.
Miami, FL 33176

20.97
Orange County Extension Center
2350 E. Michigan St.
Orlando, FL 32806

20.98
Pensacola Junior College
Biology Dept.
1000 College Blvd.
Pensacola, FL 32504
Cross-reference: 12.95

20.99
Walt Disney World
Tree Farm Nursery
Lake Buena Vista, FL 32830
Cross-references: 17.92, 20.9

GA
20.100
State Botanical Garden of Georgia
2450 S. Milledge Ave.
Athens, GA 30605
Cross-reference: 16.110

20.101
Callaway Gardens
Highway 27
Pine Mountain, GA 31822
Cross-references: 16.104, 20.10, 20.45

20.102
Dept. of Horticulture
Georgia Experiment Station
Experiment, GA 30223

20.103
Georgia Institute of Technology
915 Atlantic Dr., N.W.
Atlanta, GA 30318

20.104
Oak Hill Gardens
Berry College
Mount Berry, GA 30149

HI
20.105
University of Hawaii
Maui Agricultural Research Ctr.
located at end of Mauna Place
Kula, Maui, HI 96790

Cross-reference: 11.118

IA
20.106
Des Moines Botanical Center
909 East River Dr.
Des Moines, IA 50316

Cross-reference: 16.126

20.107
Dubuque Arboretum & Botanical Gardens
3125 West 32nd St.
Dubuque, IA 52001

Cross-reference: 16.128

20.108
Earl May Seed & Nursery
208 North Elm Street
Shenandoah, IA 51603

Cross-references: 16.129, 20.12, 20.46

20.109
Iowa State University
Agriculture Experiment Station
Ames, IA 50011

Cross-references: 16.133, 20.276

20.110
Noelridge Park
4900 Council St., N.E.
LeMars, IA 51031

Cross-reference: 16.134

20.111
Westmar College
1002 3rd Ave., S.E.
LeMars, IA 51031

ID
20.112
Ricks College
500 S. Center St.
Rexburg, ID 83440

Cross-reference: 12.123

20.113
University of Idaho
Cooperative Extension Service
Parma, ID 83660

IL
20.114
Belleville Junior College
2500 Carlyle Rd.
Belleville, IL 62221

Cross-reference: 12.125

20.115
Cantigny Gardens
1S151 Winfield Rd.
Wheaton, IL 60187

Cross-reference: 17.102

20.116
Chicago Botanic Garden
Lake Cook Rd., ½-Mi. E. of Edens
Glencoe, IL 60022

Cross-reference: 16.138

20.117
Illinois Central College
Agriculture Land Laboratory
East Peoria, IL 61635

Cross-reference: 12.131

20.118
Lake of the Woods Botanic Gardens
Champaign City, Forest Preserve
Mahomet, IL 61853

Cross-reference: 17.105

20.119
Triton Botanical Garden
Triton College
2000 5th Ave.
River Grove, IL 60171

Cross-reference: 12.139

20.120
Washington Park Botanical Garden
Fayette & Chatham Rds.
Springfield, IL 62704

Cross-reference: 16.148

IN
20.121
Brown County Master Garden
Brown County 4-H
Fairgrounds Gardens
Nashville, IN 47448

20.122
Foster Gardens
3900 Broadway
Fort Wayne, IN 46807

20.123
Parson's Patch
2003 E. Pleasant St.
Noblesville, IN 46060

KS
20.124
Botanica, The Wichita Gardens
701 Amidon
Wichita, KS 67203

Cross-reference: 16.157

20.125
NWK Research-Extension Center
Rt. 2
Colby, KS 67701

KY
20.126
Kentucky Fair & Exposition Center
Phillips Ln. & Freedom Way
Louisville, KY 40213

LA
20.127
Hodges Gardens
Many, LA 71449

Cross-reference: 16.165

20.128
Louisiana State University
Hill Farm
Baton Rouge, LA 70803

Cross-references: 12.157, 12.417, 15.89

20.129
New Orleans Botanical Garden
Victory Avenue City Park
New Orleans, LA 70119

MA
20.130
Berkshire Garden Center
Corner of Route 102/183
Stockbridge, MA 01262

Cross-reference: 16.176

20.131
University of Massachusetts
Durfee Conservatory
French Hall
Amherst, MA 01003

Cross-reference: 16.179

MD
20.132
Brookside Gardens
1500 Glenallen Ave.
Wheaton, MD 20902

Cross-reference: 16.184

20.133
Cylburn Wild Flower Preserve and Garden Center
4915 Greenspring Ave.
Baltimore, MD 21209

Cross-reference: 16.185

ME
20.134
Ornamental Trial Gardens
University of Maine
Rangely Rd.
Orono, ME 04469

Cross-references: 12.182, 15.103, 16.193

MI
20.135
Dow Gardens
Corner of Eastman (US-10) and West St. Andrews
Midland, MI 48640

Cross-reference: 16.200

20.136
Fernwood Garden and Nature Center
13988 Range Line Rd.
Niles, MI 49120

Cross-reference: 16.201

20.137
Hidden Lake Gardens
Route M-50
Tipton, MI 49287

Cross-reference: 16.202

20.138
Kalamazoo Pedestrian Mall
128 N. Kalamazoo Mall
Kalamazoo, MI 49007

20.139
Michigan State University
Horticulture Bldg.
East Lansing, MI 48824

Cross-reference: 16.204

20.140
Professional Plant Growers Association
1989 N. College Rd.
Mason, MI 48854

Cross-reference: 1.110

MN
20.141
Lyndale Park Gardens
4125 E. Lake Harriet Parkway
Minneapolis, MN 55409

Cross-reference: 16.207

20.142
Minnesota Landscape Arboretum
3675 Arboretum Dr.
Chanhassen, MN 55317

Cross-reference: 16.208

MO
20.143
Danforth FFA Garden
Charleston High School
Thorn St./Voc.-Agri. Bldg.
Charleston, MO 63834

20.144
Flower & Garden Magazine
4251 Pennsylvania Ave.
Kansas City, MO 64113

Cross-reference: 16.211

20.145
Lincoln University Gardens
1120 Chestnut St.
Jefferson City, MO 65101

Cross-reference: 12.201

20.146
Missouri Botanical Garden
4344 Shaw Blvd.
St. Louis, MO 63110

Cross-reference: 16.214

20.147
Powell Gardens
Route 1
Kingsville, MO 64061

Cross-reference: 16.215

20.148
Southeast Missouri State University
University Horticulture Display Gardens
New Madrid Dr.
Cape Girardeau, MO 63701

Cross-reference: 12.205

20.149
University of Missouri Campus
Ashland Gravel Road Greenhouses
Columbia, MO 65211

Cross-references: 12.207, 15.118, 16.217, 20.51

MS
20.150
Jackson State University
Botanical Garden
1400 John R. Lynch St.
Jackson, MS 39209

20.151
Mynelle Gardens
4736 Clinton Blvd.
Jackson, MS 39209

Cross-reference: 16.220

NC
20.152
Blue Ridge Technical Institute
Route 2
Flat Rock, NC 28731

Cross-reference: 12.214

20.153
North Carolina State University
4301 Beryl Rd.
Raleigh, NC 27695-7609

Cross-reference: 16.233

20.154
Reynolda Gardens of Wake Forest University
100 Reynolda Village
Winston-Salem, NC 27106

Cross-reference: 16.234

20.155
WPCC Formal Gardens
1001 Burkemont Ave.
Morgantown, NC 28655

ND
20.156
Agassiz Gardens
4201 S. University Dr.
Fargo, ND 58103

20.157
International Peace Garden
Route 1
Dunseith, ND 58329

Cross-reference: 16.242

20.158
North Dakota State University
Horticulture Bldg.
Fargo, ND 58105

Cross-references: 12.231, 12.436, 15.127

NE
20.159
Bluebird Nursery, Inc.
515 N. Cherry St.
Clarkson, NE 68629

Cross-reference: 20.21

20.160
State Fair Park Arboretum
1800 State Fair Park Dr.
Lincoln, NE 68510

NH
20.161
Fuller Gardens
10 Willow Ave.
North Hampton, NH 03862

Cross-reference: 17.177

20.162
Prescott Park
105 Marcy St.
Portsmouth, NH 03801
Cross-reference: 16.250

NJ
20.163
Colonial Park
Mettler's Rd.
East Millstone, NJ 08873
Cross-reference: 16.252

20.164
Cook College, Rutgers University
Ryders Ln.
New Brunswick, NJ 08903
Cross-references: 12.240, 12.437, 15.133, 16.261, 20.52

20.165
Deep Cut Park Horticultural Center
352 Red Hill Rd.
Middletown, NJ 07748
Cross-reference: 16.254

20.166
George Griswold Frelinghuysen Arboretum
53 E. Hanover Ave.
Morristown, NJ 07960
Cross-reference: 16.256

20.167
Gloucester County Vocational School
Tanyard Rd.
Sewell, NJ 08080

20.168
Hunterdon County Arboretum
R.D. 1, Rt. 31
Lebanon, NJ 08833

20.169
Ramapo College of New Jersey
505 Ramapo Valley Rd.
Mahwah, NJ 07430

20.170
Skylands Botanical Garden
Sloatsburg Rd.
Ringwood, NJ 07456
Cross-reference: 16.262

NM
20.171
New Mexico State University
University & S. Main Sts.
Las Cruces, NM 88003

NY
20.172
Brooklyn Botanic Garden
1000 Washington Ave.
Brooklyn, NY 11225
Cross-reference: 16.271

20.173
Buffalo and Erie County Botanical Gardens
South Park Ave & McKinley Pkwy.
Buffalo, NY 14218
Cross-reference: 16.272

20.174
Cutler Botanic Garden
840 Front St.
Binghamton, NY 13905
Cross-reference: 16.276

20.175
Holtsville Ecology Site
249 Buckley Rd.
Holtsville, NY 11742
Cross-reference: 16.281

20.176
Liz Christy Garden
Bowery & Hudson Sts.
New York, NY 10012

20.177
Mohonk Mountain House
Mountain Rest Rd.
New Paltz, NY 12561
Cross-reference: 17.202

20.178
Nassau County Cooperative Extension
1425 Old County Rd.
Plainview, NY 11803

20.179
Queens Botanical Garden
45-50 Main St.
Flushing, NY 11355
Cross-reference: 16.288

20.180
Sonnenberg Gardens
151 Charlotte St.
Canandaigua, NY 14424
Cross-reference: 17.209

20.181
Wild Winds Farms
Hunts Hollow Rd.
Naples, NY 14512

OH
20.182
Ashland High School
1440 King Rd.
Ashland, OH 44805

20.183
Crosby Gardens Botanical & Cultural Center
5403 Elmer Dr.
Toledo, OH 43615

20.184
Dawes Arboretum
7770 Jacksontown Road, S.E.
Newark, OH 43055

Cross-reference: 16.298

20.185
Delhi Flower & Garden Centers
135 Northland Blvd.
Cincinnati, OH 45206

20.186
Franklin Park Conservatory and Garden Center
1777 E. Broad St.
Columbus, OH 43203

Cross-reference: 16.301

20.187
Gardenview Horticultural Park
16711 Pearl Rd.
Strongsville, OH 44136

Cross-reference: 16.303

20.188
Kingwood Center
900 Park Avenue W.
Mansfield, OH 44906

Cross-reference: 16.307

20.189
Ohio State University
Agricultural Technical Institute
Route 250 & U.S. 83
Wooster, OH 44691

Cross-reference: 12.264

20.190
Schmidt Nursery
3001 Innis Rd.
Columbus, OH 43224

20.191
Spring Grove Cemetery & Arboretum
4521 Spring Grove Ave.
Cincinnati, OH 45232

Cross-reference: 18.131

20.192
Summit County Master Gardens
Summit County Fairgrounds
Howe Road at 91
Tallmadge, OH 44278

20.193
Wegerzyn Garden Center
1301 East Siebenthaler Ave.
Dayton, OH 45414

Cross-reference: 7.48

OK
20.194
Cherokee Gardens
Tahlequa, OK 74464

20.195
Oklahoma State University Technical Branch
400 N. Portland
Oklahoma City, OK 73107

Cross-reference: 12.270

OR
20.196
Clackamas Community College
19600 South Molalla Ave.
Oregon City, OR 97045

Cross-reference: 12.271

20.197
Douglas County Demonstration Farm
238 River Forks Rd.
Roseburg, OR 97470

20.198
Lewis Brown Horticultural Farm
33329 Peoria Rd.
Corvallis, OR 97333

20.199
Nichols Garden Nursery
1190 N. Pacific Hwy.
Albany, OR 97321

PA
20.200
Community College of Allegheny County/South Campus
1750 Clairton Rd.
W. Mifflin, PA 15122

20.201
Longwood Gardens
Rt. 1, P.O. Box 501
Kennett Square, PA 19348-0501

Cross-reference: 16.348

20.202
Peddlers Village
Rt. 263
Lahaska, PA 18931

20.203
Pennsylvania State University
Dept. of Horticulture
103 Tyson Bldg.
University Park, PA 16802

Cross-references: 12.278, 12.447, 15.169, 20.25, 20.56, 20.288

20.204
Rodale Research Center
R.D. 1
Kutztown, PA 18011

20.205
W. Atlee Burpee Co.
300 Park Ave.
Warminster, PA 18974
Cross-references: 20.26, 20.57

RI
20.206
Blithewold Gardens and Arboretum
101 Ferry Rd.
Bristol, RI 02809
Cross-reference: 17.234

20.207
Green Animals Topiary Garden
380 Cory's Ln.
Portsmouth, RI 02871
Cross-reference: 16.356

20.208
University of Rhode Island
Plant Science Dept.
Kingston, RI 02881
Cross-references: 12.285, 12.450

SC
20.209
Clemson University
Hwy. 76-28 & Perimeter Rd.
Clemson, SC 29634-0375
Cross-references: 12.286, 12.451, 15.176, 16.366

20.210
Weekend Gardener Journal, The
Route 6, New Bridge Rd.
Aiken, SC 29801
Cross-reference: 21.106

SD
20.211
McCrory Gardens
6th St. & 20th Ave.
Brookings, SD 57007
Cross-reference: 16.368

TN
20.212
Memphis Botanic Garden
750 Cherry Rd.
Memphis, TN 38117
Cross-reference: 16.370

20.213
Opryland Hotel
2800 Opryland Dr.
Nashville, TN 37214
Cross-reference: 16.371

20.214
University of Tennessee Agriculture Campus
Neyland Dr.
Knoxville, TN 37901
Cross-references: 12.299, 12.453, 15.181, 16.375, 20.28

20.215
University of Tennessee Botanic Gardens Maintenance
    Center
Martin, TN 38238
Cross-reference: 12.300

TX
20.216
Dallas Arboretum and Botanical Garden
8617 Garland Rd.
Dallas, TX 75218
Cross-reference: 16.380

20.217
Dallas Civic Garden Center
Fair Park
Martin Luther King, Jr. Entrance
Dallas, TX 75226
Cross-reference: 16.381

20.218
Fort Worth Botanic Garden
3220 Botanic Garden Dr.
Forth Worth, TX 76107
Cross-reference: 16.382

20.219
Houston Garden Center
1500 Hermann Dr.
Houston, TX 77004
Cross-reference: 16.385

20.220
Lubbock Memorial Arboretum
Texas A&M Agricultural Extension Service
4215 University
Lubbock, TX 79456

20.221
Porter & Son
1510 E. Washington St.
Stephenville, TX 76401

20.222
San Antonio Botanical Gardens
555 Funston Pl.
San Antonio, TX 78209
Cross-reference: 16.389

20.223
Texas A&M University
Houston Street on campus
College Station, TX 77843

Cross-references: 12.315, 12.454, 13.49, 15.188, 16.378,
20.29

UT
20.224
Brigham Young University
950 E. 820 N
Provo, UT 84602

Cross-references: 12.320, 15.196, 20.302

20.225
Utah Botanical Garden
1817 N. Main
Farmington, UT 84025

Cross-reference: 16.394

VA
20.226
American Horticultural Society
7931 East Boulevard Dr.
Alexandria, VA 22308

Cross-reference: 1.16

20.227
Busch Gardens
Williamsburg, VA 23185

20.228
Hampton Roads Agriculture Experiment Station
1444 Diamond Springs Rd.
Virginia Beach, VA 23455

20.229
J. Sargeant Reynolds Community College
Western Campus Rt. 6
Goochland, VA 23063

20.230
Kings Dominion
I-95 and Rt. 30
Doswell, VA 23047

20.231
Norfolk Botanical Gardens
Airport Rd.
Norfolk, VA 23518

Cross-reference: 16.400

20.232
Virginia Polytechnic Institute and State University
Washington St.
Blacksburg, VA 24061

Cross-references: 12.326, 12.460, 13.52, 15.201, 16.402

20.233
Virginia Western Community College
3095 Colonial Ave., S.W.
Roanoke, VA 24015

VT
20.234
University of Vermont Extension Service
Horticultural Research Center
Rt. 7 South
S. Burlington, VT 05401

20.235
Vermont Bean Seed Co.
Route 22A
Fair Haven, VT 05743

WA
20.236
Pacific Herbal Institute
902 15th Ave., S.W.
Puyallup, WA 98371

20.237
Skagit Gardens Botanical Display
1695 Johnson Rd.
Mount Vernon, WA 98273

20.238
W. W. Seymour Botanical Conservatory
316 South G St.
Tacoma, WA 98405

Cross-reference: 16.416

WI
20.239
ANR Garden Plots
Vincent High School
7501 N. Granville Rd.
Milwaukee, WI 53224

20.240
Boerner Botanical Gardens
5879 S. 92nd St.
Hales Corners, WI 53130

Cross-references: 16.419, 20.30

20.241
Horticultural Research Farms
University of Wisconsin
Arlington, WI 53911

20.242
Jung Seed Co.
335 S. High St.
Randolph, WI 53957

20.243
Olbrich Botanical Gardens
3330 Atwood Ave.
Madison, WI 53704

Cross-reference: 16.423

20.244
Shake Rag Valley
18 Shake Rag St.
Mineral Point, WI 53565

20.245
University of Wisconsin, Eau Claire
Science Hall Courtyard
Eau Claire, WI 54701

20.246
University of Wisconsin Extension
9668 Watertown Plank Rd.
Milwaukee, WI 53226

WY
20.247
Park County Fairgrounds
655 E. 5th St.
Powell, WY 82435

CANADA

AB
20.248
Alberta Special Crops and Horticultural Research Centre
3 miles East of Brooks on
Hwy. #1
Brooks, AB T0J 0J0

Cross-references: 16.429, 20.31

20.249
Bud Miller All Seasons Park
59th St. & 29th Ave.
Lloydminster, AB S9V 0T8

20.250
City of Red Deer
4914 48th Ave.
Red Deer, AB T4N 3T4

20.251
Muttart Conservatory
98th Ave. & 96 A St.
Edmonton, AB T5J 2R7

Cross-reference: 16.434

20.252
Olds College of Agricultural Technology
Plant Science Dept.
Olds, AB T0M 1P0

Cross-references: 12.354, 15.221

BC
20.253
University of British Columbia Botanical Garden
6501 Northwest Marine Drive
Vancouver, BC V6T 1W5

Cross-reference: 16.449

MB
20.254
Assiniboine Park
2355 Corydon Ave.
Winnipeg, MB R3P 0R5

Cross-reference: 16.452

NB
20.255
New Brunswick Horticulture Centre
Hoyt, NB E0G 2B0

NS
20.256
Nova Scotia Agricultural College
Horticulture & Biological Services
Truro, NS B2N 5E3

Cross-reference: 12.364

ON
20.257
Centennial Greenhouse
Elmcrest Rd.
Etobicoke, ON M9C 2Y2

20.258
Cullen Gardens and Miniature Village
300 Taunton Rd. W.
Whitby, ON L1N 5R5

Cross-reference: 17.292

20.259
Agriculture Canada
Ornamental Gardens
Central Experimental Farm
Ottawa, ON K1A 0C6

Cross-reference: 16.459

20.260
Edwards Gardens
755 Lawrence Ave. E.
Don Mills, ON M3C 1P2

Cross-reference: 16.460

20.261
Germain Gardens
242 East St.
Sarnia, ON N7S 1T5

20.262
Hendrie Park
Plains Rd. West
Burlington, ON

20.263
Humber Arboretum
205 Humber College Blvd.
Rexdale, ON M9W 5L7

Cross-reference: 16.464

20.264
Jackson Park
Tecumseh & Ouellette
Windsor, ON N8X 3N6

Cross-reference: 16.467

20.265
Niagara Parks Commission
Niagara Pkwy.
5 miles North at Falls, 1 mile South
Queenstown-Lewiston Bridge
Niagara Falls, ON L2E 6T2

Cross-references: 12.376, 16.469

20.266
Pinafore Park
89 Elm St.
St. Thomas, ON N5R 1H7

PQ
20.267
Jardin Roger Van den Hende
Université Laval
Ste.-Foy, PQ 61K 7P4

Cross-references: 12.389, 15.247, 16.478

20.268
MacDonald College of Agriculture
MacDonald Campus of McGill University
Ste. Anne de Bellevue
Quebec, PQ H9X 1C0

Cross-references: 12.391, 15.248, 16.480

20.269
W.H. Perron & Co.
2000 rue DuBois
Bois Briand
Quebec, PQ H7V 2T3

Cross-references: 20.36, 20.62

SK
20.270
4th Avenue Park
1450 B 4th Ave.
Regina, SK S4P 3C8

ALL-AMERICA ROSE SELECTIONS TEST AND DEMONSTRATION GARDENS

The All-America Rose Selections organization tests and determines the relative merits of rose cultivars before they are offered to the public. AARS is a nonprofit organization that does not breed, produce, sell, or price roses. Its sole function is to test new cultivars, give recognition to those that prove to be outstanding, and distribute information on award-winning plants to the press and the gardening public in the United States and Canada.

TEST GARDENS

CA
20.271
The Huntington
San Marino, CA

Cross-reference: 17.58

CO
20.272
Denver Botanic Garden
Denver, CO

Cross-reference: 16.65

FL
20.273
Walt Disney World
Lake Buena Vista, FL
Location: I-4 and Florida Turnpike.

Cross-reference: 17.92

GA
20.274
Fernbank Science Center
Atlanta, GA

Cross-reference: 16.106

20.275
Thomasville Nurseries, Inc.
P.O. Box 7
Thomasville, GA 31799
Location: 182 Smith Ave.
912-226-5568

IA
20.276
Iowa State University Horticultural Garden
Ames, IA

Cross-references: 16.133, 20.109

IL
20.277
Chicago Botanic Garden
Glencoe, IL

Cross-reference: 16.138

IN
20.278
Stanley W. Hayes Research Foundation, Inc.
Richmond, IN 47374

KS
20.279
E.F.A. Reinisch Rose and Test Gardens
Topeka, KS

Cross-reference: 16.159

LA
20.280
University of Southern Louisiana
Lafayette, LA 70501

Cross-reference: 16.166

MA
20.281
Stanley Park of Westfield, Inc.
Westfield, MA

Cross-reference: 16.181

MN
20.282
Lyndale Park Gardens
Minneapolis, MN

Cross-reference: 16.207 '

MO
20.283
Missouri Botanical Gardens
St. Louis, MO

Cross-reference: 16.214

OH
20.284
Chadwick Arboretum
Columbus, OH

Cross-reference: 16.295

OK
20.285
Woodward Park
Tulsa, OK 74114

OR
20.286
International Rose Test Garden
Portland, OR

Cross-reference: 16.327

PA
20.287
Hershey Gardens
Hershey, PA

Cross-reference: 16.345

20.288
Pennsylvania State University Arboretum
Dept. of Horticulture
101 Tyson Bldg.
University Park, PA 16802
814-865-7517

Cross-references: 12.278, 12.447, 20.25, 20.56, 20.203

SC
20.289
Edisto Memorial Gardens
Orangeburg, SC

Cross-reference: 16.361

TX
20.290
Tyler Municipal Rose Garden
Tyler, TX

Cross-reference: 16.390

WA
20.291
Woodland Park Zoological Gardens
Seattle, WA

Cross-reference: 18.59

WI
20.292
Boerner Botanical Gardens
Hales Corners, WI

Cross-reference: 16.419

DEMONSTRATION GARDENS

CA
20.293
Bear Creek Gardens, Inc.
Wasco, CA 93280

70.294
DeVor Nurseries, Inc.
Freedom, CA 95019

20.295
Garden Valley Ranch
Petaluma, CA 94952

CT
20.296
Elizabeth Park
Hartford, CT

Cross-reference: 16.75

MI
20.297
Michigan State University
Horticulture Gardens
East Lansing, MI

Cross-reference: 16.204

NE
20.298
Doane College
Crete, NE 68333

NY
20.299
City of Buffalo Joan Fuzak Memorial Garden
Buffalo, NY

20.300
Old Westbury Gardens
Old Westbury, NY

Cross-reference: 17.205

OR
20.301
Jackson & Perkins Co.
1 Rose Ln.
Medford, OR 97501

UT
20.302
Brigham Young University
275 WIDB
Provo, UT 84602

Cross-references: 12.320, 15.196, 20.224

~ 21 ~

Horticultural Periodicals

There are scores of horticultural magazines, journals, and newsletters published in the United States and Canada. Most national horticultural organizations, plant societies, native plant societies, scholarly organizations, garden club associations, state and provincial horticultural societies, arboreta, and conservation organizations publish some type of periodical. Please consult the appropriate chapter for full information on organization-specific periodicals.

Included below are select titles of general interest plus select university and governmental periodicals and commercial and trade periodicals.

GENERAL INTEREST

21.1
Acres U.S.A.
P.O. Box 9547
Kansas City, MO 64133

Actinidia Enthusiasts Newsletter
Friends of the Trees Society
Chelan, WA

Annual.
Cross-reference: 8.33

Aliso
Rancho Santa Ana Botanic Garden
Claremont, CA

Cross-reference: 16.48

21.2
Amaranth Today
Rodale Press
33 E. Minor St.
Emmaus, PA 18049

Quarterly.

American Horticulturist
American Horticultural Society
Alexandria, VA

Bimonthly.
Cross-reference: 1.16

American Horticulturist New Edition
American Horticultural Society
Alexandria, VA

Bimonthly.
Cross-reference: 1.16

21.3
Amicus Journal, The
122 E. 42nd St.
Room 4500
New York, NY 10168

21.4
AppAltCrop: A Newsletter on Appalachian Alternative Crops
Technical Outreach Program
Rt. 5, Box 423
Livingston, KY 40445

Quarterly.

Arnoldia
Arnold Arboretum of Harvard University
Jamaica Plain, MA

Quarterly.
Cross-reference: 16.173

21.5
Avant Gardener, The
Box 489
New York, NY 10028

Monthly.

B.U.G.S. Flyer
Biological Urban Gardening Services
Citrus Heights, CA
Cross-reference: 1.38

21.6
Backyard Gardener Idea-Letter
James J. Martin
P.O. Box 605
Winfield, IL 60190
Bimonthly.

21.7
Baer's Garden Newsletter
John Baer's Sons
Box 328
Lancaster, PA 17603
Quarterly.

21.8
Bartlett Tree Topics
F. A. Bartlett Tree Expert Co.
P.O. Box 3067
Stamford, CT 06905
Quarterly.

21.9
Better Homes and Gardens
Meredith Corp.
Locust at 17th
Des Moines, IA 50336
Monthly.

21.10
Bev Dobson's Rose Letter
Beverly R. Dobson
215 Harriman Rd.
Irvington, NY 10533
Bimonthly.

Biodynamics
Bio-Dynamic Farming and Gardening Association
Kimerton, PA
Cross-reference: 1.37

21.11
Bonsai Today
Stone Lantern Publishing Co.
P.O. Box 816
Sudbury, MA 01776
Bimonthly.

21.12
Botanical and Herb Reviews
Steven Foster
P.O. Box 106
Eureka Springs, AR 72632
Quarterly.

21.13
Cacti & Other Succulents
Marina J. Welham
8591 Lochside Dr.
Sidney, BC V8L 1M5
Bimonthly.

21.14
California Garden Magazine
San Diego Floral Association
Casa del Prado
Balboa Park
San Diego, CA 92101-1619
Bimonthly.
Cross-reference: 9.105

21.15
Canadian Garden News
514 Kortright Rd. W.
Guelph, ON N1G 3Y6

Canadian Horticultural History
Centre for Canadian Historical Horticultural Studies
Hamilton, ON
Quarterly.
Cross-reference: 17.39

21.16
Carolina Gardener
Tri-Star Communications, Inc.
P.O. Box 13269
Greensboro, NC 27415
8/year.

21.17
Coltsfoot
Rt. 1
Box 313A
Shipman, VA 22971
Bimonthly.

Common Sense Pest Control Quarterly
Bio Integral Resource Center
Berkeley, CA
Cross-reference: 1.36

21.18
Country Journal
Historical Times, Inc.
P.O. Box 392
Mt. Morris, IL 61054
Bimonthly.

Desert Plants
Boyce Thompson Southwestern Arboretum
Superior, AZ
Annual.
Cross-reference: 16.9

21.19
Design With Flowers
E. 17th St., Suite 202
Costa Mesa, CA 92627

Bimonthly.

21.20
Dwarf Conifer Notes
Theophrastus
P.O. Box 458
Little Compton, RI 02837-0458

Irregular.

21.21
Elizabeth
Elizabeth Harris
Rt. 1, Box 1746AA
Davidson, NC 28036

Bimonthly.

21.22
Endangered Species Update
School of Natural Resources
University of Michigan
Ann Arbor, MI 48109-1115

10/yr.

21.23
Euphorbia Journal
Strawberry Press
227 Strawberry Dr.
Mill Valley, CA 94941

Annual.

Fall Harvest Edition
Seed Savers Exchange
Decorah, IA

Cross-reference: 17.32

21.24
Fine Gardening
The Taunton Press
63 South Main St.
P.O. Box 355
Newtown, CT 06470-9989

Bimonthly.

Flora of North America Newsletter
Flora of North America
St. Louis, MO

Bimonthly.
Cross-reference: 4.12

21.25
Flower & Garden
4251 Pennsylvania Ave.
Kansas City, MO 64111

Bimonthly.
Cross-reference: 16.211

Four Seasons, The
Regional Parks Botanic Garden
Berkeley, CA

Quarterly.
Cross-reference: 16.49

Fremontia
California Native Plant Society
Sacramento, CA

Quarterly.
Cross-reference: 3.6

Friends of the Trees Yearbook
Friends of the Trees Society
Chelan, WA

Cross-reference: 8.33

21.26
Frontier Facts
Box 299
Norway, IA 52318

Bimonthly.

Garden Design
American Society of Landscape Architects, Inc.
Washington, DC

Quarterly.
Cross-reference: 1.25

21.27
Garden Doctor, The
1684 Willow
Denver, CO 80220

Bimonthly.

21.28
Garden Network Monthly
Benedyk and Gabella
P.O. Box 302
Villa Park, IL 60181-0302

Monthly.

21.29
Garden Railways
Sidestreet Bannerworks, Inc.
P.O. Box 61461
Denver, CO 80206

Bimonthly.

21.30
Gardener's Advocate
P.O. Box 1106
Columbia, MD 21044

Quarterly.

21.31
Gardener's Eye
Harvey Childs, Hurrican Publishing Co.
P.O. Box 22382
Denver, CO 80222

Monthly.

21.32
Gardener's Guide
6111 N. Marina Pacifica Dr., Key 15
Long Beach, CA 90803
Monthly.

21.33
Gardener's Index
CompuDel Press
P.O. Box 27041
Kansas City, MO 64110
Annual.

Gardener, The
Men's Garden Club of America
Johnston, IA
Bimonthly.
Cross-reference: 5.2

21.34
Gardeners Share
P.O. Box 243
Columbus, IN 47273
Bimonthly.

21.35
Gardening
Conde Nast Publications, Inc.
Gardening Magazine
350 Madison Ave.
New York, NY 10017
Annual.

21.36
Gardening Newsletter by Bob Flagg
P.O. Box 2306
Houston, TX 77001
Monthly.

21.37
Gardening from the Heartland
P.O. Box 57
Taylor Ridge, IL 61284
Bimonthly.

21.38
Gardens & Countrysides; A Journal of Picturesque Travel
Travel Publications, Inc.
401 Austin Hwy., Suite 209
San Antonio, TX 78209
10/year.

21.39
Gardens West
Cornwall Publishing Co., Ltd.
4962 Granville St.
Vancouver, BC V6M 3B2
7/year.

21.40
Garlic Times; The Newsletter of Lovers of the Stinking Rose
Harris Publishing
1621 Fifth St.
Berkeley, CA 94710-1714
Biannual.

21.41
Gil's Garden
Gil Whitton
1300 Casa Vista Dr.
Palm Harbor, FL 34683
Monthly.

21.42
Green Prints
P.O. Box 1355
Fairview, NC 28730
Quarterly.

Green Scene, The
Pennsylvania Horticultural Society
Philadelphia, PA
Bimonthly.
Cross-reference: 6.32

21.43
Green Thumb Companion
P.O. Box 67
Prescott, IA 50859
Bimonthly.

21.44
Green Thumb Gardening Newsletter
Spray-N-Grow
8500 Commerce Park Dr., Suite 102
Houston, TX 77036
Biannual.

21.45
Greener Gardening, Easier
E. Dexter Davis
Horticulturist
26 Norfolk St.
Holliston, MA 01746
Monthly.

21.46
Growing Edge Magazine, The
P.O. Box 1027
Corvallis, OR 97339
Quarterly.

21.47
Growing from Seed
Thompson & Morgan
P.O. Box 1308
Jackson, NJ 08527
Quarterly.

21.48
Hardy Enough
351 Pleasant St., Suite 259
Northampton, MA 01060
5/year.

21.49
Harrowsmith
7 Queen Victoria Rd.
Camden East, ON K0K 1J0
Bimonthly.
Cross-reference: 19.102

21.50
Harrowsmith
Camden House Publishing
Ferry Rd.
Charlotte, VT 05445
Bimonthly.

21.51
Helping Each Other
Judy Huber
HCR
Box 275
Bowdle, SD 67428
Monthly.

21.52
Herb Basket, The
Practical Press
P.O. Box 548
Boiling Springs, PA 17007
Quarterly.

21.53
Herb Companion, The
Interweave Press, Inc.
306 North Washington Ave.
Loveland, CO 80537
Bimonthly.

21.54
Herb Newsletter, The
P.O. Box 42236
Tacoma, WA 98442
10/year.

21.55
Herb Quarterly, The
Long Mountain Press, Inc.
P.O. Box 548
Boiling Springs, PA 17007
Quarterly.

21.56
Herb, Spice and Medicinal Plant Digest, The
L. E. Craker
Dept. of Plant and Soil Sciences
Stockbridge Hall
University of Massachusetts
Amherst, MA 01003
Quarterly.
Cross-reference: 12.171

21.57
Herbal Kitchen, The
Diane Lea Mathews
Box 134
Salisbury Center, NY 13454
5/year.

21.58
Herbal Seeker, The
P.O. Box 299
Battle Ground, IN 47920
Bimonthly.

21.59
Herbal Thymes
39 Reed St.
Marcellus, NY 13108
Quarterly.

21.60
Herbal Lifestyles
Stone Acre Press
84 Carpenter Rd.
New Hartford, CT 06057
Bimonthly.

Herbarist, The
Herb Society of America, Inc.
Boston, MA
Cross-reference: 1.60

Heritage Seed Program
Heritage Seed Program
Uxbridge, ON
3/year.
Cross-reference: 17.40

21.61
HortIdeas
G. and P. Williams
Rt. 1, Box 302
Gravel Switch, KY 40328
Monthly.

21.62
Horticulture
Horticulture Limited Partnership
P.O. Box 2595
Boulder, CO 80323
Monthly.

21.63
Hortline
Tom's World Horticulture Consulting
P.O. Box 5238
Charleston, WV 25361
Monthly.

21.64
Houseplant Forum/A Fleur de Pot
HortiCom Inc.
1449 Avenue William
Sillery, PQ G1S 4G5
Bimonthly.

IPM Practitioner, The
The Bio Integral Resource Center
Berkeley, CA
10/year.
Cross-reference: 1.36

21.65
In House Hort
P.O. Box 30961
Port Authority Station
New York, NY 10011
Monthly.

Indoor Garden, The
Indoor Gardening Society of America, Inc.
Portland, OR
Bimonthly.
Cross-reference: 1.67

21.66
International Bonsai
1070 Martin Rd.
West Henrietta, NY 14586-9623
Quarterly.

21.67
Island Grower, The
Greenhart Publications
R.R. 4
Sooke, BC V0S 1N0
11/year.

21.68
Japanese Flower Arranging
c/o S. C. Pototsky
44 Lane Park, No. 26
Brighton, MA 02135
Bimonthly.

21.69
John E. Bryan Gardening Newsletter
John E. Bryan, Inc.
300 Valley St., Suite 206
Sausalito, CA 94965
Monthly.

Journal of Community Gardening
American Community Gardening Association
Philadelphia, PA
Quarterly.
Cross-reference: 1.11

Journal of Environmental Horticulture
Horticultural Research Institute
Washington, DC
Quarterly.
Cross-reference: 1.63

21.70
Journal of Garden History
Taylor and Francis, Inc.
242 Cherry St.
Philadelphia, PA 19106-1906
Quarterly.

21.71
Kew Magazine, The
Basil Blackwell, Ltd.
2 Cambridge Center
Cambridge, MA 02142
Quarterly.

21.72
Landscape Architectural Review
Pleasance Crawford, Editor
39 Macpherson Ave.
Toronto, ON M5R 1W7
Monthly.

Landscape Architecture
American Society of Landscape Architects, Inc.
Washington, DC
Monthly.
Cross-reference: 1.25

21.73
Living Off the Land, Subtropic Newsletter
P.O. Box 2131
Melbourne, FL 32902-2131
5/year.

21.74
Living With Herbs
71 Little Fresh Pond Rd.
Box 1332
Southampton, NY 11968
Bimonthly.

21.75
*Llewellyn's Moon Sign Book; Gardening Guide and Lunar
    Almanac*
Llewellyn Publications
213 E. 4th St., Box 64383
St. Paul, MN 55164
Monthly.

Magnolia
Southern Garden History Society
Winston-Salem, NC

Quarterly.
Cross-reference: 17.35

Memo on Endangered Plants
Natural Resources Defense Council
Washington, DC

Every six weeks.
Cross-reference: 8.50

21.76
Metropolitan Gardener
P.O. Box 20120
London Terrace Station
New York, NY 10011

Bimonthly.

Minnesota Horticulturalist
Minnesota State Horticultural Society
St. Paul, MN

9/year.
Cross-reference: 6.20

National Gardening Survey
National Gardening Association
Burlington, VT

Annual.
Cross-reference: 1.89

21.77
Native Notes
Bluebird Nursery
Rt. 2, Box 550
Heiskell, TN 37754

Quarterly.

New Alchemy Quarterly, The
New Alchemy Institute
East Falmouth, MA

Quarterly.
Cross-reference: 8.52

21.78
New England Farm Bulletin & Garden Gazette
Jacob Meadow, Inc.
P.O. Box 147
Cohasset, MA 02025

24/year.

21.79
New England Gardener
P.O. Box 2699
Nantucket, MA 02584

Monthly.

21.80
Northland Berry News
Minnesota Berry Growers Association
19060 Manning Trail North
Marine on St. Croix, MN 55047

Bimonthly.
Cross-reference: 9.461

21.81
Orchid Digest
c/o Mrs. Norman K. Atkinson
P.O. Box 916
Carmichael, CA 95609

Quarterly.

21.82
Orchid Hunter
2250 Beulah Rd.
Pittsburgh, PA 15235

Monthly.

21.83
Orchid Information Exchange
1230 Plum Ave.
Simi Valley, CA 93065

Monthly.

21.84
Organic Gardening
33 E. Minor St.
Emmaus, PA 18098

Monthly.

21.85
Over the Garden Fence
Over the Garden Fence, Inc.
Box 386
Lake Dallas, TX 75065

Quarterly.

PAPPUS
Royal Botanical Gardens
Hamilton, ON

Quarterly.
Cross-reference: 16.473

21.86
Pacific Horticulture
Pacific Horticultural Foundation
P.O. Box 680
Berkeley, CA 94701

Quarterly.

People Plant Connection
American Horticultural Therapy Association
Gaithersburg, MD

11/year.
Cross-reference: 1.17

21.87
Permaculture with Native Plants
Curtin Mitchell
P.O. Box 38
Lorane, OR 97451
Quarterly.

21.88
Phalaenopsis Fancier
1230 Plum Ave.
Simi Valley, CA 93065
Monthly.

21.89
Plant Lore
16 Oak St.
Genesse, NY 14454
Biannual.

Plant Wise
Botanical Dimensions
Occidental, CA
Cross-reference: 8.17

Plants and Gardens
Brooklyn Botanic Garden
Brooklyn, NY
Quarterly.
Cross-reference: 16.271

21.90
Play Dirt
1102 Floradale Dr.
Austin, TX 78753-3924
Monthly.

Pomona
North American Fruit Explorers
Chapin, IL
Cross-reference: 1.99

21.91
Potpourri Party-Line
Berry Hill Press
7336 Berry Hill
Palos Verdes Peninsula, CA 90274
Quarterly.

21.92
Potpourri from Herbal Acres
Pine Row Publications
Box 428
Washington Crossing, PA 18977
Quarterly.

21.93
Quebec Vert
1320 boulevard St.-Joseph
Quebec, PQ G2K 1G2
Monthly.

21.94
Rosy Outlook Magazine
"Rosy" McKenney
1014 Enslen St.
Modesto, CA 95350
Quarterly.

21.95
San Diego Garden Digest
1516 W. Redwood St., Suite 106
San Diego, CA 92101
Monthly.

Solanaceae
Solanaceae Enthusiasts
Santa Clara, CA
Quarterly.
Cross-reference: 2.88

21.96
Southern Gardens
Wing Publications, Inc.
Box 11268
Columbia, SC 29202
Bimonthly.

21.97
Southern Herbs
Eve Elliot
400 S. Hawthorne Circle
Box 3722
Winter Springs, FL 32708
Quarterly.

21.98
Southern Living
P.O. Box 523
Birmingham, AL 35201
Monthly.

21.99
Spice and Herb Arts
5091 Muddy Ln.
Buckingham, FL 33905
Bimonthly.

21.100
Sunset Magazine
Lane Publishing Co.
80 Willow Rd.
Menlo Park, CA 94025-3691
Monthly.
Cross-reference: 16.58

21.101
TLC . . . for plants
Gardenvale Publishing Co., Ltd.
1 Pacifique
Ste. Anne de Bellevue, PQ H9X 1C5
Quarterly.

21.102

Texas Gardener
P.O. Box 9005
Waco, TX 76714
Bimonthly.

21.103

Thymes, The
2219 Long Hill Rd.
Gulford, CT 06437
Monthly.

21.104

Today's Herbs
P.O. Box 1422
Provo, UT 84603
Monthly.

21.105

Twenty-First Century Gardener, The
Growers Press, Inc.
P.O. Box 189
Princeton, BC V0X 1W0
Quarterly.

Urban Forest Forum
American Forestry Association
Washington, DC
Bimonthly.
Cross-reference: 8.11

21.106

Weekend Gardener Journal, The
P.O. Drawer 1607
Aiken, SC 29802
7/year.
Cross-reference: 20.210

21.107

Westscape
Rick Hassett
369 E. 900 S.
Salt Lake City, UT 84111
Quarterly.

21.108

Whole Chili Pepper, The
Out West Publishing Co.
P.O. Box 4278
Albuquerque, NM 87196
Monthly.

Wild Flower Notes
New England Wild Flower Society, Inc.
Framingham, MA
Annual.
Cross-reference: 3.22

Wildflower
Canadian Wildflower Society
North York, ON
Quarterly.
Cross-reference: 1.137

Wildflower
National Wildflower Research Center
Austin, TX
Biannual.
Cross-reference: 1.97

Winter Yearbook
Seed Savers Exchange
Decorah, IA
Cross-reference: 17.32

21.109

Woman's Day Gardening and Outdoor Living Ideas
Diamandis Comm.
Hachette Publications, Inc.
1515 Broadway
New York, NY 10036
Annual.

21.110

Yard and Garden
Johnson Hill Press, Inc.
1233 Janesville Ave.
Fort Atkinson, WI 53538
Monthly.

UNIVERSITY AND GOVERNMENTAL PERIODICALS

21.111

Annual Letter
U.S. Institute of Tropical Forestry
Southern Forest Experiment Station
Call Box 25000
Rio Piedras, PR 00928-2500
Annual.

21.112

Aquaphyte
University of Florida
Institute of Food and Agricultural Sciences
2183 McCarty Hall
Gainesville, FL 32611
Biannual.

21.113
Arid Lands Newsletter
Office of Arid Lands Studies
University of Arizona
845 North Park Ave.
Tucson, AZ 85719

21.114
California Plant Pathology
University of California
Dept. of Plant Pathology
Davis, CA 95616
Quarterly.

21.115
Canadian Plant Disease Survey
Agriculture Canada Research Program Services
Neatby Bldg., Rm. 1133
Ottawa, ON K1A 0C6
Quarterly.

21.116
Cornell Recommendations for Commercial Tree-Fruit Production
Cornell University
Media Services, 7-8 Research Park
Ithaca, NY 14850
Annual.

21.117
Cornell Recommendations for Commercial Florist Crops
Cornell University
Media Services, 7-8 Research Park
Ithaca, NY 14850
Annual.

21.118
Cultivar, The
USSC Agroecology Program
College Eight
Santa Cruz, CA 95064
Biannual.

21.119
E S F (Environmental Science and Forestry)
State University of New York
College of Environmental Science and Forestry
Public Relations Office
Bray Hall, Rm. 123
Syracuse, NY 13210
Monthly.

21.120
Frontiers of Plant Science
Agricultural Experiment Station
Box 1106
New Haven, CT 06504-1106
Biannual.
Cross-reference: 11.93

21.121
Fungi Canadensis
Agriculture Canada
Biosystematics Research Centre
Ottawa, ON K1A 0C6
Irregular.

21.122
Garden FAX
Dept. of Agriculture
Printmedia Branch
7000 113th St.
Edmonton, AB T6H 5T6
Irregular.

21.123
General Technical Report
U.S. Dept. of Agriculture
Intermountain Research Station
324 25th St.
Ogden, UT 84401
Irregular.

21.124
Horticultural Products Review
U.S. Dept. of Agriculture/Foreign Agricultural Service
Room 4644 South Building
Washington, DC 20250
Monthly.

21.125
Horticultural Review
Dept. of Horticulture
Pennsylvania State University
102 Tyson Bldg.
University Park, PA 16802
Quarterly.
Cross-reference: 12.278

21.126
Insect and Disease Conditions in Canada
Canadian Forestry Service
Place Vincet Massey
3rd Floor
Ottawa, ON K1A 1G5
Annual.

21.127
Ornamentals Northwest Newsletter
Oregon State University
Cooperative Extension Service
Dept. of Horticulture
Corvallis, OR 97331
Bimonthly.
Cross-reference: 11.42

21.128
Plant Pathology Circular
Florida Dept. of Agriculture and Consumer Services
Division of Plant Industry
1911 S.W. 34th St., Box 1269
Gainesville, FL 32602

Monthly.

21.129
Recent Reports
U.S. Dept. of Agriculture
Intermountain Research Station
324 25th St.
Ogden, UT 84401

Quarterly.

21.130
Seasonal Fruit and Vegetable Report
Ontario Ministry of Agriculture and Food Economics and
   Policy Coordination Branch
Parliament Building
Toronto, ON M7A 1B6

Monthly.

21.131
Tennessee Economic Pest Report
Dept. of Agriculture
Division of Plant Industries, Insect Survey
Box 40627, Melrose Station
Nashville, TN 37204

Annual.

21.132
Tri-ology Technical Report
Florida Dept. of Agriculture and Consumer Services
Division of Plant Industry
1911 S.W. 34th St., Box 1269
Gainesville, FL 32602

Monthly.

21.133
U.S. Department of Agriculture: Home and Garden Bulletin
U.S.D.A.
Washington, DC 20250

Irregular.

21.134
Undercover
Cooperative Extension Service
U.S.D.A.
University of Missouri
Clark Hall
Columbia, MO 65211

Monthly.
Cross-reference: 11.29

21.135
Vegetables Specialist
Dept. of Agriculture & Rural Development
Plant Industry Branch
Fredericton, NB E3B 5H1

Irregular.

21.136
Virginia Gardener, The
Virginia Cooperative Extension Service
Virginia Polytechnic Institute and State
Blacksburg, VA 24061

Monthly.
Cross-reference: 11.50

COMMERCIAL AND TRADE PERIODICALS

21.137
American Fruit Grower
Meister Publishing Co.
37841 Euclid Ave.
Willoughby, OH 44094

Monthly.

21.138
American Lawn Applicator
Group Interest Enterprises, Inc.
4012 Bridge Ave.
Cleveland, OH 44113

Monthly.

21.139
American Nurseryman
111 N. Canal St.
Chicago, IL 60606

24/year.

21.140
Business of Herbs, The
Northwind Farm Publications
Rt. 2, Box 246
Shevlin, MN 56676

Bimonthly.

21.141
Canadian Fruit Growers
Cash Crop Farming Publications, Ltd.
222 Argyle Ave.
Delhi, ON N4B 2Y2
9/yr.

21.142
Florafacts
Florafax International, Inc.
Box 45745
Tulsa, OK 74145
Monthly.

21.143
Floral Marketing Directory and Buyer's Guide
Produce Marketing Association
1500 Casho Mill Rd., Box 6036
Newark, DE 19714-6036
Annual.

21.144
Floral Mass Marketing
Cenflo, Inc.
549 W. Randolph St.
Chicago, IL 60606
Monthly.

21.145
Floral and Nursery Times
XXX Publishing Enterprises Ltd.
629 Green Bay Rd.
Wilmette, IL 60091
24/year.

21.146
Florist
29200 Northwestern Highway
P.O. Box 2227
Southfield, MI 48037
Monthly.

21.147
Florists' Review
Florists' Enterprises, Inc.
Suite 105, 2231 Wanamaker
P.O. Box 4368
Topeka, KS 66614
Monthly.

21.148
Flower News
Central Flower News
549 West Randolph St.
Chicago, IL 60606
Weekly.

21.149
Flower Shop
Cash Crop Farmings Publications Ltd.
22 Argyle Ave.
Delhi, ON N4B 2Y2
Monthly.

21.150
Flowers & Teleflora Inc.
Teleflora Plaza, Suite 260
12233 West Olympic Blvd.
Los Angeles, CA 90064
Monthly.

21.151
Garden Supply Retailer
Miller Publishing Co.
12400 Whitewater Dr.,
Suite 160
Minnetonka, MN 55343
Monthly.

21.152
Green Markets
McGraw-Hill Publications Co.
1221 Ave. of the Americas
New York, NY 10020
Weekly.

21.153
Greenhouse Canada
Cash Crop Farming Publications Ltd.
222 Argyle Ave.
Delhi, ON N4B 2Y2
Monthly.

21.154
Greenhouse Grower
Meister Publishing Co.
37841 Euclid Ave.
Willoughby, OH 44094
Monthly.

21.155
Greenhouse Manager
Branch-Smith Publishing
120 St. Louis Ave.
Fort Worth, TX 76104
Monthly.

21.156
Grounds Maintenance
Intertec Publishing Corp.
9221 Quivira Rd.
Overland Park, KS 66215
Monthly.

21.157
Grower Talks
P.O. Box 532
1 North River Lane
Suite 206
Geneva, IL 60134
Monthly.

21.158
Herb Market Report, The; For the Herb Farmer and Forager
525 S.E. H St.
Grants Pass, OR 97526
Monthly.

21.159
Landscape Contractor
J K Communications
799 Roosevelt Rd.
Bldg. 6-112
Glen Ellyn, IL 60137
Monthly.

21.160
Landscape Management Golf Daily
Edgell Communications
7500 Old Oak Blvd.
Cleveland, OH 44130
Annual.

21.161
Lawn Care Industry
7500 Old Oak Blvd.
Cleveland, OH 44130
Monthly.

21.162
Lawn Servicing
Intertec Publishing Corp.
Box 12901
9221 Quivira Rd.
Overland Park, KS 66212
10/year.

21.163
Nursery Business
Brantwood Publications, Inc.
Northwood Plaza Station
Clearwater, FL 33519-0360
Monthly.

21.164
Nursery Digest
Betrock Publishers, Inc.
10400 Griffin Rd.,
Suite # 301
Cooper City, FL 33328
Monthly.

21.165
Nursery Manager
Branch-Smith Publishing
120 St. Louis St.
Ft. Worth, TX 76104
Monthly.

21.166
Outdoor Power Equipment
Quinn Publications, Inc.
Box 1570
Fort Worth, TX 76101
Monthly.

21.167
Pro
Johnson Hill Press, Inc.
1233 Janesville Ave.
Ft. Atkinson, WI 53538
Bimonthly.

21.168
Professional Floral Designer
American Floral Services
3737 N.W. 34th St.
Oklahoma City, OK 73112
Bimonthly.

21.169
Seed Trade News
Dean Enterprises, Inc.
7535 Office Ridge Circle
Eden Prairie, MN 55434
14/year.

21.170
Southwest Lawn and Landscape
R/K Communications Group, Inc.
P.O. Box 94857
Las Vegas, NV 89193-4857
Monthly.

21.171
Turf
NEF Publishing Co.
50 Bay St., Box 391
St. Johnsbury, VT 05819
Monthly.

21.172
Western Fruit Grower
Meister Publishing Co.
37841 Euclid Ave.
Willoughby, OH 44094
Monthly.

Index of Organizations

New Jersey Small Fruits Council, Inc., 102
New Jersey State Florists' Association, Inc., 102
New Jersey State Horticultural Society, 44
New Jersey State Soil Conservation Committee, 102
New Jersey State Sweet Potato Industry Association, 102
New Jersey Tree Farm Committee, 102
New Jersey Turfgrass Association, 102
New Medico Highwatch Learning Center, 213
New Mexico Agricultural Chemical and Plant Food Association, 103
New Mexico Agricultural Experiment Station, 154
New Mexico Association of Nurserymen, 103
New Mexico Garden Clubs, Inc., 103
New Mexico Organic Growers Association, 103
New Mexico Peanut Commission, 103
New Mexico Peanut Growers Association, 103
New Mexico Pistachio Association, 103
New Mexico Soil and Water Conservation Commission, 103
New Mexico State University, 183, 293, 381
New Mexico State University Herbarium, 250
New Mexico Vine and Wine Society, 103
New Orleans Botanical Garden, 379
New Orleans Garden Society, 89
New Orleans Retail Florists Association, 90
Newton Cemetery, 360
New York Agricultural Experiment Station, 155
New York Apple Research Association, 105
New York Aquarium, 355
New York Botanical Garden, 65, 104, 184, 208, 295
New York Botanical Garden Herbarium, 251
New York Cherry Growers Association, 105
New York Christmas Tree Association, 105
New York City Department of Parks and Recreation, 367
New York City Technical College, 184
New York Corn Growers Association, 105
New York Florists Club, 105
New York Parks Council, 368
New York School of Interior Design, 184
New York Seed Improvement Cooperative, 105
New York State Association for Retarded Children, Inc., 215
New York State Flower Industries Research & Education Fund, Inc., 105
New York State Forest Owners Association, 105
New York State Green Council, 105
New York State Horticultural Society, 44
New York State Hudson River Psychiatric Center, 215
New York State Museum Herbarium, 251
New York State Nurserymen's Association, 105

New York State Office of Community Department of Agriculture & Markets, 367
New York State Seed Association, 105
New York State Small Fruit Growers Association, 105
New York State Soil and Water Conservation Commission, 105
New York State Turfgrass Association, 105
New York State Vegetable Growers Association, 105
New York State Wine Grape Growers, 105
New York Turf & Landscape Association, Inc., 105
New York Zoological Park (Bronx Zoo), 355
Niagara College of Applied Arts and Technology, 191
Niagara County Community College, 184
Niagara Frontier Botanical Society, 32
Niagara Frontier Retail Florists, 105
Niagara Parks Commission, 386
Niagara Parks Commission School of Horticulture, 191, 314
Niagara Parks Commission School of Agriculture Library, 235
Nichols Garden Nursery, 382
Nichols State University, 178
Nikka Yuko Japanese Garden, 311
916 Area Vocational Technical Institute, 180
Nitrogen Fixing Tree Association, 60
Noelridge Park, 280, 378
Norfolk Botanical Gardens, 209, 233, 307, 384
Norfolk County Agricultural School, 179
Norfolk Garden Center, 51
Norfolk School of Horticulture and Landscape Design, 188
Normandin Experimental Farm, 164
Norristown State Hospital, 216
North American Blueberry Council, 16
North American Fruit Explorers, 10, 23, 396
North American Gladiolus Council, 24, 309
North American Heather Society, 24
North American Horticulture Supply Association, 10
North American Lily Society, Inc., 24, 227
North American Mycological Association, 10, 24
North Carolina Agricultural and Technical State University, 182, 196
North Carolina Agricultural Research Service, 153
North Carolina Apple Growers Association, 98
North Carolina Arboretum, 290
North Carolina Association of Nurserymen, 98
North Carolina Botanical Garden, 65, 207, 213, 290
North Carolina Christmas Tree Growers Association, Inc., 98
North Carolina Commercial Flower Growers Association, 98
North Carolina Co-op Bulb Growers Association, 98
North Carolina Fresh Vegetable Growers Association, Inc., 98
North Carolina Grape Growers Association, 98

North Carolina Greenhouse Growers Association, Inc., 98
North Carolina Herb Association, 98
North Carolina Landscape Contractors Association, Inc., 98
North Carolina Peach Growers Society, Inc., 98
North Carolina Peanut Growers Association, 98
North Carolina Potato Association, 98
North Carolina Primrose Growers Association, 99
North Carolina State Museum of Natural Sciences, 336
North Carolina State University, 182, 196, 290, 380
North Carolina State University Herbarium, 249
North Carolina Trellised Tomato Growers Association, 99
North Carolina Watermelon Association, 99
North Carolina Wild Flower Preservation Society, 32
North Carolina Yam Commission, 99
North Carolina Zoological Park, 354
North Central Florists Association, 95
North Central Regional Plant Introduction Station, 159, 160
North Dakota Agricultural Experiment Station, 153–154
North Dakota Corn Growers Association, 99
North Dakota Dry Bean Seed Growers Association, 99
North Dakota Edible Bean Council, 95
North Dakota Federation of Garden Clubs, 99
North Dakota Natural Science Society, 60, 99
North Dakota Nursery and Greenhouse Association, 99
North Dakota State Horticultural Society, 43–44
North Dakota State Potato Council, 99
North Dakota State Seed Commission, 99
North Dakota State Soil Conservation Committee, 99
North Dakota State University, 182, 183, 196, 380
North Dakota State University Herbarium, 249
North Dakota Weed Control, 99
Northeastern Junior College, 173
Northeastern Oklahoma A&M College, 185
Northeastern Ontario Arboretum, 314
Northeastern Regional Plant Introduction Station, 159, 160
Northeast Indiana Flower Growers Association, 86
Northeast Louisiana University Herbarium, 246
Northern Alberta Institute of Technology, 190
Northern California Flower Growers and Shippers, 73
Northern California Turfgrass Council, 73
Northern Forestry Centre, 165
Northern Kentucky University Herbarium, 245
Northern Minnesota Blue Grass Growers Association, 95
Northern Nevada Native Plant Society, 32

Index of States